Advanced Structured Materials

Volume 142

Common engineering materials reach in many applications their limits and new developments are required to fulfil increasing demands on engineering materials. The performance of materials can be increased by combining different materials to achieve better properties than a single constituent or by shaping the material or constituents in a specific structure. The interaction between material and structure may arise on different length scales, such as micro-, meso- or macroscale, and offers possible applications in quite diverse fields.

This book series addresses the fundamental relationship between materials and their structure on the overall properties (e.g. mechanical, thermal, chemical or magnetic etc.) and applications.

The topics of *Advanced Structured Materials* include but are not limited to

- classical fibre-reinforced composites (e.g. glass, carbon or Aramid reinforced plastics)
- metal matrix composites (MMCs)
- micro porous composites
- micro channel materials
- multilayered materials
- cellular materials (e.g., metallic or polymer foams, sponges, hollow sphere structures)
- porous materials
- truss structures
- nanocomposite materials
- biomaterials
- nanoporous metals
- concrete
- coated materials
- smart materials

Advanced Structured Materials is indexed in Google Scholar and Scopus.

More information about this series at http://www.springer.com/series/8611

Jan Awrejcewicz · Anton V. Krysko ·
Maxim V. Zhigalov · Vadim A. Krysko

Mathematical Modelling and Numerical Analysis of Size-Dependent Structural Members in Temperature Fields

Regular and Chaotic Dynamics of Micro/Nano Beams, and Cylindrical Panels

Springer

Jan Awrejcewicz
Department of Automation, Biomechanics and Mechatronics
Lodz University of Technology
Łódź, Poland

Maxim V. Zhigalov
Department of Mathematics and Modeling
Saratov State Technical University
Saratov, Russia

Anton V. Krysko
Applied Mathematics and Systems Analysis
Saratov State Technical University
Saratov, Russia

Vadim A. Krysko
Department of Mathematics and Modeling
Saratov State Technical University
Saratov, Russia

ISSN 1869-8433 ISSN 1869-8441 (electronic)
Advanced Structured Materials
ISBN 978-3-030-55992-2 ISBN 978-3-030-55993-9 (eBook)
https://doi.org/10.1007/978-3-030-55993-9

This Springer imprint is published by the registered company Springer Nature Switzerland AG
The registered company address is: Gewerbestrasse 11, 6330 Cham, Switzerland

Preface

This book is devoted to researchers and teachers, as well as graduate students, undergraduates and bachelors in mechanical engineering, nanomechanics, nanomaterials, nanostructures and applied mathematics. It serves as a research monograph which collects the latest developments in the field of nonlinear (chaotic) dynamics of mass distributed-parameter nanomechanical structures. It can be helpful for scientists and specialists interested in a rigorous and comprehensive study of modelling nonlinear phenomena governed by PDEs. The monograph has a unique pedagogical style that is particularly suitable for independent study and self-education for many researchers and specialists who do not have time to attend classes and lectures on the subject of the monograph. In addition, the book contents stand for a good balance between Western and Eastern extensive studies of the mathematical problems of nonlinear vibrations of structural members.

The authors of the proposed book work for many years in the theoretical aspects of nonlinear dynamics of mechanical macroscale and nanoscale structures and mathematical methods for solving problems governed by nonlinear PDEs. In spite of numerous published papers, the following companion books dealing with similar problems have been recently published:

1. J. Awrejcewicz, V.A. Krysko, I.V. Papkova, A.V. Krysko, *Deterministic Chaos in One-Dimensional Continuous Systems*. World Scientific, New Jersey, 2016;
2. V.A. Krysko, J. Awrejcewicz, M.V. Zhigalov, V.F. Kirichenko, A.V. Krysko, *Mathematical Models of Higher Orders: Shells in Temperature Fields*. Springer, Switzerland, 2019.

The authors employ the accumulated experience and knowledge of the long years of their research in the proposed monograph. This book will allow readers to obtain their own new results in the field of nonlinear dynamics of continuous mass (distributed-parameter) mechanical nanostructures, based on employing the ideas contained in this monograph to design novel elements of micro/nanomechanical systems, which are challenging in development of the modern engineering world. The book material offers guidelines for the development of many sensory and executive algorithms/functions used throughout, from the largest cargo ships to the smallest handheld electronic devices and from the most advanced scientific and medical equipment to the simplest household items. This monograph is one of the first on the mechanical engineering market because the authors analyse vibrations of nanostructures based on various theories of elasticity of the higher order including the modified coupled stress theory, surface theory, nonlocal theory, gradient theory and their modifications.

The bulk of the literature on nano-objects is devoted to research in nanomechanics, nanocomposites, the theory of dislocation mechanics, etc. The issues of strength, durability and time-dependent deterioration of mechanical properties, which are the main problems for design engineers, are considered. However, majority of the dynamical problems reported in the available literature are presented for strongly order reduced systems, i.e. for the governing equations of one degree of freedom systems of the Duffing type. Moreover, there are no books on nonlinear dynamics, in particular, chaotic dynamics, for nanomechanical structures in which systems with an infinite number of degrees of freedom are studied. The issues of the "truth of chaos" are not analysed, and the scenarios of the transition from periodic to chaotic vibrations exhibited by nanomechanical systems are not satisfactorily investigated. There are also very few studies regarding dynamics of nanomechanical structures based on wavelet analysis and analysis of the largest Lyapunov exponents, while there are practically no results supported by consideration of a spectrum of Lyapunov exponents. The literature state of the art shows that there are no works devoted to the study of nanoeffects and the effect of temperature action, which play a crucial role in obtaining a reliable picture of nanostructural nonlinear dynamical systems embedded into temperature fields.

The book offers a rigorous mathematical approach and verifies numerous theory-based modelling of structural members with an emphasis on microelectromechanical structures (MEMS) and nanoelectromechanical structures (NEMS), whose nonlinear dynamics plays an important role in current research observed in applied physics and engineering.

There are no competing publications on the market for the book (except perhaps the two already mentioned), and therefore the book may have a significant impact on both theoretical- and application-oriented researchers interested in nonlinear features of nanostructural members.

The authors of this monograph are intended to fill gaps in the above-mentioned problems.

We would like to acknowledge that a part of the book material has been already published in the form of papers. We have obtained permission to reuse the mentioned material in our book.

In the case of the papers: J. Awrejcewicz, A. V. Krysko, N. P. Erofeev, V. Dobriyan, M. A. Barulina, V. A. Krysko, "Quantyfying chaos by various computational methods. Part 1: Simple systems", *Entropy*, 20(3), 2018, 175 and J. Awrejcewicz, A. V. Krysko, N. P. Erofeev, V. Dobriyan, M. A. Barulina, V. A. Krysko, "Quantyfying chaos by various computational methods. Part 2: Vibrations of the Bernoulli-Euler beam subjected to periodic and colored noise",

Entropy, 20(3), 2018, 170, the permission is not required, since they were published under full open access (Chaps. 3 and 5).

The same holds for the paper J. Awrejcewicz, V. A. Krysko, S. Pavlov, M. V. Zhigalov, L. A. Kalutsky, A. V. Krysko, "Thermoelastic vibrations of a Timoshenko microbeam based on the modified couple stress theory", *Nonlinear Dynamics*, 2020, 99, 919–943, material of which has been used in Chap. 8.

Chapter 4 is based on the paper V. A. Krysko, J. Awrejcewicz, I. V. Papkova, O. A. Saltykova, A. V. Krysko, "On reliability of chaotic dynamics of two Euler-Bernoulli beams with a small clearance", *International Journal of Non-Linear Mechanics*, 104, 2018, 8–18 (licence 4858731296328).

Chapter 6 is based on the papers: A. V. Krysko, J. Awrejcewicz, M. V. Zhigalov, S. P. Pavlov, V. A. Krysko, "Nonlinear behaviour of different flexible size-dependent beams models based on the modified couple stress theory. Part 1. Governing equations and static analysis of flexible beams", *International Journal of Non-Linear Mechanics*, 93, 2017, 96–105 (licence 4858731357453) and A. V. Krysko, J. Awrejcewicz, M. V. Zhigalov, S. P. Pavlov, V. A. Krysko, "Nonlinear behaviour of different flexible size-dependent beams models based on the modified couple stress theory. Part 2. Chaotic dynamics of flexible beams", *International Journal of Non-Linear Mechanics*, 93, 2017, 106–121 (licence 4858731418376).

Chapter 7 is based on the papers: J. Awrejcewicz, A. V. Krysko, S. P. Pavlov, M. V. Zhigalov, V. A. Krysko, "Chaotic dynamics of size-dependent Timoshenko beams with functionally graded properties along their thickness", *Mechanical Systems and Signal Processing*, 93, 2017, 415–430 (licence 4865190257673), J. Awrejcewicz, A. V. Krysko, S. P. Pavlov, M. V. Zhigalov, V. A. Krysko, "Stability of the size-dependent and functionally graded curvilinear Timoshenko beams", *Journal of Computational and Nonlinear Dynamics*, 12(4), 2017, 041018 (licence 1047522-1), A. V. Krysko, J. Awrejcewicz, I. E. Kutepov, V. A. Krysko, "Stability of curvilinear Euler-Bernoulli beams in temperature fields", *International Journal of Non-Linear Mechanics*, 94, 2017, 207–215 (licence 4858620136357) and A. V. Krysko, J. Awrejcewicz, S. P. Pavlov, M. V. Zhigalov, V. A. Krysko, "Mathematical model of a three-layer micro- and nanobeams based on the hypotheses of the Grigolyuk-Chulkov and the modified couple stress theory", *International Journal of Solids and Structures*, 117, 2017, 39–50 (licence 4858611485608).

Chapter 9 is based on the papers: A. V. Krysko, J. Awrejcewicz, S. P. Pavlov, K. S. Bodyagina, V. A. Krysko, "Topological optimization of thermoelastic composites with maximized stiffness and heat transfer", *Composites Part B*, 2019, 158, 319–327 (licence 4860611503624) and A. V. Krysko, J. Awrejcewicz, S. P. Pavlov, K. S. Bodyagina, M. V. Zhigalov, V.A. Krysko, "Non-linear dynamics of size-dependent Euler-Bernoulli beams with topologically optimized microstructure and subjected to temperature field", *International Journal of Non-Linear Mechanics*, 104, 2018, 75–86 (licence 4858620075449).

We greatly appreciate the help and support of the Editors Holm Altenbach and Andreas Öchsner.

In addition, the help of Mr. Marek Kaźmierczak in the final book preparation is highly appreciated.

Finally, we would like to acknowledge the support of the Russian Science Foundation RSF No, grant and Polish National Science Centre under the grant OPUS 14 No. 2017/27/B/ST8/01330.

Lodz, Poland Jan Awrejcewicz
Saratov, Russia Anton V. Krysko
Saratov, Russia Maxim V. Zhigalov
Saratov, Russia Vadim A. Krysko

Introduction

Currently, in connection with the rapid development of technology and instrumentation, an important issue stands for the creation of devices as small as possible. Here, we should recall the words of the great American Nobel laureate in Physics, Richard Feynman, who was the first to predict the fabrication of devices at the nanolevel. He presented a lecture given at the annual American Physical Society meeting at Caltech on December 29, 1959, entitled *"There's Plenty of Room at the Bottom: An Invitation to Enter a New Field of Physics"*. The word "bottom" in this phrase means a world of very small sizes, which occupies the area defined by nanometers (1 nm = 10^{-9} m). The mechanics of a deformable solid can describe the elements of small-sized devices, but already here theories of higher order should be taken into account. At this stage in the development of continuum mechanics, a number of elasticity theories have emerged to overcome the problem. These include the modified couple stress theory of elasticity, the surface theory of elasticity, the nonlocal theory of elasticity, the gradient elasticity theory and their modifications.

The first chapter (Chap. 1) of this monograph gives an overview of studies on the dynamics of nanoshells, nanoplates and nanobeams in natural/thermal/electric/magnetic fields, obtained on the basis of some of the above-listed theories. The review of existing papers and monographs shows that the problem of analysing the statics and dynamics of MEMS/NEMS devices is only at the initial stage of development. Namely, most of the research is limited to the analysis of Duffing-type equations, which are obtained using the Bubnov-Galerkin variational method in the first approximation, thus searching for solutions of strongly reduced order models with one or two degrees of freedom. At present, when studying the problems of statics and dynamics of plates and shells, kinematic models of the first and second approximations are mainly employed. Moreover, the majority of the approaches are based on a linear formulation and deal with a small number of degrees of freedom. Often, in the problems under consideration, the solution is not entirely reliable. Furthermore, since it is obtained by a limited number of methods, their convergence is not shown or proved. Thus, the nonlinear dynamics of nanostructures has not been sufficiently investigated, and no reliable scenarios of the transition from periodic to chaotic vibrations have been revealed. The use of such a tool for studying nonlinear dynamics as wavelet analysis and the spectrum of Lyapunov exponents is at the initial stage of its development. Furthermore, chaotic vibrations, hyperchaotic vibrations, hyper-hyperchaotic and other types of vibrations are not sufficiently studied. The critical overview of the latest nonlinear dynamics of micro/nanostructures presented in this chapter demonstrates the need for innovative theoretical and numerical tools to achieve reliable and validated results.

The second chapter (Chap. 2) concerns the latest literature review devoted to micro/nano size-dependent mathematical models of beams, plates and shells. The review includes nonlocal theory of elasticity, surface theory of elasticity, modified couple stress theory and modified theory of deformation gradient taking into account higher order shear deformation theory. Particular emphasis is placed on nonlocal models of the Euler-Bernoulli and Timoshenko beams, Kirchhoff plates and Kirchhoff-Love shells. Both the review and the real-world vibrational behaviour of the size-dependent structural member imply the need to consider physical, geometric, material and design nonlinearity. It is expected that the novel approach based on employment of the earlier developed concepts of nonlinear dynamics like the Fourier and wavelet spectra, phase portraits, Poincaré maps, the Lyapunov exponents' computation (at least the largest one), the autocorrelation functions, etc. will improve mathematical models of partial differential equations (PDEs) to be in a good fit with experimentally observed nonlinear phenomena exhibited by the size-dependent structural members.

Chapter 3, since the book is devoted to study the nonlinear phenomena exhibited by the size-dependent structural members including bifurcations and chaotic processes, provides an overview of one of the main tools for identifying the nonlinear dynamics of these objects. Namely, the concept of Lyapunov exponents is briefly revisited, which allows us to distinguish between regular (periodic or quasi-periodic) and chaotic vibrations of the size-dependent beams, plates and shells studied in this book. In particular, the methods of Benettin, Wolf, Rosenstein and Kantz that based on Jacobian estimation and the neural network method are presented and discussed. As noted above, an important issue in solving problems of nonlinear dynamics, especially at the nanolevel, is the question of the reliability of chaotic oscillations. This problem was first identified by René Lozi in 2013. In this monograph, in order to obtain reliable results, it is proposed to achieve a coincidence not only of the basic functions during chaotic oscillations, but also of their second derivatives with respect to time. This question was formulated by the authors of the monograph in the book "Deterministic Chaos in One Dimensional Continuous Systems, World Scientific, Singapore, 2016". In addition, various types of definitions of chaos are given, and a methodology for identifying the truth of chaos is presented. Moreover, this chapter is devoted to the identification of truth of chaos and the reliability of the results using various methods for determining Lyapunov exponents. This question was investigated using numerical experiments based on classical simple nonlinear systems: Hénon map, hyperchaotic Hénon map, logistic map, as well as the Rössler and Lorenz systems. The case studies are analysed using the Fourier spectrum and wavelets of various types (Morlet,

Mexican hat, Haar, Daubechies and Gauss of various orders). Preference is given to the Morlet and Gaussian wavelet 32. It was shown that when analysing Lyapunov exponents, neural network method (this method was proposed by the authors of this monograph) makes it possible to calculate the spectrum of Lyapunov exponents and hence to identify phenomena such as the transition of the system into chaos, hyper-chaos, etc. in a fast and reliable way.

The fourth chapter deals with the methodology for detecting true chaos (in terms of nonlinear dynamics) and is developed on the example of a structure composed of two beams with a small clearance. The Euler-Bernoulli hypothesis is employed, and the contact interaction between beams follows the Kantor model. The complex nonlinearity results from the von Kármán geometric nonlinearity as well as the nonlinearity implied by the contact interaction. The governing PDEs are reduced to ODEs by the second-order finite difference method (FDM). The obtained system of equations is solved by Runge-Kutta methods of different accuracy. To purify the signal from errors introduced by numerical methods, the principal component analysis is employed and the sign of the first Lyapunov exponent is estimated by the Kantz, Wolf and Rosenstein methods.

Analysis of nonlinear oscillations of classical systems is carried out in Chap. 5. It is devoted to feasible methods for computation of Lyapunov exponents, since there is no universal, verified and general method to compute their exact (in numerical sense) values. This observation leads to the conclusion that there is a need to employ qualitatively different methods while checking the reliability of "true chaotic results". Furthermore, the analysis carried out in this chapter stands for a helpful tool for studying systems with infinite dimensions. We show that the most perspective and useful is the modified method of neural networks. It gives excellent convergence to the original results and, as the only one (besides the Benettin method), allows us to compute the spectrum of all Lyapunov exponents. In addition, very good results were obtained by the Rosenstein method for all studied systems. However, the latter approach can be used to estimate only the largest Lyapunov exponents.

In Chap. 6, mathematical models of nonlinear micro- and nanocylindrical panels in temperature fields are introduced and studied. First, the application of the modified couple stress theory of thermoelastic curvilinear panels based on the third-order hypotheses has been described. Then, a technical theory of the Sheremetev-Pelekh, Timoshenko and Bernoulli-Euler models is presented. The method of solving static problems is outlined in Sect. 6.4. Chaotic dynamics of the size-dependent flexible Bernoulli-Euler, Timoshenko and Sheremetev-Pelekh beams based on the modified couple stress theory of elasticity is investigated in Sect. 6.5. The two last sections are devoted to the construction of the so-called charts of vibration character with regard to amplitude and frequency of the external excitation and their study with the use of the first, second and third kinematic hypotheses.

Chapter 7 is devoted to the analysis of nonlinear functionally graded material (FGM) straight and curved beams' behaviour based on the modified couple stress theory. Defining the deflection curve in order to simplify the governing equations, we investigate the influence of scale length parameter, and non-homogeneity coefficient on the dynamic characteristics and the scenario of transition from periodic to chaotic beam vibrations. First, properties of various FGM materials are revisited (Sect. 7.2) with an emphasis on the dependence of material properties on temperature. Then, regular and chaotic vibrations of size-dependent Timoshenko beams with functionally graded properties along their thickness are illustrated and analysed (Sect. 7.3). The following new properties regarding the research topic under consideration can be derived.

(i) We have considered the dynamics of nonlinear FG Timoshenko beams on the basis of the modified couple stress theory, using a novel concept of the bending line.
(ii) The influence of the size-dependent coefficient and the grading parameter on the load-deflection dependence is investigated for the static problem. In order to get solutions for nonlinear static problem, the relaxation method was employed. It has been shown that the minimum deflection is achieved by the beam when the functional grading process is taken into account and the stiffer layer is located on the upper side in both cases, i.e. with/without the size-dependent behaviour.
(iii) The functionally graded beam with the stiffer layer on the upper side is suitable for application to carry dynamic loads for a given frequency and amplitude of the harmonic excitation. This conclusion coincides with that formulated for static problems. In the case of the homogeneous beam and the beam with the stiffer layer located on the bottom side, the essential dependence of the obtained results on the size-dependent coefficient is observed.
(iv) In order to validate the reliability of the largest Lyapunov exponents (LLEs) computed with Wolf's algorithm, three qualitatively different methods have been applied, i.e. Rosenstein's, Kantz's and the neural network one. The analysis implies that all methods give qualitatively the same result, i.e. positive or negative values of the LLEs, over the whole time interval studied.
(v) For the considered values of the size-dependent and material grading parameters, the universal route to chaos, following the classical Ruelle-Takens-Newhouse scenario, has been detected.

In Sect. 7.4, the size-dependent model based on a modified coupled stress theory has been constructed for the geometrically nonlinear curvilinear functionally graded Timoshenko beams. To build the model, the concept of a reference line was used. Compared to previous models for functionally graded beams, this model is significantly simpler. In the case of the straight-line beams, the size-dependent effect decreases the value of the beam deflection for the same coefficient of non-homogeneity. It has been reported that the localization of the most rigid layer on the upper side of the beam essentially decreases the deflections of the beam for the same value of the size-dependent coefficients. In the case of the curvature, the most loading ability is observed for the functionally graded beam with the most rigid layer located on the beam top. In the case of the LE computations, in all studied cases we have observed a transition from the negative to positive LE value, which is associated with the transition from the pre-critical to post-critical beam state. In the variants 2 and 8, there are two positive LE. It means that in these two cases, stiffer stability loss is exhibited. In Sect. 7.5, the investigation of stability of flexible curvilinear Euler-Bernoulli beams in a temperature field has been carried out without any restrictions regarding the temperature field distribution. It has been shown that the occurrence of imperfections due to either beam curvature or external load implies a different form of beam stability loss while increasing the temperature intensity. The type of temperature field has an essential impact on the beam stability loss regarding the temperature intensity and external loading. Inclusion of the beam curvature in the heat transfer equation yields an increase of the critical load responsible for the stability loss as well as changes of the beam form regarding its pre-critical state. Section 7.6 is devoted to the study of the mathematical model of a three-layer micro- and nanobeams. Based on both Grigolyuk-Chulkov and modified couple stress theories, the new model validated by both static and dynamic analyses of the three-layer microbeams including only one scalar/length parameter has been constructed, which takes into account the size effect. The employed Hamilton principle yielded the governing equation of motion as well as general boundary and initial conditions regarding displacements formulated for the microbeams. The proposed model of the microbeam deformation is one of the most simple models, and it includes the only one scalar length parameter. However, it allows us to take into account the microstructural effects in both external and internal beam layers for any boundary conditions. The finally formulated boundary value problem is of the sixth order, and in the case of the static problem it is solved analytically. The numerical results show that the studied beam model can explain the scale effect exhibited by the microbeams. The obtained deflections and stresses based on the introduced modified couple stress model are smaller compared to the classical three-layer Grigolyuk-Chulkov beam model while increasing beam thickness.

Thermoelastic vibrations of the Timoshenko microbeams based on the modified couple stress theory are studied in Chap. 8. In particular, the dependence of the quality factor of nonlinear microbeam resonators under thermoelastic damping for Timoshenko beams with regard to geometric nonlinearity is analysed. The constructed mathematical model is based on the modified couple stress theory which implies prediction of size-dependent effects in microbeam resonators. The Hamilton principle yields coupled nonlinear thermoelastic PDEs governing dynamics of the Timoshenko microbeams for both plane stresses and plane deformations. Nonlinear thermoelastic vibrations are investigated analytically and numerically, and quality factors of the resonators versus geometric and material microbeam properties are estimated. Results are presented for gold microbeams for different ambient temperatures and different beam thicknesses, and they are compared with results yielded by the classical theory of elasticity in linear/nonlinear cases. The most important conclusions of our study are summarized in the following three points.

1. Results reported in Figs. 8.4, 8.6 and Figs. 8.5, 8.7 with size-dependent behaviour and in the case of linear vibrations exhibit increase of the eigenfrequencies and increase of the quality factor of the beam resonator.
2. The occurrence of nonlinearity in both tested boundary (simple-simple) and (clamped-clamped) conditions implies a significant increase in the frequency of the fundamental vibration mode as well as the quality factor of the resonator (Figs. 8.8–8.11).
3. The analysis of the obtained results shows that in the case of the thermoelastic damping, it is necessary to take into account the nonlinear behaviour as well as size-dependent effects of the microbeams. Both mentioned features are crucial for the quality factor of the beam resonators.

In Chap. 9, problems associated with multifunctional requirements with respect to effective characteristics of composites consisting of two components as well as composites with holes or technological inclusions have been studied. In the process of investigation, a strong dependence of the optimal topology of the distribution of materials in the microstructure of composites on the form of the target functions has been detected. The study of transformations of the optimal topology of the composite microstructure with a change in the weight coefficient from $\omega = 0$ (maximization of the heat transfer) up to $\omega = 1$ (maximization of the mechanical moduli) has been conducted. Moreover, a set of alternatives optimal in the Pareto sense has been constructed. The considered examples clearly indicate the inability to achieve the best/required properties simultaneously in both cases, which is caused by conflicting criteria in the target function. In addition, nonlinear dynamics of the size-dependent Euler-Bernoulli beams embedded into temperature field with topologically optimized microstructure is studied. The following new results are presented:

(i) We have developed a mathematical model of the size-dependent nonlinear beam, taking into account the topological optimization under the criterion of maximum stiffness. The mathematical model is based on the Bernoulli-Euler, von Kármán and Duhamel-Newmann hypotheses. Also, an algorithm and a computer program for numerical computations of the optimized beam microstructure for the given boundary conditions, the form of external load and the temperature have been developed (both static and dynamic problems have been considered).

(ii) Reliability of the results was confirmed by investigating the convergence along the spatial variable as well as by examining the solution to the Cauchy problems, and investigating Lyapunov exponents and time evolution of the frequency obtained using wavelet analysis.

(iii) The analysis of the reliability of the results obtained for different numbers of spatial partitions was carried out based on the analysis of the power spectrum for chaotic system states. The reliability of chaos was validated by computing LLEs using four different methods.

(iv) The Cauchy problem was solved by numerous methods and the fourth-order Runge-Kutta method was chosen as the most efficient. The optimal step was chosen using the Runge principle.

(v) Based on the tests carried out for numerous wavelets (Daubechies, Gauss, Haar and Morlet), the Morlet wavelets were chosen as the most feasible for our problem.

(vi) The static analysis was carried out for three values of temperature and two values of size-dependent parameter. The investigated "frequency-deflection" dependency exhibits different results for homogeneous and non-homogeneous (optimized) beams for all values of the length-dependent parameter.

(vii) The use of beams with the optimized microstructure allows for an increase in the range of working loads regimes compared to homogeneous beams for which the vibration regimes are either periodic or quasi-periodic.

(viii) The analysis of the scenarios of transition from periodic to chaotic vibrations was carried out. In all cases, the transition into chaotic vibrations followed a scenario similar to the classical Pomeau-Manneville scenario (a few exceptions were observed for some values of temperature and the length-dependent parameter).

Contents

1 Nanostructural Members in Various Fields: A Literature Review 1
1.1 Introduction 1
1.1.1 Nanobeams 2
1.1.2 Nanoplates 7
1.1.3 Nanoshells 12
References 19

2 Size-Dependent Theories of Beams, Plates and Shells 25
2.1 Introduction 25
2.2 Literature Review 25
2.3 Non-classical (Size-Dependent) Models of Beams, Plates and Shells 28
2.3.1 Nonlocal Theory 28
2.3.2 Modified Couple Stress Theory 30
2.3.3 Modified Theory of a Gradient of Deformations 31
2.3.4 Surface Theory of Elasticity 33
2.3.5 Size-Dependent Theory of Beams, Plates and Shells 35
2.3.6 Size-Dependent Theory of Beams, Plates and Shells Based on the Shear Deformations of the First Order Under the Timoshenko Theory (TBT) 45
2.3.7 Size-Dependent Theory of Beams, Plates and Shells Based on the Shear Deformation Model of the Third Order 51
2.3.8 Nonlocal HSDT-Based Models 53
References 58

3 Lyapunov Exponents and Methods of Their Analysis 79
3.1 Introduction 79
3.2 Largest Lyapunov Exponent (LLE) 80

3.3 Spectrum of Lyapunov Exponents 81
3.4 Benettin's Method 82
3.5 Wolf's Method 84
3.6 Rosenstein's Method 85
3.7 Kantz Method 86
3.8 Method Based on Jacobian Estimation 86
3.9 Modification of the Neural Network Method 88
References 90

4 Reliability of Chaotic Vibrations of Euler-Bernoulli Beams with Clearance 93
4.1 Introduction 93
4.2 Literature Review 94
4.3 Mathematical Model 95
4.4 Principal Component Analysis (PCA) 100
4.5 Numerical Experiment 101
4.6 Application of the Principal Component Analysis (PCA) 108
References 111

5 Analysis of Simple Nonlinear Dynamical Systems 113
5.1 Introduction 113
5.2 Gauss Wavelets 113
5.3 Logistic Map 114
5.4 Hénon Map 114
5.5 Hyperchaotic Generalized Hénon Map 116
5.6 Rössler Attractor 122
5.7 Lorenz Attractor 124
References 129

6 Mathematical Models of Micro- and Nano-cylindrical Panels in Temperature Field 131
6.1 Introduction 131
6.2 Literature Review 131
6.3 Modified Couple Stress Theory of Thermoelastic Curvilinear Panels Based on the Third-Order Sheremetev–Pelekh, Timoshenko and Bernoulli–Euler Hypotheses 134
6.4 Technical Theory for the Sheremetev–Pelekh, Timoshenko and Bernoulli–Euler Models 147
6.5 Static Solutions 161
6.6 Chaotic Dynamics of the Size-Dependent Flexible Bernoulli–Euler, Timoshenko and Sheremetev–Pelekh Beams Within the Modified Couple Stress Theory of Elasticity 169
6.6.1 Analysis of Dynamic Characteristics of the Bernoulli–Euler, Timoshenko and Sheremetev–Pelekh Models for the Size-Dependent Beams Versus the Size Length 172

6.7 Construction of the Charts of Vibration Characters Verusus Amplitude and Frequency of the Exciting Load and Their Analysis (First-, Second- and Third-Order Kinematic Hypotheses) 184
6.7.1 Analysis of the Charts of the Vibration Regimes for the Euler–Bernoulli, Timoshenko and Sheremetev–Pelekh Modes Versus λ 184
6.7.2 Comparison of the Charts of the Vibration Regimes for One Chosen Model Versus the Relative Length λ with Account of the Size-Dependent Parameter l/h 190
6.8 Influence of a Type of Kinematic Models of the Zero-, First- and Third-Order Approximations on the Scenario of Transition from Periodic to Chaotic Vibrations 192
References 193

7 Mathematical Models of Functionally Graded Beams in Temperature Field 197
7.1 Introduction 197
7.2 Literature Review 199
7.3 Laws of Properties Change of FGM 202
7.3.1 Properties of Material P-FGM 202
7.3.2 Properties of Material E-FGM 204
7.3.3 Properties of S-FGM Material 205
7.3.4 Properties of Porous Materials 206
7.3.5 Homogenization of Properties of Graded Material Based on Mori-Tanaka and Self-consistent Methods 207
7.3.6 Dependence of Material Properties on Temperature 208
7.4 Chaotic Dynamics of Size-Dependent Timoshenko Beams with Functionally Graded Properties Along Their Thickness 210
7.4.1 Introduction 210
7.4.2 Mathematical Background 212
7.4.3 Derivation of the Equations of Motion 215
7.4.4 Statement of the Problem 217
7.4.5 Results and Discussions 219
7.5 Stability of the Size-Dependent and Functionally Graded Curvilinear Timoshenko Beams 231
7.5.1 Introduction 231
7.5.2 Theoretical Background 233
7.5.3 Derivation of Equations of Motion 235
7.5.4 The Methods of Analysis 238
7.5.5 Numerical Results 239

7.6 Stability of Curvilinear Euler-Bernoulli Beams in Temperature Fields 246
7.6.1 Introduction 246
7.6.2 Mathematical Model 248
7.6.3 Numerical Solution 251
7.6.4 Influence of Imperfections 255
7.7 Mathematical Model of Three-Layer Micro- and Nano-Beams Based on the Hypotheses of the Grigolyuk-Chulkov and the Modified Couple Stress Theory 261
7.7.1 Introduction 261
7.7.2 Theory of Bending Including Shear Effects and the Modified Couple Stress Theory 267
7.7.3 Equations of Motion and Boundary Conditions for the Three-Layer Beam 273
7.7.4 Static Transversal Bending of the Three-Layer Beam 276
7.7.5 Vibrations of a Three-Layer Beam 279
7.7.6 Numerical Results and Their Validation 282
References 285

8 Thermoelastic Vibrations of Timoshenko Microbeams (Modified Couple Stress Theory) 295
8.1 Introduction 295
8.2 Literature Review 296
8.3 Formulation of the Problem 301
8.3.1 Stress and Deformation Fields 301
8.3.2 Beam Equation of Motion Based on the Modified Couple Stress Theory 304
8.3.3 Coupled Problem of Thermoelasticity for Timoshenko Beams 306
8.4 Quality Factor of the Resonator with Thermoelastic Damping 308
8.5 Numerical and Analytical Results for a Rectangular Cross-Sectional Beam Resonator 314
8.5.1 Analytical Solution for the Linear Size-Dependent Beam 316
8.5.2 Analytical Solution for the Nonlinear Size-Dependent Beam 324
References 328

9 Vibrations of Size-Dependent Beams Under Topologic Optimization and Temperature Field 333
9.1 Introduction 333
9.2 Literature Review 334

9.3 Topological Optimization Problem 338
9.3.1 Methods Based on Density 338
9.3.2 Regularization 340
9.4 Applying Topological Cell Optimization 342
9.4.1 Flexibility Minimization 342
9.4.2 Thermoelastic Structures 343
9.4.3 Topological Optimization of Stress State 346
9.4.4 Topological Optimization in the Field of Thermoelasticity 348
9.4.5 Interpolating Schemes 351
9.5 Reactions on Boundary 353
9.5.1 Reaction on Boundary and Their Sensitivity 353
9.5.2 Formulating Problems of Topological Optimization 354
9.6 Topological Optimization of Thermoelastic Composites with Maximized Stiffness and Heat Transfer 357
9.6.1 Introduction 357
9.6.2 Determination of the Effective Tensor of Elastic Properties 360
9.6.3 Determination of the Effective Heat Transfer Coefficient 364
9.6.4 Topological Optimization 365
9.6.5 Numerical Results 366
9.7 Non-Linear Dynamics of Size-Dependent Euler-Bernoulli Beams with Topologically Optimized Microstructure and Subjected to Temperature Field 373
9.7.1 Introduction 374
9.7.2 Mathematical Background 377
9.7.3 Results and Discussions 384
References 396

Chapter 1
Nanostructural Members in Various Fields: A Literature Review

1.1 Introduction

An overview of studies on the dynamics of nanoshells, nanoplates and nanobeams in natural/thermal/electric/magnetic fields, obtained on the basis of some of the above-listed theories of higher order, is given. The review of existing papers and monographs shows that the problem of analysing the statics and dynamics of MEMS/NEMS devices is only at the initial stage of development. Namely, most of the research is limited to the analysis of Duffing-type equations, which are obtained using the Bubnov-Galerkin variational method in the first approximation, thus searching for solutions of strongly reduced order models with one degree of freedom. At present, when studying the problems of statics and dynamics of plates and shells, kinematic models of the first and second approximations are mainly employed. Moreover, the majority of the approaches are based on a linear formulation and deal with a small number of degrees of freedom. Often, in the problems under consideration, the solution is not entirely reliable. Furthermore, since it is obtained by a limited number of methods, their convergence is not shown or proved. Thus, the nonlinear dynamics of nanostructures has not been sufficiently investigated, and no reliable scenarios of the transition from periodic to chaotic vibrations have been revealed. The use of such a tool for studying nonlinear dynamics as wavelet analysis and the spectrum of Lyapunov exponents is at the initial stage of its development. Furthermore, chaotic vibrations, hyperchaotic vibrations, hyper-hyperchaotic and other types of vibrations are not sufficiently studied. The critical overview of the latest nonlinear dynamics of micro/nanostructures presented in this chapter demonstrates the need for innovative theoretical and numerical tools to achieve reliable and proven results.

J. Awrejcewicz et al., *Mathematical Modelling and Numerical Analysis of Size-Dependent Structural Members in Temperature Fields*, Advanced Structured Materials 142,
https://doi.org/10.1007/978-3-030-55993-9_1

1.1.1 Nanobeams

1.1.1.1 Natural Environment

Guo and Zhao [1] developed a theoretical background aimed at a study of the size-dependent bending elastic properties of nanobeams with surface effects.

Lim and Wang [2] employed the exact variational nonlocal stress modelling with asymptotic higher order strain gradients for nanobeams.

Wang et al. [3] analysed vibrations of initially stressed micro- and nanobeams.

Challamel and Wang [4] solved a paradox associated with the small-scale effect for a nonlocal cantilever beam. Lim et al. [5] illustrated the stiffness strengthening effects of nonlocal stress and axial tension on free vibration of cantilever nanobeams.

Aydogdu [6] proposed a general nonlocal beam theory to analyse bending, buckling and free vibration of nanobeams based on the classical Euler-Bernoulli, Timoshenko, Reddy and Levinson theories.

Zhang et al. [7] analysed bending and vibration of hybrid nonlocal beams.

Murmu and Adhikari [8] studied nonlocal transverse vibration of double-nanobeam systems within the framework of Eringen's nonlocal elasticity theory. The following results were reported: nonlocal natural frequencies are smaller than the corresponding local frequencies, small-scale effects are higher with increasing values of nonlocal parameter, whereas increase of stiffness of coupling springs reduces the nonlocal effects.

Li et al. [9] derived a sixth-order PDE governing dynamics of simply supported nanobeams under initial axial force based on nonlocal elasticity theory. The effects of the nonlocal nanoscale and dimensionless axial force on the first twice mode frequencies were presented and discussed.

Hosseini-Hashemi et al. [10] investigated the surface effects (elasticity, stress and density) of free vibrations of the Euler-Bernoulli and Timoshenko nanobeams employing the nonlocal elasticity theory. The governing PDEs were studied with regard to three different boundary conditions, i.e. simple-simple, clamped-simple and clamped-clamped. In particular, it was shown that rotary inertia and shear deformation had more effects on the surface than the nonlocal parameter.

Simsek [11] proposed a novel size-dependent beam model for nonlinear free vibration of a functionally graded nanobeam by matching the nonlocal strain gradient theory and the Euler-Bernoulli beam theory with an account of the von Kármán's nonlinearity. Hamilton's principle yielded the governing PDEs and boundary conditions. A few case studies were supplemented pointing out the important features of the strain gradient length scale, the nonlocal parameters, vibration amplitude and various material compositions.

Hashemi and Khaniki [12] studied free vibrations of a Timoshenko nanobeam with variable cross section in frame of the nonlocal elasticity theory. The small-scale effects were modelled by Eringen's nonlocal elasticity theory. They derived an analytical solution with regard to the Timoshenko nanobeams for three different boundary conditions. They also illustrated how the nanoscale effects change the Timoshenko beam natural frequencies.

Sahmani and Aghdam [13] investigated the size-dependent nonlinear free vibration response of multi-layer functionally graded graphene platelet-reinforced composite nanobeams. Both of the hardening/softening stiffness were taken into considerations within the framework of the third-order shear deformation beam theory and nonlocal strain gradient elasticity theory. The non-classical governing PDEs were obtained, and next a perturbation technique in conjunction with the Galerkin method yielded an explicit analytical solution for nonlocal strain gradient nonlinear frequency of the studied nanobeams.

Jena and Chakraverty [14] studied the free vibration of nanobeams based on nonlocal Euler-Bernoulli theory and the developed differential transform method. The latter allowed to transform the governing differential equations to the algebraic equations. The numerical results for different scaling parameters and four boundary conditions were presented and discussed.

Khaniki [15] studied vibrations of nanobeams based on the modified Eringen's two-phase local/nonlocal integral model. Three different examples including in-phase vibration, out-phase vibration and fixation of the under heath beam layer were analysed. It was shown, among others, that the elastic coupling term and nonlocal parameters had a significant effect on the natural frequencies of the studied nanobeams.

1.1.1.2 Thermal Environment

Jiang et al. [16] determined the thermal extension coefficient of carbon nanotubes using an analytical approach.

Yan and Han [17] investigated the torsional and axially compressed buckling of multi-walled and double-walled carbon nanotubes versus temperature change.

Lee and Chang [18] found a closed-form solution while investigating the critical buckling temperature of single-walled carbon nanotubes under a uniform temperature rise.

Tounsi et al. [19] studied the small size effects on wave propagation in double-walled carbon nanotube subjected to temperature.

Lim and Yang [20] employed the variational principle and integrated the straining energy density of a nanobeam under thermal field based on originally developed higher order differential equation and the corresponding boundary conditions. The effects of the nonlocal nanoscale and temperature on the nanobeams transverse deflection were illustrated and analysed. It was concluded that at low and room temperature the nanobeams' transverse deflection decreased with increase in temperature difference, whereas at higher temperature the transverse deflections increased as the temperature difference increased.

Chen et al. [21] predicted the damping behaviour of the nanobeams including thermal fluctuations and the paddling effect.

Youssef et al. [22] developed the model of vibration of gold nanobeam induced by laser pulse heating in framework of two-temperature generalized thermoelasticity and non-Fourier heat conduction. It was shown through numerical results the effects of the two-temperature parameter and the laser pulse parameters on the damping of energy accumulated inside the beams.

Leijssen and Verhagen [23] demonstrated experimentally optomechanical interactions in a sliced photonic crystal nanobeam. In particular, they showed how the large interaction enables detection of the thermal motion with detection noise below that associated with standard quantum limit.

Hoang [24] studied thermoelastic damping depending on vibration nanobeam modes by using the finite element method. He showed that a properly carried out optimization of the resonant beam dimensions may essentially increase its quality factor.

Ebrahimi and Salari [25] studied the thermal action on buckling and free vibration characteristics of functionally graded and size-dependent Timoshenko nanobeams under in-plane thermal loading. The scale effect was studied based on Eringen's nonlocal elasticity theory. The effects of a few parameters including thermal effect, material distribution profile, small size effects, beam thickness and mode number on the critical buckling temperature and normalized natural frequencies of the temperature-dependent nanobeams were analysed.

Ebrahimi and Barati [26] derived an analytical model of inhomogeneous functionally graded nanobeam in thermal environment on a basis of nonlocal strain gradient theory. The temperature across the nanobeam thickness was distributed in a nonlinear way. The reported numerical examples allowed to observe how the characteristics of the wave propagation of nanobeams depend on the nonlocality parameter length scale parameter, gradient index and temperature changes.

Ghadiri et al. [27] derived the equation of motion for a rotating nanocantilever based on the Euler-Bernoulli beam model. The effects of temperature, angular velocity and small scale were studied. Increase of the non-dimensional frequency of the first mode was implied by an increase of the nonlocal parameter.

Ebrahimi and Barati [28] studied thermal effects on the buckling of functionally graded nanobeams subjected to different types of thermal loading. The derived PDEs were solved analytically. In particular, the effects of the power-law index, nonlocal parameter slenderness ratio and thermal loading were illustrated and discussed.

Ebrahimi and Barati [29] carried out the analysis of surface and thermal effects on the vibration characteristics of viscoelastic foundation by utilizing the nonlocal strain gradient elasticity theory, the Euler-Bernoulli beam model and the Gurtin-Murdoch elasticity theory. The Hamilton principle yielded the governing equations, which were solved analytically for simple-simple and clamped-clamped boundary conditions. Effects of linear, shear and viscous layers of foundation, structural damping coefficient, surface elasticity, length scale parameter, nonlocal parameter, temperature change, and slenderness ratio of the nanobeam frequencies were exhibited.

Abouelregal and Zenkour [30] studied the vibrational response of thermoelastic nanobeam resonators subjected to ramp-type heating and exponential decaying time-varying load based on the Euler-Bernoulli beam theory. The small-scale effects were captured by a nonlocal parameter. The effects of nonlocal, point load and ramping-time parameters on the nanobeam vibrations were investigated.

Shafiei et al. [31] analysed transverse vibration of rotary functionally graded size-dependent tapered Euler-Bernoulli nanobeam in thermal environment. Nonlocal equations of motions were yielded by Hamilton's principle, and they were solved by the differential quadrature method. The following important parameters influence on the nanobeams flapwise bending vibration were considered: angular velocity, material distribution profile, boundary conditions, small-scale parameter and rate of cross-sectional change.

Shahabinejad et al. [32] analysed free vibrations of rotating functionally graded nanobeams under in-plane thermal loading. The Euler-Bernoulli beam theory, Hamilton's principle and the small-scale effect based on the Eringen elasticity theory were used. Free vibration frequencies were estimated for cantilever and proposed cantilever boundary conditions.

Arefi and Zenkour [33] employed the analytical approach for estimation of the thermal stresses and deformations of a curved nanobeam resting on Pasternak's foundation. Influence of the following important parameters on the thermal stresses was carried out: spring and shear parameters, thermal loads, nonlocal parameter, and beam curvature radius.

1.1.1.3 Electric Field

McCutcheon et al. [34] demonstrated experimentally high-quality factor of dual-polarized photonic crystal nanobeam cavities.

Hu et al. [35] studied experimentally self-heating and external strain coupling induced phase transition in a nanobeam. They demonstrated the accompanied gigantic change in resistivity and optical transmittance.

Liang and Shen [36] analysed the effect of an electrostatic force on an Euler-Bernoulli piezoelectric nanobeam dynamics. Influence of the electrostatic force on the first four natural frequencies was demonstrated, and a possibility of adjusting the natural frequency of a nanobeam by using voltage control was shown.

Stabile et al. [37] presented experimental results of the equilibrium/non-equilibrium transport properties of vanadium oxide nanobeams near the metal-insulator transition. It was shown a crucial role of both temperature and electric fields in the transitional process, and that both fields can be separated.

Wang et al. [38] investigated vibration of nanowires based on the Timoshenko beam model embedded in electric field via molecular dynamics simulation. It was demonstrated an increase/decrease of the natural frequencies by increasing of positive/negative electric field in polarization direction. The vibration frequencies of the cantilever Timoshenko beam with axial force were estimated versus the employed electric field.

Phunpeng et al. [39] studied piezoelectric and flexoelectric effects of nanobeam by using finite element method.

Arefi et al. [40] investigated free vibration responses of piezoelectric sandwich curved nanobeams resting on Winkler-Pasternak foundation in the framework of the nonlocal elastic theory and higher order shear deformation theories. The Hamilton principle yielded the governing equations of motion and boundary conditions. The carried out numerical analysis allowed to study the influence of important parameters like nonlocal parameter, the volume fraction, different boundary conditions, the external electric field and dimensionless geometric parameters of the dynamic characterizes of the nanobeams.

Jasulaneca et al. [41] reviewed research devoted to switch architectures and structural elements in the field of electrostatically actuated nanobeam-based nanoelectromechanical structures. Experimental results overview was presented focused on reliability issues and of the operating environment.

Arefi et al. [42] considered nonlocal magneto-electro-thermo-elastic behaviour of the functionally graded nanobeams subjected to magneto-electro-elastic loads. The governing equations were derived based on the third-order shear deformation theory of beams, the principle of virtual work and the nonlocal magneto-electro-thermo-elastic relations. The nanobeams were under transverse loads and electric/magnetic potentials. They reported electric and magnetic potential distributions through the nanobeam thickness as well as the influence of the chosen parameters including inhomogeneous parameter, electric and magnetic potential, nonlocal parameter and thermal load.

Arefi [43] analysed the thickness stretching effect in the framework on the shear and normal deformation for magneto-electro-elastic vibrations of a three-layered curved nanobeam with nanocore and two piezomagnetic layers. Eringen's nonlocal elasticity theory was utilized to emphasize the size dependency in the governing PDEs. Both analytical and numerical studies were employed. In particular, the influence of the following parameters was investigated, electro-magneto-mechanical load, size-dependent parameter, opening angle, Pasternak's foundation parameter and core thickness.

1.1.1.4 Magnetic Field

Firouz-Abadi and Hosseinian [44] analysed the resonance frequency and stability of the nanobeams embedded into a longitudinal magnetic field with an account of the small-scale effect. The study includes the Lorentz forces and thermal stress effects. The governing equations were solved using the Galerkin method.

Karličić et al. [45] studied vibration of a cracked nanobeam in an elastic Winkler-type medium with an account of the effects of longitudinal magnetic field and temperature change. The considerations were based on the Euler-Bernoulli beam theory and the nonlocal elasticity as well as on the Maxwell classical equation. The influence of the nonlocal parameter, stiffness of rotational spring, temperature change and magnetic field on the vibration frequencies were investigated. The crack position versus boundary conditions was also analysed.

Baghani et al. [46] illustrated and discussed the effects of magnetic field, surface energy and compressive axial load on the nanobeam dynamics and stability. The vibration frequencies and critical buckling loads of the nanobeam were estimated using the differential quadrature method. It was demonstrated that the magnetic field, surface energy and angular velocity play an important role in the dynamic and stability analysis of the nanobeams.

Du et al. [47] reported the design, fabrication and characterization of a resonant Lorentz force magnetic field sensor based on dual-coupled photonic crystal nanobeam cavities. The resonance wavelength shift of a selected supermode of the coupled cavities caused by the Lorentz force-induced displacement was employed to achieve the optical transmission variation.

Alibeigi et al. [48] investigated the buckling response of nanobeams based on the Euler-Bernoulli model, the von Kármán geometric nonlinearity, and the modified couple stress theory under action of thermal, electric and magnetic loadings. The governing equation and boundary conditions were obtained by using the minimum potential energy principle. The problem was solved using the Galerkin approach with an account of size effect as well as length and thickness influence on the critical buckling temperature.

Kerid et al. [49] investigated the magnetic field, thermal loads and small-scale effects on vibrations of a nanobeam structure using the Euler-Bernoulli and Timoshenko beam theories. The resonance frequency change, the magnetic field intensity, the thermal load and small-scale effects were presented and discussed.

1.1.2 Nanoplates

1.1.2.1 Natural Environment

Alibeigloo [50] analysed dynamics of a nanoplate by employing 3D theory of elasticity and nonlocal continuum mechanics. A closed-form solution was proposed based on the state-space method in the thickness direction and Fourier series in the nanoplate directions. In particular, the effects of the nonlocal parameter, aspect ratio, thickness-to-length ratio and half wavenumbers on the frequencies were illustrated and discussed.

Yan and Jiang [51] investigated the surface effects on the vibration and buckling of a simply supported piezoelectric nanoplate employing a modified Kirchhoff plate model. The effects of the applied electric potential, the mode number, the plate aspect ratio and the plate thickness on the vibration frequencies were illustrated numerically. The authors detected a critical transition point where the combined surface effects on the critical electric voltage may disappear.

The vibration response of a double-piezoelectric-nanoplate system subjected to external electric voltage was studied by Asemi and Farajpour [52] where two nanoplates were coupled by a polymer matrix. The governing PDEs were derived based on the Hamilton principle, and both natural frequencies and critical electric voltages were estimated.

Salehipour et al. [53] used converted couple stress and three-dimensional elasticity conception to derive a model for static and vibration of functionally graded microplates and nanoplates. Developed model included small size effect. Hamilton's principle was used to create the equations of motion and boundary conditions. For in-plane and out-of-plane free vibrations of simply supported plates, analytical closed-form solutions were described. In order to achieve the analytical solutions, the elasticity modulus and mass density were assumed to differ exponentially by the thickness of the plate.

Jafari et al. [54] used nonclassical constitutive equations consisting of the first/second-order strain gradients. Navier and Galerkin methods were employed to solve the governing PDEs and obtain the approximate system outputs, respectively. The author studied the influence of different parameters, boundary conditions and the plate size on the natural frequencies of the nanoplates.

Ghassabi et al. [55] derived governing PDEs and the associated boundary conditions with an account of a nonlocal parameter based on Hamilton's principle. The Kirchhoff, Mindlin and the third-order shear deformation theories were employed. The analysed case studies included simply supported and cantilever nanoplates with an emphasis put to demonstrate a role of the dimensionless plate length, plate theory, power-low index and nonlocal parameter ratio on the vibration of the functionally graded rectangular nanoplates.

Mehdi Zarei et al. [56] studied free vibration and buckling of a round tapered nanoplate subjected to in-plane forces. Assumption of nonlocal resilience was used to demonstrate effects depending on the size. In order to obtain frequency equations for simply supported and clamped nanoplates, the Raleigh-Ritz method and differential transformation approach were used. The final results showed that incrementing the parameter of taper yielded rise of the buckling load and natural frequencies.

Shahrbabaki [57] employed the Ritz and Galerkin methods to study 3D nonlocal elasticity of rectangular nanoplate. Two simple cases of 3D free vibrations of simply supported nanoplate and wave propagation in 3D infinite nonlocal solid were studied. In particular, the author utilized novel trigonometric series as approximating functions while using Galerkin approach. Effects of length to thickness ratio, aspect ratio, nonlocal parameter and different boundary conditions influence on the natural frequencies of the nanoplate vibrations were analysed.

Despotovic [58] investigated the problem of stability and vibrations of square single-layer graphene sheet employing Eringen's approach. Natural frequencies of transverse vibrations versus the body length and nonlocality features were estimated based on the Galerkin method and the classical and nonlocal elasticity theories. Critical values of the body load parameter and the mode shapes were determined.

Singh et al. [59] carried out the vibration study of a nanoplate supported by Winkler foundations in the framework of the classical/ Eringen's elasticity theory. Effects of various nanoplates' parameters on the non-dimensional frequencies were illustrated.

Ajri et al. [60] studied the non-stationary free vibration and nonlinear behaviour of the viscoelastic nanoplates by employing the consistent couple stress theory. The plate material obeyed the Leaderman integral constitutive relation. The governing nonlinear second-order integral-partial differential equations were yielded by the Hamilton principle. The frequency and nanofrequency responses were presented. It was also demonstrated the amplitude-dependent damping mechanism.

1.1.2.2 Thermal Environment

Pelevic and Meer [61] investigated numerically heat transfer performed by flow over a plate surface with carbon nanofibers. The lattice Boltzmann model was employed. The obtained results exhibited a substantial heat transfer enhancement for a densely covered surface with carbon nanofibers of varying length.

Khorshidi et al. [62] determined the free vibration assessment of functionally graded rectangular nanoplates. The concept of nonlocal elasticity predicated on the idea of exponential shear deformation was applied to acquire the nanoplate's natural frequencies. In the principle of exponential shear deformation, exponential functions have been used to provide the effect of transverse shear deformation and rotating inertia in terms of thickness coordinates. The concept of nanolocal elasticity was employed to scrutinize the effect of the small scale on the natural frequency of the orthogonal nanoplate graded functionally. By enforcing the Hamilton concept, the governing equations and the corresponding boundary conditions were extracted.

Hosseini and Jamalpoor [63] used Eringen's nonlocal elasticity theory to illustrate the dynamic characteristics of a viscoelastic nanoplate-system subjected to temperature change. Two Kirchhoff nanoplates were combined with an internal viscoelastic medium of Kelvin-Voight and were restricted to the external elastic Pasternak's foundation. In addition, the effects of viscoelastic structural damping, higher order modes and damping coefficient of the viscoelastic medium on vibration characteristics were analysed. Numerical results suggested that surface elastic modulus and leftover surface stress significantly influenced natural frequency.

Mahmoud et al. [64] presented a new series of nanoplate nanocomposites made from polyazomethine/graphene in the form of PAMs/GNP. Pure PAMS and PAMs/GNP nanocomposites were also defined by available characterization methods, such as X-ray diffraction, scanning electron microscopy and conductivity tests. Pure PAMs and PAMs/GNP nanocomposites were thermally degraded in two steps. The degradation steps relied on the nature of the needed nanocomposites that was primarily connected to the decomposition of the content of pure PAMS. In addition, SEM images provided more convincing evidence for the structure of nanocomposites.

Malikan [65] studied the buckling analysis of the orthogonal nanoplate with different boundary conditions using the couple stress continuum. The simplified principle of first-order shear deformation was employed. The governing differential equations were yielded by the Hamilton principle. The results received were compared with molecular dynamic simulation.

Barati [66] proved that sort of vibration, temperature and moisture increase, strain gradient and nonlocal parameter, elastic foundation, material graduation and proportion of side to thickness have a significant impact on the vibration behaviour of double-layer nanoscale plates. A double-layered nanoplate was for the first time constructed by means of nonlocal strain gradient theory, which matched both rigidity-softening and rigidity-hardening effects. Consequences of magnetic and hydrothermal actions on double-layered nanoplates were also explored.

Barati [67] illustrated a nonlocal strain gradient plate framework for the vibration assessment of double-layered nanoplates in hygrothermal areas with linearly variable mechanical loads in the plane. The suggested hypothesis included two parameters of scale connected to the effects of nonlocal and strain gradients. The Hamilton concept was employed to derive governing equations of a nonlocal strain gradient nanoplate on an elastic medium. The consequences of various factors such as load factor, nonlocal and length scale parameters, increase in the percentage of moisture and temperature or boundary conditions on vibration characteristics of a double-layered nanoplate were illustrated.

1.1.2.3 Electric Fields

Pugno [68] analysed nanoelectromechanical (NEMS) three-dimensional systems. A general formula based on free energy for the treatment of statics and dynamics of three-dimensional NEMS was extracted and described in compliance with a classical/quantum mechanics. Nanoplates and nanowires were studied. The structural destabilization, which would result from the so-called pull-in voltage, would pertain to the device switch. The amplitude and frequency of the nanoplate thermal vibrations were assessed according to the voltage employed. The impact of the forces of van der Waals on the dynamics of nanoelectromechanical three-dimensional systems was presented. Based on the concept of uncertainty, the amplitude and frequency of oscillations were estimated.

Wenjun Yang et al. [69] studied the influence of flexoelectricity on the electromechanical coupling behaviour of a piezoelectric nanoplate that was simply supported by the Kirchhoff theory. The governing equations and the corresponding boundary conditions were derived from the principle of Hamilton, and the analytical solutions for deflection and natural frequency were obtained. The results showed that the deflections predicted by the current model were smaller than those calculated by the classical model. For thinner plates, the flexoelectric effect was more prominent. Differences in deflections or frequencies between the two models were gradually decreasing with an increase of the plate thickness. This paper can help to understand the mechanism of electromechanical higher order coupling.

Sobhy [70] described a functionally graded material (FGM) and its thermomechanical bending. Winkler springs with a variable modulus constituted one of the layers. Moreover, the plates were the other way round. By solving the one-dimensional heat conduction equation, the temperature was estimated. The plate's material features were established to be graded across the thickness of the panel. Numerous numerical outcomes were described, such as the impact of boundary conditions, power factor, plate aspect ratio and side-to-thickness bending ratio of FGM plates.

Park and Han [71] presented buckling of nonlocal magneto-electro-elastic nanoplates, which was examined on the basis of the theory of higher order shear deformation. The variation of the magneto-electro-elastic plate was established in compliance with the Maxwell equation and the magneto-electric boundary condition. Furthermore, Eringen's nonlocal differential theory was employed. The governing equations of nonlocal principle were derived by using the variation principle. Numerical results revealed the relationship between local and nonlocal hypotheses. The consequences of aspect ratio, nonlocal parameters and in-plate load directions on the reaction to buckling were analysed.

1.1.2.4 Magnetic Fields

Fujiwara et al. [72] investigated magnetic orientation of carbon nanotubes, which can be used in the engineering of nanometers. Nanotube alignment is essential for their anisotropic behaviour. The orientation of paramagnetic and diamagnetic substances and proteins in magnetic fields was examined. The exposure emerges from the energy of magnetic anisotropy and obeys the thermal equilibrium distribution of Boltzmann. The method could be applied for nanotubes in an isolated condition and therefore can be used to position gas and liquid phases in production and process.

Kiani [73] investigated behaviour of free in-plane and out-of-plane vibration of rectangular nanoplates, which were submitted to unidirectional in-plane magnetic fields of concern. The body forces on the nanoplate were acquired based on the theories of Kirchhoff, Mindlin and higher order plate. The couching described demonstrated the magnetic field's small scale. The frequencies of the nanoplates were assessed for the suggested models. The role of dimensional relations and magnetic field strength in both in-plane and out-of-plane frequencies was discussed.

Karlicić et al. [74] developed nanostructure of the graphene sheet. The impact of in-plane magnetic field on the viscoelastic orthotropic multi-nanoplate system (VOMNPS) inserted in a viscoelastic medium using nonlocal principle was investigated. The system of partial differential equations characterizing the free transverse vibration of VOMNPS under the uniaxial in-plane magnetic field applying the nonlocal elasticity of Eringen principle and the plate hypothesis of Kirchhoff taking into account the viscoelastic and orthotropic properties of nanoplates was derived and studied.

Amiri et al. [75] analysed the free vibration of circular magneto-electro-elastic (MEE) nanoplates. Those nanoplates were modelled using Kirchhoff's assumption in the context of the theory of nonlocal elasticity, in order to take into account the small-scale effect. The MEE nanoplate explored here was deemed to be completely clamped under external electrical end magnetic potentials. The influence of the magneto-electric potential on the system's destabilization was examined, and critical values of the potential employed were measured. A comprehensive numerical investigation was carried out to study the effects of the nanoplate and piezoelectric volume fraction of the MEE material on the natural frequencies of the nanoplate with regard to small scale, thickness and radius.

In order to analyse the mechanical behaviour of a single-layer graphene sheet as an orthotropic nanoplate, Karličić et al. [76] applied the nonlocal theory of Kirchhoff-Love. Using the classical equations of Maxwell, the equation of motion of a simply supported nanoplate was derived. The analysis suggested that the sensitivity of nanomechanical detectors can be enhanced with the help of magnetic field.

Arani et al. [77] used for the first time a control feedback system to analyse the free vibration response of magnetic material (MsM). Due to the consideration of both normal and shear modulus, the Pasternak foundation was chosen to model the elastic medium. Nonlocal equations of motion were derived using the Hamilton principle and resolved using a differential quadrature method (DQM) taking into account various boundary conditions. The outcomes showed the impact of different parameters of MsNP's vibration behaviour, in particular, the control effect of velocity feedback gains to minimize the frequency.

Karami et al. [78] focused on the assessment of nanoplate propagation, made of functionally graded porous (FG) temperature-reliant materials based on the Winkler-Pasternak foundation in the magnetic in-plate field. The porosity distribution of nanoplates was regarded in accordance with the power-law rule as an even pattern. The Hamilton principle was used in connecting to the theory of nonlocal strain gradients to derive the governing equations based on the hypothesis of second-order shear deformation.

1.1.3 Nanoshells

1.1.3.1 Vibration

Arman et al. [79] studied the spectrum of vibrational radial modes in composite metal nanostructures. Metal nanoshells with dielectric core in an environment and bimetallic core-shell particles were investigated. For all of these nanostructures, the frequencies and damping rates of essential (breathing) modes with those of two higher order modes were measured together. For metal nanoshells, the frequency of breathing mode was invariably lower than the frequency of solid molecules of equal dimension, while damping was higher and enhanced with a reduction in the thickness of the shell. Two regimes that can be defined in the appearance of an external medium as weakly damped and overdamped vibrations were classified. For bimetallic particles, the frequency and damping rate depended periodically on the thickness of the shell with the period defined by the number of the mode. The frequency of higher modes was almost independent of the environment for both forms of nanostructures, while the damping rate demonstrated a powerful sensitivity to the external excitation.

Zaera et al. [80] analysed the free axisymmetric vibrations of a closed spherical nanoshell applying the hypothesis of nonlocal elasticity of Eringen. Bearing in mind the theories of thin shells, the motion equations were adequately derived, and the solution was acquired using the classical separation of variables. Compared to their local counterparts, the impact of the nonlocal parameter on natural frequencies and modal forms was debated. The carried out research may be helpful in biomedical and bioengineering-oriented nanotechnology fields.

Ke et al. [81] investigated the thermo-electro-mechanical vibration of piezoelectric cylindrical nanoshells by applying nonlocal hypothesis and the concept of the thin shell of Love. From the principle of Hamilton, the boundary conditions and governing equations were derived. For the simply supported piezoelectric nanoshell, an analytical solution was provided by employing the double Fourier series of the displacement components. The method of differential quadrature (DQ) was used to acquire numerical solutions for piezoelectric nanoshells under different boundary conditions. The effect of the nonlocal parameter, increase in temperature, external electric voltage, radius-to-thickness and radius-to-radius ratio on the natural frequencies of piezoelectric nanoshells were extensively analysed. The nonlocal effect and thermoelectric loading were discovered to have a major impact on the natural frequencies of piezoelectric nanoshells.

Rouhi et al. [82] observed that when dimensions were in the order of nanometer, the surface stress effect played a key role in the mechanical behaviour of nanostructures. The free vibration behaviour of simply supported cylindrical nanoshells was investigated in the context of the surface elasticity principle, taking into account the aforementioned effect using an exact solution method. Energy-based approach was employed to receive the governing equations of motion and boundary conditions. In compliance with the Gurtin-Murdoch hypothesis, the effect of surface stress was estimated. The nanoshell was modelled using the principle of shell deformation of the first-order shear. The problem of free vibration was resolved due to the dimensionless form of governing equations and then explained using a Navier-type solution under the simply supported boundary conditions. Chosen numerical outcomes on the effects of surface stress and surface material features on the natural frequencies of different radii and lengths of nanoshells were illustrated. The findings indicated that the surface energies have a major impact on the vibrational behaviour of nanoshells with small thickness magnitudes. It was noted that the nanoshell's natural frequency depends on the properties of the surface material.

Mehralian et al. [83] investigated the size-dependent conical shell formulation. It was derived on the basis of the modified couple stress theory and the first-order shear deformation model in order to confirm the free vibration of functionally graded conical shell inserted in an elastic Pasternak medium and subjected to thermal conditions. The material properties were deemed temperature-dependent and classified according to the distribution of power law in the thickness direction. By using the principle of Hamilton, the boundary conditions and governing equations were derived. The size effect was assessed applying the modified hypothesis of couple stress, and the free vibration of simply supported conical nanoshell truncated by FG was examined as a special case. The consequences of various parameters such as dimensionless length parameter or change in temperature were studied on the basis of the modified couple stress and classical continuum theories.

Rouhi et al. [84] presented an analytical method to examine the geometrically nonlinear free vibrations of cylindrical nanoshells. The Gurtin-Murdoch continuum model was used to recreate the impact of surface stress. The equations governing the shell's nonlinear vibrations were derived applying energy-based technique, taking into account the effect of surface stress. To get the frequency-amplitude curves of nanoshells, a perturbation method was employed. In order to examine the vibrational behaviour of nanoshells with various geometric and surface material properties, different numerical outcomes were afforded. When the surface stress was extremely thin, it has been shown that the nonlinear free vibration behaviour of the nanoshells was greatly impacted. It was also proved that if the surface residual stress is negative, the impact of geometric nonlinearity is more apparent.

Razavi et al. [85] derived the electromechanical vibration equations of the cylindrical nanoshell produced of functionally graded piezoelectric material (FGPM) applying the consistent hypothesis of torque stress and the cylindrical shell model. By the energy technique and the concept of Hamilton, the governing equations and boundary conditions were defined. The free vibration of a unique FGPM nanoshell model under various boundary conditions was subsequently examined using Navier and Galerkin theories. The length-to-radius, radius-to-thickness ratio and dimensionless length scale parameter have been proved to play an important role in the vibration behaviour of the FGPM cylindrical nanoshell build on the size-dependent principle.

Barati [86] studied the free vibrational behaviour of porous nanoshells graded functionally applying the hypothesis of nonlocal strain gradient. A nonlocal parameter and a strain gradient parameter were used to define nanoshell stiffness reduction and stiffness improvement. Porositics were allocated evenly and unevenly thoroughly by nanoshell thickness. The power-law function was defined as the gradation of material properties with porosities. The nanoshell was constructed by the hypothesis of first-order shear deformation, and the technique of Galerkin was used to acquire vibration frequencies. In nonlocal strain gradient principle, shape functions that achieved the available classical and nonclassical boundary conditions have been presented. The vibrational behaviour of the nanoshell was influenced by the nonlocal and strain gradient coefficients, boundary conditions, fraction of the porosity volume, porosity distribution and radius-to-thickness ratio.

Sahmani et al. [87] proposed, in the context of the surface elasticity hypothesis, a size-dependent shell model that considered for geometrical imperfection sensitivity of the axial post-buckling characteristics of a cylindrical nanoshell created of functionally graded material (FGM). On the basis of the theory of virtual work, the nonclassical differential equations were derived and presumed from boundary layer-type. A perturbation-based solving methodology was then used to deduce the size dependence on the nonlinear instability of perfect and imperfect axially loaded FGM nanoshells with different shell thickness values, index of material property gradients and different uniform variations in temperature. In the case of thicker FGM nanoshells, in which the consequence of surface-free energy decrease, the impact of the initial geometric imperfection on the critical buckling load was higher than its impact on the minimum load in the post-buckling domain.

Barati [88] developed the free vibrational behaviour of nanoshells of porous nanocrystalline silicon applying the hypothesis of strain gradients. Nanocrystalline materials based on multi-phase composites in which nanopores, nanograins and interface phases contribute. The nanoshell was constructed by strain gradient hypothesis due to the experimental analysis of strain gradients close to the interface phase. To integrate the dimension of nanograins/nanopores and their surface energies, a micromechanical solution consisting of the Mori-Tanaka scheme was used. The nanoshell was modelled by the hypothesis of first-order shear deformation, and the technique of Galerkin was illustrated to acquire vibration frequencies. The vibrational behaviour of the nanoshell affected by the porosity proportion of nanograin size, the strain gradient ratio, boundary conditions and the nanograin/nanopore surface phase.

Razavi [89] studied free vibration of a simply supported magneto-electro-elastic doubly curved nanoshell in the appearance of a rotating inertia effect on the basis of the first-order shear deformation hypothesis. Gauss' electrostatic and magnetostatic principles were applied to model the electric and magnetic behaviours of the nanoshell. An analytical relationship was then acquired for the natural frequency of the double-curved magneto-electro-elastic nanoshell. In addition, the consequences of the electric and magnetic potential, the increase in temperature, nonlocal parameters, Pasternak foundation parameters and the geometry of the nanoshell on the natural frequencies of double-curved magneto-electro-elastic nanoshells were examined. It has been proved that natural frequency of described nanoshell reduces with rising temperature and electrical potential or reducing magnetic potential.

Shojaeefard et al. [90] analysed free vibration of a functionally graded piezomagnetic material cylindrical nanoshell. The nanoshell was nested in viscoelastic media under external electric, rotational and magnetic loadings. The nanoshell's equations were derived from the nonlocal hypothesis of Eringen. The structure's magnetic and piezoelectric properties exponentially changed in thickness. Due to the initial hoop tension, the rotational loading was measured. The outcomes were achieved by applying a generalized technique of differential quadrature to the governing equations and related boundary conditions. Angular velocity, external amperage and voltage parameter, viscoelastic media parameters and functionally graded power index impacted to the free vibration characteristics of nanoshell.

Karami et al. [91] investigated the porous nanoshell's wave propagation. In accordance with a higher order shear deformation shell hypothesis, the Bi-Helmholtz nonlocal strain gradient principle was used to involve size-dependent influence. The nanoshells were created of a functionally graded material (P-FGM), which varies continuously in the direction of thickness. The sensitivity of the wave reaction was studied for different fractions of the porosity volume, material features, nonlocal parameters, humidity, temperature, length scales of the strain gradient and wavenumbers. Depending on the outcome, the reaction size dependence was confirmed to be nearly the same as that of beams, plates and tubes.

1.1.3.2 Thermal Field

Seong-Kyun Cheong et al. [92] presented a new technique of calculation to estimate the thermal reaction of gold nanoshells nested in a tissue-like medium when lighted by a laser with a near-infrared (NIR). The heat formed by several gold nanoshells as a result of the photothermal effect was figured and connected with the outcomes for the medium without gold nanoshells in order to estimate the global temperature increase within the gold nanoshell-laden medium. After adjusting the model criteria to correctly provide for these differences, the computational outcomes and experimental data corresponded decently well to the average percentage difference of 10%.

Changhong Liu et al. [93] analysed the optical spectrum and near-field augmentation of a multi-layered gold nanoshell to discover its potential biological application. The mathematical model has been created in the context of a multi-layered concentrated sphere expansion. The analysis suggested that compared to a traditional single-layer Au-SiO_2 nanoshell, a multi-layer Au-SiO_2-Au nanoshell has the asset of achieving a concentrated surface plasmon resonance at a wavelength of 1300 nm or longer, which is thought to be more useful for coherent optical imagery with ultra-high resolution. With a single-layer nanoshell, an incredibly thin gold layer was needed for long-wave resonance and creating such a thin layer in the latest synthesis methods would be almost impossible.

Avetisyan [94] developed a strategy to analyse the temperature field of nanoparticles, taking into account the absorbed local intensity of composite spherical nanoparticles (nanoshells) with pulse laser radiation. This strategy enabled to examine the spatial inhomogeneities of the light field diffracted into a nanoshell and the resultant distribution of the absorption energy and to guarantee a mathematical solution for the equation of time-dependent heat conduction, taking into account the correlating spatially inhomogeneous distribution of heating sources. The detected influence had potential uses for monitored cell optoporation and nanosurgery in cell biology, as well as cancer cell killing.

Sahmani [95] studied the nonlinear destabilization of piezoelectric cylindrical nanoshells under the coupled radial of compression and electrical load, along with the influence of surface-free energy. The Gurtin-Murdoch elasticity hypothesis and the classical shell concept were used to create an efficient size-dependent shell model to take into account the surface effects. A linear modification of normal stress is presumed by the thickness of the bulk to achieve the balance conditions on the surfaces of nanoshells. In the transverse direction, electric field was also used. Employing the virtual work theory, nonlinear differential equations depending on the size were derived. The boundary layer hypothesis was subsequently used to incorporate the influence of surface-free energy in accordance with nonlinear pre-buckling deformation, initial geometric imperfection and large deflections in the post-buckling regime. Eventually, a singular two-step perturbation method was used to acquire the size-dependent critical buckling pressure and the consociated post-buckling equilibrium path for alternative electrical loadings. The electrical load enhances or reduces the critical buckling pressure and critical nanoshell end-shortening, which relied on the sign of the interpreted voltage. In addition, it was discovered that the influence of electrical load on the post-buckling behaviour of nanoshell rises by considering the effects of surface-free energy.

1.1.3.3 Electric Field

Fan and Huang [96] used a class of nonlinear optical materials consisting of single-domain ferromagnetic nanoparticles coated by a non-magnetic nanoshell with an inherent sensitivity to the second harmonic generation (SHG) in a non-magnetic host fluid. The SHG of these components had magnetic-field controllability, i.e. magnetic-field-controlled anisotropy, redshift and enhancement, which were induced by shift of the resonant plasmon frequency by the formation of the chains of the coated nanoparticles.

Tanabe [97] showed that the metal nanoparticles and nanoshells composed of metal shells and dielectric cores vastly improve the electromagnetic fields around them because of surface plasmons. Field enhancement coefficients for spherical metal nanoparticles and nanoshells in the quasistatic limit were measured using empirical dielectric constants depending on the wavelength. It prosecuted the relationship between the field enhancement factor and different parameters, such as distance from the nanoparticle/nanoshell, wavelength, dielectric core material surrounding medium, metal element and diameter ratio between the core and the shell. The peak field enhancement factor was the strongest for the value around 0.9 of a core-to-shell diameter ratio. Due to optimization of the parameter, it was discovered that an Ag nanoshell with a Teflon core and with a core-to-shell diameter ratio of 0.88 puts a peak field enhancement factor of 1400 in the area of water.

Weber et al. [98] developed an SiO_2/Au nanoshell's field improvement behaviour in a strong-field physics context. Concentrated plasmonic areas caused enhance in local electric fields with the potential influence on a strong-field regiment without the necessity to apply of expensive amplified lasers. Electrons were ionized from the nanoshell and speeded up by the local field, in which spectral and polarization properties were spatially inhomogeneous. By expanding the volume ratio between core and particles, the localized reaction to ultra-short femtosecond pulses could be examined. For applying ultra-short pulses centred at 800 nm, optimal geometric parameters of the nanoshell were chosen. Where femtosecond oscillators performed, the phase and amplitude of the remodifying of the incident pulse by the ultra-short response of the incident pulses can be mainly rectified using the active leasing of the shape to the ultra-short reaction of the medium. Classical free electron trajectories were determined to illustrate the inhomogeneous essence of a strong-field picture of local enhancement.

1.1.3.4 Magnetic Field

Moran et al. [99] analysed capacitive coupled shortwave radiofrequency fields, which resistively warmed up low concentrations (-1 ppm) of gold nanoparticles. Gold nanoparticles with a diameter under 50 nm heat at almost twice the rate of nanoparticle with over 50 nm diameter. The reason is the higher resistivity of smaller gold nanostructures. To illustrate this phenomenon, a Joule heating model was acquired. The model affords important insights into the logical design and architecture of nanoscale materials for non-invasive cancer therapy.

Zabow et al. [100] investigated different types of tunable magnetic resonance imaging agent founded on accurately sized cylindrical magnetic nanoshells. The nanoshells were manufactured using top-down prepatterned substrates. The redeposition of material back-sputtered during ion-milling was successively applied. The well-resolved magnetic resonance peaks of the consequent nanostructures confirmed the control of nanoscale manufacturing and the general viability of such sputter redeposition for the manufacture of a multitude of self-supporting, highly monodisperse nanoscale systems.

Vallecchi et al. [101] developed symmetric and antisymmetric polarization resonances at optical frequencies in tightly correlated nanoshell particles produced of either a metallic core and a dielectric shell or a dielectric core and a metal shell. The study was carried out using the single-dipole approximation (SDA) including all dynamically retarded field terms. In addition, for the first time, analytical equations for the four acceptable resonances were derived by maintaining only the static (non-retarded) term in the dipolar field expression. The image concept was used to differentiate a priori between symmetric and antisymmetric variants and to verify the verification of resonances achieved by the SDA for full-wave simulations. The resonance frequencies of a couple of nanoshells can be adjusted over a wide spectrum of wavelengths/frequencies by varying the core and shell's relative dimensions. This makes these particle pairs very suitable for use as components of metamaterials or in order to improve local fields when operating frequencies range from the visible to the infrared spectral areas.

Shu-Min et al. [102] investigated the electric field enhancement features of an active gold nanoshell with gain structure inside by using Mie hypothesis. As the inner core gain factor enhances to a critical value, a super-resonance occurs in the act gold nanoshell and enormous improvements in the electric fields can be located near the particle surface. The critical value of the gain coefficient for the super-resonance of the active gold nanoshell reduces and then enhances with increasing shell thickness, and the improved Raman scattering factor (G factor) also first enhances and then reduces with the corresponding surface. This optimized active gold nanoshell had a high-efficiency SERS effect and can be helpful for the detection of single molecule.

Gladilin et al. [103] applied the time-dependent formalization of Ginzburg-Landau, which was used to examine the vortex states and dynamics of vortex in superconducting spherical nanoshells. The latter were exposed to strong dc and weak ac magnetic fields. Shell thickness non-uniformity can have a significant impact on the ac magnetic reaction of a 3D array of superconducting nanoshells. Substantially, the reaction was greatly influenced not only by the pertinent geometric, material parameters and the frequency of the ac field, but also by the magnitude of the field used. By modifying the field, the real part of the effective ac magnetic permeability may be adjusted from positive values considerably larger than one down to negative values.

References

1. Guo, J.G., Zhao, Y.P.: The size-dependent bending elastic properties of nanobeams with surface effects. Nanotechnology **18**, 295701 (2007)
2. Lim, C.W., Wang, C.M.: Exact variational nonlocal stress modelling with asymptotic higher-order strain gradients for nanobeams. J. Appl. Phys. **101**, 054312 (2007)
3. Wang, C.M., Zhang, Y.Y., Kitipornchai, S.: Vibration of initially stressed micro- and nano-beams. Int. J. Struct. Stab. Dyn. **7**, 555–570 (2007)
4. Challamel, N., Wang, C.M.: The small length scale effect for a non-local cantilever beam: a paradox solved. Nanotechnology **19**, 345703 (2008)
5. Lim, C.W., Li, C., Yu, J.L.: The effects of stiffness strengthening nonlocal stress and axial tension on free vibration of cantilever nanobeams. Interact. Multiscale Mech. **2**(3), 223–233 (2009)
6. Aydogdu, M.: A general nonlocal beam theory: its application to nanobeam bending, buckling and vibration. Physica E **41**, 1651–1655 (2009)
7. Zhang, Y.Y., Wang, C.M., Challamel, N.: Bending and vibration of hybrid nonlocal beams. J. Eng. Mech. ASCE J. Eng. Mech. **136**(5), 562–574 (2010)
8. Murmu, T., Adhikari, S.: Nonlocal transverse vibration of double-nanobeam-systems. J. Appl. Phys. **108**, 083514 (2010)
9. Li, C., Lim, C.W., Yu, J.L., Zeng, Q.C.: Analytical solutions for vibration of simply supported nonlocal nanobeams with an axial force. Int. J. Struct. Stab. Dyn. **11**(2), 257–271 (2011)
10. Hosseini-Hashemi, S., Fakher, M., Nazemnezhad, R.: Surface effects on free vibration analysis of nanobeams using nonlocal elasticity: a comparison between Euler-Bernoulli and Timoshenko. J. Solid Mech. **5**(3), 290–304 (2013)
11. Simsek, M.: Nonlinear free vibration of a functionally graded nanobeam using nonlocal strain gradient theory and a novel Hamiltonian approach. Int. J. Eng. Sci. **105**, 12–27 (2016)
12. Hashemi, S.H., Khaniki, H.B.: Vibration analysis of the Timoshenko non-uniform nanobeam based on nonlocal theory: an analytical solution. Int. J. Nano Dimens. **8**(1), 70–81 (2017)
13. Sahmani, S., Aghdam, M.M.: Small scale effects on the large amplitude nonlinear vibrations of multilayer functionally graded composite nanobeams reinforced with graphene-nanoplatelets. Int. J. Nanosci. Nanotechnol. **14**(3), 207–227 (2018)
14. Jena, S.K., Chakraverty, S.: Free vibration analysis of Euler-Bernoulli nanobeam using differential transform method. Int. J. Comput. Mater. Sci. Eng. **7**(3), 1850020 (2018)
15. Khaniki, H.B.: On vibrations of nanobeam systems. Int. J. Eng. Sci. **124**, 85–103 (2018)
16. Jiang, H., Liu, R., Huang, Y., Hwang, K.C.: Thermal expansion of single wall carbon nanotubes. J. Eng. Mater. Technol. **126**, 265–270 (2004)
17. Yao, X., Han, Q.: Buckling analysis of multiwalledcarbon nanotubes under torsional load coupling with temperature change. J. Eng. Mater. Technol. **128**, 419–427 (2006)
18. Lee, H.L., Chang, W.J.: A closed-form solution for critical buckling temperature of a single-walled carbon nanotube. Physica E **41**, 1492–1494 (2009)
19. Tounsi, A., Heireche, H., Beerabah, H.M., Benzair, A., Boumia, L.: Effect of small size on wave propagation in double-walled carbon nanotubes under temperature field. J. Appl. Phys. **104**, 104301 (2007)
20. Lim, C.W., Yang, Q.: Nonlocal thermal-elasticity for nanobeam deformation: exact solutions with stiffness enhancement effects. J. Appl. Phys. **110**, 013514 (2011)
21. Chen, Ch., Ma, M., Liu, J.Z., Zheng, Q., Xu, Z.: Viscous damping of nanobeam resonators: humidity, thermal noise, and a paddling effect. J. Appl. Phys. **110**, 034320 (2011)
22. Youssef, H.M., El-Bary, A.A., Elsibai, K.A.: Vibration of gold nano beam in context of two-temperature generalized thermoelasticity subjected to laser pulse. Lat. Am. J. Solids Struct. **11**(13), 2460–2482 (2014)
23. Leijssen, R., Verhagen, E.: Strong optomechanical interactions in a sliced photonic crystal nanobeam. Sci. Rep. **5**, 15974 (2015)
24. Hoang, C.M.: Thermoelastic damping depending on vibration modes of nano beam resonator. Commun. Phys. **25**(4), 317–325 (2015)

25. Ebrahimi, F., Salari, E.: Thermal buckling and free vibration analysis of size dependent Timoshenko FG nanobeams in thermal environments. Compos. Struct. **128**, 363–380 (2015)
26. Ebrahimi, F., Barati, M.R.: Thermal environment effects on wave dispersion behaviour of inhomogeneous strain gradient nanobeams based on higher order refined beam theory. J. Therm. Stress. **39**(12), 1560–1571 (2016)
27. Ghadiri, M., Hosseini, S.H.S., Shafiei, N.: A power series for vibration of a rotating nanobeam with considering thermal effect. Mech. Adv. Mater. Struct. **23**(12), 1414–1420 (2016)
28. Ebrahimi, F., Barati, M.R.: Thermal buckling analysis of size-dependent FG nanobeams based on the third-order shear deformation beam theory. Acta Mech. Solida Sin. **29**(5), 547–554 (2016)
29. Ebrahimi, F., Barati, M.R.: Vibration analysis of viscoelastic inhomogeneous nanobeams resting on a viscoelastic foundation based on nonlocal strain gradient theory incorporating surface and thermal effects. Acta Mech. **228**, 1197–1210 (2017)
30. Abouelregal, A.E., Zenkour, A.M.: Thermoelastic response of nanobeam resonators subjected to exponential decaying time varying load. J. Theor. Appl. Mech. **55**(3), 937–948 (2017)
31. Shafiei, N., Ghadiri, M., Mahinzare, M.: Flapwise bending vibration analysis of rotary tapered functionally graded nanobeam in thermal environment. Mech. Adv. Mater. Struct. 1–17 (2017)
32. Shahabinejad, E., Shafiei, N., Ghadiri, M.: Influence of temperature change on modal analysis of rotary functionally graded nano-beam in thermal environment. J. Solid Mech. **10**(4), 779–803 (2018)
33. Arefi, M., Zenkour, A.M.: Thermal stress and deformation analysis of a size-dependent curved nanobeam based on sinusoidal shear deformation theory. Alex. Eng. J. **57**, 2177–2185 (2018)
34. McCutcheon, M.W., Deotare, P.B., Zhang, Y., Loncar, M.: High-Q transverse-electric/transverse-magnetic photonic crystal nanobeam cavities. Appl. Phys. Lett. **98**, 111–117 (2011)
35. Hu, B., Zhang, Y., Chen, W., Xu, Ch., Wang, Z.L.: Self-heating and external strain coupling induced phase transition of VO_2 nanobeam as single domain switch. Adv. Mater. **23**, 3536–3541 (2011)
36. Liang, X., Shen, S.-P.: Dynamic analysis of Bernoulli-Euller piezoelectric nanobeam with electrostatic force. Sci. China Phys. Mech. Astron. **56**, 1930–1937 (2013)
37. Stabile, A.A., Singh, S.K., Wu, T.-L., Whittaker, L., Banerjee, S., Sambandamurthy, G.: Separating electric field and thermal effects across the metal-insulator transition in vanadium oxide nanobeams. Appl. Phys. Lett. **107**, 013503 (2015)
38. Wang, L., Liu, S., Yi, J.: Vibration of Zinc oxide nanowires in electric field. In: ASME International Mechanical Engineering Congress and Exposition, vol. 4B, p. V04BT05A036 (2017)
39. Phunpeng, V., Aphirakmethawong, J., Baiz, P.M.: Numerical analysis of piezoelectric nanobeams with flexoelectric effect. In: Proceedings of 5th IIAE International Conference on Industrial Application Engineering 2017, pp. 378–381 (2017)
40. Arefi, M., Pourjamshidian, M., Arani, G.: Free vibration analysis of a piezoelectric curved sandwich nano-beam with FG-CNTRCs face-sheets based on various high-order shear deformation and nonlocal elasticity theories. Eur. Phys. J. Plus **133**, 193–200 (2018)
41. Jasulaneca, L., Kosmaca, J., Meija, R., Andzane, J., Erts, D.: Review: electrostatically actuated nanobeam-based nanoelectromechanical switches materials solutions and operational conditions. Beilstein J. Nanotechnol. **9**, 271–300 (2018)
42. Arefi, M., Arani, A.H.S.: Higher order shear deformation bending results of a magneto-electrothermoelastic functionally graded nanobeam in thermal, mechanical, electrical, and magnetic environments. Mech. Based Des. Struct. Mach. **46**(6), 669–692 (2018)
43. Arefi, M.: Magneto-electro-mechanical size-dependent vibration analysis of three-layered nanobeam with initial curvature considering thickness stretching. Int. J. Nano Dimens. **10**(1), 48–61 (2019)
44. Firouz-Abadi, R.D., Hosseinian, A.R.: Resonance frequencies and stability of a current-carrying suspended nanobeam in a longitudinal magnetic field. Theor. Appl. Mech. Lett. **2**, 031012 (2012)

45. Karlicić, D., Jovanović, D., Kozić, P., Cajić, M.: Thermal and magnetic effects on the vibration of a cracked nanobeam embedded in a elastic medium. J. Mech. Mater. Struct. **10**(1), 43–62 (2015)
46. Baghani, M., Mohammadi, M., Farajpour, A.: Dynamic and stability analysis of the rotating nanobeam in a nonuniform magnetic field considering the surface energy. Int. J. Appl. Mech. **8**(4), 1650048 (2016)
47. Du, H., Zhou, G., Zhao, Y., Chen, G., Chau, F.S.: Magnetic field sensor based on coupled photonic crystal nanobeam cavities. Appl. Phys. Lett. **110**, 061110 (2017)
48. Alibeigi, B., Beni, Y.T., Mehralian, F.: On the thermal buckling of magneto-electro-elastic piezoelectric nanobeams. Eur. Phys. J. Plus **133**, 133 (2018)
49. Kerid, R., Bourouina, H., Yahiaoui, R., Bounekhla, M., Aissat, A.: Magnetic field effect on nonlocal resonance frequencies of structure-based filter with periodic square holes network. Physica E **105**, 83–89 (2019)
50. Alibeigloo, A.: Free vibration analysis of nano-plate using three-dimensional theory of elasticity. Acta Mech. **222**, 149–159 (2011)
51. Yan, Z., Jiang, Y.: Vibration and buckling analysis of a piezoelectric nanoplate considering surface effects and in-plane constraints. Proc. R. Soc. A **468**, 3458–3475 (2012)
52. Asemi, S.R., Farajpour, A.: Vibration characteristics of double-piezoelectric-nanoplate-systems. Micro Nano Lett. **9**, 280–285 (2014)
53. Salehipour, H., Nahvi, H., Shahidi, A.R.: Exact closed-form free vibration analysis for functionally graded micro/nano plates based on modified couple stress and three-dimensional elasticity theories. Compos. Struct. **124**, 283–291 (2015)
54. Jafari, A., Shah-Enayati, S.S., Atai, A.A.: Size dependency in vibration analysis of nano plates; one problem, different answers. Eur. J. Mech.-A/Solids **59**, 124–139 (2016)
55. Ghassabi, A.A., Dag, S., Cigeroglu, E.: Free vibration analysis of functionally graded rectangular nanoplates considering spatial variation of the nonlocal parameter. Arch. Mech. **69**, 105–130 (2017)
56. Zarei, M., Faghani, G.R., Ghalami, M., Rahimi, G.H.: Buckling and vibration analysis of tapered circular nano plate. J. Appl. Comput. Mech. **4**, 40–54 (2018)
57. Shahrbabaki, E.A.: On three-dimensional nonlocal elasticity: free vibration of rectangular nanoplate. Eur. J. Mech.-A/Solids **71**, 122–133 (2018)
58. Despotovic, N.: Stability and vibration of a nanoplate under body force using nonlocal elasticity theory. Acta Mech. **229**, 273–284 (2018)
59. Singh Pratap, P., Azam, M.S., Ranjan, V.: Analysis of free vibration of nano plate resting on Winkler Foundation. Vibroeng. Procedia **21**, 65–70 (2018)
60. Ajri, M., Fakhrabadi, M.M.S., Rastgoo, A.: Analytical solution for nonlinear dynamic behavior of viscoelastic nano-plates modeled by consistent couple stress theory. Lat. Am. J. Solids Struct. **15**, e113 (2018)
61. Pelevic, N., van der Meer, T.: Numerical investigation of heat transfer enhancement by carbon nano fibers deposited on a flat plate. Comput. Math. Appl. **65**, 914–923 (2013)
62. Khorshidi, K., Asgari, T., Fallah, A.: Free vibrations analysis of functionally graded rectangular nano-plates based on nonlocal exponential shear deformation theory. Mech. Adv. Compos. Struct. **2**, 79–93 (2015)
63. Hosseini, M., Jamalpoor, A.: Analytical solution for thermomechanical vibration of double-viscoelastic nanoplate-systems made of functionally graded materials. J. Therm. Stress. **38**, 1428–1456 (2015)
64. Mahmound, A.H., El-Shishtawy, R.M., Obaid, A.Y.: The impact of graphene nano-plates on the behavior of novel conducting polyazomethine nanocomposites. R. Soc. Chem. **7**, 9998–10008 (2017)
65. Malikan, M.: Analytical predictions for the buckling of a nanoplate subjected to non-uniform compression based on the four-variable plate theory. J. Appl. Comput. Mech. **3**, 218–228 (2017)
66. Barati, M.R.: Magneto-hygro-thermal vibration behavior of elastically coupled nanoplate systems incorporating nonlocal and strain gradient effects. J. Braz. Soc. Mech. Sci. Eng. **9**, 4335–4352 (2017)

67. Barati, M.R.: Nonlocal stress-strain gradient vibration analysis of heterogeneous double-layered plates under hygro-thermal and linearly varying in-plane loads. J. Vib. Control **24**, 4630–4647 (2017)
68. Pugno, N.: Non-linear statics and dynamics of nanoelectromechanical systems based on nanoplates and nanowires. Proc. Inst. Mech. Eng. **219**, 29–40 (2005)
69. Yang, W., Liang, X., Shen, S.: Electromechanical responses of piezoelectric nanoplates with flexoelectricity. Acta Mech. **226**, 3097–3110 (2015)
70. Sobhy, M.: Thermoelastic Response of FGM plates with temperature-dependent properties resting on variable elastic foundations. Int. J. Appl. Mech. **7**, 1550082 (2015)
71. Park, W.-T., Han, S.-C.: Buckling analysis of nano-scale magneto-electro-elastic plates using the nonlocal elasticity theory. Adv. Mech. Eng. **10**, 1–16 (2018)
72. Fujiwara, M., Oki, E., Hamada, M., Tonimoto, Y.: Magnetic orientation and magnetic properties of a single carbon nanotube. J. Phys. Chem. A **105**, 4383–4386 (2001)
73. Kiani, K.: Free vibration of conducting nanoplates exposed to unidirectional in-plane magnetic fields using nonlocal shear deformable plate theories. Physica E **57**, 179–192 (2014)
74. Karličić, D., Cajić, M., Adhikari, S., Kozić, P., Murmu, T.: Vibrating nonlocal multi-nanoplate system under inplane magnetic field. Eur. J. Mech.-A/Solids **64**, 29–45 (2017)
75. Amiri, A., Fakhari, S.M., Pournaki, I.J., Rezazadeh, G., Shabani, R.: Vibration analysis of circular magneto-electro-elastic nano-plates based on Eringen's nonlocal theory. Int. J. Eng. **28**, 1808–1817 (2015)
76. Karlici, D., Cajic, M., Adhikari, S., Kozi, P., Murmu, T., Lazarevic, M.: Nonlocal mass-nanosensor model based on the damped vibration of single-layer graphene sheet influenced by in-plane magnetic field. Int. J. Mech. Sci. **96–97**, 132–142 (2015)
77. Arani, G.A., Maraghi, K.Z., Arani, K.H.: Smart vibration control of magnetostrictive nano-plate using nonlocal continuum theory. J. Solid Mech. **8**, 300–314 (2016)
78. Karami, B., Shahsavari, D., Li, L.: Temperature-dependent flexural wave propagation in nanoplate-type porous heterogeneous material subjected to in-plane magnetic field. J. Therm. Stress. **41**, 483–499 (2018)
79. Arman, S., Kirakosyan, S., Tigran Shahbazyan, V.: Vibrational modes of metal nanoshells and bimetallic core-shell nanoparticles. J. Chem. Phys. **129**, 034708 (2008)
80. Zaera, Z., Fernandez-Saez, J., Loya, J.A.: Axisymmetric free vibration of closed thin spherical nano-shell. Compos. Struct. **104**, 154–161 (2013)
81. Ke, L.L., Wang, Y.S., Reddy, J.N.: Thermo-electro-mechanical vibration of size-dependent piezoelectric cylindrical nanoshells under various boundary conditions. Compos. Struct. **116**, 626–636 (2014)
82. Rouhi, H., Ansari, R., Darvizeh, M.: Exact solution for the vibrations of cylindrical nanoshells considering surface energy effect. J. Ultrafine Grained Nanostructured Mater. **48**, 113–124 (2015)
83. Mehralian, F., Tadi Beni, Y.: Thermo-mechanical vibration of size dependent shear deformable functionally graded conical nanoshell resting on elastic foundation. Int. J. Eng. Appl. Sci. **8**, 68–86 (2016)
84. Rouhi, H., Ansari, R., Darvizeh, M.: Size-dependent large amplitude vibration analysis of nanoshells using the Gurtin-Murdoch model. Int. J. Nanosci. Nanotechnol. **13**, 241–252 (2017)
85. Razavi, H., Babadi, A.F., Beni, Y.T.: Free vibration analysis of functionally graded piezo-electric cylindrical nanoshell based on consistent couple stress theory. Compos. Struct. **160**, 1299–1309 (2017)
86. Barati, M.R.: Vibration analysis of multi-phase nanocrystalline material nanoshells using strain gradient elasticity. Mater. Res. Express **4**, 105021 (2017)
87. Sahmani, S., Aghdam, M.M.: Imperfection sensitivity of the nonlinear axial buckling behavior of FGM nanoshells in thermal environments based on surface slasticity theory. Int. J. Comput. Mat. Sci. Eng. **6**, 1750002 (2017)
88. Barati, M.R.: Vibration analysis of porous FG nanoshells with even and uneven porosity distributions using nonlocal strain gradient elasticity. Acta Mech. **229**, 1183–1196 (2018)

89. Razavi, S.: Magneto-electro-thermo-mechanical response of a multiferroic doubly-curved nano-shell. J. Solid Mech. **10**, 130–141 (2018)
90. Shojaeefard, M.H., Googarchin, H.S., Mahinzare, M.: Vibration and buckling analysis of a rotary functionally graded piezomagnetic nanoshell embedded in viscoelastic media. J. Intell. Mater. Syst. Struct. **29**, 2344–2361 (2018)
91. Karami, B., Shahsavari, D., Janghorban, M., Dimitri, R., Tornabene, F.: Wave propagation of porous nanoshells. Nanomaterials **9**, 9010022 (2019)
92. Cheong, S.-K., Krishnan, S., Cho, S.H.: Modeling of plasmonic heating from individual gold nanoshells for near-infrared laser-induced thermal therapy. Med. Phys. **36**, 4664–71 (2009)
93. Liu, Ch., Chunting, ChM, Li, B.Q.: The plasmon resonance of a multilayered gold nanoshell and its potential bioapplications. IEEE Trans. Nanotechnol. **10**, 797–805 (2011)
94. Avetisyan, Y.A., Yakunin, A.N., Tuchin, V.V.: Novel thermal effect at nanoshell heating by pulsed laser irradiation: hoop-shaped hot zone formation. J. Biophotonics **5**, 734–744 (2012)
95. Sahmani, S., Aghdam, M.M., Akbarzadeh, A.H.: Surface stress effect on nonlinear instability of imperfect piezoelectric nanoshells under combination of hydrostatic pressure and lateral electric field. AUT J. Mech. Eng. **2**, 177–190 (2018)
96. Fan, C.Z., Huang, J.P.: Second-harmonic generation with magnetic-field controllabilities. Appl. Phys. Lett. **89**, 141906 (2006)
97. Tanabe, K.: Field enhancement around metal nanoparticles and nanoshells: a systematic investigation. J. Phys. Chem. C **112**, 15721–15728 (2008)
98. Weber, S.J., Des Francs, C.G., Girard, C.: Spatiotemporal properties of nanoshell plasmonic response for strong-field experiments. Phys. Rev. B. **91**, 205419–1 (2018)
99. Moran, ChH., Wainerdi, S.M., Cherukuri, T.K., Kittrell, C., Wiley, B.J., Nicholas, N.W., Curley, S.A., Kanzius, J.S., Cherukuri, P.: Size-dependent Joule heating of gold nanoparticles using capacitively coupled radiofrequency fields. Nano Res. **2**, 400–405 (2009)
100. Zabow, G., Dodd, S.J., Moreland, J., Koretsky, A.P.: The fabrication of uniform cylindrical nanoshells and their use as spectrally tunable MRI contrast agents. Nanotechnology **20**, 385301 (2009)
101. Vallecchi, A., Campione, S., Capolino, F.: Symmetric and antisymmetric resonances in a pair of metal-dielectric nanoshells: tunability and closed-form formulas. J. Nanophotonics **4**, 041577 (2010)
102. Jiang, S.-M., Wu, D.-J., Wu, X.-W., Liu, X.-J.: Enormous enhancement of electric field in active gold nanoshells. Chin. Phys. B **23**, 047807 (2014)
103. Gladilin, V.N., Tempere, J., Devreese, J.T., Moshchalkov, V.V.: Negative-μ regime in the ac magnetic response of superconductor nanoshells. Solid State Commun. **5** (2018)

Chapter 2
Size-Dependent Theories of Beams, Plates and Shells

2.1 Introduction

This chapter concerns the latest literature review devoted to micro/nano-size-dependent mathematical models of beams, plates and shells. The review includes nonlocal theory of elasticity, surface theory of elasticity, modified couple stress theory and modified theory of deformation gradient taking into account higher order shear deformation theory. Particular emphasis is placed on nonlocal models of the Euler-Bernoulli and Timoshenko beams, Kirchhoff plates and Kirchhoff-Love shells. Both the review and the real-world vibrational behaviour of the size-dependent structural member implies the need to consider physical, geometric, material and design nonlinearity. It is expected that the novel approach based on employment of the earlier developed concepts of nonlinear dynamics like the Fourier and wavelet spectra, phase portraits, Poincaré maps, the Lyapunov exponents computation (at least the largest one), the autocorrelation functions, etc. will improve mathematical models of partial differential equations (PDEs) to be in a good fit with experimentally observed nonlinear phenomena exhibited by the size-dependent structural members.

2.2 Literature Review

Micro- and nanosize-dependent beams, plates and shells are widely employed in micro- and nanoelectromechanical systems (MEMS and NEMS) serving as vibration sensors [1], micro-drives [2] and micro-switchers [3]. The mentioned objects exhibited the size-dependent effects with regard to other mechanical properties [4–8]. The classical solid mechanics cannot give a proper interpretation of the size-dependent behaviour occurring in structures exhibiting scale effects. In the last years, many novel theories have been proposed allowing for modelling of the scale effects in continuum.

J. Awrejcewicz et al., *Mathematical Modelling and Numerical Analysis of Size-Dependent Structural Members in Temperature Fields*, Advanced Structured Materials 142,
https://doi.org/10.1007/978-3-030-55993-9_2

The mentioned size effects can be explained with a use of the modelling of molecular dynamics (MD) or mechanics of continuum of a higher order. Though the MD method can yield accurate and reliable results, it is expensive from a point of view of computations. Therefore, the methods of higher orders of mechanics of a solid medium are widely employed while modelling of nanoscale structures. Development of theory of continuum of a higher order begun with the work of Dell'Isola in nineteenth century [9, 10] as well as the pioneering works of Cosserat and Cosserat [11] published in 1909. However, the ideas proposed by the Cosserat brothers attracted researchers at the beginning of 1960 yielding a large number of theories of continuum of higher order.

In general, the developed theories can be divided into three parts: (i) theory of gradient deformations, (ii) theory of micro-continuum and (iii) nonlocal theories of elasticity.

The class of theory of gradient deformations consists of the modified couple stress theory and the modified theory of deformations gradient. In the class of the deformation gradients, the deformation energy consists of the stress and deformation gradients, and hence the size effect can be accounted with the help of size parameters characterizing the material length. In the stress couple theory worked out by Toupin [12], Mindlin and Tiersten [13], and Koiter [14], in the deformation energy, only a vector of rotation gradient is employed, and hence only two size-dependent length parameters are used. Yang et al. [15], based on the change of the couple stress theory, proposed a modified couple stress theory. In the latter case, the number of the size-dependent parameters is decreased from two to one. The first theory of deformation gradient has been proposed by Mindlin [16] who considered only the first gradient of deformations. One year later, Mindlin [17] developed the second theory of deformation gradient, where the first and the second gradients of deformations have been taken into account. Lam et al. [18] proposed the modified couple stress theory based on three size-dependent length parameters. The latter approach was based on the modification of the Mindlin theory with the help of approach analogous to that of Yang et al. [15].

Theory of micro-continuum has been worked out by Eringen [19–21], and it includes micropolar, microcracks and micromorphic (3M) theories. The micropolar theory, practically proposed by Cosserat brothers [11], belongs to the most simple among 3M theories, whereas the micromorphic theory stands for the most general among 3M theories. In the 3M theories, each particle can rotate and undergoes deformation independently on the centroid of motion of a particle (see [22–28]).

Nonlocal theory used of elasticity has been initially proposed by Kroner [29] and then improved by Eringen [30, 31] and Eringen and Edelen [32]. In the latter theory, stress in a control point of a continuum depends on deformation of all solid body points, and therefore the size-dependent effect is accounted with the help of state equations using a nonlocal parameter. The nonlocal theory of elasticity has been first formulated in the integral form, and later on [33] reformulated into its counterpart differential form with the help of a defined kernel function. Owing to simplicity of the differential models, the differential theory has been widely used in modelling of nanostructures. Besides, more recently one more class of theories of higher order named as the nonlocal theory of deformations gradient has been developed. It is based on unification of the theory of elasticity and theory of deformations gradient (see [34–36]).

Though a big achievement has been obtained for the magneto-electro-elastic material (MEU) with microscopic dimension, the application of the traditional theory of elasticity being based on microscopic characteristics does not allow to achieve reliable results in the nanoscale. Namely, the influence of size-dependent effects on the mechanical properties of materials is observed on the nanosize level what has been confirmed by atomic modelling and experiments [37–41]. In the nanoscale, a surface has remarkable influence on the general size-dependent mechanical properties of materials. From the mathematical point of view, the materials can be presented as a surface layer with zero-order thickness and volume kernel with mechanical properties being different from the surface layer. Based on the surface theory of elasticity, many different important problems have been investigated including influence of the surface on the bending [42–44], stability [45–47] and propagation of waves [48–51] as well as free vibrations [47, 52–60] of nanomaterials and nanostructures. The mentioned works pointed out the importance of account of the surface elasticity and surface stress into the fundamental governing equation for getting structural characteristics of nanomaterials. Based on the theory of surface energy, it is assumed that the surface properties cannot be ignored while studying nanostructures and nanomaterials due to large ratio of the surface area and the amount of volume in the nanosize structures [61]. The size-dependent behaviour in nanostructures has been investigated in [43, 62].

The size-dependent models are widely employed for prognosis of global behaviour of beam, plate and shell nanostructures such as carbon nanotubes (CNTs) and graphene leaves. CNTs have been first discovered by Iijima [63]. Depending on the fabrication technology, one may produce various types of CNTs, such as single-wall nanotubes (SWCNTs), double-wall nanotubes (DWCNTs) and multi-wall nanotubes (MWCNTs). They can be obtained by twisting of single-layer graphene sheets (SLGS), double-layer graphene sheets (DLGS) and multi-layer graphene sheets (MLGS) in all areas of nanotechnology [64–70].

One of the important aspects of the use of the size-dependent models relies on their application to the problems of statics and dynamics of beams. There is now available a palette of various size-dependent models in theory of beams and plates. The simplest models are based on theory of Euler-Bernoulli (EBT) and the classical theory of plates (CPT). Those models are applicable only for modelling of thin beams and thin plates, since they do not include effects of shear deformation. In order to overcome limitation of EBT and CPT, a series of theories according to shear deformation have been proposed. Models including shear deformations of the first order are based on the theory of Timoshenko beams (TBT) and theory of shear deformation of the first order (FSDT). Since in those models plane displacements undergo changes along thickness, there is a need to introduce correction coefficient of the shear effect. In order to remove the mentioned correction coefficient and to get more exact estimation of deflection in thick beams and plates, a few theories of higher order shear deformation (HSDT) are proposed including Reddy's beam theory (RBT) and theory of shear deformation of third order (TSDT) [71]. The full review of the state of the art of theory of plates can be found in the work of Thai and Kim [72].

The resolving equations yielded by the described size-dependent models can be solved with the help either of analytical or numerical methods. However, application of analytical approaches is available for rather simple geometry of nanostructures as well as for simple loads and boundary conditions. In the case of practical problems dealing with a general geometry, load and boundary conditions, a search for analytical solutions is almost impossible due to complexity of the size-dependent models in comparison with classical ones. Therefore, the most suitable methods are based on the finite element method, differential-integral method, meshless method, Ritz method, Galerkin method, etc. The key role in the numerical approaches is played by the finite difference method being most suitable to analyse size-dependent structures.

In the recent dozen of years, there were carried out large investigations of microbeams, microplates and microshells but without account of the size-dependent effects. This is why, we briefly describe the state of the art of development of size-dependent models aimed at the forecast of dynamical behaviour of micro/nanobeam, plate and shell structures. It includes mainly models of beams, plates and shells developed based on the nonlocal theory of elasticity [33], the surface theory of elasticity [61], the modified couple stress theory [15] and the modified theory of deformation gradient [18].

2.3 Non-classical (Size-Dependent) Models of Beams, Plates and Shells

2.3.1 Nonlocal Theory

Nonlocal theory of elasticity has been formulated by Eringen [30–32] with the help of integral equation

$$\sigma_{ij} = \int_{\bar{x}} k(|x - \bar{x}|, \kappa)\sigma_{ij}^{L} dx, \tag{2.1}$$

where σ_{ij} and σ_{ij}^{L} stand for components of the nonlocal and local tensors of stress, respectively, k is the kernel function defined by the nonlocal parameter κ in the neighbourhood of $|x - \bar{x}|$, where $\kappa = e_0 a$ and a is the material constant and internal characteristic length dimension or a molecular parameter, respectively. The value e_0 can be estimated based on experiment or obtained via appropriate modelling or based on the static analysis of bending of single-layer graphene sheets by Huang et al. [73]. Arash and Ansari [74] also estimated the value of nonlocal parameter for the case of free vibrations of single-walled carbon nanotubes by comparison of the SS values obtained via nonlocal shell model based on the first-order shear deformation theory and the model of molecular dynamics. Duan et al. [75] proposed the model of microbeam to find e_0 by monitoring free vibrations of nonlocal beams. They obtained analytical formula for e_0 based on the geometric properties and modes of vibration. Zhang et al. [76–78] proposed the microstructural beam mesh model to find e_0 under consideration of free vibrations of nonlocal beams [76], stability loss and free vibrations of nonlocal plates [77]. It has been found that the value e_0 depends on the initial stress, rotation inertia, vibration modes and a ratio of the rectangular plate sides. In the general case, owing to the conservative estimations, the value of the nonlocal parameter of the single-walled carbon nanotubes is $e_0 < 2.0$ nm [79].

We consider a particular kernel function k [33] transformed the nonlocal state equation into the differential counterpart form

$$(1 - \mu \nabla^2)\sigma_{ij} = \sigma_{ij}^L, \tag{2.2}$$

where $\mu = \kappa^2$ and ∇^2 stands for the Laplacian. Explicit form of equation (2.2) can be derived for three case studies dealing with an isotropic material.

In the case of 1D problems, we have

$$\sigma_{xx} - \mu \frac{d^2\sigma_{xx}}{dx^2} = E\varepsilon_{xx}, \tag{2.3}$$

$$\sigma_{xz} - \mu \frac{d^2\sigma_{xz}}{dx^2} = E\varepsilon_{xz}. \tag{2.4}$$

In the case of (2D) problems, one gets

$$\sigma_{xx} - \mu \left(\frac{d^2\sigma_{xx}}{dx^2} + \frac{d^2\sigma_{xx}}{dy^2} \right) = \frac{E}{1 - \nu^2}(\varepsilon_{xx} + \nu\varepsilon_{yy}), \tag{2.5}$$

$$\sigma_{xx} - \mu \left(\frac{d^2\sigma_{yy}}{dx^2} + \frac{d^2\sigma_{yy}}{dy^2} \right) = \frac{E}{1 - \nu^2}(\nu\varepsilon_{xx} + \nu\varepsilon_{yy}), \tag{2.6}$$

$$\sigma_{xy} - \mu \left(\frac{d^2\sigma_{xy}}{dx^2} + \frac{d^2\sigma_{xy}}{dy^2} \right) = \frac{E}{1 + \nu}\varepsilon_{xy}, \tag{2.7}$$

$$\sigma_{xz} - \mu \left(\frac{d^2\sigma_{xz}}{dx^2} + \frac{d^2\sigma_{xz}}{dy^2} \right) = \frac{E}{1 + \nu}\varepsilon_{xz}, \tag{2.8}$$

$$\sigma_{yz} - \mu \left(\frac{d^2\sigma_{yz}}{dx^2} + \frac{d^2\sigma_{yz}}{dy^2} \right) = \frac{E}{1 + \nu}\varepsilon_{yz}. \tag{2.9}$$

Finally, in the case of 3D problems, the following formulas hold:

$$\sigma_{xx} - \mu \left(\frac{d^2\sigma_{xx}}{dx^2} + \frac{d^2\sigma_{xx}}{dy^2} + \frac{d^2\sigma_{xx}}{dz^2} \right) = \frac{E}{1 - \nu^2}(\varepsilon_{xx} + \nu\varepsilon_{yy} + \nu\varepsilon_{zz}), \tag{2.10}$$

$$\sigma_{yy} - \mu\left(\frac{d^2\sigma_{yy}}{dx^2} + \frac{d^2\sigma_{yy}}{dy^2} + \frac{d^2\sigma_{yy}}{dz^2}\right) = \frac{E}{1-\nu^2}(\varepsilon_{xx} + \nu\varepsilon_{yy} + \nu\varepsilon_{zz}), \tag{2.11}$$

$$\sigma_{zz} - \mu\left(\frac{d^2\sigma_{zz}}{dx^2} + \frac{d^2\sigma_{zz}}{dy^2} + \frac{d^2\sigma_{zz}}{dz^2}\right) = \frac{E}{1-\nu^2}(\varepsilon_{xx} + \nu\varepsilon_{yy} + \nu\varepsilon_{zz}), \tag{2.12}$$

$$\sigma_{xy} - \mu\left(\frac{d^2\sigma_{xy}}{dx^2} + \frac{d^2\sigma_{xy}}{dy^2} + \frac{d^2\sigma_{xy}}{dz^2}\right) = \frac{E}{1+\nu}\varepsilon_{xy}, \tag{2.13}$$

$$\sigma_{xz} - \mu\left(\frac{d^2\sigma_{xz}}{dx^2} + \frac{d^2\sigma_{xz}}{dy^2} + \frac{d^2\sigma_{xz}}{dz^2}\right) = \frac{E}{1+\nu}\varepsilon_{xz}, \tag{2.14}$$

$$\sigma_{yz} - \mu\left(\frac{d^2\sigma_{yz}}{dx^2} + \frac{d^2\sigma_{yz}}{dy^2} + \frac{d^2\sigma_{yz}}{dz^2}\right) = \frac{E}{1+\nu}\varepsilon_{xz}. \tag{2.15}$$

Here, ε_{ij} are components of the deformation tensor and E is the Young modulus, whereas ν is Poisson's coefficient. The differential model is simple and is widely employed for nanostructures investigation. However, the latter model can give sometimes paradoxal results, for example, in the problems of bending and vibration of cantilever beams. The detailed information about paradoxes of the differential models can be found in [80–83].

2.3.2 *Modified Couple Stress Theory*

The modified couple stress theory has been proposed by Yang et al. [15], in result of modification of the classical couple stress theory developed by Toupin [12], Mindlin and Tiersten [84] and Koiter [14]. Introducing the additional condition of balance of stress couples in order to guarantee symmetry of the couple stress tensor implied decrease of a number of additional internal material length parameters from two to one. The latter result improved efficiency of the modified couple stress theory, since estimation of the internal parameter of material length does not belong to easy task. Energy of deformation U, owing to that theory, is a function of both deformation and curvature [15] and takes the form

$$U = \frac{1}{2}\int_V (\sigma_{ij}\varepsilon_{ij} + m_{ij}\chi_{ij})dV, \tag{2.16}$$

where $m_{i,j}$ are components of deviation part of the symmetric tensor of moments, whereas $\chi_{i,j}$ are components of the tensor of symmetric curvature defined by the following formulas:

$$\chi_{xx} = \frac{\partial\theta_x}{\partial x}, \tag{2.17}$$

$$\chi_{yy} = \frac{\partial \theta_y}{\partial x}, \tag{2.18}$$

$$\chi_{zz} = \frac{\partial \theta_z}{\partial x}, \tag{2.19}$$

$$\chi_{xy} = \frac{1}{2}\left(\frac{\partial \theta_x}{\partial x} + \frac{\partial \theta_y}{\partial y}\right), \tag{2.20}$$

$$\chi_{xz} = \frac{1}{2}\left(\frac{\partial \theta_x}{\partial x} + \frac{\partial \theta_z}{\partial z}\right), \tag{2.21}$$

$$\chi_{yz} = \frac{1}{2}\left(\frac{\partial \theta_y}{\partial y} + \frac{\partial \theta_z}{\partial z}\right). \tag{2.22}$$

Here, the vector of rotation θ is defined via displacements field (u_x, u_y, u_z) in the following way:

$$\theta_x = \frac{1}{2}\left(\frac{\partial u_z}{\partial z} - \frac{\partial u_y}{\partial y}\right), \tag{2.23}$$

$$\theta_y = \frac{1}{2}\left(\frac{\partial u_x}{\partial z} - \frac{\partial u_z}{\partial x}\right), \tag{2.24}$$

$$\theta_x = \frac{1}{2}\left(\frac{\partial u_y}{\partial x} - \frac{\partial u_x}{\partial y}\right). \tag{2.25}$$

In the case of linear elastic material, we have

$$m_{ij} = \frac{E}{1+\nu}\chi_{ij} l^2, \tag{2.26}$$

where l stands for the internal length material parameter. Estimation of the parameter l can be found in Refs. [85, 86].

2.3.3 Modified Theory of a Gradient of Deformations

In this theory [18], the deformation energy includes two additional parts of the gradient, i.e. the gradient of dilatation γ and the deviatory gradient of extension η as a supplement to the symmetric curvature χ_{ij}. Therefore, energy of deformation is [18]

$$U = \frac{1}{2}\int_V (\sigma_{ij}\varepsilon_{ij} + p_i\gamma_i + \tau_{ijk}\eta_{ijk} + m_{ij}\chi_{ij})dV, \tag{2.27}$$

where the symmetric tensor of curvature χ_{ij} is defined by Eq. (2.16). The vector of the dilatation gradient γ_i and the deviatory tensor of the extension gradient η_{ijk} are defined via Eqs. (2.28)–(2.30) and (2.31)–(2.40) using the following relations:

$$\gamma_x = \frac{\partial \varepsilon_{xx}}{\partial x} + \frac{\partial \varepsilon_{yy}}{\partial x} + \frac{\partial \varepsilon_{zz}}{\partial x}, \tag{2.28}$$

$$\gamma_y = \frac{\partial \varepsilon_{xx}}{\partial y} + \frac{\partial \varepsilon_{yy}}{\partial y} + \frac{\partial \varepsilon_{zz}}{\partial y}, \tag{2.29}$$

$$\gamma_z = \frac{\partial \varepsilon_{xx}}{\partial z} + \frac{\partial \varepsilon_{yy}}{\partial z} + \frac{\partial \varepsilon_{zz}}{\partial z}, \tag{2.30}$$

$$\eta_{xxx} = \frac{\partial \varepsilon_{xx}}{\partial x} - \frac{1}{5}\left[\gamma_x + 3\left(\frac{\partial \varepsilon_{xx}}{\partial x} + \frac{\partial \varepsilon_{xy}}{\partial y} + \frac{\partial \varepsilon_{xz}}{\partial z}\right)\right], \tag{2.31}$$

$$\eta_{yyy} = \frac{\partial \varepsilon_{yy}}{\partial y} - \frac{1}{5}\left[\gamma_y + 3\left(\frac{\partial \varepsilon_{xy}}{\partial x} + \frac{\partial \varepsilon_{yy}}{\partial y} + \frac{\partial \varepsilon_{yz}}{\partial z}\right)\right], \tag{2.32}$$

$$\eta_{zzz} = \frac{\partial \varepsilon_{zz}}{\partial z} - \frac{1}{5}\left[\gamma_z + 3\left(\frac{\partial \varepsilon_{xz}}{\partial x} + \frac{\partial \varepsilon_{yz}}{\partial y} + \frac{\partial \varepsilon_{zz}}{\partial z}\right)\right], \tag{2.33}$$

$$\eta_{yyx} = \eta_{yxy} = \eta_{xyy} = \frac{1}{3}\left[\eta_{xxx} + 2\frac{\partial \varepsilon_{xy}}{\partial y} + \frac{\partial\left(\varepsilon_{yy} - \varepsilon_{xx}\right)}{\partial x}\right], \tag{2.34}$$

$$\eta_{zzx} = \eta_{zxz} = \eta_{xzz} = \frac{1}{3}\left[\eta_{xxx} + 2\frac{\partial \varepsilon_{xz}}{\partial z} + \frac{\partial\left(\varepsilon_{zz} - \varepsilon_{xx}\right)}{\partial x}\right], \tag{2.35}$$

$$\eta_{xxy} = \eta_{xyx} = \eta_{yxx} = \frac{1}{3}\left[\eta_{yyy} + 2\frac{\partial \varepsilon_{xy}}{\partial x} + \frac{\partial\left(\varepsilon_{xx} - \varepsilon_{yy}\right)}{\partial x}\right], \tag{2.36}$$

$$\eta_{zzy} = \eta_{zyz} = \eta_{yzz} = \frac{1}{3}\left[\eta_{yyy} + 2\frac{\partial \varepsilon_{yz}}{\partial z} + \frac{\partial\left(\varepsilon_{zz} - \varepsilon_{yy}\right)}{\partial y}\right], \tag{2.37}$$

$$\eta_{xxz} = \eta_{xzx} = \eta_{zxx} = \frac{1}{3}\left[\eta_{zzz} + 2\frac{\partial \varepsilon_{xz}}{\partial x} + \frac{\partial\left(\varepsilon_{xx} - \varepsilon_{zz}\right)}{\partial z}\right], \tag{2.38}$$

$$\eta_{yyz} = \eta_{yzy} = \eta_{zyy} = \frac{1}{3}\left[\eta_{zzz} + 2\frac{\partial \varepsilon_{yz}}{\partial y} + \frac{\partial\left(\varepsilon_{yy} - \varepsilon_{zz}\right)}{\partial z}\right], \tag{2.39}$$

$$\eta_{xyz} = \eta_{yxz} = \eta_{zxy} = \eta_{xzy} = \eta_{zyx} = \eta_{yxz} = \frac{1}{3}\left[\frac{\partial \varepsilon_{yz}}{\partial x} + \frac{\partial \varepsilon_{xz}}{\partial y} + \frac{\partial \varepsilon_{xz}}{\partial z}\right]. \quad (2.40)$$

In the case of linear elastic material, the stresses of higher order $(p_i, \tau_{ijk}, m_{ij})$ are as follows:

$$p_i = 2\eta l_0^2 \varepsilon_{mm,j}, \quad \tau_{ijk}^{(1)} = 2\eta l_1^2 \eta_{ijk}^{(1)}, \quad m_{ij}^s = 2\eta l_2^2 \chi_{ij}^{(s)}$$

$$p_i = \frac{E}{1+\nu} l_0^2 \varepsilon_{mm,j}, \quad (2.41)$$

$$\tau_{ijk}^{(1)} = \frac{E}{1+\nu} l_1^2 \eta_{ijk}^{(1)}, \quad (2.42)$$

$$m_{ij}^s = \frac{E}{1+\nu} l_2^2 \chi_{ij}^{(s)}, \quad (2.43)$$

where l_0, l_1, l_2 are internal parameters of the material length coupled with the dilatation gradient, the deviatory gradient of extension and the gradient of symmetric curvature, respectively.

2.3.4 Surface Theory of Elasticity

It is known that a nature of the chemical coupling and equilibrium atomic localization of the surface atoms is different from the nature of internal atom localization (see [87, 88]). In result, material properties in vicinity of a surface principally differ from the atoms located inside the solid body volume.

The surface energy being necessary for isothermic creation of a unit part of the new surface can be used as measure of chemical coupling in the surface vicinity [89]. In the case of fluid, the surface energy is equal to surface tension, but it is different for the case of solid body. In the case when the surface energy does not depend on deformation, Shuttleworth [90] and Herring [91] proposed the following formula (see also [92, 93]):

$$\tau_{\alpha\beta} = \Gamma \delta_{\alpha\beta} + \frac{\partial \Gamma}{\partial \varepsilon_{\alpha\beta}}, \quad (2.44)$$

where Γ—surface energy; $\tau_{ab}, \varepsilon_{\alpha\beta}, \delta_{\alpha\beta}$—surface tension, surface deformation and Kronecker's symbol, respectively. In formula (2.44) and later, we employ standard notation of indices together with Einstein summation concept where each Greek index takes the value from 1 to 2, and each Latin symbol changes from 1 to 3.

In the theory of surface elasticity, a surface is considered as a thin membrane [61, 94] exhibiting bending stiffness [95, 96], which is ideally coupled with volume material. It allows to include the elastic constant parameters for the surface and to satisfy various governing relations. Owing to the linear theory of surface elasticity [94], a volume body still obeys boundary conditions of the classical theory of elasticity, whereas the surface properties include equations with surface constant parameters and surface residual stresses. Namely, the volume body satisfies the equations of equilibrium and resolving equations of the classical linear theory of elasticity, and they can be presented in the following forms:

$$\sigma_{ij,j} + b_i = 0, \tag{2.45}$$

$$\sigma_{ij} = \lambda \varepsilon_{kk} \delta_{ij} + 2\mu \varepsilon_{ij}, \tag{2.46}$$

where σ_{ij} stand for the stress Cauchy components; b_i are force components; λ and μ are the standard Lame constants; and ε_{ij} stand for components of infinitely small deformations governed by the following relations:

$$\varepsilon_{ij} = \frac{1}{2}(u_{i,j} + u_{j,i}), \tag{2.47}$$

where u_i are components of displacement.

On the body surface, the resolving equations have the following form ([37, 97, 98]):

$$\begin{aligned} \sigma_{\beta\alpha} n_\beta &= -\tau_{\beta\alpha,\beta}, \\ \sigma_{ij} n_i n_j &= \tau_{\beta\alpha} \kappa_{\beta\alpha}, \end{aligned} \tag{2.48}$$

where $\kappa_{\alpha\beta}$—components of surface curvature tensor; n_i—components of external unit normal, n_β component of vector $\mathbf{n} = n_i e_i$ on the surface; $\tau_{\alpha\beta}$—components of a tensor of surface stresses [94], and

$$\tau = \left[\tau_0 + (\lambda_0 + \tau_0) tr(\varepsilon^s)\right] \Big/ \mathbf{I}^s + 2\mu_0 \varepsilon^s - \tau_0 (\nabla s u)^T. \tag{2.49}$$

Here μ_0 and λ_0 are surface elastic constants; τ_0 stands for residual surface stress (i.e. surface stress for zero deformation); $\mathbf{I}^s$ is projection tensor with $\mathbf{I}^s = \mathbf{I} - n \otimes n$; ∇s stands for gradient surface operator defined as $\nabla s(\cdot) = (\mathbf{I} - n \Theta n)\, \nabla(\cdot)$, where the superscript T denotes transposition matrix; and ε^s is the tensor of surface deformation of the following form:

$$\varepsilon^s = \frac{1}{2}\left[\nabla s u + (\nabla s u)^T\right] = \frac{1}{2}(\mathbf{I} - n \otimes n)\left[\nabla u + (\nabla u)^T\right], \tag{2.50}$$

where I is the unit tensor of the second order.

Observe that there are three surface constants μ_0, λ_0 and τ_0 (see [37, 99]).

2.3.5 Size-Dependent Theory of Beams, Plates and Shells

2.3.5.1 Nonlocal Models of Euler-Bernoulli Beams (EBT), the Kirchhoff Plates (CPT) and the Kirchhoff-Love Shells (KLT)

The first models of nonlocal beams based on EBT have been developed by Peddieson et al. [100] and Sudak [101]. Peddieson et al. [100] employed their own model to estimate the influence of size dependence on the bending behaviour of isotropic nanoshells, while Sudak [101] employed his model to investigate stability loss of MWCNT. Then numerous articles appear devoted to modelling of nanobeams and CNT with the help of nonlocal EBT model. For instance, Zhang et al. [102] investigated free vibrations of the DWCNT. In order to study the size effect of the characteristics of the DWCNT vibrations, the analytical solutions have been derived for the eigen frequencies of simply supported DWCNT. Wang et al. [103] obtained a general solution in a closed form for stability loss of CNT with various boundary conditions. Aydogdu [104] investigated the influence of size dependence on a basis of vibration of nanorods for various boundary conditions and obtained explicit formulas for eigenfrequencies. Murmu and Pradhan [105] introduced thermal effects into analysis of free vibrations under boundary conditions of SWCNTs with the help of method of differential quadratures (DQ). The similar approach was also used by Civalek and Demir [106] for getting bending moments and deflections of nanobeams for different boundary conditions. Mustapha and Zhong [107] investigated free vibrations of axially loaded non-prismatic SWCNTs employing the Bubnov-Galerkin method. Li et al. [108] obtained closed-form solutions for eigenfrequencies of axially loaded simply supported nanobeams. Ghannadpour et al. [109] used the Ritz method to solve the governing equations of nonlocal EBT, i.e. model of bending under critical load and eigenfrequencies for different boundary conditions. Ansari et al. [110] worked out the nonlocal EBT taking into account the geometric von Kármán nonlinearity based on the MWCNT model with an account of temperature, whereas Fang et al. [111] proposed the nonlinear nonlocal EBT model for nonlinear vibrations with DWCNT. Nonlinear free vibrations and excited vibrations of nanobeams with different boundary conditions were analysed by Simsek [112] and Bagdatli [113].

Nonlocal model of EBT has been also proposed for the nanobeams made from functionally graded (FG) materials. Simsek [114] investigated nonlocal effect under axial vibrations of FG nanorods with changeable transversal cross section. It has been assumed that the elasticity modulus and mass density of the nanorods are changed in the axial direction. Nguyen et al. [115] proposed analytical solutions of the nonlocal EBT model for the static analysis of bending of FG beams with different boundary conditions. Elasticity modulus of FG nanobeams can be changed in both longitudinal and transverse directions. Galerkin, Niknam and Akhdam [116] obtained solutions to the nonlocal EBT model for the eigenfrequencies and critical loads of the FG nanobeams lying on an elastic foundation. Ebrahimi and Salari [117, 118] studied the influence of temperature on the free vibrations of FG nanobeams with different boundary conditions with the help of semi-analytical approach. Nejad and Hadi [119] and Nejad et al. [120] investigated bending and stability loss of FG nanobeams with an account of changes of the elasticity modulus in the axial and transverse directions. The governing equations were solved by the DQ method. Nonlinear free vibrations of the FG nanobeams were investigated by Nazemnezhad and Hosseini-Hashemi [121] with the help of nonlinear theory of von Kármán and using the nonlocal EBT model. Multiple scales method yielded solutions for the case of eigenfrequencies of nonlinear nonlocal EBT model and for the case of free and excited vibrations of the simply supported beams. El-Borgi et al. [122] proposed the nonlocal EBT model for nonlinear free and excited vibrations of the FG nanobeams. The multiple-scale method was employed to obtain nonlinear frequencies of the simply supported beams. Shafiei et al. [123] also worked out the nonlinear nonlocal EBT model for investigation of nonlinear free vibrations of FG beams with changeable transversal cross sections. Nonlinear frequency of beams was obtained for various boundary conditions with the help of generalized DQ methodology.

Zhang et al. [124] developed one of the earlier nonlocal shell models for analysis of stability loss of MWCNT under axial compression on the basis of classical theory of shells. In order to solve the problems with an account of the nonlocal effect of the axial buckling of simply supported DWCNT, the solutions were derived in the closed form. Li and Kardomateas [125, 126] derived non-classical nonlocal models of shells for a study of thermal bending and free vibration of MWCNT. The model of nonlocal classical shell was proposed by Wang and Varadan [127] and Hu et al. [128] and employed to investigate the wave propagation in CNTs. The accuracy of nonlocal classical shell model under prognosis of buckling of axially loaded SWCNTs was quantified by Zhang et al. [129] in comparison with the results obtained by MD modelling. It was shown that for the long SWCNT the local KLT (Kirchhoff-Love theory) model can yield the results compared with the results obtained with the help of nonlocal KLT model and MD modelling. However, for short SWCNT, only nonlocal shell model may guarantee the results compared with MD. Rouhi and Ansari [130] also proposed the nonlocal classical shell model for the case of axial compression of DWCNT validated for different boundary conditions. More recently, Sarvestani [131] proposed the nonlocal classical shell model to analyse stability loss of the bounded MWCNT under axial comparison.

One of the most earlier models of the nonlocal CPT plate was given by Lu et al. [132] based on classical CPT. The latter model was employed for the study of influence of the size-dependent bending behaviour and stability of the isotropic nanoplates. Duan and Wang [133] derived exact solutions of nonlocal CPT model while analysing axially symmetric of circle nanoplates under arbitrary load. The influence of the nonlocal parameter regarding deviation, radial moment, etc. has been investigated.

Aksencer and Aydogdu [134] obtained Levy's solution for the nonlocal CPT model in the stability problems and free vibrations of the rectangular nanoplates with two edges simply supported and two arbitrary supported. Shakouri et al. [135] employed Galerkin's approach for solving the resolving equations of the nonlocal CPT model for estimation of the eigenfrequencies of the isotropic nanoplates with various boundary conditions.

The nonlocal CPT model was also used for computation of the size-dependent SLGS and MLGS behaviours. For instance, Pradhan and Murmu [136] and Pradhan and Kumar [137] investigated the nonlocal effect influenced on the SLGS stability using the DQ method, whereas Babaei and Shahidi [138] and Farajpour et al. [139] considered the influence of size dependence on bending of rectangular SLGS [138] and SLGS with variable thickness [139] employing the Galerkin method. Free vibrations of MLGS embedded into a polymer matrix were investigated by Pradhan and Phadikar [140]. It was shown how the size effect increases with the increase of the layer number. Shen et al. [141] extended application of the nonlocal CPT model for investigation of vibrations of the simple support mass sensor of SLGS. Ansari et al. [142] derived analytical formulas for the SLGS eigenfrequencies with arbitrary boundary conditions based on construction of the atomic potential while estimating elastic properties of SLGS. More recently, Zhang et al. [143–145] employed equations of the nonlocal CPT model for derivation of the eigenfrequencies [143], nonlinear bending [144] and critical loads [145] of SLGS for various boundary conditions.

Employment of the nonlocal CPT in the above-mentioned investigations was limited to consideration of graphene sheets made from isotropic materials. However, numerical results of MD modelling carried out by Ni et al. [146] showed that mechanical properties of the graphene sheets are anisotropic due to hexagonal structure of the elementary graphene cells [146]. Therefore, nonlocal orthotropic CPT models were carried out to account of influence of the anisotropic mechanical properties of the graphene sheets. Pouresmaeeli et al. [147] proposed the nonlocal CPT model for vibrations of the orthotropic DLGS, embedded into an elastic matter. Mohammadi et al.[148, 149] worked out the nonlocal CPT model for free vibrations of orthotropic SLGS embedded into the temperature field. Navier and Levy solutions of the eigenfrequencies of the rectangular SLGS were obtained. Sari and Al-Kouz [150] developed the nonlocal CPT model to analyse free vibrations of orthotropic SLGS with an account of the changeable thickness. Anjomshoa [151] and Anjomshoa et al. [152] proposed nonlocal CPT models for study stability loss [151] and free vibrations [152] of orthotropic circled and elliptic SLGS embedded into elastic matter. The nonlocal CPT model was also proposed by Mohammadi et al. [153] for the investigation of stability loss and shear effects of orthotropic SLGS inclusions subjected to temperature fields. Ashoori et al. [154] worked out the nonlocal CPT model suitable for prognosis of thermal stability of ring-type FG nanoplates under various kinds of thermal loads. Exact solutions for the critical temperature responsible for stability loss were also obtained for ring-clamped FG nanoplates.

2.3.5.2 Beam (EBT), Plate (CPT) and Shell (KLT) Models Based on the Modified Couple Stress Theory

Park and Gao [155] belong to the first who presented the modified couple stress EBT model for isotropic microbeams. They employed the developed model for the investigation of influence of the length parameter on deflections and bending stiffness of a cantilever epoxidal beam with a discrete load on its free end. It was demonstrated that an account of the length parameter of material implies an increase of stiffness under bending of the cantilever beam. The latter effect played an important role when the beam thickness was small but it decreased with increase of the beam thickness. These observations were validated by experimental investigations. The modified couple stress theory of EBT was extended by Kong et al. [156, 157] for the problem of free vibrations [156] and buckling of isotropic microbeams [157].

The model of nonlinear modified couple stress theory of EBT was first worked out by Xia et al. [158] for the case of nonlinear bending behaviour, nonlinear analysis of stability and free vibrations of isotropic microbeams on a basis of the von Kármán nonlinearity. The obtained results showed the importance of account of nonlinearity and size effects during the study of microscale devices and systems such as biosensors, atomic-force microscopes and MEMS [158]. Simsek [159] also proposed nonlinear EBT model for fitting nonlinear deflections and free vibrations of isotropic microbeams lying an elastic foundation. Nonlinear EBT model has been widely employed for analysing the influence of the size effects on nonlinear deflections [160], nonlinear vibrations [161–165] and stability [164, 165] of isotropic microbeams. It should be noted that Farokhi et al. [162] considered the initial geometric imperfections while studying nonlinear beam vibrations, whereas Togun and Bagdatli [163] took into account axial extension of the microbeam vibrations. Wang et al. [160, 164] considered the thermal effects within nonlinear microbeam deflections [160] as well as stability problems and vibrations of the beams [164]. Ansari et al. [165] obtained analytical solutions while investigating both vibrations and stability of microbeams under various boundary conditions.

The developed EBT model was also applied to FG microbeams. For example, it was employed by Asghari et al. [166] to study bending and free vibrations of the FG microbeams. Free vibrations of FG cantilevers with changeable properties of material in the longitudinal direction were considered by Akgoz and Civalek [167] and Shafiei et al. [168]. It should be mentioned that Akgoz and Civalek [167] analysed only cantilever, whereas Shafiei et al. [168] considered beams with account of the geometric nonlinearity with various boundary conditions. Simsek [169] also investigated nonlinear free vibrations of microbeam functionally graded in the longitudinal direction. He employed the Galerkin methods and his variational method to get the approximate solutions for beams with simple support and clamping. Dehrouyeh-Semnani et al. [170] include the initial geometric imperfections while studying vibrations of the FG microbeams.

The modified couple stress theory under CPT model was first proposed by Tsiatas [171] for analysis of bending of isotropic microplates of arbitrary form. Yin et al. [172] and Jomehzadeh et al. [173] expended that theory to include free vibrations of simply supported microplates [172] and the Levi-type microplates [173]. Akgoz and Civalek [174] proposed the CPT model within the modified couple stress theory in order to investigate the influence of the size dependence on free vibrations of simply supported SLGS included in an elastic matrix. It was shown that the size effect was observed for higher vibration modes. Akgoz and Civalek [175] also included elastic medium into the CPT model for the static bending, the stability problem and analysis of free vibrations of isotropic microplates. Askari and Tahani [176] obtained analytical solutions for the eigenfrequencies of the clamped CPT microplates with the use of the extended Kantorovich method. Simsek et al. [177] adapted the CPT model to study the influence of the size on the excited vibrations of isotropic microplates under action of moving load. The dynamic behaviour of microplates for various boundary conditions was quantified with the use of the Newmark implicit integration method. Zhou et al. [178] developed the modified couple stress model for free vibrations of isotropic microbeams based on the classical theory of shells. It was found that the size effect plays an important role when the size of the characteristic radius is compared with the length scale material parameter [178].

Asghari [179] proposed the CPT model and the nonlinear modified couple stress theory for the geometric nonlinear analysis of microplates of an arbitrary form. Wang et al. [180, 181] worked out the CPT model within the nonlinear modified couple stress theory in order to investigate the size effects influenced on nonlinear free vibrations [180] and nonlinear deflections [181] of circle microplates. Farokhi and Ghayesh [182] also employed CPT model and nonlinear modified couple stress theory to carry out nonlinear dynamic analysis of microplates with geometric imperfections.

The similar approaches were employed also to the FG microplates. Ke et al. [183] studied the influence of size effects on the deflections, critical loads and eigenfrequencies of ring FG microplates for different boundary conditions. Asghari and Taati [184] investigated free vibrations of the FG microbeams of the arbitrary form. Ashoori and Sadough Vanini [185] presented the modified CPT model to study stability loss of the FG microplates including thermal effects and interaction between plate and elastic matter. Ashoori and Sadough Vanini [186] extended their earlier work [185] in order to include the geometric nonlinearity under thermal buckling of circle FG microplates. Taati [187] obtained analytical formulas of the CPT model with the help of modified couple stress theory for stability of the FG microplates with various boundary conditions under shear in plane, two-side compression and uniform transversal load. Based on the classical theory of KLT shells, Beni et al. [188] developed the modified couple stress theory for the investigation of influence of the size effects on free frequencies of the simply supported cylindrical FG microplates. Tsiatas and Yiotis [189] also employed the modified couple stress theory of KLT model in order to investigate the size-dependent influence on the state deflections and characteristics of longitudinal and free vibrations of the microplates. In comparison with nonlocal KLT regimes, it was shown that the influence of scale length parameter on the critical loads and eigenfrequencies stands in contrary to the nonlocal models.

2.3.5.3 Beams (EBT), Plates (CPT) and Shells (KLT) Models Based on the Modified Gradient Theory

Kong et al. [190] proposed one of the earlier EBT models with gradient of deformations in order to study the size-dependent influence on deflections and eigenfrequency of isotropic cantilever microbeams. Accuracy of results yielded by the theory of gradient deformations was compared with the accuracy of modified couple stress theory and classical theory. The comparison results showed that the model of gradient deformations predicts the size effect in a better way than the method of modified couple stress theory, since it takes in addition into account the additional tensor of the dilatation gradient, deviator tensor of extension gradient and a supplement to the tensor of rotations gradient. Akgoz and Civalek [191] extended the EBT model of deformations gradient into analysis of stability loss of isotropic microbeams for cantilever and simply supported beams. Akgoz and Civalek [192–195] also employed EBT model of the deformation gradient for the study of influence of the size effect on stability of SWCNT [192], static bending SWCNT [193], stability of linearly narrow microbeams [194] and longitudinal vibrations of microbeams [195]. Zhao et al. [196] developed the nonlinear EBT model of deformation gradient for nonlinear stability loss and nonlinear analysis of free vibrations of isotropic microbeams. They pointed out the importance of geometric nonlinearity and size-dependent effects while getting reliable results. Rajabi and Ramezani [197] also developed the model of nonlinear gradient of EBT deformations for the isotropic microbeams but mainly considered static bending and free vibrations. The model of nonlinear EBT beam was considered by Mohammadi and Mahzoon [198] on a basis of deformation gradient where also the influence of temperature on stability of isotropic microbeams was taken into account. They derived analytical solutions of microbeams with various boundary conditions. Vatankhah et al. [199] used the model of nonlinear gradient of deformations of EBT for a study of nonlinear vibrations of isotropic microbeams.

Kahrobaiyan et al. [200] extended application of the EBT model with gradient of deformations to quantify stability and free vibrations of the FG microbeams. Further extension of this model into problems of stability of FG microbeams was carried out by Akgoz and Civalek [201]. For FG microscopes for various boundary conditions, the solutions regarding estimation of critical loads have been derived. Akgoz and Civalek [202] employed the EBT model of the gradient of deformations for a study of longitudinal vibrations of the FG microbeams. The method of Rayleigh-Ritz was used to solve the problem regarding eigenfrequencies of the FG beams with clamped ends and one free end. Rahaeifard et al. [203] developed the EBT model of nonlinear gradient of deformations for study of influence of the geometric nonlinearity and scale length material parameter on deflections and eigenfrequencies of FG simply supported beam.

Wang et al. [204] belong to the first who developed the CPT model with an account of the deformations gradient to predict the dependence of the characteristics of isotropic size-dependent microplates. A comparison between the model of deformation gradient and the model based on the modified couple stress theory showed that the first one yields better results [204]. Deflections of CPT model within theory of gradient of deformations were obtained by Ashoori and Mahmoodi [205] for the microplates and for various boundary conditions with the help of extended Kantorovich method. Besides, the analytical buckling solutions were obtained by Mohammadi and Fooladi [206] and Mohammadi et al. [207] for the Levi microplates. Wang et al. [208] derived the CPT model within gradient of deformations to analyse bending of microplates for various boundary conditions. Zeighampour and Tadi Beni

[209], Allahbakhshi and Allahbakhshi [210], Li et al. [211], Hosseini et al. [212] and Zhang et al. [213] extended the CPT model based on theory of deformations gradient into SWCNT [209], MLGS [210], two-layer isotropic microplates [211], multi-layer orthotropic microplates [212] and isotropic microplates lying on elastic basis [213].

2.3.5.4 Beams (EBT), Plates (CPT) and Shells (KLT) Models Based on the Surface Theory of Elasticity

In order to achieve high accuracy models of size-dependent behaviour in nanostructures, both surface and volume effects are taken into account. Theory of surface energy and theory of elasticity of deformation gradient are introduced to fit both surface and volume effects, respectively. There are proposed novel models for the Bernoulli-Euler and Timoshenko beams. The fundamental equations, initial and boundary conditions are introduced simultaneously using the Hamilton principle. New models include the Poisson's effect and have three parameters of material length and three constants of the surface elasticity to the account of scale material length as well as three constants of the surface elasticity in order to quantify size effects in the surface and volume of the nanostructure, as well as the modified couple stress models and classical models in order to illustrate features of novel models regarding the problems of static bending and free vibrations of simply supported Euler-Bernoulli and Timoshenko nanobeams. Numerical results show that the differences in deflections and eigenfrequency predicted by the mentioned model and other models are large for small beam thickness. The differences are decreased when the beam size increases. There are considered Euler-Bernoulli and Timoshenko models based on the continuum theory of Gurtin-Merdok to study thin and thick nanobeams with arbitrary cross section [214]. Gao et al. [60, 215, 216] also proposed models of beams and plates including the size effect of the surface energy while studying size-dependent mechanical properties. Theory of surface energy characterizing the surface effect takes into account only the influence of surface layer. The modified couple stress theory and theory of deformations gradient characterizing the size effect take into account the influence of volume material. In other words, the mentioned theories characterize the size effect either for the volume part or for the surface.

In the reviewed literature, majority of the works characterize the size effect in nanostructures either from the point of view of volume or surface [206, 217]. There is a limited number of works to study size effect including both features [60, 215].

In Ref. [218], there is proposed a universal beam model taking into account both surface theory of elasticity and the volume theory. There are obtained formulation of nanosized Bernoulli-Euler and Timoshenko beams under the theory of gradient elasticity and theory of surface energy. They yielded the governing equations, initial conditions and all possible boundary conditions. There are solved problems of static bending and free vibrations of Bernoulli-Euler and Timoshenko beams with simply supported ends.

The size-dependent characteristics of piezoelectric nanomaterials which are coupled with surface effects and flexoelectricity are investigated for the Euler-Bernoulli beam [219]. The surface effects are taken into account in the modified theory of beams via the model regarding the surface piezoelectricity and the generalized Young-Laplace equation, whereas flexoelectricity effects are considered with the help of theory of piezoelectricity of higher order. The results of modelling of dynamics of piezoelectric nanobeams show that the influence of surface effects and flexoelectricity effects are changed with respect to the beam thickness and beam size ratio.

It should be emphasized that due to large ratio of the surface area to volume of typical nanostructures, the size effects are coupled with surface effects. Gurtin and Merdok [61] proposed the model of surface elasticity for elastic materials, where the surface is modelled as thin layer with small thickness attached to the volume. The governing and equilibrium equations for the surface layer differ from those inside the volume of the rigid body. Using that model, the size-dependent properties of nanomaterials were reported in [42, 46, 216]. However, the model of surface elasticity is not sufficient for investigation of mechanical behaviour of piezoelectric nanostructure with electromechanical couplings. In order to solve these problems for the piezoelectric nanostructures, the modified continuum model with surface-dependent effects depended on electric field was developed in [220]. In that model, the surface stresses depended on surface piezoelectricity and played a role of supplement to the surface elasticity. In that new surface piezoelectric model, both static and dynamic behaviours of various piezoelectric nanostructures were investigated [220–223]. The modelling results showed that the surface effects exhibit essential influence on properties of the static bending, vibrations and stability of the mentioned piezoelectric nanostructures.

It is also recognized that the flexoelectricity is responsible for size-dependent properties of piezoelectric nanomaterials which concerns a spontaneous polarization of electric materials due to either deformation gradient or non-homogeneous deformation field. Maranganti et al. [224] proposed a theoretical basis for dielectrics with an account of flexoelectricity and clarified mechanism of electromechanical coupling owing to gradients of deformations and/or polarization effects. Majdoub et al. [225] demonstrated strong increase of the effective piezoelectric coefficients on the size which depends on the size and included the flexoelectric phenomena into theory of piezoelectricity. Eliseev et al. [226] investigated renormalization process in properties of nanoferroids due to spontaneous flexoelectric effect. More recently, Lu et al. [227] investigated the influence of flexoelectricity on the electrostatic potential in the piezoelectric nanotube. The described so far investigations imply necessity of the account of flexoelectric effect while characterizing properties of piezoelectric nanostructures.

In Ref. [228], the theoretical model for investigation of the size-dependent elastic properties of nanobeam with an account of surface relaxation effects and surface tension was studied. The surface layer and its thickness were estimated in a unique way. The 3D model of nanofilm with a relaxation layer of atoms was analysed. Four non-zero elastic constants of the nanomembrane were obtained, and then Young modulus for straight extension was derived. Employing the ratio of energetic equilibrium, the dependences of effective elasticity modulus and effective bending stiffness of the nanobeam for two types of the transverse cross sections were obtained. Their dependence on the surface relaxation and surface tension were reported.

Yet, there was no theoretical model created which could adequately explain all phenomena of the size dependence. Villain et al. [229] detected a strong decrease of Young modulus with the help of atomic modelling with account of the surface tension. Employing the method of molecular statics based on the atomic potential Wolf [230] and Liang [231] showed that the modulus of elasticity can increase or decrease on the nano-level. The similar results were also obtained by Zhou and Huang [232] with the use of combinations of the molecular statics theory and computations. Miller and Shenoy [37] developed the model consisting of surface tension in order to forecast elastic properties of the nanobeams and nanoplates. They concluded that the effective stiffness is closely related to the values of elastic surface constants. Streitz and Cammarata [233] analysed the influence of surface energy on the elastic properties of the nanomaterials and forecasted increase of the elastic modulus of nanomembranes. Dingreville et al. [234] worked out a structural scheme in order to include the surface energy in the theory of continuum and demonstrated how the general model of elasticity of nanostructures can either increase or decrease. Guo and Zhao [235] also developed a 3D model with a layer composed of relaxed atoms which served as a basis of investigation of size-dependent elastic constants of nanomembranes and showed that moduli of elasticity of a nanomembrane may either increase or decrease while decreasing of its thickness.

Non-classical model of Mindlin plate located on elastic foundation was studied with the help of modified theory of couple stresses, theory of surface elasticity and two-parametric model of the Winkler-Pasternak foundation [236]. The model includes five kinematic parameters, and the governing equations and boundary conditions were yielded by the Hamilton principle. Microstructure effects, surface energy and fundamental effects are taken into account. It included one material scale parameter useful for descriptions of the microstructure effects, three elasticity constants of the surface needed to include surface energy and two parameters of foundation effect. The derived non-classical model can be reduced to its classical counterpart form based on elasticity theory assuming that the microstructure effects, surface energy and foundation effects are removed. Besides, this new model includes the Mindlin plate model which takes into account microstructural effects, surface energy influence and influence of the foundation being treated as separate cases in the carried out investigations.

Thin beams and plates supported by elastic foundation are widely employed in nano- and microscale devices and systems. There were presented various types of the elastic foundation models including those derived by Winkler [237], Filonenko-Borodich [238], Pasternak [239], Kerr [240] and Vlasov [241]. Winkler's model consists of only one parameter and it is the simplest one, whereas the Winkler-Pasternak model consists of two parameters and may exhibit the influence of foundation effects [242–244]. Since the size effects play an important role in applications in the nano- and microscales while the classical continuum theories cannot demonstrate the latter effects, more recently continuum theories of higher orders were worked out. The latter includes the parameters of material length and may include effects of microstructure and surface energy. Khajeansari et al. [244] investigated the bending deformation of the Euler-Bernoulli beam lying on the Winkler-Pasternak foundation employing theory of surface elasticity (see, for instance, [94, 245]). Simsek and Reddy [246] proposed the functionally gradient model of a microbeam embedded into an elastic medium using the modified couple stress theory (see [15, 247]) and the Winkler-Pasternak foundation. Limkatanyu et al. [248] proposed the Euler-Bernoulli beam model supported by the Winkler-Pasternak foundation and included the microstructure and surface energy effects, which extended the non-classical model proposed by Gao and Mahmuda [218]. More recently, Gao and Zhang [249] proposed the non-classical model which takes into account the microstructure, surface energy and influence of the foundation for the Kirchhoff plates with the use of variational formulation.

Novel non-classical model of Kirchhoff plates with the use of the modified couple stress theory, the surface theory of elasticity and two-parameter model of elastic foundation was presented in [249]. The variational formulation based on the Hamilton principle which yielded the derived equations of motion and the boundary conditions was used. New plate model consisted of the scale material length parameter to account of the microstructure effect, three elastic surface constants for description of the surface energy, and two elastic moduli of foundation. In order to demonstrate model abilities, the problem of static bending and free vibrations of the rectangular plate simply supported was solved analytically. In the case of the static bending, the numerical results indicate that in the considered cases deflection of the plate lying on the elastic foundation is less than that forecasted by the classical models. Besides, a difference in deflection forecast by new and classical models of plates is of large amount if the plate thickness is efficiently small (it is decreased with increase of a plate thickness). In the case of free vibrations, it was found that the eigenfrequencies obtained for new plate models under elastic foundation or without are higher than those obtained by the classical plate models. The so far described size effects in microscale are validated by laboratory experiments. Besides, it was shown both analytically and numerically that an occurrence of elastic foundation implies decrease of a plate deflection and increase of fundamental plate frequency, what was to be expected. In the case of a rigid body with large ratio of the surface to its volume, the size effects play a crucial role [37]. The size effects can be understood with the help of theory of surface elasticity where locations of the atoms and material properties are different in the surface and inside the body volume, and one surface effect is included into artificially constructed member with a negligible thickness [87, 95, 96].

Theory of surface elasticity [61, 94] was employed for analysis of thin plates coupled with surface effects. For instance, Miller and Shenoy [37] developed a model suitable for the description of the dependence of effective stiffness of nanosized structural element (rod, beam or plate) on the size. Lim and He [250] proposed a geometrically nonlinear model of nanoplates and nanomembranes based on Kirchhoff's hypothesis and von Kármán deformations. Lu et al. [251] developed the model of a thin plate dependent on the size with inclusion of normal stresses out and inside of the surface of volume layer. Lu et al. [252] worked out the nonlinear plastic model for functionally graded membrane with the help of Kirchhoff's kinematic relations and nonlinear von Kármán deformation of a volume material. Wang and Wang [253] proposed the model of nonlinear vibrations of the Kirchhoff plate and a Mindlin plate with the help of von Kármán model. However, in general, there is a small amount of models for thin plates with an account of both microstructure and surface energy. One of the non-classical models of thin Kirchhoff's plates using theory of elasticity of deformation gradient was proposed in [254]. It consisted of two additional scale length parameters: one was associated with energy volume deformation and the second represented energy of surface [216].

Influence of the surface stresses on rigidity of the cantilever plates within the employment of 3D model was investigated in [255]. A link between surface stress and stiffness of the plate was proposed with an account of the plate sizes. However, in general, the use of the short nanosized plates stands for the most challenging future investigations.

Rouhi et al. [256] proposed an analytical approach for analysis of geometrically nonlinear free vibrations of cylindrical nanoshells. The continuum model of Gurtin-Merdok was employed for the account of surface stresses. The governing PDEs of the shell with the surface stresses were derived based on the energetic method. The amplitude-frequency characteristics were reported. There were given numerous numerical results for the investigation of dynamic behaviour of nanosized shells with various geometric and surface material properties. It was shown that the surface stress depends essentially on the behaviour of nonlinear free vibrations of nanosized shells being very thin. Besides, it was detected that the effect of geometric nonlinearity was more exhibited when the residual stress was negative.

2.3.6 *Size-Dependent Theory of Beams, Plates and Shells Based on the Shear Deformations of the First Order Under the Timoshenko Theory (TBT)*

2.3.6.1 Nonlocal Timoshenko Models

The earliest nonlocal TBT model was proposed by Wang [257] to study wave propagation in CNT. The latter models account for the shear deformation which becomes essential for short and thick CNT. Wang and Varadan [258] also developed the nonlocal TBT model which was employed for the investigation of free vibrations either as SWCNTs or DWCNTs. The closed-form solutions for eigenfrequencies of simply supported CNT were reported. Wang et al. [259–262] obtained solutions in closed form for the critical loads [259], eigenfrequencies [260] and deflections [261, 262] with the help of nonlocal TBT model with four different boundary conditions including simply supported clamped cantilever and propped cantilever. In those models, the transversal shear stress was estimated based on the local theory. Lu et al. [263] included the nonlocal effect into both normal and transverse shear effects.

The compatibility nonlocal model was worked out also by Reddy and Pang [264] who changed EBT and TBT with the use of nonlocal governing Eringen's equations. They presented solutions for deflections, critical loads and eigenfrequencies for the nanobeams with four different boundary conditions. It should be mentioned that closed-form solutions obtained by Reddy and Pang [87] differ from those proposed by Wang et al. [259–262], since they were based on various TBT models.

The compatibility nonlocal TBT model has been widely employed to study nonlocal effects in CNT. For instance, Murmu and Pradhan [265] investigated the influence of the nonlocal parameter and transverse shear buckling OCNT embedded in an elastic medium. The latter work was extended by Ansari et al. [266], where the influence of high temperature was included. Pradhan and Murmu [267] investigated nonlocal TBT models based on the DQ method. Numerical solutions to the carbon nanotubes using nonlocal TBT and small-scale effects on vibration analysis of TBT nanomodel were presented by Roque et al. [268] based on meshless method with the global and local collocation technique and the use of radial basic functions. Vibrations of OCNT were also investigated by Wu and Lai [269] using RMVT-based nonlocal TBT models worked out on the basis of the variational Reissner's principle as well the principle of virtual displacements. Amirian et al. [270] and Zidour et al. [271] had taken into account the influence of temperature of the OCNT vibrations, while Ansari et al. [272] investigated the influence of thermal environment on stability of MWCNT. Ansari et al. [273] worked out a nonlocal TBT model for nonlinear excited vibrations of the magneto-electro-thermo-elastic nanobeams.

There were also proposals for nonlocal TBT model for FG nanobeams. Simsek and Yurtcu [274] proposed the nonlocal EBT models and TBT models to study bending and stability of FG beams. The compatibility nonlocal TBT model was extended by Rahmani and Pedram [275] for analysis of simply supported FG nanobeams. Ebrahimi and Salari [276, 277] also developed nonlocal TBT model to study the stability and free vibrations of FG nanobeams where thermal effects were taken into account. The first proposal for nonlocal first-order shear deformation theory (FSDT) model was proposed by Lu et al. [132] for the isotropic nanoplates. Then, it was employed to study the size effect in the problems of bending and problems of eigenfrequencies of simply supported isotropic nanoplates. Pradhan and Phadikar [278, 279] proposed nonlocal CPT and FSDT models of free vibrations [278] and analysed stability loss of [279] SLGS and MLGS. In the MLGS models, the interaction between two graphene sheets was modelled using the Winkler model. Influence of the size dependence, shear deformations, elasticity modulus and stiffness of the Winkler foundation on the eigenfrequencies and critical loads of the simply supported graphene sheets were carried out by Kananipour [280]. Ansari et al. [281, 282] investigated vibration of SLGS [281] and MLGS [282] with an account of different boundary conditions and using the nonlocal FSDT and the DQ method. Nonlocal FSDT was also employed by Samaei et al. [283] and Bedroud et al. [284] in order to study buckling of single SLGS [283] and circular nanoplates [284]. Arani et al. [285] considered electro-thermo-torsional buckling of nanotubes with double walls based on the nonlocal FSDT shell model. Naderi and Saidi [286] modified the nonlocal FSDT model to study the stability of nanoplates without the nonlocal effects for the transverse shear stresses. Nonlinear nonlocal CPT and FSDT models were worked out by Reddy [287] for nonlinear analysis of bending of isotropic nanoplates on a basis of von Kármán nonlinearity with the help of variational approaches.

Nonlocal FSDT models were also proposed for nanoplates made from FG and orthotropic materials. Hosseini-Hashemi et al. [288] proposed nonlocal FSDT for the circular/annular FG nanoplates. The closed-form solutions for the eigenfrequencies of circular nanoplates for various boundary conditions were also reported. Anjomshoa and Tahani [289] developed nonlocal FSDT model for free vibrations of orthotropic circular and elliptical SLGS embedded into an elastic medium. Golmakani and Rezatalab [290] proposed a nonlocal nonlinear FSDT model for nonlinear analysis of bending of orthotropic SLGS with the use of the von Kármán nonlinear deformations. Dastjerdi et al. [291] and Dastjerdi and Jabbarzadeh [292] developed nonlinear nonlocal model for geometric nonlinear analysis of the annular/circular orthotropic SLGS [291] and MLGS [292], where the temperature influence was taken into account.

2.3.6.2 Modified Couple Stress Timoshenko Theory

Ma et al. [293] developed the modified couple stress TBT model [155] in order to the account of the shear deformation effect. The model was employed to the investigation of the influence of length scale parameter and shear deformation on deflections and eigenfrequencies of simply supported isotropic microbeams. The closed form of solutions of the TBT model was proposed by Asghari et al. [294] for beam bending for various boundary conditions, whereas Dos Santos and Reddy [295] employed the Ritz method for the input loads and eigenfrequencies of the beams for various boundary conditions. Dehrouyeh-Semnani and Nikkhah-Bahrami [296] employed the EBT and TBT models for a study of the Poisson's effect in the isotropic microbeams. It was shown that inclusion of Poisson's effect into the model of the modified couple stress theory yields under estimated deflection of epoxidal cantilever. Liu and Reddy [297] worked out the modified coupled stress theory of the TBT model for the isotropic bended microbeams, and employed it to the problems of bending and free vibrations of the simply supported beams. Taati et al. [298] also worked out the TBT model for investigation of heat effects in the isotropic microbeams. Asghari et al. [299] proposed a nonlinear TBT model for analysis of deflections and free vibrations of isotropic microbeams. Ghayesh et al. [300, 301] also proposed a nonlinear TBT model for nonlinear dynamic problem of isotropic microbeams.

The TBT model was employed to study FG microbeams made from laminated composite materials. Reddy [302] developed the EBT and TBT models for FG microbeams with an account of a geometric nonlinearity. There were also derived analytical formulas for the critical loads and eigenfrequencies of the simply supported microbeams. Ke et al. [303] employed the nonlinear TBT model for a study of the influence of a size length parameter on nonlinear characteristics of vibrating FG microbeams. Asghari et al. [304] proposed the TBT model to investigate an influence of the size effects of the FG cantilevers on their deflection and rotations as well as on the eigenfrequency of simply supported FG. However, geometric nonlinearity was not taken into account. Ke and Wang [305] employed the TBT model to study free vibrations, static bending and dynamic stability of the FG microbeams under different boundary conditions and using the DQ method. Stability and free vibrations of the FG microbeams in the problem of thermoelasticity were investigated by Nateghi and Salamat-talab [306] with the help of modified couple stress theory and the TBT model. The DQ method was implemented for estimation of the critical load and eigenfrequencies of the FG microbeams with various boundary conditions. Numerical results showed that the influence of the temperature is of more importance for higher values of the ratio of the beam thickness to the scale length parameter. Simsek et al. [307] chose the TBT model for investigation of the size effect on deflection of the simply supported FG microbeams subjected to discrete and/or continuous loads.

Chen et al. [308], Chen and Li [309], Roque et al. [310] and Mohammad-Abadi and Daneshmehr [311] proposed suitable models for static bending [308, 310], free vibrations [309] and buckling [311] of the laminated composite microbeams. Thai et al. [312] considered application of the TBT model for static bending, stability and free vibrations of the sandwich FG microbeams. Most recently, Krysko et al. [313] developed the TBT model for static bending and free vibrations of the three-layer microbeams based on the Grigoluk-Chulkov theory.

One of the earlier models of the FSDT modified theory was developed by Ma et al. [314] and Ke et al. [315] for isotropic microplates. It should be mentioned that the FSDT model [314] takes into account both longitudinal and bending deformations, whereas Ke et al. [315] considered only the bending deformation in their model. Besides, Ma et al. [314] obtained solutions in the closed form for the problem of bending and vibrations of the simply supported plates, while Ke et al. [315] obtained the numerical solutions for the eigenfrequencies of plates with clamped sides with the help of the Ritz method. Roque et al. [316] presented the numerical solutions of the FSDT model of modified couple stress theory for analysis of the static bending of the isotropic microplates using the meshless method of collocation with radial basic functions. Zhou and Gao [317] worked out the modified couple stress theory of the FSDT model to carry out the axially symmetric analysis of bending of the isotropic circular microplates. Alinaghizadeh et al. [318] developed the modified couple stress theory of the FSDT model for the static analysis of bending of the FG annular microplates sector. In order to solve the bending problem of the microplates, the DQ method was used for the different boundary conditions. He et al. [319] extended the FSDT model to analyse static bending of the laminated composites microplates, whereas Simsek and Aydin [320] extended the FSDT model to solve the static bending and vibrations of the externally loaded FG microplates under action of moving load.

Reddy and Berry [321] generalized the axially symmetric model of FSDT in order to explain the influence of the geometric nonlinearity, temperature and non-homogeneous behaviour of the FG materials and the axially symmetric analysis of nonlinear bending of the circle microplates. Ke et al. [322, 323] also developed the axially symmetric FSDT model for nonlinear free vibrations [322] and stability [323] of annular FG microplates. Thai and Choi [324] proposed nonlinear CPT and FSDT models for the FG microplates. They obtained analytical formulas for linear/nonlinear deflections, critical loads and the eigenfrequencies of the simply supported microplates and quantified the influence of the size effect on bending, stability and dynamic characteristics of the FG microplates. Jung et al. [325, 326] took into consideration interaction of a plate with an elastic medium in the FSDT model while investigating the size-dependent bending effect, vibration [325] and stability [326] of simply supported FG microplates and nanoplates. Nonlinear models were employed by Ansari et al. [327, 328] for the case of nonlinear vibration [327], nonlinear bending and stability analysis [328] of the FG microplates. It was illustrated and discussed that in the nonlinear FSDT model worked out in [328] the physically neutral surface of the FG plate was considered, and the bending due to extension was removed. Ansari et al. [329] used nonlinear FSDT model to investigate the size effect on stability and frequencies of the FG microplates. Lou and He [330] also developed nonlinear models of CPT and FSDT for the case of nonlinear analysis of bending and free vibrations of the FG microplates. In their models, an interaction between the plate and elastic medium, as well as physical neutral plane position of the FG plate was taken into account.

Based on the FSDT, Zeighampour and Beni [331] and Hosseini-Hashemi et al. [332] proposed the models of shells to analyse free vibrations of the isotropic cylindrical micro/nanoshells [331] and spherical micro/nanoshells [332]. Gholami et al. [333] also developed the FSDT shell model which was used to study axial buckling and dynamic stability loss of the FG microshell. Tadi Beni et al. [334] proposed the FSDT shell model for the cylindrical FG microshells and employed it to the problems of free vibrations. Lou et al. [335] worked out the nonlinear FSDT shell model for estimation of the pre-buckling and buckling deformation and studied the influence and a value of the length size parameter on the critical loads of the stability loss of the cylindrical FG shells. Those models included a concept of a physical neutral plane position of the FG shells.

2.3.6.3 Modified Gradient Timoshenko Theory

Wang et al. [220] belong to the first who developed the TBT model with an account of the deformation gradient for the static bending and analysis of free vibrations of isotropic size-dependent and simply supported microbeams. Models of TBT nonlinear gradient of deformation were developed by Ansari et al. [336] and Asghari et al. [337] for isotropic microbeams using the von Kármán nonlinearity. It should be noted that Ansari et al. [336] employed their model for a study of free nonlinear vibrations, whereas Asghari et al. [337] considered in addition to nonlinear bending. Ansari et al. [338] extended the TBT model with the deformations gradient onto FG microbeams. Solutions in closed forms for the eigenfrequencies were obtained of the simply supported microbeams in order to follow the influence of index of material gradient and size-dependent length material parameter on free vibration of the FG beams. Ansari et al. [339] generalized their earlier work [338] into study of free vibrations of curvilinear FG microbeams. Ansari et al. [340] also developed the TBT model based on the modified strain gradient theory for the case of thermal bending of the FG microbeams with different boundary conditions. More recently, Ansari et al. [341] extended the TBT model of the strain gradient theory to study both linear and nonlinear viscoelastic beams. It should be emphasized that Gholami et al. [342] worked out the TBT model with deformations gradient in order to study nonlinear stability and dynamics of the micro-FG switches followed by the work of Mindlin [17]. Small-scale effects and Casimir forces were considered. Xie et al. [343] employed the collocation method based on the indirect radial basics functions to solve the EBT and TBT models regarding bending, stability loss and computation of the eigenfrequencies of the FG microbeams under various boundary conditions. It should be emphasized that in the previous work devoted to FG microbeams the size length parameters were chosen constant along the microbeams thickness. Therefore, Tajalli et al. [344] improved the previous models based on the strain gradient theory through account of the changeable length parameter along beam thickness. Practical investigations of the static and bending and free vibrations validated correctness of the approach [344].

Nonlinear TBT model based on the modified strain gradient theory was worked out by Ansari et al. [345], in order to study the influence of size length-dependent parameter and initial geometric interpretations on stability of the FG microbeams. Approximate solutions to the problems of stability loss of the FG microbeam for different boundary conditions were obtained with the help of the DQ method. Ansari et al. [346] supplemented their previous work [345] by considering the influence of thermal effects.

One of the earlier FSDT models with the gradient of deformations was proposed by Sahmani and Ansari [347] and Ansari et al. [348] for a study of free vibrations and thermal stability loss of the FG microplates. Sahmani and Ansari [349] considered only simply supported plates, whereas Ansari et al. [350] considered microplates with various boundary conditions and with the use of the DQ method. Ansari et al. [351] developed the model of nonlinear modified gradient FSDT theory to investigate stability of the annular FG microplates under thermal load. Ansari et al. [352] extended their previous work [350] to include the influence of temperature on bending, stability and free vibrations of the FG microplates for various boundary conditions. Shenas and Malekzadeh [351] also studied the temperature influence on free vibrations of the FG microplates for various boundary conditions. However, they employed Chebyshev-Ritz method instead of the DQ approach as it was done in the work of Ansari et al. [350]. Ansari et al. [352] developed the FSDT model for circular/annular FG microplates for different boundary conditions with the use of the DQ method.

Gholami et al. [353] worked out the FSDT shell model based on the modified strain gradient elasticity theory for the cylindrical FG microshells. Analytical solutions were reported for the critical load of simply supported cylindrical FG microshells under axial compression. Zhang et al. [354] also worked out the FSDT shell model based on the modified strain gradient elasticity theory for cylindrical FG microshells, but it was spanned on four FSDT unknowns proposed by Thai et al. [355–357]. Therefore, their model was simpler than that proposed by Gholami et al. [348], which included five unknowns.

2.3.6.4 Surface Theory of Timoshenko

A complex Timoshenko beam model to study static bending problems of nanowires/nanobeams with surface effects was used in [43]. There were obtained explicit solutions of the nonlinear problem with various boundary conditions for investigation of the combined influence of the residual surface stress, surface elasticity and shear deformation on the effective stiffness and Young modulus. It was shown that stiffness depends on the size, which is important for thin beams. The residual surface stress has a tendency to increase or decrease stiffness of a nanobeam versus boundary conditions. Shear deformation always makes the nanobeam more soft in comparison with the Euler-Bernoulli model. The obtained solutions well coincided with the experimental measurement of the Young modulus estimation, in particular, for a beam with small ratio of its length to thickness.

2.3.7 Size-Dependent Theory of Beams, Plates and Shells Based on the Shear Deformation Model of the Third Order

2.3.7.1 Nonlocal Theory of RBT, TSDT, HSDT

Based on the nonlocal constitutive relations, Reddy [358] modified theory of beams (EBT, TBT, RBT) in order to take into account a nonlocal effect. There were obtained variational equations for four models which were used to account of nonlocal models based on finite elements. For the case of simplified beams, there were obtained solutions in closed forms of the deflections, longitudinal loads and the eigenfrequencies. Ebrahimi and Salari [359] introduced thermal effects into nonlocal RBT to study the influence of temperature and nonlocal parameter on the characteristics of free vibration of OCNT. Emam [360] proposed a unified nonlinear nonlocal model for analysis of stability of isotropic nanobeams. There were obtained analytical solutions for critical load estimation for simply supported and clamped nanobeams.

Rahmani and Jandaghian [361] extended the nonlocal RBT model into the FG nanobeams. Analytical solutions for the critical loads were obtained for the FG nanobeams for various boundary conditions with the use of the Rayleigh-Ritz method. Ebrahimi and Reza Barati [362] also worked out the nonlocal RBT model for FG nanobeams where thermal effects and interactions between nanobeam and elastic medium were considered.

One of the earlier nonlocal HSDT (higher order shear deformation theory) models was proposed by Aydogdu [363, 364] for isotropic nanobeams based on general exponential theory of shear deformations. The latter theory generalized the exponential shear theory proposed by Karama et al. [365].

Thai [366] also proposed a nonlocal HSDT model for the isotropic nanobeams but it was based on the refined theory of plates developed by Shimpi [367]. In this theory, the field of displacement was separated into shear and bending parts. Tounsi et al. [368] and Zemri et al. [369] extended the nonlocal HSDT model by inclusion of thermal effects [368] and non-homogeneous FG material distribution [369]. Thai and Vo [370] developed the nonlocal HSDT model for isotropic nanobeams based on the sinusoidal theory of shear deformations. Touratier [371] and Tounsi et al. [372] proposed the nonlocal quasi-3D-model for isotropic nanobeams based on the quasi-3D sinusoidal theory proposed by Thai and Kim [373].

It should be noted that in contrary to the HSDT the quasi-3D-model enables to describe an effect of thickness change which plays a crucial role in thick and short beams. The general sinusoidal model [370] and quasi-3D sinusoidal model [372] for the FG nanobeams were more extended by Ahouel et al. [374] and Chaht et al. [375], respectively. The Thai and Vo [370] model was also revised by Pour et al. [376] and Sadatshojaei and Sadatshojaie [377] for the prognosis of nonlinear dynamic reactions of SWCNTs of carbon and boron nitride nanotube. Berrabah et al. [378] compared the accuracy of various nonlocal HSDT models for the prognosis of deflections, critical loads and eigenfrequency of nanobeams.

The fields of displacements of the nonlocal HSDT models were developed based on the simple HSDT proposed by Thai and Choi [379], where tangential and transverse displacements have influenced the bending and shear components. Ebrahimi and Barati [380] worked out the unified nonlocal HSDT model for heterogeneous FG beams based on the simple HSDT model introduced by Thai and Choi [379]. This model was employed to study the influence of temperature on vibration characteristics of the FG nanobeams. Mashat et al. [381] investigated vibrations and thermal buckling of nanobeams embedded in an elastic medium for various boundary conditions with the use of a unified nonlocal HSDT model including EBT, TBT, RBT and the sinusoidal theory. More recently, Thai et al. [382] proposed the simple nonlocal HSDT model for isotropic nanobeams which consisted of only one unknown. In the case of deflections and eigenfrequencies, there were obtained closed-form solutions for the studied boundary conditions. The carried out simulations showed that accuracy of the mentioned theory is similar to the TBT nonlocal model though it is built only on one unknown.

Nonlocal TSDT model was first proposed by Aghababaei and Reddy [383] for isotropic nanoplates by transforming the TSDT proposed by Reddy [71] with the use of nonlocal Eringen's relations. In the case of freely supported nanoplates, there were also obtained solutions in closed forms for the deflections and eigenfrequencies. The last model was employed by Pradhan [384] and Pradhan and Sahu [385] to study nonlocal effect under stability loss [384] and estimation of the eigenfrequencies [385] of simply supported SLGS. The bending of SLGS was also considered by Ansari and Sahmani [386] with the use of the unified nonlocal modes including various theories like CPT, FSDT and TSDT. Hosseini-Hashemi et al. [387] obtained Levi's solutions for the critical loads and eigenfrequencies of the isotropic nanoplates. Daneshmehr et al. [388, 389] extended employment of the nonlocal TSDT into stability [388] and analysis of free vibrations [389] of the FG nanoplates.

2.3.8 Nonlocal HSDT-Based Models

Narendar [390] proposed a nonlocal HSDT model for the analysis of stability loss for isotropic micro/nanoscale plates based on the improved theory of plates proposed by Shimpi [367] (two-variable refined plate theory). The latter model was extended by Malekzadeh and Shojaee [391], Narendar and Gopalakrishnan [392] to study free vibrations of the nanoplates [391] and to analyse stability loss of orthotropic nanoplates [392]. This model was also employed by Sobhy [393] to study free vibrations of the orthotropic DLGS under hygrothermal deformations. Sobhy [394] proposed the general model of HSDT and MLGS based on the simple HSDT proposed by Thai and Choi [379]. Analytical solutions with regard to eigenfrequencies, critical loads, temperature and stability loss were also obtained for MLGS for various boundary conditions. Levi's solutions of the nonlocal HSDT model introduced by Narendar [390] were also obtained by Sobhy [395, 396] while analysing bending of isotropic SLGS in the thermal environment [395] and orthotropic nanoplates in hygrothermal environment [396]. Zenkour and Sobhy [397], Alzahrani et al. [398], Thai et al. [399] and Sobhy [400, 401] developed the nonlocal sinusoidal models of thermal buckling of nanoplates [397], hygro-thermomechanical bending of nanoplates [398], isotropic nanoplates [399], thermomechanical bending of SLGS [400] and buckling of orthotropic nanoplates [401] based on the sinusoidal theory of Touratier [371]. It should be noted that the nonlocal sinusoidal model proposed by Sobhy [402] for the FGM nanoplates was based on the simple theory of Thai and Vo [403], and it is more simple in comparison to nonlocal sinusoidal models proposed in [397–401]. Belkorissat et al. [404] also worked out a simple nonlocal HSDT model for the FG nanoplates based on the work of Sobhy [402], but it was based on the hyperbolic function proposed by Soldatos [405]. Khorshidi and Fallah [406] reformulated the exponential theory of Karama et al. [365] for the nanoplates. Bessaim et al. [407] worked out the nonlocal quasi-3D model for the analysis of free vibrations of the isotropic nanoplates based on the quasi-3D sinusoidal theory of Thai and Kim [373], which is spanned on five unknowns. In the last years, Sobhy and Radwan [408] also developed a nonlocal quasi-3D theory for free vibrations and bending of the FG nanoplates. The latter consists of five unknown and is analogous to that proposed in [407] but being based on the new hyperbolic function.

2.3.8.1 Modified Coupled Stress Theory RBT, TSDT, NSDT, HSDT

Model of the modified RBT couple stress theory was first proposed by Ma et al. [409] for the isotropic microbeams. It was used for investigation of the influence of size effects on static deflections and free vibrations of the simply supported microbeams. Application of the RBT models was extended by Mohammad-Abadi and Daneshmehr [410] for investigation of the size material length parameter on stability of microbeams based on the models of EBT and TBT. In order to compare the obtained results, analytical solutions for the simply supported beams were developed. Salamat-talab et al. [411] extended the application of the RBT model into the FG simply supported microbeams. Nateghi et al. [412] and Aghazadeh et al. [413] proposed the unified model of stability [412] of deflection and free vibrations [413] of the FG beams. Their model includes three different theories of beams: EBT, FBT and TBT. The DQ method yielded the values of critical load, deflections and eigenfrequencies of the FG microbeams for various boundary conditions. Chen et al. [414] worked out the model RBT for laminated composite microbeams on the basis of novel introduced relations for anisotropic materials. The model was used for investigation of the size effects on deflection of the simply supported microbeams under uniformed load. Mohammad-Abadi and Daneshmehr [415] and Mohammad-Abadi et al. [416] extended their isotropic model presented in [229] for a study of the free vibrations [232] and thermal bending [417] for the laminated microbeams for various boundary conditions.

Darijani and Mohammadabadi [417] proposed the modified couple stress HSDT model for the isotropic microbeams by the separation of axial and transverse displacements into two parts related to shear and bending. The function of the form of shear part is based on observation that the transversal shear stress and couple stress of a higher order vanish on the upper and lower beam surfaces. More recently, Noori et al. [418] proposed the HSDT model for free vibrations of the isotropic microbeams on a basis of description of the axial displacement along thickness by a polynomial of fifth order. In order to define the eigenfrequencies of nanobeams for various boundary conditions, the DQ method was employed. Simsek and Reddy [419] developed the unified HSDT model for the FG microbeams by spanning seven various theories of beams including EBT, TBT and RBT. Touratier [371] sinusoidal theory, hyperbolic theory of Soldatos [420], exponential theory of Karama et al. [365] and general exponential theory of Aydogdu [364]. This model was employed to the problems of bending and free vibrations of simply supported FG microbeams. That model was extended by Simsek and Reddy [419] and Akbarzadeh et al. [421] for the problem FG microbeams and the problem of post-buckling stability of the FG microbeams with general boundary conditions [421]. Trinh et al. [422] also proposed the modified couple stress FG microbeam model based on both HSDT and the quasi-2D theory of beams. The field of displacement of their model was reconsidered by Thai et al. [423] based on the assumption of separating component of bending, shear and thickness change.

Based on the sinusoidal theory of Touratier [371], Akgoz and Civalek [424] worked out the modified couple stress sinusoidal model for investigation of thermomechanical buckling behaviour of the FG microbeams. The results showed that the influence of temperature on the bending loads of the FG microbeams becomes essential when a ratio of the length scale parameter to material thickness increases [424]. Al-Basyouni et al. [425] also proposed a modified couple stress sinusoidal model for the FG microbeams. However, it was based on the simple sinusoidal theory proposed by Thai and Vo [403] and included the physical plane position surface of the FG beams.

A TSDT model based on modified couple stress theory was first developed by Gao et al. [426] to study isotropic plates. The model was used for investigation of the size of simply supported microplates on their eigenfrequencies. The model was validated by Thai and Kim [420] and Chen et al. [427] for the FG microplates and the laminated composite materials, respectively. Jung and Han [428] also proposed the TSDT model for the FG microplates, but the authors used other law for computation of the equivalent mechanical properties of the FG microplates. Eshraghi et al. [429] worked out the TSDT model for the FG microplates with annular and circular forms. Field of displacements was unified with the inclusion of three different plate theories, i.e. CPT, FSDT and TSDT. In order to solve the problem of static bending and free vibrations, the DQ method was used by Eshraghi et al. [430]. Ghayesh and Farokhi [431] employed their previous work [429] and developed the nonlinear TSDT model to study nonlinear characteristics of vibrations of isotropic microplates. Based on the TSDT, Sahmani et al. [432] worked out the modified couple stress model of shells to analyse the dynamic instability of cylindrical FG microshells. They obtained closed-form solutions for the simply supported cylindrical microshells.

Thai and Vo [433] proposed the HSDT model based on the modified couple stress theory of the FG microplates. The field of displacements was based on the sinusoidal theory of Touratier [371]. For the case of simply supported microplates, there were obtained solutions of deflections and eigenfrequencies. Darijani and Shahdadi [434] proposed a simple HSDT model for the isotropic microplates via separating shear and bending components. Function of the form of shear component in plane was obtained from a condition of zeroth value of the transversal shear and of a higher order couple stresses on the upper/lower plate boundary. He et al. [435] reformulated theory of plates of Shimpi [367] for account of size effects in the FG microplates with the use of the modified couple stress theory. Though the work of Lou et al. [436] is analogous to that of He et al. [435], however, Lou et al. [437] used various forms of the Thai and Choi [379]. Lou et al. [436] more recently validated the work of He et al. [435] and included a geometric nonlinearity, and took into account the interaction between the plate and an elastic medium. Trinh et al. [438] also presented the unified and modified couple stress model for the FG microplates with mechanical and thermal loads based on the quasi-3D theory.

Reddy and Kim [439] worked out the nonlinear quasi-3D model for the FG microplates with an account of thermal effects. The latter was based on the von Kármán nonlinearity and a general quasi-3D theory an account cubic and square variations of plane and transversal displacement along thickness. The field of displacement of the general quasi-3D theory contained 11 unknowns. The CPT, FSDT and TSDT are yielded by the general theory as a particular case. The analytical formulas of this model were obtained by Kim and Reddy [440] for the simply supported plates. Lei et al. [441] proposed a simple quasi-3D theory to study static bending and free vibrations of the FG microplates which include only five unknowns and cubic form nonlinearity.

2.3.8.2 Modified Gradient Theories of RBT, TSDT, NSDT, HSDT

Based on a strain gradient elasticity theory of Lam et al. [18], Wang et al. [442] formulated the RBT model for the account of influence of the size on static bending and free vibrations of the isotropic microbeams. Sahmani and Ansari [443] improved the model RBT of the deformations gradient by inclusion of the thermal effects of the FG materials to study the stability of FG microbeams. Ansari et al. [444] investigated the influence of size-dependent parameter on free vibrations of the simply supported FG beams. Zhang et al. [445] proposed the RBT model for the FG beams based on the improved RBT model of Shi [446]. Sahmani et al. [447] proposed the RBT nonlinear model on the basis of the modified gradient theory of free vibrations of the FG microbeams. In addition to the RBT, the HSDT models were also proposed for the microbeams on the basis of the modified gradient strain theories such as the sinusoidal theory of Touratier [371], hyperbolic theory of Soldatos [405] and theory of shear n-th order deformation proposed by Xiang et al. [448]. For instance, Akgoz and Civalek [449] and Lei et al. [450] proposed the sinusoidal models of the deformation gradient for the analysis of bending and free vibrations of microplates made from isotropic materials [449] and FG materials [450] on the basis of Touratier theory [371]. Akgoz and Civalek [451] employed their previous theory [449] to study problems of buckling of isotropic microbeams. Akgoz and Civalek [452] also worked out the sinusoidal model of deformation gradient for the FG microbeams, as it was done by Lei et al. [450]. They also presented a new equation useful for computation of the shear correction coefficient of TBT model. In the mentioned equation, the coefficient of the shear correction stands for the function of the scale length material parameter. Akgoz and Civalek [453] extended their previous work [452] in order to clarify the interaction of FG microbeam and the Winkler elastic foundation. Based on the hyperbolic theory of Soldatos [405], Akgoz and Civalek [454] proposed the hyperbolic model of the deformation gradient to study bending and stability of the microbeams. Akgoz and Civalek [455] proposed unified HSDT model for analysis of bending of simply supported carbon nanotubes resting on an elastic foundation. The field of displacements followed the work of Simsek and Reddy [419], which accounts of theories of beams including EBT, TBT, RBT, sinusoidal theory of Touratier [371], the hyperbolic theory of Soldatos [420], exponential theory of Karama et al. [365] and the general exponential theory proposed by Aydogdu [364]. Zhang et al. [456] proposed the HSDT model to study bending and vibrations of curved FG microbeams based on the theory of shear deformations of n-th order (see [448]).

Sahmani and Ansari [347] developed the TSDT model with a gradient of deformations for free vibrations analysis of the FG microplates. In the case of simply supported plates, there were also presented closed solutions for the eigenfrequencies. Zhang et al. [457] proposed a simple TSDT model for circular/annular FG microplates earlier introduced by Thai and Kim [458], which included only four unknowns. The DQ method was used to compute deflection, critical values and eigenfrequencies of circular/annular FG microplates with various boundary conditions. Zhang et al. [459] worked out a simple HSDT model with deformation gradient for the FG microplates based on the simple HSDT proposed by Thai and Choi [460–463], which possesses only four unknowns (the authors considered an interaction of the plate and elastic foundations). Akgoz and Civalek [464] developed the sinusoidal model of the deformation gradient for analysis of bending, stability and free vibrations of the isotropic microplates based on the sinusoidal theory of Touratier [371].

2.3.8.3 Surface Theory: RBT, TSDT, NSDT, HSDT

In work [255], an influence of the surface stresses on rigidity of the cantilever plates, using full 3D model, was studied. It allowed to describe a link between the surface stresses and stiffness of the cantilever, and then the cantilever behaviour versus the size effects was studied.

In spite of the large account of the works published within the book research topic, mainly linear models were used for computational analysis, though the experimental results imply a need of account of the nonlinear features of the micro/nanomechanical/thermomechanical systems [465].

This opens a novel challenging research field aimed on development of more precise nonlinear deformation models of the size-dependent beams subjected to static and dynamic loads. To study dynamics of the size-dependent beams, there is a need to use the apparatus of nonlinear dynamics including the Fourier and wavelet spectra, phase portraits, Poincaré maps, the Lyapunov exponent's computation (at least the largest one), the autocorrelation functions, etc. The mentioned approach was earlier employed by the authors of the given monograph to study nonlinear dynamics of continuous systems in the form of beams, plates and shells [466–472]. There were considered three types of nonlinearity, i.e. physical, geometric and design one (contact interaction in time). However, the mentioned results were obtained based on the classical theory of elasticity, i.e. without considering the size-dependent behaviour of constructions made from homogeneous materials.

The presented so far state of the art of literature shows that the investigations of the size-dependent beams modelled via the Euler-Bernoulli, Timoshenko and Sheremetev-Pelekh theories are based on the Duffing-type equations which are yielded by the employment of Bubnov-Galerkin method in the first approximation to the governing PDEs. There were considered linear problems aimed at estimation of eigenfrequencies and static problems focused on analysing the influence of parameters responsible for the size effects.

In all cases, investigation of nonlinear static and dynamics, and in particular chaotic dynamics, was not taken into account.

References

1. Fu, Y., Zhang, J.: Electromechanical dynamic buckling phenomenon in symmetric electric fields actuated microbeams considering material damping. Acta Mech. **212**, 29–42 (2010)
2. Moghimi, Z.M., Ahmadian, M.T.: Static pull-in analysis of electrostatically actuated microbeams using homotopy perturbation method. Appl. Math. Model. **34**, 1032–1046 (2010)
3. Jia, X.L., Yang, J., Kitipornchai, S.: Pull-in instability of geometrically nonlinear micro-switches under electrostatic and Casimir forces. Acta Mech. **218**, 161–174 (2011)
4. Stolken, J.S., Evans, A.G.: Microbend test method for measuring the plasticity length scale. Acta Mater. **46**(14), 5109–5115 (1998)
5. Nix, W.D.: Mechanical properties of thin films. Metall. Trans. A **20**, 2217–2245 (1989)
6. Ma, Q., Clarke, D.R.: Size dependent hardness of silver single crystals. J. Mater. Res. **10**(4), 853–863 (1995)
7. Fleck, N.A., Muller, G.M., Ashby, M.F., Hutchinson, J.W.: Strain gradient plasticity: theory and experiments. Acta Metall. Mater. **42**, 475–487 (1994)
8. Chong, A.C.M., Lam, D.C.C.: Strain gradient plasticity effect in indentation hardness of polymers. Mater. Res. **14**(10), 4103–4110 (1999)
9. Dell'Isola, F., Andreaus, U., Placidi, L.: At the origins and in the vanguard of peridynamics, non-local and higher-gradient continuum mechanics: An underestimated and still topical contribution of Gabrio Piola. Math. Mech. Sol. **20**, 887–928 (2015)
10. Dell'Isola, F., Della Corte, A., Esposito, R., Russo, L.: Some cases of unrecognized transmission of scientific knowledge: from antiquity to Gabrio Piola's peridynamics and generalized continuum theories. In: Altenbach, H., Forest, S. (eds.) Generalized Continua as Models for Classical and Advanced Materials, pp. 77–128. Springer, Berlin (2016)
11. Cosserat, E., Cosserat, F.: Théorie des Corps Déformables. Hermann et Fils, Paris (1909)
12. Toupin, R.A.: Elastic materials with couple-stresses. Arch. Ration. Mech. Anal. **11**, 385–414 (1962)
13. Mindlin, R.D., Tiersten, H.F.: Effects of couple-stresses in linear elasticity. Arch. Ration. Mech. Anal. **11**, 415–448 (1962)
14. Koiter, W.T.: Couple-stresses in the theory of elasticity, I and II. Proc. K. Ned. Akad. Wet. B **67**, 17–44 (1964)
15. Yang, F., Chong, M., Lam, D.C.C., Tong, P.: Couple stress based strain gradient theory for elasticity. Int. J. Solids Struct. **39**, 2731–2743 (2002)
16. Mindlin, R.D.: Micro-structure in linear elasticity. Arch. Ration. Mech. Anal. **16**, 51–78 (1964)
17. Mindlin, R.D.: Second gradient of strain and surface-tension in linear elasticity. Int. J. Sol. Struct. **1**, 417–38 (1965)
18. Lam, D.C.C., Yang, F., Chong, A.C.M., Wang, J., Tong, P.: Experiments and theory in strain gradient elasticity. J. Mech. Phys. Soli **51**, 1477–1508 (2003)
19. Eringen, A.C.: Simple microfluids. Int. J. Eng. Sci. **2**, 205–217 (1964)
20. Eringen, A.C.: Linear theory of micropolar elasticity. J. Math. Mech. **15**, 909–923 (1966)
21. Eringen, A.C.: Micropolar fluids with stretch. Int. J. Eng. Sci. **7**, 115–127 (1969)
22. Altenbach, H., Eremeyev, V.A.: Strain rate tensors and constitutive equations of inelastic micropolar materials. Int. J. Plast. **63**, 3–17 (2014)
23. Ansari, R., Bazdid-Vahdati, M., Shakouri, A., Norouzzadeh, A., Rouhi, H.: Micromorphic prism element. Math. Mech. Sol. **22**(6), 17–29 (2016)
24. Ansari, R., Bazdid-Vahdati, M., Shakouri, A., Norouzzadeh, A., Rouhi, H.: Micromorphic first-order shear deformable plate element. Mecca. **51**, 1797–809 (2016)
25. Ansari, R., Shakouri, A.H., Bazdid-Vahdati, M., Norouzzadeh, A., Rouhi, H.: A nonclassical finite element approach for the nonlinear analysis of micropolar plates. J. Comput. Nonlin. Dyn. **12**, 011019–12 (2016)
26. Chen, Y., Lee, J.D.: Determining material constants in micromorphic theory through phonon dispersion relations. Int. J. Eng. Sci. **41**, 871–886 (2003)
27. Eringen, A.C.: Microcontinuum Field Theories: I Foundations and Solids. Springer, Berlin (1999)

28. Eringen, A.C.: Microcontinuum Field Theories: II Fluent Media. Springer, Berlin (2001)
29. Kroner, E.: Elasticity theory of materials with long range cohesive forces. Int. J. Sol. Struct. **3**, 731–742 (1967)
30. Eringen, A.C.: Linear theory of nonlocal elasticity and dispersion of plane waves. Int. J. Eng. Sci. **10**, 425–435 (1972)
31. Eringen, A.C.: Nonlocal polar elastic continua. Int. J. Eng. Sci. **10**, 1–16 (1972)
32. Eringen, A.C., Edelen, D.G.B.: On nonlocal elasticity. Int. J. Eng. Sci. **10**, 233–248 (1972)
33. Eringen, A.C.: On differential equations of nonlocal elasticity and solutions of screw dislocation and surface waves. J. Appl. Phys. **54**, 4703–4710 (1983)
34. Aifantis, E.C.: On the gradient approach Relation to Eringen's nonlocal theory. Int. J. Eng. Sci. **49**, 1367–1377 (2011)
35. Lim, C., Zhang, G., Reddy, J.: A higher-order nonlocal elasticity and strain gradient theory and its applications in wave propagation. J. Mech. Phys. Sol. **78**, 298–313 (2015)
36. Narendar, S., Gopalakrishnan, S.: Ultrasonic wave characteristics of nanorods via nonlocal strain gradient models. J. Appl. Phys. **107**, 084312–8 (2010)
37. Miller, R.E., Shenoy, V.B.: Size-dependent elastic properties of nanosized structural elements. Nanotechnology **11**, 139–147 (2000)
38. Gavan, K.B., Westra, H.J.R., van der Drift, E.W.J.M., Venstra, W.J., van der Zant, H.S.J.: Size-dependent effective Young's modulus of silicon nitride cantilevers. Appl. Phys. Lett. **94**, 233108 (2009)
39. Olsson, P.A.T., Park, H.S.: On the importance of surface elastic contributions to the flexural rigidity of nanowires. J. Mech. Phys. Sol. **60**, 2064–2083 (2012)
40. Li, C., Guo, W.L., Kong, Y., Gao, H.J.: Size-dependent piezoelectricity in zinc oxide nanofilms from first-principles calculations. Appl. Phys. Lett. **90**, 033108–033103 (2007)
41. Majdoub, M.S., Sharma, P., Cagin, T.: Enhanced size-dependent piezoelectricity and elasticity in nanostructures due to the flexoelectric effect. Phys. Rev. B **77**, 125424 (2008)
42. He, J., Lilley, C.M.: Surface effect on the elastic behavior of static bending nanowires. Nano Lett. **8**, 1798–1802 (2008)
43. Jiang, L.Y., Yan, Z.: Timoshenko beam model for static bending of nanowires with surface effects. Phys. E **42**, 2274–2279 (2010) 1572
44. Chiu, M.S., Chen, T.Y.: Effects of high-order surface stress on buckling and resonance behavior of nanowires. Acta. Mech. **223**, 1473–1484 (2012)
45. Wang, G.F., Feng, X.Q.: Surface effects on buckling of nanowires under uniaxial compression. Appl. Phys. Lett. **94**, 141913 (2009)
46. Wang, G.F., Feng, X.Q.: Timoshenko beam model for buckling and vibration of nanowires with surface effects. J. Phys. D Appl. Phys. **42**, 155411 (2009)
47. Yan, Z., Jiang, L.Y.: Surface effects on the vibration and buckling of piezoelectric nanoplates. Europhy. Lett. **99**, 27007 (2012)
48. Zhang, C.L., Chen, W.Q., Zhang, C.: On propagation of anti-plane shear waves in piezoelectric plates with surface effect. Phys. Lett. A **376**, 3281–3286 (2012)
49. Huang, G.Y., Kang, Y.L.: Acoustic vibrations of a circular nanowire by considering the effect of surface. J. Appl. Phys. **110**, 023526 (2011)
50. Assadi, A., Farshi, B.: Size-dependent longitudinal and transverse wave propagation in embedded nanotubes with consideration of surface effects. Acta Mech. **222**, 27–39 (2011)
51. Chen, W.Q., Wu, B., Zhang, C.L., Zhang, C.: On wave propagation in anisotropic elastic cylinders at nanoscale: surface. elasticity and its effect. Acta Mech. **225**, 2743–2760 (2014)
52. Wang, G.F., Feng, X.Q.: Effects of surface elasticity and residual surface tension on the natural frequency of microbeams. Appl. Phys. Lett. **90**, 231904 (2007)
53. He, J., Lilley, C.M.: Surface stress effect on bending resonance of nanowires with different boundary conditions. Appl. Phys. Lett. **93**, 263108 (2008)
54. Abbasion, S., Rafsanjani, A., Avazmohammadi, R., Farshidianfar, A.: Free vibration of microscaled Timoshenko beams. Appl. Phys. Lett. **95**, 143122 (2009)
55. Farshi, B., Assadi, A., Alinia-ziazi, A.: Frequency analysis of nanotubes with consideration of surface effects. Appl. Phys. Lett. **96**, 093105 (2010)

56. Assadi, A.: Size dependent dynamic analysis of nanoplates. J. Appl. Phys. **107**, 124310 (2010)
57. Lu, P., Lee, H.P., Lu, C., O'Shea, S.J.: Surface stress effects on the resonance properties of cantilever sensors. Phys. Rev. B **72**, 085405 (2005)
58. He, Q., Lilley, C.M.: Resonant frequency analysis of Timoshenko nanowires with surface stress for different boundary conditions. J. Appl. Phys. **112**, 074322 (2012)
59. Yan, Z., Jiang, Y.: Vibration and buckling analysis of a piezoelectric nanoplate considering surface effects and in-plane constraints. Proc. R. Soc. A. **468**, 3458–3475 (2012)
60. Gao, X.L.: A new Timoshenko beam model incorporating microstructure and surface energy effects. Acta Mech. **226**, 457–474 (2015)
61. Gurtin, M.E., Murdoch, A.I.: A continuum theory of elastic material surfaces. Arch. Ration. Mech. Anal. **57**(4), 291–323 (1975)
62. Koochi, A., Hosseini-Toudeshky, H., Ovesy, H.R., Abadyan, M.: Modeling the influence of surface effect on instability of nano-cantilever in presence of van der Waals force. Int. J. Struct. Stab. Dyn. **13**(4), 1250072 (2013)
63. Iijima, S.: Helical microtubules of graphitic carbon. Nature **354**, 56–58 (1991)
64. Ansari, R., Norouzzadeh, A., Gholami, R., Faghih Shojaei, M., Darabi, M.A.: Geometrically nonlinear free vibration and instability of fluid-conveying nanoscale pipes including surface stress effects. Microflu. Nanoflu. **20**, 28–42 (2016)
65. Ansari, R., Gholami, R., Norouzzadeh, A., Darabi, M.A.: Wave characteristics of nanotubes conveying fluid based on the non-classical Timoshenko beam model incorporating surface energies. Arab. J. Sci. Eng. **41**, 4359–4369 (2016)
66. Ansari, R., Gholami, R., Norouzzadeh, A.: Size-dependent thermo-mechanical vibration and instability of conveying fluid functionally graded nanoshells based on Mindlin's strain gradient theory. Thin Wall. Struct. **105**, 172–184 (2016)
67. Ansari, R., Gholami, R., Norouzzadeh, A., Sahmani, S.: Size-dependent vibration and instability of fluid-conveying functionally graded microshells based on the modified couple stress theory. Microflu. Nanoflu. **19**, 509–522 (2015)
68. Ansari, R., Gholami, R., Norouzzadeh, A., Darabi, M.: Surface stress effect on the vibration and instability of nanoscale pipes conveying fluid based on a size-dependent Timoshenko beam model. Acta Mech. Sin. **31**, 708–719 (2015)
69. Ansari, R., Norouzzadeh, A., Gholami, R., Shojaei, M.F., Hosseinzadeh, M.: Size-dependent nonlinear vibration and instability of embedded fluid-conveying SWBNNTs in thermal environment. Phys E. **61**, 148–157 (2014)
70. Wang, L.: Vibration analysis of fluid-conveying nanotubes with consideration of surface effects. Phys E. **43**, 437–439 (2010)
71. Reddy, J.N.: A simple higher-order theory for laminated composite plates. J. Appl. Mech. **51**, 745–752 (1984)
72. Thai, H.T., Kim, S.E.: A review of theories for the modeling and analysis of functionally graded plates and shells. Compos. Struct. **128**, 70–86 (2015)
73. Huang, L.Y., Han, Q., Liang, Y.J.: Calibration of nonlocal scale effect parameter for bending single-layered graphene sheet under molecular dynamics. Nano. **7**, 125003–8 (2012)
74. Arash, B., Ansari, R.: Evaluation of nonlocal parameter in the vibrations of single-walled carbon nanotubes with initial strain. Phys E. **42**, 2058–2064 (2010)
75. Duan, W.H., Challamel, N., Wang, C.M., Ding, Z.: Development of analytical vibration solutions for microstructured beam model to calibrate length scale coefficient in nonlocal Timoshenko beams. J. Appl. Phys. **114**, 104312–11 (2013)
76. Zhang, Z., Wang, C.M., Challamel, N., Elishakoff, I.: Obtaining Eringen's length scale coefficient for vibrating nonlocal beams via continualization method. J. Sound Vib. **333**, 4977–4990 (2014)
77. Zhang, Z., Wang, C.M., Challamel, N.: Eringen's length scale coefficient for buckling of nonlocal rectangular plates from microstructured beam-grid model. Int. J. Sol. Struct. **51**, 4307–4315 (2014)
78. Zhang, Z., Wang, C.M., Challamel, N.: Eringen's length-scale coefficients for vibration and buckling of nonlocal rectangular plates with simply supported edges. J. Eng. Mech. **141**, 04014117–10 (2015)

79. Wang, Q., Wang, C.M.: The constitutive relation and small scale parameter of nonlocal continuum mechanics for modelling carbon nanotubes. Nanotechnology **18**, 075702–4 (2007)
80. Norouzzadeh, A., Ansari, R., Rouhi, H.: Pre-buckling responses of Timoshenko nanobeams based on the integral and differential models of nonlocal elasticity: an isogeometric approach. Appl. Phys. A. **123**, 330–341 (2017)
81. Norouzzadeh, A., Ansari, R.: Finite element analysis of nano-scale Timoshenko beams using the integral model of nonlocal elasticity. Phys E. **88**, 194–200 (2017)
82. Fernandez-Saez, J., Zaera, R., Loya, J.A., Reddy, J.N.: Bending of Euler Bernoulli beams using Eringen's integral formulation: A paradox resolved. Int. J. Eng. Sci. **99**, 107–116 (2016)
83. Challamel, N., Wang, C.M.: The small length scale effect for a non-local cantilever beam: a paradox solved. Nanotech. **19**, 345703 (2008)
84. Mindlin, R.D., Tiersten, H.F.: Effects of couple-stresses in linear elasticity. Arch. Ration. Mech. Anal. **11**, 415–448 (1962)
85. Dehrouyeh-Semnani, A.M., Nikkhah-Bahrami, M.: A discussion on evaluation of material length scale parameter based on micro-cantilever test. Compos. Struct. **122**, 425–429 (2015)
86. Khajueenejad, F., Ghanbari, J.: Internal length parameter and buckling analysis of carbon nanotubes using modified couple stress theory and Timoshenko beam model. Mater. Res. Exp. **2**, 105009–10 (2015)
87. Cammarata, R.C.: Surface and interface stress effects in thin films. Prog. Surf. Sci. **46**, 1–38 (1994)
88. Thomson, R., Chuang, T.-J., Lin, I.-H.: The role of surface stress in fracture. Acta Metall. **34**, 1133–1143 (1986)
89. Maugis, D.: Contact, Adhesion, and Rupture of Elastic Solids. Springer, Berlin (2000)
90. Shuttleworth, R.: The surface tension of solids. Proc. Phys. Soc. A **63**, 444–457 (1950)
91. Herring, C.: Some theorems on the free energies of crystal surfaces. Phys. Rev. **82**, 87–93 (1951)
92. Nix, W.D., Gao, H.: An atomistic interpretation of interface stress. Scr. Mater. **39**, 1653–1661 (1998)
93. Spaepen, F.: Interfaces and stresses in thin films. Acta Mater. **48**, 31–42 (2000)
94. Gurtin, M.E., Murdoch, A.I.: Surface stress in solids. Int. J. Solids Struct. **14**, 431–440 (1978)
95. Steigmann, D.J., Ogden, R.W.: Plane deformation of elastic solids with intrinsic boundary elasticity. Proc. R. Soc. A **453**, 853–877 (1997)
96. Steigmann, D.J., Ogden, R.W.: Elastic surface-substrate interactions. Proc. R. Soc. A **455**, 437–474 (1999)
97. Wang, G.F., Feng, X.Q.: Effects of surface stresses on contact problems at nanoscale. J. Appl. Phys. **101**, 013510-1–013510-6 (2007)
98. Zhao, X.J., Rajapakse, R.K.N.D.: Analytical solutions for a surface-loaded isotropic elastic layer with surface energy effects. Int. J. Eng. Sci. **47**, 1433–1444 (2009)
99. Shenoy, V.B.: Atomistic calculations of elastic properties of metallic fcc crystal surfaces. Phys. Rev. B **71**, 094104-1–094104-11 (2005)
100. Peddieson, J., Buchanan, G.R., McNitt, R.P.: Application of nonlocal continuum models to nanotechnology. Int. J. Eng. Sci. **41**, 305–312 (2003)
101. Sudak, L.J.: Column buckling of multiwalled carbon nanotubes using nonlocal continuum mechanics. J. Appl. Phys. **94**, 7281–7287 (2003)
102. Zhang, Y.Q., Liu, G.R., Xie, X.Y.: Free transverse vibrations of double-walled carbon nanotubes using a theory of nonlocal elasticity. Phys. Rev. B. **71**, 195404–7 (2005)
103. Wang, Q., Varadan, V.K., Quek, S.T.: Small scale effect on elastic buckling of carbon nanotubes with nonlocal continuum models. Phys. Lett. A. **357**, 130–135 (2006)
104. Aydogdu, M.: Axial vibration of the nanorods with the nonlocal continuum rod model. Phys. E. **41**, 861–864 (2009)
105. Murmu, T., Pradhan, S.C.: Thermo-mechanical vibration of a single walled carbon nanotube embedded in an elastic medium based on nonlocal elasticity theory. Comput. Mater. Sci. **46**, 854–859 (2009)

106. Civalek, O., Demir, C.: Bending analysis of microtubules using nonlocal Euler-Bernoulli beam theory. Appl. Math. Model. **35**, 2053–2067 (2011)
107. Mustapha, K.B., Zhong, Z.W.: Free transverse vibration of an axially loaded non-prismatic single-walled carbon nanotube embedded in a two-parameter elastic medium. Comput. Mater. Sci. **50**, 742–751 (2010)
108. Li, C., Lim, C.W., Yu, J.L., Zeng, Q.C.: Analytical solutions for vibration of simply supported nonlocal nanobeams with an axial force. Int. J. Struct. Stab. Dyn. **11**(2), 257–271 (2011)
109. Ghannadpour, S.A.M., Mohammadi, B., Fazilati, J.: Bending, buckling and vibration problems of nonlocal Euler beams using Ritz method. Compos. Struct. **96**, 584–589 (2013)
110. Ansari, R., Ramezannezhad, H., Gholami, R.: Nonlocal beam theory for nonlinear vibrations of embedded multiwalled carbon nanotubes in thermal environment. Nonlinear Dyn. **67**, 2241–2254 (2012)
111. Fang, B., Zhen, Y.X., Zhang, C.P., Tang, Y.: Nonlinear vibration analysis of double-walled carbon nanotubes based on nonlocal elasticity theory. Appl. Math. Model. **37**, 1096–1107 (2013)
112. Simsek, M.: Large amplitude free vibration of nanobeams with various boundary conditions based on the nonlocal elasticity theory. Compos. B Eng. **56**, 621–628 (2014)
113. Bagdatli, S.M.: Non-linear vibration of nanobeams with various boundary condition based on nonlocal elasticity theory. Compos. B Eng. **80**, 43–52 (2015)
114. Simsek, M.: Nonlocal effects in the free longitudinal vibration of axially functionally graded tapered nanorods. Comput. Mater. Sci. **61**, 257–265 (2012)
115. Nguyen, N.T., Kim, N.I., Lee, J.: Analytical solutions for bending of transversely or axially FG nonlocal beams. Steel. Compos. Struct. **17**, 639–663 (2014)
116. Niknam, H., Aghdam, M.M.: A semi analytical approach for large amplitude free vibration and buckling of nonlocal FG beams resting on elastic foundation. Compos. Struct. **119**, 452–462 (2014)
117. Ebrahimi, F., Salari, E.: Thermo-mechanical vibration analysis of nonlocal temperature-dependent FG nanobeams with various boundary conditions. Compos. B Eng. **78**, 272–290 (2015)
118. Ebrahimi, F., Salari, E.: Nonlocal thermo-mechanical vibration analysis of functionally graded nanobeams in thermal environment. Acta Astronaut. **113**, 29–50 (2015)
119. Nejad, M.Z., Hadi, A.: Eringen's non-local elasticity theory for bending analysis of bi-directional functionally graded Euler-Bernoulli nano-beams. Int. J. Eng. Sci. **106**, 1–9 (2016)
120. Nejad, M.Z., Hadi, A., Rastgoo, A.: Buckling analysis of arbitrary two-directional functionally graded Euler-Bernoulli nano-beams based on nonlocal elasticity theory. Int. J. Eng. Sci. **103**, 1–10 (2016)
121. Nazemnezhad, R., Hosseini-Hashemi, S.: Nonlocal nonlinear free vibration of functionally graded nanobeams. Compos. Struct. **110**, 192–199 (2014)
122. El-Borgi, S., Fernandes, R., Reddy, J.N.: Non-local free and forced vibrations of graded nanobeams resting on a non-linear elastic foundation. Int. J. Non-Linear Mech. **77**, 348–363 (2015)
123. Shafiei, N., Kazemi, M., Safi, M., Ghadiri, M.: Nonlinear vibration of axially functionally graded non-uniform nanobeams. Int. J. Eng. Sci. **106**, 77–94 (2016)
124. Zhang, Y.Q., Liu, G.R., Wang, J.S.: Small-scale effects on buckling of multiwalled carbon nanotubes under axial compression. Phys. Rev. B. **70**, 205430–6 (2004)
125. Li, R., Kardomateas, G.A.: Thermal buckling of multi-walled carbon nanotubes by nonlocal elasticity. J. Appl. Mech. Trans. ASME. **74**, 399–405 (2007)
126. Li, R., Kardomateas, G.A.: Vibration characteristics of multiwalled carbon nanotubes embedded in elastic media by a nonlocal elastic shell model. J. Appl. Mech. Trans. ASME. **74**, 1087–1094 (2007)
127. Wang, Q., Varadan, V.K.: Application of nonlocal elastic shell theory in wave propagation analysis of carbon nanotubes. Smart. Mater. Struct. **16**, 178–191 (2007)
128. Hu, Y.G., Liew, K.M., Wang, Q., He, X.Q., Yakobson, B.I.: Nonlocal shell model for elastic wave propagation in single- and double-walled carbon nanotubes. J. Mech. Phys. Sol. **56**, 3475–3485 (2008)

129. Zhang, Y.Y., Wang, C.M., Duan, W.H., Xiang, Y., Zong, Z.: Assessment of continuum mechanics models in predicting buckling strains of single-walled carbon nanotubes. Nanotech. **20**, 395707–8 (2009)
130. Rouhi, H., Ansari, R.: Nonlocal analytical Flugge shell model for axial buckling of double-walled carbon nanotubes with different end conditions. Nano. **7**, 1250018–10 (2012)
131. Sarvestani, H.Y.: Buckling analysis of curved nanotube structures based on the nonlocal shell theory. Int. J. Multiscale Comput. Eng. **14**, 45–54 (2016)
132. Lu, P., Zhang, P.Q., Lee, H.P., Wang, C.M., Reddy, J.N.: Non-local elastic plate theories. Proc. R Soc. A Math. Phys. Eng. Sci. **463**, 3225–3240 (2007)
133. Duan, W.H., Wang, C.M.: Exact solutions for axisymmetric bending of micro/nanoscale circular plates based on nonlocal plate theory. Nanotech. **18**, 385704–5 (2007)
134. Aksencer, T., Aydogdu, M.: Levy type solution method for vibration and buckling of nanoplates using nonlocal elasticity theory. Phys. E. **43**, 954–959 (2011)
135. Shakouri, A., Ng, T.Y., Lin, R.M.: Nonlocal plate model for the free vibration analysis of nanoplates with different boundary conditions. J. Comput. Theor. Nanosci. **8**, 2118–2128 (2011)
136. Pradhan, S.C., Murmu, T.: Small scale effect on the buckling of single-layered graphene sheets under biaxial compression via nonlocal continuum mechanics. Comput. Mater. Sci. **47**, 268–274 (2009)
137. Pradhan, S.C., Kumar, A.: Buckling analysis of single layered graphene sheet under biaxial compression using nonlocal elasticity theory and DQ method. J. Comput. Theor. Nanosci. **8**, 1325–1334 (2011)
138. Babaei, H., Shahidi, A.R.: Small-scale effects on the buckling of quadrilateral nanoplates based on nonlocal elasticity theory using the Galerkin method. Arch. Appl. Mech. **81**, 1051–1062 (2011)
139. Farajpour, A., Danesh, M., Mohammadi, M.: Buckling analysis of variable thickness nanoplates using nonlocal continuum mechanics. Phys. E. **44**, 719–727 (2011)
140. Pradhan, S.C., Phadikar, J.K.: Small scale effect on vibration of embedded multilayered graphene sheets based on nonlocal continuum models. Phys. Lett. A. **373**, 1062–1069 (2009)
141. Shen, Z.B., Tang, H.L., Li, D.K., Tang, G.J.: Vibration of single-layered graphene sheet-based nanomechanical sensor via nonlocal Kirchhoff plate theory. Comput. Mater. Sci. **61**, 200–205 (2012)
142. Ansari, R., Shahabodini, A., Rouhi, H.: A nonlocal plate model incorporating interatomic potentials for vibrations of graphene with arbitrary edge conditions. Curr. Appl. Phys. **15**, 1062–1069 (2015)
143. Zhang, Y., Lei, Z.X., Zhang, L.W., Liew, K.M., Yu, J.L.: Nonlocal continuum model for vibration of single- layered graphene sheets based on the element-free kp-Ritz method. Eng. Anal. Bound. Elem. **56**, 90–97 (2015)
144. Zhang, Y., Zhang, L.W., Liew, K.M., Yu, J.L.: Nonlocal continuum model for large deformation analysis of SLGSs using the kp-Ritz element-free method. Int. J. Non-Linear Mech. **79**, 1–9 (2016)
145. Zhang, Y., Zhang, L.W., Liew, K.M., Yu, J.L.: Buckling analysis of graphene sheets embedded in an elastic medium based on the kp-Ritz method and non-local elasticity theory. Eng. Anal. Bound. Elem. **70**, 31–39 (2016)
146. Ni, Z., Bu, H., Zou, M., Yi, H., Bi, K., Chen, Y.: Anisotropic mechanical properties of graphene sheets from molecular dynamics. Phys. B Condens. Matter. **405**, 1301–1306 (2010)
147. Pouresmaeeli, S., Fazelzadeh, S.A., Ghavanloo, E.: Exact solution for nonlocal vibration of double- orthotropic nanoplates embedded in elastic medium. Compos. B Eng. **43**, 3384–3390 (2012)
148. Mohammadi, M., Farajpour, A., Goodarzi, M., Heydarshenas, R.: Levy type solution for nonlocal thermo- mechanical vibration of orthotropic mono-layer graphene sheet embedded in an elastic medium. J. Solid Mech. **5**, 116–132 (2013)
149. Mohammadi, M., Moradi, A., Ghayour, M., Farajpour, A.: Exact solution for thermo-mechanical vibration of orthotropic mono-layer graphene sheet embedded in an elastic medium. Lat. Am. J. Sol. Struct. **11**, 437–458 (2014)

150. Sari, M.S., Al-Kouz, W.G.: Vibration analysis of non-uniform orthotropic Kirchhoff plates resting on elastic foundation based on nonlocal elasticity theory. Int. J. Mech. Sci. **114**, 1–11 (2016)
151. Anjomshoa, A.: Application of Ritz functions in buckling analysis of embedded orthotropic circular and elliptical micro/nano-plates based on nonlocal elasticity theory. Meccanica **48**, 1337–1353 (2013)
152. Anjomshoa, A., Shahidi, A.R., Shahidi, S.H., Nahvi, H.: Frequency analysis of embedded orthotropic circular and elliptical micro/nano-plates using nonlocal variational principle. J. Sol. Mech. **7**, 13–27 (2015)
153. Mohammadi, M., Farajpour, A., Moradi, A., Ghayour, M.: Shear buckling of orthotropic rectangular graphene sheet embedded in an elastic medium in thermal environment. Compos. B Eng. **56**, 629–635 (2014)
154. Ashoori, A.R., Salari, E., Sadough Vanini, S.A.: Size-dependent thermal stability analysis of embedded functionally graded annular nanoplates based on the nonlocal elasticity theory. Int. J. Mech. Sci. **119**, 396–411 (2016)
155. Park, S.K., Gao, X.L.: Bernoulli-Euler beam model based on a modified couple stress theory. Micromech. Microeng. **16**(11), 2355–2359 (2006)
156. Kong, S., Zhou, S., Nie, Z., Wang, K.: The size-dependent natural frequency of Bernoulli Euler micro-beams. Int. J. Eng. Sci. **46**, 427–437 (2008)
157. Kong, S., Zhou, S., Nie, Z., Wang, K.: Size effect on the buckling loads of slender columns based on a modified couple stress theory. J. Mech. Streng. **31**, 136–139 (2009)
158. Xia, W., Wang, L., Yin, L.: Nonlinear non-classical microscale beams: Static bending, post-buckling and free vibration. Int. J. Eng. Sci. **48**, 2044–2053 (2010)
159. Simsek, M.: Nonlinear static and free vibration analysis of microbeams based on the nonlinear elastic foundation using modified couple stress theory and He's variational method. Compos. Struct. **112**, 264–272 (2014)
160. Wang, Y.G., Lin, W.H., Liu, N.: Nonlinear bending and post-buckling of extensible microscale beams based on modified couple stress theory. Appl. Math. Model. **39**, 117–127 (2015)
161. Wang, Y.G., Lin, W.H., Liu, N.: Nonlinear free vibration of a microscale beam based on modified couple stress theory. Phys. E. **47**, 80–85 (2013)
162. Farokhi, H., Ghayesh, M.H., Amabili, M.: Nonlinear dynamics of a geometrically imperfect microbeam based on the modified couple stress theory. Int. J. Eng. Sci. **68**, 11–23 (2013)
163. Togun, N., Bagdatli, S.M.: Size dependent nonlinear vibration of the tensioned nanobeam based on the modified couple stress theory. Compos. B Eng. **97**, 255–262 (2016)
164. Wang, Y.G., Lin, W.H., Zhou, C.L., Liu, R.X.: Thermal postbuckling and free vibration of extensible microscale beams based on modified couple stress theory. J. Mech. **31**, 37–46 (2014)
165. Ansari, R., Ashrafi, M.A., Arjangpay, A.: An exact solution for vibrations of postbuckled microscale beams based on the modified couple stress theory. Appl. Math. Model. **39**, 3050–3062 (2015)
166. Asghari, M., Ahmadian, M.T., Kahrobaiyan, M.H., Rahaeifard, M.: On the size-dependent behavior of functionally graded micro-beams. Mater. Des. **31**, 2324–2329 (2010)
167. Akgoz, B., Civalek, O.: Free vibration analysis of axially functionally graded tapered Bernoulli-Euler microbeams based on themodified couple stress theory. Compos. Struct. **98**, 314–322 (2013)
168. Shafiei, N., Kazemi, M., Ghadiri, M.: Nonlinear vibration of axially functionally graded tapered microbeams. Int. J. Eng. Sci. **102**, 12–26 (2016)
169. Simsek, M.: Size dependent nonlinear free vibration of an axially functionally graded (AFG) microbeam using He's variational method. Compos. Struct. **131**, 207–214 (2015)
170. Dehrouyeh-Semnani, A.M., Mostafaei, H., Nikkhah-Bahrami, M.: Free flexural vibration of geometrically imperfect functionally graded microbeams. Int. J. Eng. Sci. **105**, 56–79 (2016)
171. Tsiatas, G.C.: A new Kirchhoff plate model based on a modified couple stress theory. Int. J. Sol. Struct. **46**(13), 2757–2764 (2009)

172. Yin, L., Qian, Q., Wang, L., Xia, W.: Vibration analysis of microscale plates based on modified couple stress theory. Acta Mech. Solida. Sin. **23**, 386–393 (2010)
173. Jomehzadeh, E., Noori, H.R., Saidi, A.R.: The size-dependent vibration analysis of micro-plates based on a modified couple stress theory. Phys E. **43**, 877–883 (2011)
174. Akgoz, B., Civalek, O.: Free vibration analysis for single-layered graphene sheets in an elastic matrix via modified couple stress theory. Mater Des. **42**, 164–171 (2012)
175. Akgoz, B., Civalek, O.: Modeling and analysis of micro-sized plates resting on elastic medium using the modified couple stress theory. Mecc. **48**, 863–873 (2013)
176. Askari, A.R., Tahani, M.: Analytical determination of size-dependent natural frequencies of fully clamped rectangular microplates based on the modified couple stress theory. J. Mech. Sci. Technol. **29**, 2135–2145 (2015)
177. Simsek, M., Aydn, M., Yurtcu, H.H., Reddy, J.N.: Size-dependent vibration of a microplate under the action of a moving load based on the modified couple stress theory. Acta Mech. **226**, 3807–3822 (2015)
178. Zhou, X., Wang, L., Qin, P.: Free vibration of micro- and nano-shells based on modified couple stress theory. J. Comput. Theor. Nanosci. **9**, 814–828 (2012)
179. Asghari, M.: Geometrically nonlinear micro-plate formulation based on the modified couple stress theory. Int. J. Eng. Sci. **51**, 292–309 (2012)
180. Wang, Y.G., Lin, W.H., Liu, N.: Large amplitude free vibration of size-dependent circular microplates based on the modified couple stress theory. Int. J. Mech. Sci. **71**, 51–57 (2013)
181. Wang, Y.G., Lin, W.H., Zhou, C.L.: Nonlinear bending of size-dependent circular microplates based on the modified couple stress theory. Arch. Appl. Mech. **84**, 391–400 (2014)
182. Farokhi, H., Ghayesh, M.H.: Nonlinear dynamical behaviour of geometrically imperfect microplates based on modified couple stress theory. Int. J. Mech. Sci. **90**, 133–144 (2015)
183. Ke, L.L., Yang, J., Kitipornchai, S., Bradford, M.A.: Bending, buckling and vibration of size-dependent functionally graded annular microplates. Compos. Struct. **94**, 3250–3257 (2012)
184. Asghari, M., Taati, E.: A size-dependent model for functionally graded micro-plates for mechanical analyses. J. Vib. Control. **19**, 1614–1632 (2013)
185. Ashoori, A.R., Sadough Vanini, S.A.: Thermal buckling of annular microstructure-dependent functionally graded material plates resting on an elastic medium. Compos. B Eng. **87**, 245–253 (2016)
186. Ashoori, A.R., Sadough Vanini, S.A.: Nonlinear thermal stability and snap-through behavior of circular microstructure-dependent FGM plates. Eur. J. Mech. A Solids. **59**, 323–332 (2016)
187. Taati, E.: Analytical solutions for the size dependent buckling and postbuckling behavior of functionally graded micro-plates. Int. J. Eng. Sci. **100**, 45–60 (2016)
188. Beni, Y.T., Mehralian, F., Zeighampour, H.: The modified couple stress functionally graded cylindrical thin shell formulation. Mech. Adv. Mater. Struct. **23**, 791–801 (2016)
189. Tsiatas, G.C., Yiotis, A.J.: Size effect on the static, dynamic and buckling analysis of orthotropic Kirchhoff-type skew micro-plates based on a modified couple stress theory: comparison with the nonlocal elasticity theory. Acta Mech. **226**, 1267–1281 (2015)
190. Kong, S., Zhou, S., Nie, Z., Wang, K.: Static and dynamic analysis of micro beams based on strain gradient elasticity theory. Int. J. Eng. Sci. **47**, 487–498 (2009)
191. Akgoz, B., Civalek, O.: Strain gradient elasticity and modified couple stress models for buckling analysis of axially loaded micro-scaled beams. Int. J. Eng. Sci. **49**, 1268–1280 (2011)
192. Akgoz, B., Civalek, O.: Buckling analysis of cantilever carbon nanotubes using the strain gradient elasticity and modified couple stress theories. J. Comput. Theor. Nanosci. **8**, 1821–1827 (2011)
193. Akgoz, B., Civalek, O.: Investigation of size effects on static response of single-walled carbon nanotubes based on strain gradient elasticity. Int. J. Comput. Meth. **9**, 1240032–19 (2012)
194. Akgoz, B., Civalek, O.: Buckling analysis of linearly tapered micro-Columns based on strain gradient elasticity. Struct. Eng. Mech. **48**, 195–205 (2013)
195. Akgoz, B., Civalek, O.: Longitudinal vibration analysis for microbars based on strain gradient elasticity theory. J. Vib. Control. **20**, 606–616 (2014)

196. Zhao, J., Zhou, S., Wang, B., Wang, X.: Nonlinear microbeam model based on strain gradient theory. Appl. Math. Model. **36**, 2674–2686 (2012)
197. Rajabi, F., Ramezani, S.: A nonlinear microbeam model based on strain gradient elasticity theory. Acta Mech. Solida Sin. **26**, 21–34 (2013)
198. Mohammadi, H., Mahzoon, M.: Thermal effects on postbuckling of nonlinear microbeams based on the modified strain gradient theory. Compos. Struct. **106**, 764–776 (2013)
199. Vatankhah, R., Kahrobaiyan, M.H., Alasty, A., Ahmadian, M.T.: Nonlinear forced vibration of strain gradient microbeams. Appl. Math. Model. **37**, 8363–8382 (2013)
200. Kahrobaiyan, M.H., Rahaeifard, M., Tajalli, S.A., Ahmadian, M.T.: A strain gradient functionally graded Euler-Bernoulli beam formulation. Int. J. Eng. Sci. **52**, 65–76 (2012)
201. Akgoz, B., Civalek, O.: Buckling analysis of functionally graded microbeams based on the strain gradient theory. Acta Mech. **224**, 2185–2201 (2013)
202. Akgoz, B., Civalek, O.: Longitudinal vibration analysis of strain gradient bars made of functionally graded materials (FGM). Compos. B Eng. **55**, 263–268 (2013)
203. Rahaeifard, M., Kahrobaiyan, M.H., Ahmadian, M.T., Firoozbakhsh, K.: Strain gradient formulation of functionally graded nonlinear beams. Int. J. Eng. Sci. **65**, 49–63 (2013)
204. Wang, B., Zhou, S., Zhao, J., Chen, X.: A size-dependent Kirchhoff micro-plate model based on strain gradient elasticity theory. Eur. J. Mech. A Sol. **30**, 517–524 (2011)
205. Ashoori, M.A., Mahmoodi, M.J.: A micro-scale modeling of Kirchhoff plate based on modified strain-gradient elasticity theory. Eur. J. Mech. A Sol. **40**, 50–59 (2013)
206. Mohammadi, M., Fooladi, M.M.: An analytical solution for buckling analysis of size-dependent rectangular micro-plates according to the modified strain gradient and couple stress theories. Acta Mech. **226**, 3477–3493 (2015)
207. Mohammadi, M., Fooladi, M., Darijani, H.: Exact boundary conditions for buckling analysis of rectangular micro-plates based on the modified strain gradient theory. Int. J. Multiscale Comput. Eng. **13**, 265–280 (2015)
208. Wang, B., Huang, S., Zhao, J., Zhou, S.: Reconsiderations on boundary conditions of Kirchhoff micro-plate model based on a strain gradient elasticity theory. Appl. Math. Model. **40**, 7303–7317 (2016)
209. Zeighampour, H., Beni, Y.T.: Cylindrical thin-shell model based on modified strain gradient theory. Int. J. Eng. Sci. **78**, 27–47 (2014)
210. Allahbakhshi, A., Allahbakhshi, M.: Vibration analysis of nano-structure multilayered graphene sheets using modified strain gradient theory. Front Mech. Eng. **10**, 187–197 (2015)
211. Li, A., Zhou, S., Wang, B.: A size-dependent model for bi-layered Kirchhoff micro-plate based on strain gradient elasticity theory. Compos. Struct. **113**, 272–280 (2014)
212. Hosseini, M., Bahreman, M., Jamalpoor, A.: Using the modified strain gradient theory to investigate the size-dependent biaxial buckling analysis of an orthotropic multi-microplate system. Acta Mech. **227**, 1621–1643 (2016)
213. Zhang, L., Liang, B., Zhou, S., Wang, B., Xue, Y.: An application of a size-dependent model on microplate with elastic medium based on strain gradient elasticity theory. Mecc. **52**, 251–262 (2016)
214. Chang, L., Rajapakse, R.K.N.D.: Continuum models incorporating surface energy for static and dynamic response of nanoscale beams. IEEE Trans. Nanotechnol. **9**(4), 422–431 (2010)
215. Gao, X.L., Mahmoud, F.F.: A new Bernoulli-Euler beam model incorporating microstructure and surface energy effects. Z. Angew. Math. Phys. **65**(2), 393–404 (2014)
216. Shaat, M., Mahmoud, F.F., Gao, X.L., Faheem, A.F.: Size dependent bending analysis of Kirchhoff nano-plates based on a modified couple-stress theory including surface effects. Int. J. Mech. Sci. **79**, 31–37 (2014)
217. Wang, B., Zhao, J., Zhou, S.: A micro scale Timoshenko beam model based on strain gradient elasticity theory. Eur. J. Mech. A/Sol. **29**, 591–599 (2010)
218. Zhang, L., Wang, B., Zhou, S., Xue, Y.: Modeling the size-dependent nanostructures: Incorporating the bulk and surface effects. J. Nanomech. Micromech. **7**(2), 04016012 (2017)
219. Yan, Z., Jiang, L.: Influence of surface effects and flexoelectricity on vibration of piezoelectric nanobeams. In: 13th International Conference on Fracture, Beijing, China (2013). Accesssed June 16–21, 2013

220. Huang, G.Y., Yu, S.W.: Effect of surface piezoelectricity on the electromechanical behaviour of a piezoelectric ring. Phys. Stat. Sol. B-Basic **243**, R22–R24 (2006)
221. Yan, Z., Jiang, L.Y.: Surface effects on the electromechanical coupling and bending behaviours of piezoelectric nanowires. J. Phys. D-Appl. Phys. **44**, 075404 (2011)
222. Yan, Z., Jiang, L.Y.: The vibrational and buckling behaviors of piezoelectric nanobeams with surface effects. Nanotech. **22**, 245703 (2011)
223. Li, Y.H., Fang, B., Zhang, J.H., Song, J.Z.: Surface effects on the wrinkling of piezoelectric films on compliant substrates. J. Appl. Phys. **110**, 114303 (2011)
224. Maranganti, R., Sharma, N.D., Sharma, P.: Electromechanical coupling in nonpiezoelectric materials due to nanoscale size effects: Green's function solutions and embedded inclusions. Phys. Rev. B **74**, 014110 (2006)
225. Majdoub, M.S., Sharma, P., Cagin, T.: Dramatic enhancement in energy harvesting for a narrow range of dimensions in piezoelectric nanostructures. Phys. Rev. B **78**, 121407 (2008)
226. Eliseev, E.A., Morozovska, A.N., Glinchuk, M.D., Blinc, R.: Spontaneous flexoelectric/flexomagnetic effect in nanoferroics. Phys. Rev. B **79**, 165433 (2009)
227. Liu, C.C., Hu, S.L., Shen, S.P.: Effect of flexoelectricity on electrostatic potential in a bent piezoelectric nanowire. Smart Mater. Struct. **21**, 115024 (2012)
228. Guo, J.G., Zhao, Y.P.: The size-dependent bending elastic properties of nanobeams with surface effects. Nanotech. **18**, 295701 (2007)
229. Villain, P., Beauchamp, P., Badwi, K.F., Goudeau, P., Renault, P.O.: Scr. Mater. **50**, 1247 (2004)
230. Wolf, D.: Surface-stress-induced structure and elastic behavior of thin films. Appl. Phys. Lett. **58**, 2081 (1991)
231. Liang, H., Upmanyu, M.: Size-dependent elasticity of nanowires: Nonlinear effects. Phys. Rev. B **71**, 241403 (2005)
232. Zhou, L.G., Huang, H.: Are surfaces elastically softer or stiffer? Appl. Phys. Lett. **84**, 1940 (2004)
233. Streitz, F.H., Cammarata, R.C., Sieradzki, K.: Surface-stress effects on elastic properties. I. Thin metal films. Phys. Rev. B **49**, 10699 (1994)
234. Dingreville, R., Qu, J., Cherkaoui, M.: Surface free energy and its effect on the elastic behavior of nano-sized particles, wires and films. J. Mech. Phys. Sol. **53**, 1827 (2005)
235. Guo, J.G., Zhao, Y.P.: The size-dependent elastic properties of nanofilms with surface effects. J. Appl. Phys. **98**, 074306 (2005)
236. Gao, X.L., Zhang, G.Y.: A non-classical Mindlin plate model incorporating microstructure, surface energy and foundation effects. Proc. R. Soc. A: Math. Phys. Eng. Sci. **472**(2191), 20160275 (2016)
237. Winkler, E.: Die Lehre von der Elasticitaet und Festigkeit. Verlag von H. Dominicus, Prague, Czech Republic (1867)
238. Filonenko-Borodich, M.M.: Some approximate theories of the elastic foundation. Sci. Not. Moskow Nat. Univ. Mech. **46**, 3–18 (1940). (in Russian)
239. Pasternak, P.L.: On a New Method of Analysis of an Elastic Foundation by Means of Two Foundation Constants. Gosudarstvennoe Izdatelstvo Literaturi po Stroitelstvu i Arkhitekture, Moscow (1954). (in Russian)
240. Kerr, A.D.: Elastic and viscoelastic foundation models. ASME J. Appl. Mech. **31**, 491–498 (1964)
241. Vlasov, V.Z.: Beams, Plates and Shells on Elastic Foundations. Jerusalem, Israel (1966)
242. Feng, Z.H., Cook, R.D.: Beam elements on two-parameter elastic foundations. J. Eng. Mech. **109**, 1390–1402 (1983)
243. Eisenberger, M., Clastornik, J.: Beams on variable two-parameter elastic foundation. J. Eng. Mech. **113**, 1454–1466 (1987)
244. Khajeansari, A., Baradaran, G.H., Yvonnet, J.: An explicit solution for bending of nanowires lying on Winkler-Pasternak elastic substrate medium based on the Euler-Bernoulli beam theory. Int. J. Eng. Sci. **52**, 115–128 (2012)

245. Eremeyev, V.A., Lebedev, L.P.: Mathematical study of boundary-value problems within the framework of Steigmann-Ogden model of surface elasticity. Contin. Mech. Thermodyn. **28**, 407–422 (2016)
246. Simsek, M., Reddy, J.N.: A unified higher order beam theory for buckling of a functionally graded microbeam embedded in elastic medium usingmodified couple stress theory. Compos. Struct. **101**, 47–58 (2013)
247. Park, S.K., Gao, X.L.: Variational formulation of a modified couple stress theory and its application to a simple shear problem. Z. Angew Math. Phys. **59**, 904–917 (2008)
248. Limkatanyu, S., Ponbunyanon, P., Prachasaree, W., Kuntiyawichai, K., Kwon, M.: Correlation between beam on Winkler-Pasternak foundation and beam on elastic substrate medium with inclusion of microstructure and surface effects. J. Mech. Sci. Tech. **28**, 3653–3665 (2014)
249. Gao, X.L., Zhang, G.Y.: A non-classical Kirchhoff plate model incorporating microstructure, surface energy and foundation effects. Contin. Mech. Thermodyn. **28**, 195–213 (2016)
250. Lim, C.W., He, L.H.: Size-dependent nonlinear response of thin elastic films with nano-scale thickness. Int. J. Mech. Sci. **46**, 1715–1726 (2004)
251. Lu, P., He, L.H., Lee, H.P., Lu, C.: Thin plate theory including surface effects. Int. J. Solids Struct. **43**, 4631–4647 (2006)
252. Lu, C.F., Wu, D.Z., Chen, W.Q.: Nonlinear responses of nanoscale FGM films including the effects of surface energies. IEEE Trans. Nanotech. **10**, 1321–1327 (2011)
253. Wang, K.F., Wang, B.L.: Effects of residual surface stress and surface elasticity on the nonlinear free vibration of nanoscale plates. J. Appl. Phys. **112**, 013520-1–013520-6 (2012)
254. Lazopoulos, K.A.: On bending of strain gradient elastic micro-plates. Mech. Res. Commun. **36**, 777–783 (2009)
255. Michael, J.: Lachut, John, E.: Sader effect of surface stress on the stiffness of cantilever plates. Phys. Rev. Lett. **99**(20), e206102 (2007)
256. Rouhi, H., Ansari, R., Darvizeh, M.: Size-dependent large amplitude vibration analysis of nanoshells using the Gurtin-Murdoch model. Int. J. Nanosci. Nanotech. **13**, 241–252 (2017)
257. Wang, Q.: Wave propagation in carbon nanotubes via nonlocal continuum mechanics. J. Appl. Phys. **98**, 124301 (2005)
258. Wang, Q., Varadan, V.K.: Vibration of carbon nanotubes studied using nonlocal continuum mechanics. Smart Mater. Struct. **15**, 659–666 (2006)
259. Wang, C.M., Zhang, Y.Y., Ramesh, S.S., Kitipornchai, S.: Buckling analysis of micro- and nano-rods/tubes based on nonlocal Timoshenko beam theory. J. Phys. D: Appl. Phys. **39**, 3904–3909 (2006)
260. Wang, C.M., Zhang, Y.Y., He, X.Q.: Vibration of nonlocal Timoshenko beams. Nanotech. **18**, 105401–9 (2007)
261. Wang, C.M., Kitipornchai, S., Lim, C.W., Eisenberger, M.: Beam bending solutions based on nonlocal Timoshenko beam theory. J. Eng. Mech. **134**, 475–481 (2008)
262. Wang, Q., Liew, K.M.: Application of nonlocal continuum mechanics to static analysis of micro- and nano- structures. Phys. Lett. A. **363**, 236–242 (2007)
263. Lu, P., Lee, H.P., Lu, C., Zhang, P.Q.: Application of nonlocal beam models for carbon nanotubes. Int. J. Sol. Struct. **44**, 5289–52300 (2007)
264. Reddy, J.N., Pang, S.D.: Nonlocal continuum theories of beams for the analysis of carbon nanotubes. J. Appl. Phys. **103**, 023511–16 (2008)
265. Murmu, T., Pradhan, S.C.: Buckling analysis of a single-walled carbon nanotube embedded in an elastic medium based on nonlocal elasticity and Timoshenko beam theory and using DQM. Phys E. **41**, 1232–1239 (2009)
266. Ansari, R., Gholami, R., Darabi, M.A.: Thermal buckling analysis of embedded single-walled carbon nanotubes with arbitrary boundary conditions using the nonlocal timoshenko beam theory. J. Therm. Stresses. **34**, 1271–1281 (2011)
267. Pradhan, S.C., Murmu, T.: Small-scale effect on vibration analysis of single-walled carbon nanotubes embedded in an elastic medium using nonlocal elasticity theory. J. Appl. Phys. **105**, 124306–9 (2009)

268. Roque, C.M.C., Ferreira, A.J.M., Reddy, J.N.: Analysis of Timoshenko nanobeams with a nonlocal formulation and meshless method. Int. J. Eng. Sci. **49**, 976–984 (2011)
269. Wu, C.P., Lai, W.W.: Free vibration of an embedded single-walled carbon nanotube with various boundary conditions using the RMVT-based nonlocal Timoshenko beam theory and DQ method. Phys E. **68**, 8–21 (2015)
270. Amirian, B., Hosseini-Ara, R., Moosavi, H.: Thermal vibration analysis of carbon nanotubes embedded in two-parameter elastic foundation based on nonlocal Timoshenko's beam theory. Arch. Mech. **64**, 581–602 (2012)
271. Zidour, M., Benrahou, K.H., Semmah, A., Naceri, M., Belhadj, H.A., Bakhti, K.: The thermal effect on vibration of zigzag single walled carbon nanotubes using nonlocal Timoshenko beam theory. Comput. Mater. Sci. **51**, 252–260 (2012)
272. Ansari, R., Gholami, R., Sahmani, S., Norouzzadeh, A., Bazdid-Vahdati, M.: Dynamic stability analysis of embedded multi-walled carbon nanotubes in thermal environment. Acta Mech. Sol. Sin. **28**, 659–667 (2015)
273. Ansari, R., Gholami, R., Rouhi, H.: Size-dependent nonlinear forced vibration analysis of magneto-electro- thermo-elastic Timoshenko nanobeams based upon the nonlocal elasticity theory. Compos Struct. **126**, 216–226 (2015)
274. Simsek, M., Yurtcu, H.H.: Analytical solutions for bending and buckling of functionally graded nanobeams based on the nonlocal Timoshenko beam theory. Compos. Struct. **97**, 378–386 (2013)
275. Rahmani, O., Pedram, O.: Analysis and modeling the size effect on vibration of functionally graded nanobeams based on nonlocal Timoshenko beam theory. Int. J. Eng. Sci. **77**, 55–70 (2014)
276. Ebrahimi, F., Salari, E.: Thermal buckling and free vibration analysis of size dependent Timoshenko FG nanobeams in thermal environments. Compos. Struct. **128**, 363–380 (2015)
277. Ebrahimi, F., Salari, E.: Effect of various thermal loadings on buckling and vibrational characteristics of nonlocal temperature-dependent functionally graded nanobeams. Mech. Add. Mater. Struct. **23**, 1379–1397 (2016)
278. Pradhan, S.C., Phadikar, J.K.: Nonlocal elasticity theory for vibration of nanoplates. J. Sound Vib. **325**, 206–223 (2009)
279. Pradhan, S.C., Phadikar, J.K.: Nonlocal theory for buckling of nanoplates. Int. J. Struct. Stab. Dyn. **11**, 411–429 (2011)
280. Kananipour, H.: Static analysis of nanoplates based on the nonlocal Kirchhoff and Mindlin plate theories using DQM. Lat. Am. J. Sol. Struct. **11**, 1709–20 (2014)
281. Ansari, R., Sahmani, S., Arash, B.: Nonlocal plate model for free vibrations of single-layered graphene sheets. Phys. Lett. A. **375**, 53–62 (2010)
282. Ansari, R., Arash, B., Rouhi, H.: Vibration characteristics of embedded multi-layered graphene sheets with different boundary conditions via nonlocal elasticity. Compos Struct. **93**, 2419–2429 (2011)
283. Samaei, A.T., Abbasion, S., Mirsayar, M.M.: Buckling analysis of a single-layer graphene sheet embedded in an elastic medium based on nonlocal Mindlin plate theory. Mech. Res. Commun. **38**, 481–485 (2011)
284. Bedroud, M., Hosseini-Hashemi, S., Nazemnezhad, R.: Buckling of circular/annular Mindlin nanoplates via nonlocal elasticity. Acta Mech. **224**, 2663–2676 (2013)
285. Arani, A.G., Abdollahian, M., Kolahchi, R., Rahmati, A.: Electro-thermo-torsional buckling of an embedded armchair DWBNNT using nonlocal shear deformable shell model. Compos. B Eng. **51**, 291–299 (2013)
286. Naderi, A., Saidi, A.R.: Modified nonlocal mindlin plate theory for buckling analysis of nanoplates. J. Nanomech. Micromech. **4**, A4013015–8 (2014)
287. Reddy, J.N.: Nonlocal nonlinear formulations for bending of classical and shear deformation theories of beams and plates. Int. J. Eng. Sci. **48**, 1507–1518 (2010)
288. Hosseini-Hashemi, S., Bedroud, M., Nazemnezhad, R.: An exact analytical solution for free vibration of functionally graded circular/annular Mindlin nanoplates via nonlocal elasticity. Compos. Struct. **103**, 108–118 (2013)

289. Anjomshoa, A., Tahani, M.: Vibration analysis of orthotropic circular and elliptical nanoplates embedded in elastic medium based on nonlocal Mindlin plate theory and using Galerkin method. J. Mech. Sci. Technol. **30**, 2463–2474 (2016)
290. Golmakani, M.E., Rezatalab, J.: Nonlinear bending analysis of orthotropic nanoscale plates in an elastic matrix based on nonlocal continuum mechanics. Compos. Struct. **111**, 85–97 (2014)
291. Dastjerdi, S., Jabbarzadeh, M., Aliabadi, S.: Nonlinear static analysis of single layer annular/circular graphene sheets embedded in Winkler-Pasternak elastic matrix based on non-local theory of Eringen. Ain. Shams. Eng. J. **7**, 873–884 (2016)
292. Dastjerdi, S., Jabbarzadeh, M.: Non-linear bending analysis of multi-layer orthotropic annular/circular graphene sheets embedded in elastic matrix in thermal environment based on non-local elasticity theory. Appl. Math. Model. **41**, 83–101 (2017)
293. Ma, H.M., Gao, X.-L., Reddy, J.N.: A microstructure-dependent Timoshenko beam model based on a modified couple stress theory. J. Mech. Phys. Sol. **56**, 3379–3396 (2008)
294. Asghari, M., Kahrobaiyan, M.H., Rahaeifard, M., Ahmadian, M.T.: Investigation of the size effects in Timoshenko beams based on the couple stress theory. Arch. Appl. Mech. **81**, 863–874 (2011)
295. Dos Santos, J.V.A., Reddy, J.N.: Free vibration and buckling analysis of beams with a modified couple- stress theory. Int. J. Appl. Mech. **4**, 1250026–28 (2012)
296. Dehrouyeh-Semnani, A.M., Nikkhah-Bahrami, M.: A discussion on incorporating the Poisson effect in microbeam models based on modified couple stress theory. Int. J. Eng. Sci. **86**, 20–25 (2015)
297. Liu, Y.P., Reddy, J.N.: A nonlocal curved beam model based on a modified copled stress theory. Int. J. Str. Stab. Dyn. **11**, 495–512 (2011)
298. Taati, E., Molaei Najafabadi, M.M., Basirat Tabrizi, H.: Size-dependent generalized thermoelasticity model for Timoshenko microbeams. Acta Mech. **225**(7), 1823–1842 (2014)
299. Asghari, M., Kahrobaiyan, M.H., Ahmadian, M.T.: A nonlinear Timoshenko beam formulation based on the modified couple stress theory. Int. J. Eng. Sci. **48**, 1749–1766 (2010)
300. Ghayesh, M.H., Farokhi, H., Amabili, M.: Nonlinear dynamics of a microscale beam based on the modified couple stress theory. Compos. B Eng. **50**, 318–324 (2013)
301. Ghayesh, M.H., Amabili, M., Farokhi, H.: Three-dimensional nonlinear size-dependent behaviour of Timoshenko microbeams. Int. J. Eng. Sci. **71**, 1–14 (2013)
302. Reddy, J.N.: Microstructure-dependent couple stress theories of functionally graded beams. Mech. Phys. Sol. **59**, 2382–2399 (2011)
303. Ke, L.L., Wang, Y.S., Yang, J., Kitipornchai, S.: Nonlinear free vibration of size-dependent functionally graded microbeams. Int. J. Eng. Sci. **50**(1), 256–267 (2012)
304. Asghari, M., Rahaeifard, M., Kahrobaiyan, M.H., Ahmadian, M.T.: The modified couple stress functionally graded Timoshenko beam formulation. Mater. Des. **32**, 1435–1443 (2011)
305. Ke, L.L., Wang, Y.S.: Size effect on dynamic stability of functionally graded microbeams based on a modified couple stress theory. Compos. Struct. **93**, 342–350 (2011)
306. Nateghi, A., Salamat-Talab, M.: Thermal effect on size dependent behavior of functionally graded microbeams based on modified couple stress theory. Compos. Struct. **96**, 97–110 (2013)
307. Simsek, M., Kocaturk, T., Akbas, S.D.: Static bending of a functionally graded microscale Timoshenko beam based on the modified couple stress theory. Compos. Struct. **95**, 740–747 (2013)
308. Chen, W., Li, L., Xu, M.: A modified couple stress model for bending analysis of composite laminated beams with first order shear deformation. Compos. Struct. **93**, 2723–2732 (2011)
309. Chen, W.J., Li, X.P.: Size-dependent free vibration analysis of composite laminated Timoshenko beam based on new modified couple stress theory. Arch. Appl. Mech. **83**(3), 431–444 (2013)
310. Roque, C.M.C., Fidalgo, D.S., Ferreira, A.J.M., Reddy, J.N.: A study of a microstructure-dependent composite laminated Timoshenko beam using a modified couple stress theory and a meshless method. Compos. Struct. **96**, 532–537 (2013)

311. Mohammad-Abadi, M., Daneshmehr, A.R.: An investigation of modified couple stress theory in buckling analysis of micro composite laminated Euler-Bernoulli and Timoshenko beams. Int. J. Eng. Sci. **75**, 40–53 (2014)
312. Thai, H.T., Vo, T.P., Nguyen, T.K., Lee, J.: Size-dependent behavior of functionally graded sandwich microbeams based on the modified couple stress theory. Compos. Struct. **123**, 337–349 (2015)
313. Krysko, A.V., Awrejcewicz, J., Pavlov, S.P., Zhigalov, M.V., Krysko, V.A.: Mathematical model of a three-layer micro- and nano-beams based on the hypotheses of the Grigolyuk-Chulkov and the modified couple stress theory. Int. J. Sol. Struct. **117**, 39–50 (2017)
314. Ma, H.M., Gao, X.L., Reddy, J.N.: A non-classical Mindlin plate model based on a modified couple stress theory. Acta Mech. **220**, 217–235 (2011)
315. Ke, L.L., Wang, Y.S., Yang, J., Kitipornchai, S.: Free vibration of size-dependent Mindlin microplates based on the modified couple stress theory. J. Sound Vib. **331**, 94–106 (2012)
316. Roque, C.M.C., Ferreira, A.J.M., Reddy, J.N.: Analysis of Mindlin micro plates with a modified couple stress theory and a meshless method. Appl. Math. Model. **37**, 4626–4633 (2013)
317. Zhou, S.S., Gao, X.L.: A nonclassical model for circular mindlin plates based on a modified couple stress theory. J. Appl. Mech. **81**, 051014–8 (2014)
318. Alinaghizadeh, F., Shariati, M., Fish, J.: Bending analysis of size-dependent functionally graded annular sector microplates based on the modified couple stress theory. Appl. Math. Model. **44**, 540–556 (2017)
319. He, D., Yang, W., Chen, W.: A size-dependent composite laminated skew plate model based on a new modified couple stress theory. Acta Mech. Sol. Sin. **30**, 75–86 (2017)
320. Simsek, M., Aydn, M.: Size-dependent forced vibration of an imperfect functionally graded (FG) microplate with porosities subjected to a moving load using the modified couple stress theory. Compos. Struct. **160**, 408–421 (2017)
321. Reddy, J.N., Berry, J.: Nonlinear theories of axisymmetric bending of functionally graded circular plates with modified couple stress. Compos. Struct. **94**, 3664–3668 (2012)
322. Ke, L.L., Yang, J., Kitipornchai, S., Bradford, M.A., Wang, Y.S.: Axisymmetric nonlinear free vibration of size-dependent functionally graded annular microplates. Compos. B Eng. **53**, 207–217 (2013)
323. Ke, L.L., Yang, J., Kitipornchai, S., Wang, Y.S.: Axisymmetric postbuckling analysis of size-dependent functionally graded annular microplates using the physical neutral plane. Int. J. Eng. Sci. **81**, 66–81 (2014)
324. Thai, H.T., Choi, D.H.: Size-dependent functionally graded Kirchhoff and Mindlin plate models based on a modified couple stress theory. Compos. Struct. **95**, 142–153 (2013)
325. Jung, W.Y., Park, W.T., Han, S.C.: Bending and vibration analysis of S-FGM microplates embedded in Pasternak elastic medium using the modified couple stress theory. Int. J. Mech. Sci. **87**, 150–162 (2014)
326. Jung, W.Y., Han, S.C., Park, W.T.: A modified couple stress theory for buckling analysis of S-FGM nanoplates embedded in Pasternak elastic medium. Compos. B Eng. **60**, 746–756 (2014)
327. Ansari, R., Faghih, S.M., Mohammadi, V., Gholami, R., Darabi, M.A.: Nonlinear vibrations of functionally graded Mindlin microplates based on the modified couple stress theory. Compos. Struct. **114**, 124–134 (2014)
328. Ansari, R., Gholami, R., Faghih, S.M., Mohammadi, V., Darabi, M.A.: Size-dependent nonlinear bending and postbuckling of functionally graded Mindlin rectangular microplates considering the physical neutral plane position. Compos. Struct. **127**, 87–98 (2015)
329. Ansari, R., Faghih, S.M., Mohammadi, V., Gholami, R., Darabi, M.A.: Size-dependent vibrations of post-buckled functionally graded mindlin rectangular microplates. Lat. Am. J. Sol. Struct. **11**, 2351–2378 (2014)
330. Lou, J., He, L.: Closed-form solutions for nonlinear bending and free vibration of functionally graded microplates based on the modified couple stress theory. Compos. Struct. **131**, 810–820 (2015)

331. Zeighampour, H., Beni, Y.T.: A shear deformable cylindrical shell model based on couple stress theory. Arch. Appl. Mech. **85**, 539–553 (2015)
332. Hosseini-Hashemi, S., Sharifpour, F., Ilkhani, M.R.: On the free vibrations of size-dependent closed micro/nano-spherical shell based on the modified couple stress theory. Int. J. Mech. Sci. **115–116**, 501–515 (2016)
333. Gholami, R., Ansari, R., Darvizeh, A., Sahmani, S.: Axial buckling and dynamic stability of functionally graded microshells based on the modified couple stress theory. Int. J. Struct. Stab. Dyn. **15**, 1450070–24 (2015)
334. Tadi Beni, Y., Mehralian, F., Razavi, H.: Free vibration analysis of size-dependent shear deformable functionally graded cylindrical shell on the basis of modified couple stress theory. Compos. Struct. **120**, 65–78 (2015)
335. Lou, J., He, L., Wu, H., Du, J.: Pre-buckling and buckling analyses of functionally graded microshells under axial and radial loads based on the modified couple stress theory. Compos. Struct. **142**, 226–237 (2016)
336. Ansari, R., Gholami, R., Darabi, M.A.: A nonlinear Timoshenko beam formulation based on strain gradient theory. J. Mech. Mater. Struct. **7**(2), 1749–1761 (2012)
337. Asghari, M., Kahrobaiyan, M.H., Nikfar, M., Ahmadian, M.T.: A size-dependent nonlinear Timoshenko microbeam model based on the strain gradient theory. Acta Mech. **223**, 1233–1249 (2012)
338. Ansari, R., Gholami, R., Sahmani, S.: Free vibration analysis of size-dependent functionally graded microbeams based on the strain gradient Timoshenko beam theory. Compos. Struct. **94**, 221–228 (2011)
339. Ansari, R., Gholami, R., Sahmani, S.: Size-dependent vibration of functionally graded curved microbeams based on the modified strain gradient elasticity theory. Arch. Appl. Mech. **83**, 1439–1449 (2013)
340. Ansari, R., Gholami, R., Faghih, S.M., Mohammadi, V., Sahmani, S.: Buckling of FGM Timoshenko microbeams under in-plane thermal loading based on the modified strain gradient theory. Int. J. Multi. Comput. Eng. **11**, 389–405 (2013)
341. Ansari, R., Faraji, O.M., Rouhi, H.: Studying linear and nonlinear vibrations of fractional viscoelastic Timoshenko micro-/nano-beams using the strain gradient theory. Nonlinear Dyn. **87**, 695–711 (2017)
342. Gholami, R., Ansari, R., Rouhi, H.: Studying the effects of small scale and Casimir force on the non-linear pull-in instability and vibrations of FGM microswitches under electrostatic actuation. Int. J. Non-Linear Mech. **77**, 193–207 (2015)
343. Xie, X., Zheng, H., Yang, H.: Indirect radial basis function approach for bending, free vibration and buckling analyses of functionally graded microbeams. Compos. Struct. **131**, 606–615 (2015)
344. Tajalli, S.A., Rahaeifard, M., Kahrobaiyan, M.H., Movahhedy, M.R., Akbari, J., Ahmadian, M.T.: Mechanical behavior analysis of size-dependent micro-scaled functionally graded Timoshenko beams by strain gradient elasticity theory. Compos. Struct. **102**, 72–80 (2013)
345. Ansari, R., Shojaei, M.F., Mohammadi, V., Gholami, R., Darabi, M.A.: Buckling and post-buckling behavior of functionally graded Timoshenko microbeams based on the strain gradient theory. J. Mech. Mater. Struct. **7**, 931–949 (2012)
346. Ansari, R., Faghih, S.M., Gholami, R., Mohammadi, V., Darabi, M.A.: Thermal postbuckling behavior of size-dependent functionally graded Timoshenko microbeams. Int. J. Non-Linear Mech. **50**, 127–135 (2013)
347. Sahmani, S., Ansari, R.: On the free vibration response of functionally graded higher-order shear deformable microplates based on the strain gradient elasticity theory. Compos. Struct. **95**, 430–442 (2013)
348. Ansari, R., Gholami, R., Faghih, S.M., Mohammadi, V., Darabi, M.A.: Thermal buckling analysis of a mindlin rectangular FGM microplate based on the strain gradient theory. J. Therm. Stress. **36**, 446–465 (2013)
349. Ansari, R., Shojaei, M.F., Mohammadi, V., Gholami, R., Rouhi, H.: Size-dependent thermal buckling and postbuckling of functionally graded annular microplates based on the modified strain gradient theory. J. Therm. Stress. **37**, 174–201 (2014)

350. Ansari, R., Hasrati, E., Faghih, S.M., Gholami, R., Mohammadi, V., Shahabodini, A.: Size-dependent bending, buckling and free vibration analyses of microscale functionally graded mindlin plates based on the strain gradient elasticity theory. Lat. Am. J. Sol. Struct. **13**, 632–664 (2016)
351. Shenas, A.G., Malekzadeh, P.: Free vibration of functionally graded quadrilateral microplates in thermal environment. Thin Wall. Struct. **106**, 294–315 (2016)
352. Ansari, R., Gholami, R., Faghih, S.M., Mohammadi, V., Sahmani, S.: Bending, buckling and free vibration analysis of size-dependent functionally graded circular/annular microplates based on the modified strain gradient elasticity theory. Eur. J. Mech. A Sol. **49**, 251–267 (2015)
353. Gholami, R., Darvizeh, A., Ansari, R., Hosseinzadeh, M.: Size-dependent axial buckling analysis of functionally graded circular cylindrical microshells based on the modified strain gradient elasticity theory. Mecc. **49**, 1679–1695 (2014)
354. Zhang, B., He, Y., Liu, D., Shen, L., Lei, J.: Free vibration analysis of four-unknown shear deformable functionally graded cylindrical microshells based on the strain gradient elasticity theory. Compos. Struct. **119**, 578–597 (2015)
355. Thai, H.T., Choi, D.H.: A simple first-order shear deformation theory for laminated composite plates. Compos. Struct. **106**, 754–763 (2013)
356. Thai, H.T., Choi, D.H.: A simple first-order shear deformation theory for the bending and free vibration analysis of functionally graded plates. Compos. Struct. **101**, 332–340 (2013)
357. Thai, H.T., Nguyen, T.K., Vo, T.P., Lee, J.: Analysis of functionally graded sandwich plates using a new first- order shear deformation theory. Eur. J. Mech. A Sol. **45**, 211–225 (2014)
358. Reddy, J.N.: Nonlocal theories for bending, buckling and vibration of beams. Int. J. Eng. Sci. **45**, 288–307 (2007)
359. Ebrahimi, F., Salari, E.: Thermo-mechanical vibration analysis of a single-walled carbon nanotube embedded in an elastic medium based on higher-order shear deformation beam theory. J. Mech. Sci. Technol. **29**, 3797–3803 (2015)
360. Emam, S.A.: A general nonlocal nonlinear model for buckling of nanobeams. Appl. Math. Model. **37**, 6929–6939 (2013)
361. Rahmani, O., Jandaghian, A.A.: Buckling analysis of functionally graded nanobeams based on a nonlocal third-order shear deformation theory. Appl. Phys. A **119**, 1019–1032 (2015)
362. Ebrahimi, F., Barati, M.R.: Vibration analysis of nonlocal beams made of functionally graded material in thermal environment. Eur. Phys. J. Plus. **131**, 279–301 (2016)
363. Aydogdu, M.: A general nonlocal beam theory: its application to nanobeam bending, buckling and vibration. Phys. E **41**, 1651–1655 (2009)
364. Aydogdu, M.: A new shear deformation theory for laminated composite plates. Compos. Struct. **89**, 94–101 (2009)
365. Karama, M., Afaq, K.S., Mistou, S.: Mechanical behaviour of laminated composite beam by the new multi- layered laminated composite structures model with transverse shear stress continuity. Int. J. Sol. Struct. **40**, 1525–1546 (2003)
366. Thai, H.T.: A nonlocal beam theory for bending, buckling, and vibration of nanobeams. Int. J. Eng. Sci. **52**, 56–64 (2012)
367. Shimpi, R.P.: Refined plate theory and its variants. AIAA J. **40**, 137–146 (2002)
368. Tounsi, A., Semmah, A., Bousahla, A.A.: Thermal buckling behavior of nanobeams using an efficient higher-order nonlocal beam theory. J. Nanomech. Micromech. **3**, 37–42 (2013)
369. Zemri, A., Houari, M.S.A., Bousahla, A.A., Tounsi, A.: A mechanical response of functionally graded nanoscale beam: An assessment of a refined nonlocal shear deformation theory beam theory. Struct. Eng. Mech. **54**, 693–710 (2015)
370. Thai, H.T., Vo, T.P.: A nonlocal sinusoidal shear deformation beam theory with application to bending, buckling, and vibration of nanobeams. Int. J. Eng. Sci. **54**, 58–66 (2012)
371. Touratier, M.: An efficient standard plate theory. Int. J. Eng. Sci. **29**, 901–916 (1991)
372. Tounsi, A., Benguediab, S., Houari, M.S.A., Semmah, A.: A new nonlocal beam theory with thickness stretching effect for nanobeams. Int. J. Nanosci. **12**, 1350025–8 (2013)

373. Thai, H.T., Kim, S.E.: A simple quasi-3D sinusoidal shear deformation theory for functionally graded plates. Compos. Struct. **99**, 172–180 (2013)
374. Ahouel, M., Houari, M.S.A., Bedia, E.A.A., Tounsi, A.: Size-dependent mechanical behavior of functionally graded trigonometric shear deformable nanobeams including neutral surface position concept. Steel. Compos. Struct. **20**, 963–981 (2016)
375. Chaht, F.L., Kaci, A., Houari, M.S.A., Tounsi, A., Beg, O.A., Mahmoud, S.R.: Bending and buckling analyses of functionally graded material (FGM) size-dependent nanoscale beams including the thickness stretching effect. Steel. Compos. Struct. **18**, 425–442 (2015)
376. Pour, H.R., Vossough, H., Heydari, M.M., Beygipoor, G., Azimzadeh, A.: Nonlinear vibration analysis of a nonlocal sinusoidal shear deformation carbon nanotube using differential quadrature method. Struct. Eng. Mech. **54**, 1061–1073 (2015)
377. Sadatshojaei, E., Sadatshojaie, A., Fakhar, M.H.: Differential quadrature method for nonlocal nonlinear vibration analysis of a boron nitride nanotube using sinusoidal shear deformation theory. Mech. Adv. Mater. Struct. **23**, 1278–1283 (2016)
378. Berrabah, H.M., Tounsi, A.L., Semmah, A., Adda Bedia, E.A.: Comparison of various refined nonlocal beam theories for bending, vibration and buckling analysis of nanobeams. Struct. Eng. Mech. **48**, 351–365 (2013)
379. Thai, H.T., Choi, D.H.: Efficient higher-order shear deformation theories for bending and free vibration analyses of functionally graded plates. Arch. Appl. Mech. **83**, 1755–1771 (2013)
380. Ebrahimi, F., Barati, M.R.: A unified formulation for dynamic analysis of nonlocal heterogeneous nanobeams in hygro-thermal environment. Appl. Phys. A. **122**, 792–806 (2016)
381. Mashat, D., Zenkour, A., Sobhy, M.: Investigation of vibration and thermal buckling of nanobeams embedded in an elastic medium under various boundary conditions. J. Mech. **32**, 277–287 (2016)
382. Thai, S., Thai, H.T., Vo, T.P., Patel, V.I.: A simple shear deformation theory for nonlocal beams. Compos. Struct. (2017). https://doi.org/10.1016/j.compstruct.2017.03.022
383. Aghababaei, R., Reddy, J.N.: Nonlocal third-order shear deformation plate theory with application to bending and vibration of plates. J. Sound Vib. **326**, 277–289 (2009)
384. Pradhan, S.C.: Buckling of single layer graphene sheet based on nonlocal elasticity and higher order shear deformation theory. Phys. Lett. A. **373**, 4182–8 (2009)
385. Pradhan, S.C., Sahu, B.: Vibration of single layer graphene sheet based on nonlocal elasticity and higher order shear deformation theory. J. Comput. Theor. Nanosci. **7**, 1042–50 (2010)
386. Ansari, R., Sahmani, S.: Prediction of biaxial buckling behavior of single-layered graphene sheets based on nonlocal plate models and molecular dynamics simulations. Appl. Math. Model. **37**, 7338–7351 (2013)
387. Hosseini-Hashemi, S., Kermajani, M., Nazemnezhad, R.: An analytical study on the buckling and free vibration of rectangular nanoplates using nonlocal third-order shear deformation plate theory. Eur. J. Mech. A Solids. **51**, 29–43 (2015)
388. Daneshmehr, A., Rajabpoor, A., Pourdavood, M.: Stability of size dependent functionally graded nanoplate based on nonlocal elasticity and higher order plate theories and different boundary conditions. Int. J. Eng. Sci. **82**, 84–100 (2014)
389. Daneshmehr, A., Rajabpoor, A., Hadi, A.: Size dependent free vibration analysis of nanoplates made of functionally graded materials based on nonlocal elasticity theory with high order theories. Int. J. Eng. Sci. **95**, 23–35 (2015)
390. Narendar, S.: Buckling analysis of micro-/nano-scale plates based on two-variable refined plate theory incorporating nonlocal scale effects. Compos. Struct. **93**, 3093–3103 (2011)
391. Malekzadeh, P., Shojaee, M.: Free vibration of nanoplates based on a nonlocal two-variable refined plate theory. Compos. Struct. **95**, 443–452 (2013)
392. Narendar, S., Gopalakrishnan, S.: Scale effects on buckling analysis of orthotropic nanoplates based on nonlocal two-variable refined plate theory. Acta Mech. **223**, 395–413 (2012)
393. Sobhy, M.: Hygrothermal vibration of orthotropic double-layered graphene sheets embedded in an elastic medium using the two-variable plate theory. Appl. Math. Model. **40**, 85–99 (2016)
394. Sobhy, M.: Generalized two-variable plate theory for multi-layered graphene sheets with arbitrary boundary conditions. Acta Mech. **225**, 2521–2538 (2014)

395. Sobhy, M.: Levy-type solution for bending of single-layered graphene sheets in thermal environment using the two-variable plate theory. Int. J. Mech. Sci. **90**, 171–178 (2015)
396. Sobhy, M.: Hygrothermal deformation of orthotropic nanoplates based on the state-space concept. Compos. B Eng. **79**, 224–235 (2015)
397. Zenkour, A.M., Sobhy, M.: Nonlocal elasticity theory for thermal buckling of nanoplates lying on Winkler-Pasternak elastic substrate medium. Phys. E. **53**, 251–259 (2013)
398. Alzahrani, E.O., Zenkour, A.M., Sobhy, M.: Small scale effect on hygro-thermo-mechanical bending of nanoplates embedded in an elastic medium. Compos. Struct. **105**, 163–172 (2013)
399. Thai, H.T., Vo, T.P., Nguyen, T.K., Lee, J.: A nonlocal sinusoidal plate model for micro/nanoscale plates. Proc. Inst. Mech. Eng., Part C: J. Mech. Eng. Sci. **228**, 2652–2660 (2014)
400. Sobhy, M.: Thermomechanical bending and free vibration of single-layered graphene sheets embedded in an elastic medium. Phys. E. **56**, 400–409 (2014)
401. Sobhy, M.: Natural frequency and buckling of orthotropic nanoplates resting on two-parameter elastic foundations with various boundary conditions. J. Mech. **30**, 443–453 (2014)
402. Sobhy, M.: A comprehensive study on FGM nanoplates embedded in an elastic medium. Compos. Struct. **134**, 966–980 (2015)
403. Thai, H.T., Vo, T.P.: A new sinusoidal shear deformation theory for bending, buckling, and vibration of functionally graded plates. Appl. Math. Model. **37**, 3269–3281 (2013)
404. Belkorissat, I., Houari, M.S.A., Tounsi, A., Bedia, E.A.A., Mahmoud, S.R.: On vibration properties of functionally graded nano-plate using a new nonlocal refined four variable model. Steel. Compos. Struct. **18**, 1063–1081 (2015)
405. Soldatos, K.P.: A transverse shear deformation theory for homogeneous monoclinic plates. Acta Mech. **94**, 195–220 (1992)
406. Khorshidi, K., Fallah, A.: Buckling analysis of functionally graded rectangular nano-plate based on nonlocal exponential shear deformation theory. Int. J. Mech. Sci. **113**, 94–104 (2016)
407. Bessaim, A., Houari, M.S.A., Bernard, F., Tounsi, A.: A nonlocal quasi-3D trigonometric plate model for free vibration behaviour of micro/nanoscale plates. Struct. Eng. Mech. **56**, 223–240 (2015)
408. Sobhy, M., Radwan, A.F.: A new quasi 3D nonlocal plate theory for vibration and buckling of FGM nanoplates. Int. J. Appl. Mech. **9**, 1750008–29 (2017)
409. Ma, H.M., Gao, X.L., Reddy, J.N.: A nonclassical Reddy-Levinson beam model based on a modified couple stress theory. Int. J. Multiscale Comput. Eng. **8**, 167–180 (2010)
410. Mohammad-Abadi, M., Daneshmehr, A.R.: Size dependent buckling analysis of microbeams based on modified couple stress theory with high order theories and general boundary conditions. Int. J. Eng. Sci. **74**, 1–14 (2014)
411. Salamat-Talab, M., Nateghi, A., Torabi, J.: Static and dynamic analysis of third-order shear deformation FG micro beam based on modified couple stress theory. Int. J. Mech. Sci. **57**, 63–73 (2012)
412. Nateghi, A., Salamat-Talab, M., Rezapour, J., Daneshian, B.: Size dependent buckling analysis of functionally graded micro beams based on modified couple stress theory. Appl. Math. Model. **36**, 4971–4987 (2012)
413. Aghazadeh, R., Cigeroglu, E., Dag, S.: Static and free vibration analyses of small-scale functionally graded beams possessing a variable length scale parameter using different beam theories. Eur. J. Mech. A Sol. **46**, 1–11 (2014)
414. Chen, W., Chen, W., Sze, K.Y.: A model of composite laminated Reddy beam based on a modified couple- stress theory. Compos. Struct. **94**, 2599–2609 (2012)
415. Mohammad-Abadi, M., Daneshmehr, A.R.: Modified couple stress theory applied to dynamic analysis of composite laminated beams by considering different beam theories. Int. J. Eng. Sci. **87**, 83–102 (2015)
416. Mohammad-Abadi, M., Daneshmehr, A.R., Homayounfard, M.: Size-dependent thermal buckling analysis of micro composite laminated beams using modified couple stress theory. Int. J. Eng. Sci. **92**, 47–62 (2015)

417. Darijani, H., Mohammadabadi, H.: A new deformation beam theory for static and dynamic analysis of microbeams. Int. J. Mech. Sci. **89**, 31–9 (2014)
418. Noori, J., Fariborz, S.J., Vafa, J.P.: A higher-order micro-beam model with application to free vibration. Mech. Adv. Mater. Struct. **23**, 443–450 (2016)
419. Simsek, M., Reddy, J.N.: Bending and vibration of functionally graded microbeams using a new higher order beam theory and the modified couple stress theory. Int. J. Eng. Sci. **64**, 37–53 (2013)
420. Thai, H.T., Kim, S.E.: A size-dependent functionally graded Reddy plate model based on a modified couple stress theory. Compos. B Eng. **45**, 1636–1645 (2013)
421. Akbarzadeh, K.M., Shariati, M., Emam, S.A.: Postbuckling of functionally graded nanobeams based on modified couple stress theory under general beam theory. Int. J. Mech. Sci. **110**, 160–9 (2016)
422. Trinh, L.C., Nguyen, H.X., Vo, T.P., Nguyen, T.-K.: Size-dependent behaviour of functionally graded microbeams using various shear deformation theories based on the modified couple stress theory. Compos. Struct. **154**, 556–572 (2016)
423. Thai, H.T., Vo, T.P., Bui, T.Q., Nguyen, T.K.: A quasi-3D hyperbolic shear deformation theory for functionally graded plates. Acta Mech. **225**, 951–964 (2014)
424. Akgoz, B., Civalek, O.: Thermo-mechanical buckling behavior of functionally graded microbeams embedded in elastic medium. Int. J. Eng. Sci. **85**, 90–104 (2014)
425. Al-Basyouni, K.S., Tounsi, A., Mahmoud, S.R.: Size dependent bending and vibration analysis of functionally graded micro beams based on modified couple stress theory and neutral surface position. Compos. Struct. **125**, 621–630 (2015)
426. Gao, X.L., Huang, J.X., Reddy, J.N.: A non-classical third-order shear deformation plate model based on a modified couple stress theory. Acta Mech. **224**, 2699–2718 (2013)
427. Chen, W., Xu, M., Li, L.: A model of composite laminated Reddy plate based on new modified couple stress theory. Compos. Struct. **94**, 2143–2156 (2012)
428. Jung, W.Y., Han, S.C.: Static and eigenvalue problems of sigmoid functionally graded materials (S-FGM) micro-scale plates using the modified couple stress theory. Appl. Math. Model. **39**, 3506–3524 (2015)
429. Eshraghi, I., Dag, S., Soltani, N.: Consideration of spatial variation of the length scale parameter in static and dynamic analyses of functionally graded annular and circular micro-plates. Compos. B Eng. **78**, 338–48 (2015)
430. Eshraghi, I., Dag, S., Soltani, N.: Bending and free vibrations of functionally graded annular and circular micro-plates under thermal loading. Compos. Struct. **137**, 196–207 (2016)
431. Ghayesh, M.H., Farokhi, H.: Coupled size-dependent behavior of shear deformable microplates. Acta Mech. **227**, 757–775 (2016)
432. Sahmani, S., Ansari, R., Gholami, R., Darvizeh, A.: Dynamic stability analysis of functionally graded higher-order shear deformable microshells based on the modified couple stress elasticity theory. Compos. B Eng. **51**, 44–53 (2013)
433. Thai, H.T., Vo, T.: A size-dependent functionally graded sinusoidal plate model based on a modified couple stress theory. Compos. Struct. **96**, 376–383 (2013)
434. Darijani, H., Shahdadi, A.H.: A new shear deformation model with modified couple stress theory for microplates. Acta Mech. **226**, 2773–2788 (2015)
435. He, L., Lou, J., Zhang, E., Wang, Y., Bai, Y.: A size-dependent four variable refined plate model for functionally graded microplates based on modified couple stress theory. Compos. Struct. **130**, 107–115 (2015)
436. Lou, J., He, L., Du, J., Wu, H.: Nonlinear analyses of functionally graded microplates based on a general four-variable refined plate model and the modified couple stress theory. Compos. Struct. **152**, 516–527 (2016)
437. Lou, J., He, L., Du, J.: A unified higher order plate theory for functionally graded microplates based on the modified couple stress theory. Compos. Struct. **133**, 1036–1047 (2015)
438. Trinh, L.C., Vo, T.P., Thai, H.T., Mantari, J.L.: Size-dependent behaviour of functionally graded sandwich microplates under mechanical and thermal loads. Compos. B Eng. **124**, 218–241 (2017)

439. Reddy, J.N., Kim, J.: A nonlinear modified couple stress-based third-order theory of functionally graded plates. Compos. Struct. **94**, 1128–1143 (2012)
440. Kim, J., Reddy, J.N.: Analytical solutions for bending, vibration, and buckling of FGM plates using a couple stress-based third-order theory. Compos. Struct. **103**, 86–98 (2013)
441. Lei, J., He, Y., Zhang, B., Liu, D., Shen, L., Guo, S.: A size-dependent FG micro-plate model incorporating higher-order shear and normal deformation effects based on a modified couple stress theory. Int. J. Mech. Sci. **104**, 8–23 (2015)
442. Wang, B., Liu, M., Zhao, J., Zhou, S.: A size-dependent Reddy-Levinson beam model based on a strain gradient elasticity theory. Mecc. **49**, 1427–1441 (2014)
443. Sahmani, S., Ansari, R.: Size-dependent buckling analysis of functionally graded third-order shear deformable microbeams including thermal environment effect. Appl. Math. Model. **37**, 9499–9515 (2013)
444. Ansari, R., Gholami, R., Sahmani, S.: Free vibration of size-dependent functionally graded microbeams based on the strain gradient reddy beam theory. Int. J. Comput. Meth. Eng. Sci. Mech. **15**, 401–412 (2014)
445. Zhang, B., He, Y., Liu, D., Gan, Z., Shen, L.: Size-dependent functionally graded beam model based on an improved third-order shear deformation theory. Eur. J. Mech. A Sol. **47**, 211–230 (2014)
446. Shi, G.: A new simple third-order shear deformation theory of plates. Int. J. Sol. Struct. **44**, 4399–4417 (2007)
447. Sahmani, S., Bahrami, M., Ansari, R.: Nonlinear free vibration analysis of functionally graded third-order shear deformable microbeams based on the modified strain gradient elasticity theory. Compos. Struct. **110**, 219–230 (2014)
448. Xiang, S., Jin, Y.-X., Bi, Z.-Y., Jiang, S.-X., Yang, M.-S.: A n-order shear deformation theory for free vibration of functionally graded and composite sandwich plates. Compos. Struct. **93**, 2826–2832 (2011)
449. Akgoz, B., Civalek, O.: A size-dependent shear deformation beam model based on the strain gradient elasticity theory. Int. J. Eng. Sci. **70**, 1–14 (2013)
450. Lei, J., He, Y., Zhang, B., Gan, Z., Zeng, P.: Bending and vibration of functionally graded sinusoidal microbeams based on the strain gradient elasticity theory. Int. J. Eng. Sci. **72**, 36–52 (2013)
451. Akgoz, B., Civalek, O.: A new trigonometric beam model for buckling of strain gradient microbeams. Int. J. Mech. Sci. **81**, 88–94 (2014)
452. Akgoz, B., Civalek, O.: Shear deformation beam models for functionally graded microbeams with new shear correction factors. Compos. Struct. **112**, 214–225 (2014)
453. Akgoz, B., Civalek, O.: Bending analysis of FG microbeams resting on Winkler elastic foundation via strain gradient elasticity. Compos. Struct. **134**, 294–301 (2015)
454. Akgoz, B., Civalek, O.: A novel microstructure-dependent shear deformable beam model. Int. J. Mech. Sci. **99**, 10–20 (2015)
455. Akgoz, B., Civalek, O.: Bending analysis of embedded carbon nanotubes resting on an elastic foundation using strain gradient theory. Acta Astronaut. **119**, 1–12 (2016)
456. Zhang, B., He, Y., Liu, D., Gan, Z., Shen, L.: A novel size-dependent functionally graded curved mircobeam model based on the strain gradient elasticity theory. Compos. Struct. **106**, 374–392 (2013)
457. Zhang, B., He, Y., Liu, D., Lei, J., Shen, L., Wang, L.: A size-dependent third-order shear deformable plate model incorporating strain gradient effects for mechanical analysis of functionally graded circular/annular microplates. Compos. B Eng. **79**, 553–580 (2015)
458. Thai, H.T., Kim, S.E.: A simple higher-order shear deformation theory for bending and free vibration analysis of functionally graded plates. Compos. Struct. **96**, 165–173 (2013)
459. Zhang, B., He, Y., Liu, D., Shen, L., Lei, J.: An efficient size-dependent plate theory for bending, buckling and free vibration analyses of functionally graded microplates resting on elastic foundation. Appl. Math. Model. **39**, 3814–3845 (2015)
460. Thai, H.T., Choi, D.H.: A refined plate theory for functionally graded plates resting on elastic foundation. Compos. Sci. Technol. **71**, 1850–1858 (2011)

461. Thai, H.T., Choi, D.H.: A refined shear deformation theory for free vibration of functionally graded plates on elastic foundation. Compos. B Eng. **43**, 2335–2347 (2012)
462. Thai, H.T., Choi, D.H.: An efficient and simple refined theory for buckling analysis of functionally graded plates. Appl. Math. Model. **36**, 1008–1022 (2012)
463. Thai, H.T., Choi, D.H.: Analytical solutions of refined plate theory for bending, buckling and vibration analyses of thick plates. Appl. Math. Model. **37**, 8310–8323 (2013)
464. Akgoz, B., Civalek, O.: A microstructure-dependent sinusoidal plate model based on the strain gradient elasticity theory. Acta Mech. **226**, 2277–2294 (2015)
465. Krysko, A.V., Awrejcewicz, J., Zhigalov, M.V., Pavlov, S.P., Krysko, V.A.: Nonlinear behaviour of different flexible size-dependent beams models based on the modified couple stress theory. Part 1. Governing equations and static analysis of flexible beams. Int. J. Non-Lin. Mech. **93**, 96–105 (2017)
466. Krysko, V.A., Awrejcewicz, J., Vakakis, A.: Nonlinear Dynamics of Continuous Elastic Systems. Springer, Berlin (2004)
467. Krysko, V.A., Awrejcewicz, J.: Chaos in Structural Mechanics. Springer, Berlin (2008)
468. Krysko, V.A., Awrejcewicz, J.: Nonclassical Thermoelastic Problems in Nonlinear Dynamics of Shells. Springer, Berlin (2003)
469. Krysko, V.A., Awrejcewicz, J., Krysko, A.V.: Thermo-Dynamics of Plates and Shells. Springer, Berlin (2007)
470. Awrejcewicz, J.A., Krysko, V.A., Zhigalov, M.V., Saltykova, O.A., Krysko, A.V.: Chaotic vibrations in flexible multilayered Bernoulli-Euler and Timoshenko type beams. Lat. Am. J. Sol. Struct. **5**(4), 319–363 (2008)
471. Krysko, A.V., Awrejcewicz, J., Saltykova, O.A., Zhigalov, M.V., Krysko, V.A.: Investigations of chaotic dynamics of multi-layer beams using taking into account rotational inertial effects. Comm. Nonlinear Sci. Num. Simul. **19**(8), 2568–2589 (2014)
472. Awrejcewicz, J., Krysko, V.A., Papkova, I.V., Krysko, A.V.: Deterministic Chaos in One-Dimensional Continuous Systems. World Scientific, Singapore (2016)

Chapter 3
Lyapunov Exponents and Methods of Their Analysis

3.1 Introduction

The book is devoted to study the nonlinear phenomena exhibited by the size-dependent structural members including bifurcations and chaotic processes, and hence this chapter provides an overview of one of the main tools for identifying the nonlinear dynamics of these objects. Namely, the concept of Lyapunov exponents is briefly revisited, which allows us to distinguish between regular (periodic or quasi-periodic) and chaotic vibrations of the size-dependent beams, plates and shells studied in this book. In particular, the methods of Benettin, Wolf, Rosenstein, Kantz based on Jacobian estimation and the neural network method are presented and discussed. As noted above, an important issue in solving problems of nonlinear dynamics, especially at the nano-level, is the question of the reliability of chaotic oscillations. This problem was first identified by René Lozi in 2013. In this monograph, in order to obtain reliable results, it is proposed to achieve a coincidence not only of the basic functions during chaotic oscillations, but also of their second derivatives with respect to time. This question was formulated by the authors of the monograph in the book "Deterministic Chaos in One Dimensional Continuous Systems, World Scientific, Singapore, 2016". In addition, various types of definitions of chaos are given and a methodology for identifying the truth of chaos is presented. Moreover, this chapter is devoted to identification of the truth of chaos and the reliability of the results using various methods for determining Lyapunov exponents. This question was investigated using numerical experiments based on classical simple nonlinear systems: Hénon map, hyperchaotic Hénon map, logistic map, as well as the Rössler and Lorenz systems. The case studies are analyzed using the Fourier spectrum and wavelets of various types (Morlet, Mexican hat, Haar, Daubechies and Gauss of various orders). Preference is given to the Morlet and Gaussian wavelet 32. It was shown that when analyzing Lyapunov exponents, neural network method

J. Awrejcewicz et al., *Mathematical Modelling and Numerical Analysis of Size-Dependent Structural Members in Temperature Fields*, Advanced Structured Materials 142,
https://doi.org/10.1007/978-3-030-55993-9_3

(this method was proposed by the authors of the monograph) makes it possible to calculate the spectrum of Lyapunov exponents and hence to identify phenomena such as the transition of the system into chaos, hyper-chaos, etc. in a fast and reliable way [1, 2].

3.2 Largest Lyapunov Exponent (LLE)

Let there be a given dynamical system

$$\dot{x} = f(x), \tag{3.1}$$

where x stands for N-dimensional state vector.

We choose two closely located phase points x_1 and x_2, belonging to trajectories ($x_1(t)$ and $x_2(t)$), and we follow how the distance between the mentioned points

$$d(t) = |\boldsymbol{\varepsilon}\,(t)| = |x_2(t) - x_1(t)| \tag{3.2}$$

is changed during the system evolution governed by (3.1).

If nonlinear dynamics of system (3.1) is chaotic, thn $d(t)$ is expected to increase in the following exponential way

$$d(t) \approx d(0)e^{kt}. \tag{3.3}$$

Equation (3.3) yields an average velocity of the exponential divergence of the trajectory as follows

$$k \approx \frac{\ln\left[d(t)/d(0)\right]}{t}, \tag{3.4}$$

or equivalently

$$k = \lim_{\substack{d(0)\to 0\\ t\to\infty}} \frac{\ln\left[d(t)/d(0)\right]}{t}. \tag{3.5}$$

The quantity k is known as the Kolmogorov-Sinai entropy or KS-entropy [3]. Based on the definition of KS-entropy, one may quantify dynamic regimes which can be either regular or chaotic. In particular, if a system dynamics is either periodic or quasi-periodic, i.e. $d(t)$ does not increase in time, then KS-entropy equals zero ($k = 0$). If in the system a stable periodic non-movable point occurs then $d(t) \to 0$ and $k < 0$. In the case of chaotic system dynamics the KS-entropy $k > 0$.

3.3 Spectrum of Lyapunov Exponents

Le's spectrum gives a possibility for a qualitative estimation of the features of a local attractor/replier stability.

We consider the phase trajectory $x(t)$ of the dynamical system (3.1) starting from the point $x(0)$ as well as its neighbourhood trajectory

$$x_1(t) = x(t) + \boldsymbol{\varepsilon}(t). \tag{3.6}$$

Consider the function

$$\lambda\left[\boldsymbol{\varepsilon}(0)\right] = \lim_{t\to\infty} \frac{\ln\left[\frac{|\boldsymbol{\varepsilon}(t)|}{|\boldsymbol{\varepsilon}(0)|}\right]}{t} \tag{3.7}$$

defined on the vectors of initial displacement $\boldsymbol{\varepsilon}(0)$ such that $|\boldsymbol{\varepsilon}(0)| = \varepsilon$, where $\varepsilon \to 0$.

All possible rotations of the vector of initial displacement with regard to n directions in the N-dimensional phase space imply a jump like dynamics of the function (3.7) tending to the finite number of the values $\lambda_1, \lambda_2, \lambda_3, \ldots, \lambda_n$. The mentioned values of λ are known as the Lyapunov exponents. The positive (negative) values of LEs present a measure of the average exponential divergence (convergence) of the neighbourhood trajectories.

A sum of the LEs describes an average divergence of the flow of the phase trajectories which in the case of a dissipative system should be negative, i.e. the system has an attractor. However, owing to the numerical results, some dissipative systems exhibit LEs being invariant with respect to all chosen initial conditions. This means that the LEs can be used to quantify properties of attractors.

The LEs are presented in the decreased order. For instance, the symbols $(+, 0, -)$ mean that a certain attractor in the 3D state-space exhibits an exponential elongation along one of the directions, in the second direction the phase flow has a neutral stability, and finally, in the remaining third direction the trajectories are compressed in an exponential way. It should be emphasized that attractors other than stable stationary points always have one LE equal to zero. It means that in an average sense the points on a trajectory cannot be either divergent or convergent.

In what follows, we consider a correspondence of the LEs with the features and types of attractors:

(i) $n = 1$ (n stands for dimension of a phase space). Only stable non-movable point (node or focus) can be an attractor. In this case, there is only one LE $\lambda_1 = (-)$ having a negative value.

(ii) $n = 2$. In 2D dynamical systems there are two types of attractors, i.e. either stable fixed points or limiting cycles. The LEs follow

$$(\lambda_1, \lambda_2) = (-, -) - \text{stable fixed point;}$$

$$(\lambda_1, \lambda_2) = (0, -) - \text{stable limiting cycle (one of LEs is equal to zero);}$$

(iii) $n = 3$. In 3D phase space there are four types of attractors: stable fixed points, limiting cycles 2D tori and strange attractors. A possible order of the LEs is as follows:

$$(\lambda_1, \lambda_2, \lambda_3) = (-, -, -) - \text{stable fixed point;}$$

$$(\lambda_1, \lambda_2, \lambda_3) = (0, -, -) - \text{stable limiting cycle;}$$

$$(\lambda_1, \lambda_2, \lambda_3) = (0, 0, -) - \text{stable 2D tori;}$$

$$(\lambda_1, \lambda_2, \lambda_3) = (+, 0, -) \text{ strange chaotic attractor.}$$

Analytical estimation of the LEs is not possible for majority of the dynamical systems since it requires getting analytical solutions to the system of evolutionary differential equations. However, nowadays there are efficient numerical algorithms for their estimation.

KS-entropy describes the maximum (largest) Lyapunov exponent (LLE) allowing for the definition of a velocity responsible for the lack of information on the system initial state.

3.4 Benettin's Method [4]

We begin with an example of the numerical estimation of the Kolmogorov entropy for the case of the Hénon-Heiles model. The numerical computations are carried out with an accuracy up to 10^{-14} and with the help of the method of integration known as the method of central points (independently of the studies in reference [4]).

We have also investigated numerically the ergodic properties of the dissipative dynamical systems with a few degrees of freedom using examples of the Lorenz systems. The Lorenz system possesses LEs spectrum of the type $(+, 0, -)$ with the same values of all orbits originating in an arbitrary point lying on the attractor. The obtained results imply that the ergodic property of dynamical systems can be characterized properly by Les' spectra. Now we briefly described the used method of the LEs estimation.

Let the point x_0 belongs to the attractor A of a dynamical system. We follow evolution of the point x_0 along its trajectory. We choose the positive quality ε essentially less than the investigated dimension of the studied attractor. In addition, we choose an arbitrary perturbated point $\tilde{x}_0$ in a way to satisfy the condition $\|\tilde{x}_0 - x_0\| = \varepsilon$. We consider evolutions of the chosen points x_0 and $\tilde{x}_0$ within a short time interval T and we denote the new obtained points associated with T by x_1 and $\tilde{x}_1$, respectively. The vector $\Delta x_1 = \tilde{x}_1 - x_1$ stands as the perturbation vector. Hence, one may estimate the value of λ in the following way:

$$\tilde{\lambda}_1 = \frac{1}{T} \ln \frac{\|\Delta x_1\|}{\varepsilon}. \tag{3.8}$$

The time intervals T is chosen in the way to keep the perturbation amplitude less than linear dimension of the phase portrait non-homogeneity and less of the being investigated attractor dimension. Let us consider the normalized vector of perturbation $\Delta x_1' = \varepsilon \Delta x_1 / \|\Delta x_1\|$, and the associated new perturbation point $\tilde{x}_1' = x_1 + \Delta x_1'$. We prolong the so far described procedure by taking into account the points x_0 and $\tilde{x}_0$ instead of x_1 and $\tilde{x}_1$, respectively.

Repeating the mentioned procedure M times, one may estimate λ as the averaged arithmetic value of the quantities $\tilde{\lambda}_l$ estimated on each computational step, i.e. we have

$$\lambda \cong \frac{1}{M} \sum_{i=1}^{M} \tilde{\lambda}_l = \frac{1}{M} \sum_{i=1}^{M} \frac{1}{T} \ln \frac{\|\Delta x_i\|}{\varepsilon} = \frac{1}{MT} \sum_{i=1}^{M} \ln \frac{\|\Delta x_i\|}{\varepsilon}. \tag{3.9}$$

In order to achieve more accurate estimation, one may arbitrarily choose larger values of M and carry out the computations for a different initial point x_0.

The so far presented method can be employed only if the evolutionary equations governing system dynamics are known (in the case of experimental data the governing equations are typically not known).

In order to obtain numerically the Lyapunov spectrum one may use the modified algorithm generalizing the classical Benettin's approach. Instead of using one trajectory, a few of them are considered based on perturbations of a few initial points. A number of the perturbed trajectories is equal to the dimension of the phase space. In order for smooth realizations of the latter approach, a numerical method based on the governing equations derivation based on a variation procedure is employed [4]. Since LLE plays a crucial influence on the evolution of all perturbed trajectories, then each step of the algorithm requires not only normalization of the perturbation vectors but also their orthogonalization. We consider the procedure of numerical estimation of the Lyapunov spectrum of a given dynamical system. We assume for simplicity that the phase space is three-dimensional (3D).

Let r_0 be a certain point belonging to the attractor. We fix a certain relatively small (in comparison to the attractor dimension) positive number ε and we choose the perturbation points x_0, y_0 and z_0 to keep the length of perturbation vectors $\Delta x_0 = x_0 - r_0$, $\Delta y_0 = y_0 - r_0$, $\Delta z_0 = z_0 - r_0$ equal to ε, and we keep them mutually orthogonal. Now, the points r_0, x_0, y_0 and z_0 evaluate into points r_1, x_1, y_1 and z_1, respectively, within the monitored time interval T. We consider the new perturbation vectors $\Delta x_1 = x_1 - r_1$, $\Delta y_1 = y_1 - r_1$, $\Delta z_1 = z_1 - r_1$, and again we carry out their orthogonalization based on the known method of Gramm-Schmidt [5].

After the orthogonalization process, the obtained perturbation vectors $\Delta x''_1$, $\Delta y''_1$, $\Delta z''_1$ are orthonormalized, i.e. they are mutually orthogonalized and they have unit length. Now, we again employ the normalization procedure of the orthogonalized perturbation vectors to make their length equal to ε in the following way

$$\Delta x_1''' = \Delta x''_1 \cdot \varepsilon, \; \Delta y'''_1 = \Delta y''_1 \cdot \varepsilon, \; \Delta z_1''' = \Delta z''_1 \cdot \varepsilon. \tag{3.10}$$

We consider the following set of the perturbation points

$$x_1' = x_1 + \Delta x_1''', \; y_1' = y_1 + \Delta y_1''', \; z_1' = z_1 + \Delta z_1'''. \tag{3.11}$$

The process is repeated, but instead of the points r_0, x_0, y_0 and z_0, we consider the points r_1, x_1', y_1' and z_1', respectively.

Repeating the mentioned procedure M times the following sums are computed:

$$S_1 = \sum_{k=1}^{M} \ln \|\Delta x_k'\|, \quad S_2 = \sum_{k=1}^{M} \ln \|\Delta y_k'\|, \quad S_3 = \sum_{k=1}^{M} \ln \|\Delta z_k'\|, \tag{3.12}$$

Finally, the Lyapunov spectrum $\Lambda = \{\lambda_1, \lambda_2, \lambda_3\}$ is estimated using the following formula

$$\lambda_i = \frac{S_i}{MT}, \quad i = 1, 2, 3. \tag{3.13}$$

It should be emphasized that a fundamental role in getting reliable results plays the choice of the time interval T. Namely, if one takes into account a relatively large T then the perturbed trajectories will follow the direction corresponding to the LLE, and hence the obtained results will be not reliable.

3.5 Wolf's Method [3]

Wolf et al. [6] proposed the method which yields an estimation of LEs based on analysis of the time series. It was shown how LEs are associated with fast divergence or convergence in the phase space. Conceptually, the mentioned method is based on the earlier worked out methodology who can be employed only for the analytically modelled systems. The method allows to follow long time increase of the attractor elements of small volumes.

The method allows to estimate the LLE based on a choice of data from one co-ordinate time history without knowledge of the system evolutionary equations but it does not allow to measure all system phase co-ordinates.

Consider the time series $x(t)$, $t = 1, \ldots, N$ measurements regarding one co-ordinate of a chaotic process carried out in equal time intervals. Employment of the method of mutual information one may define a time delay τ, whereas the method of false neighbourhood allows to estimate dimension of the embedding space m. In result, the following set of points in R^m is obtained:

$$x_i = (x(i), x(i-\tau), \ldots, x(i-(m-1)\cdot\tau) = (x_1(i), x_2(i), \ldots, x_m(i)), \tag{3.14}$$

where $i = ((m-1)\tau + 1), \ldots, N$.

We choose a point x_0 from the series (3.14). We find a counterpart point $\widetilde{x_0}$, which satisfies the inequality $\|\widetilde{x_0} - x_0\| = \varepsilon_0 < \varepsilon$, where $\varepsilon > 0$ and is essentially less than the dimension of the reconstructed attractor (both points x_0 and $\widetilde{x_0}$ should be chosen for different time instants). Then we monitor evolution of both points on the reconstructed attractor unless a distance between them is not larger than the given quantity $\varepsilon_{\max}$. The obtained points x_1 and $\widetilde{x_1}$ refer to the time evolution T_1.

Then we consider again the series (3.14) and we look for a point $\widetilde{x_1'}$, being close to x_1, where the following formula is satisfied $\left\| \widetilde{x_1'} - x_1 \right\| = \varepsilon_1 < \varepsilon$. The vectors $\widetilde{x_1} - x_1$ and $\widetilde{x_1'} - x_1$ should possess (possibly) the same direction. The mentioned procedure is repeated for the points x_1 and $\widetilde{x_1}$.

Repeating the so far described procedure M times, the LLE is estimated based on the following formula

$$\lambda \tilde{=} \sum_{k=0}^{M-1} \ln(\varepsilon_k' / \varepsilon_k) / \sum_{k=1}^{M} T_k. \tag{3.15}$$

The given method has been employed by the authors of this monograph to the systems with known spectra of LEs such as the Hénon map, Rössler and Lorenz systems and Mackey-Glass equations [7]. Besides, the method is applicable to study the Belousov-Zhabotinsky reaction [8] and the Couette-Taylor flow [9].

Wolf et al. [6] introduced a few constraints on the choice of the embedding dimension and time delay τ while carrying out the attractor reconstruction for getting the most precise estimation of LEs.

Based on investigation of the Rössler system [10] and Belousov-Zhabotinsky reactions [11], the influence on the time delay required for attractor reconstruction, the time of system development between steps of changes, length of the exchange vector and the minimum possible vector length on the numerically obtained results for the LLE computation have been illustrated and discussed.

In addition, it has been illustrated how variation of time of the systems development between the mentioned changes from 0.5 to 1.5 of the orbit almost always implied stable estimation of the LLEs for the Lorenz system, Rössler attractor [7] and the Belousov-Zhabotinsky reaction [12].

The so far described algorithms can be used to detect chaos and to estimate the corresponding parameters based on the available experimental data while estimating the first positive LEs. Besides, the carried out numerical experiments showed that the deterministic chaos can differ from the external noise (Belousov-Zhabotinsky attractor) and topological complexity (Lorenz attractor). However, it requires a special choice of the used data, and the studied attractor should have a large dimension.

3.6 Rosenstein's Method [13]

The method proposed by Rosenstein [13] is simple in realization and allows to get results relatively fast but in fact it yields a special function [13] instead of the numerical value of λ_1 of the following form

$$y(i, \Delta t) = \frac{1}{\Delta t}\langle \ln d_j(i), d_j(i)\rangle = \min_{x_j} \left\| x_j - x_j' \right\|, \tag{3.16}$$

where x_j stands for the considered point, and x_j' refers to its neighbour point.

The carried out algorithm relies on the coupling of d_j and the Lyapunov exponents $d_j(i) \approx e^{\lambda_1(i\Delta t)}$. LLE is defined by an angle of inclination of the most linear part of the computed function.

3.7 Kantz Method [10]

The Kantz algorithm [10] allows for computation of LLE by looking for all neighbours in the vicinity of the studied trajectory and yields the average distance between neighbours and the studied trajectory as a function of time (or relative time). It is based on the following equation

$$S(\tau) = \frac{1}{T}\sum_{t=1}^{T} \ln\left(\frac{1}{|U_t|}\sum_{i\in U_t} |x_{t+\tau} - x_{i+\tau}|\right), \tag{3.17}$$

where x_t—an arbitrary signal point; U_t—neighbourhood of x_t; x_i—neighbour of x_t; τ—relative time multiplied by the frequency of the choice; T—dimension of the choice; $S(\tau)$—elongation factor in domain of the linear increase of the curve whose slope defines the Lyapunov exponent, i.e. $e^{\lambda\tau} \propto e^{S(\tau)}$. However, the requirement of the linear increase of the curve yields occurrence of new errors. Though the method is useful and enough accurated for the case of systems with known values of LLEs, the choice of parameters and regions when the mentioned linear increase occurs is realized in an arbitrary way.

3.8 Method Based on Jacobian Estimation [11, 12]

This method has been proposed in references [11, 12]. Its main idea is to use an algorithm, the scheme of which is illustrated in Fig. 3.1. A sphere of small radius ε is taken. After a few iterations m, a certain operator T^m transforms this sphere into an ellipsoid having $a_1, \ldots, a_p$ half-axes. The sphere is stretched along the axes $a_1, \ldots, a_s > \varepsilon$, where s is the number of positive LEs. For sufficiently small ε, the operator T^m is close to the sum of the shear operator and the linear operator A. The LLEs are computed as averaged eigenvalues of the operator A on the whole attractor.

A vector ζ_j is chosen, and a set $\{\zeta_{ki}\}(i = 1, \ldots, N)$ of i-th neighbourhood vectors is found. The following set of vectors $y_i \equiv \zeta_{ki} - \zeta_j$, where $||y_i|| \le \varepsilon$, is taken. After m successive iterations, the operator T^m transforms the vector ζ_j into ζ_{j+m}, and the

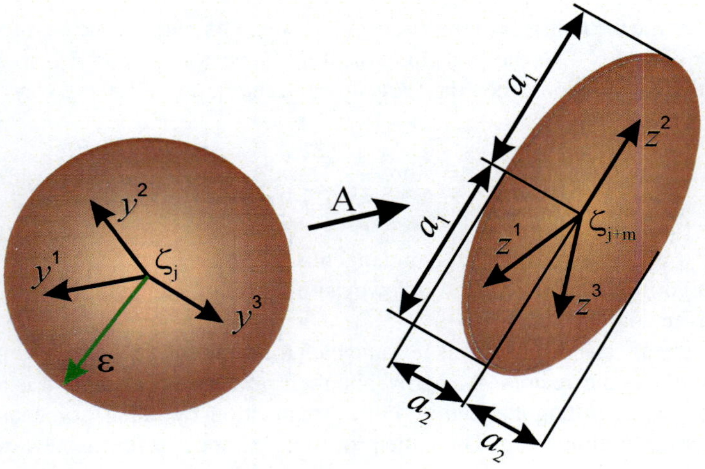

Fig. 3.1 Transformation of a sphere of small radius into a counterpart ellipsoid

vector ζ_{k_i} into $\zeta_{k_{i+m}}$. Consequently, the vectors y_i are transformed into

$$y_{i+m} = \zeta_{k_{i+m}} - \zeta_{j+m}.$$

Assuming that the radius ε is sufficiently small, one can introduce the operator A_j as follows

$$y_{i+m} = A_j y_i.$$

The operator A_j describes the system in variations. To estimate the operator A, the least-square method can be employed:

$$\min_{A_j} S = \min_{A_j} \frac{1}{N} \sum_{i=0}^{N} (y_{i+m} - A_j y_i)^2.$$

This yields the following system of equations of the dimension $n \times n$:

$$A_j V = C, \, (V)_{kl} = \frac{1}{N} \sum_{i=1}^{N} y_i^k y_i^l,$$

$$(C)_{kl} = \frac{1}{N} \sum_{i=1}^{N} y_{i+m}^k y_i^l,$$

where V, C are the matrices of the dimension $n \times n$, y_i^k stands for the k-th component of vector y_i, and y_{i+m}^k is the k-th component of the vector y_{i+m}. If A is a solution to the mentioned equations, then the LEs can be found in the following way

$$\lambda_i = \lim_{n \to \infty} \frac{1}{n\tau} \sum_{j=1}^{n} \ln A_j e_i^j,$$

where $\{e_j\}$ is a set of basic vectors in tangent space ζ_j.

The algorithm can be realized in a way similar to the computation of LEs of the ODEs given analytically.

Let us choose an arbitrary basis $\{e^s\}$ and then follow the changes in the length of the vector $A_j e^s$. As the vectors $A_j e^s$ grow and their orientations change, it is necessary to perform their orthogonalization and normalization by using, for example, the Gramm-Schmidt procedure. Then, the procedure is repeated for the new basis.

The mentioned method allows one to estimate a spectrum of nonnegative LEs. However, the method has a serious disadvantage—it is highly sensitive to noise and errors.

3.9 Modification of the Neural Network Method [14, 15]

We proposed a novel and counterpart method to compute LEs based on a modification of the neural network method (see Fig. 3.2)

To realize the neural network algorithm, the following criteria were taken into account:

(i) the network is sensitive to the input information (information is given in the form of real numbers);
(ii) the network is self-organizing, i.e. it yields the output space of solutions only based on the inputs;

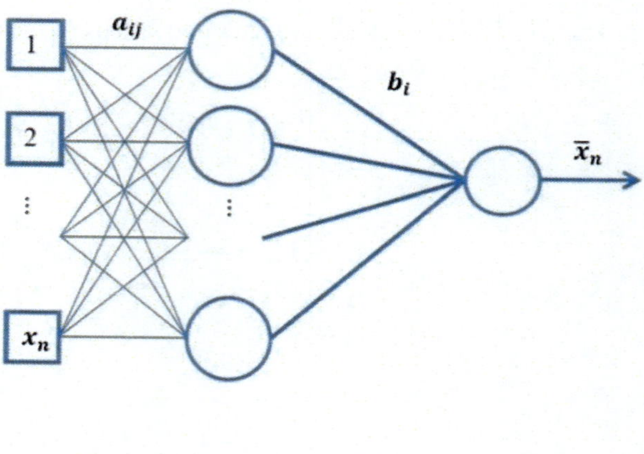

Fig. 3.2 One-layer neutral network

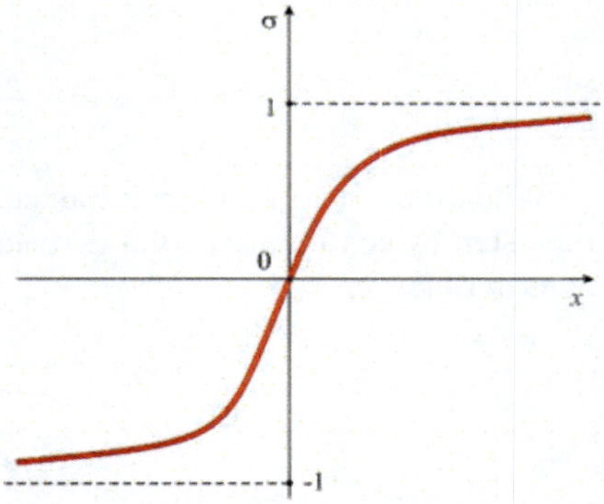

Fig. 3.3 Transition function

(iii) the neural network is a network of straight distribution (all connections are directed from input neurons to output neurons);
(iv) owing to the synapses tuning, the network exhibits dynamic couplings (in the learning process, the tuning of the synaptic coupling takes place ($dW/dt \neq 0$), where W stands for the weighted coefficients of the network).

In the network, there is a hidden layer of neurons, which contains the hyperbolic tangent playing a role of an activation function (Fig. 3.3).

A derivative of the hyperbolic tangent is described by a quadratic function, as it is in the case of a logistic function. However, in contrast to the logistic function, the space of the values of the hyperbolic tangent falls within the interval (–1;1). This results in higher convergence in comparison to the standard logistic function.

Prognosis of $\hat{x}_k$ of a scalar time series x_k is made by employing the following formula

$$\hat{x}_k = \sum_{i=1}^{n} b_i \tanh\left(a_{i0} + \sum_{j=1}^{d} a_{ij} x_{k-j}\right), \tag{3.18}$$

where n stands for the number of neurons, d is the number of the searched LE, a_{ij} stands for the $n \times (d+1)$ matrix of coefficients, and b_i is the vector of length n. The matrix a_{ij} contains the coupling forces with respect to the network input and the vector b_i is used to control the input of each neuron to the network output, whereas the vector a_{i0} is used for relatively simple learning based on data with nonzero averaged value.

Weights a and b are chosen in a probabilistic way, and the dimension of the searched solution is decreased in the process of learning. The associated Gaussian is chosen in a way to have initial standard distribution 2^{-j}, centred with respect to zero in order to promote the most recent time delays (small values of j) in the phase space. The coupling forces are chosen in a way to minimize the averaged one step mean square error of a forecast

$$e = \frac{\sum\limits_{k=d+1}^{c} (\hat{x}_k - x_k)^2}{c-d}. \quad (3.19)$$

When the network is being trained, sensitivity of the output is defined in each time step by computing partial derivatives of all averaged points of the time series in each time step x_{k-j}:

$$\hat{S}(j) = \frac{1}{c-j} \sum_{k=j+1}^{c} \left| \frac{\partial \hat{x}_k}{\partial x_{k-j}} \right|. \quad (3.20)$$

In the case of the network given by (3.18), the partial derivatives have the following form

$$\frac{\partial \hat{x}_k}{\partial x_{k-j}} = \sum_{i=1}^{n} a_{ij} b_i \sec h^2 \left(a_{i0} + \sum_{m=1}^{d} a_{im} x_{k-m} \right). \quad (3.21)$$

The largest value j is the optimal embedding dimension, and the key role is played by $\hat{S}(j)$ as in the false nearest neighbours method. The individual values of $\hat{S}(j)$ yield a quantitative estimate of the importance of each time step using the associated terms of the autocorrelation function or coefficients of the associated linear model.

The weight coefficients of the trained neural network are substituted to the matrix of solutions, and the input data are used to define the initial state. The computation of the spectrum is realized by employment of the generalized Benettin algorithm based on the obtained system of equations.

References

1. Awrejcewicz, J., Krysko, A.V., Erofeev, N.P., Dobriyan, V., Barulina, M.A., Krysko, V.A.: Quantyfying chaos by various computational methods. Part 1: Simple systems. Entropy **20**(3) 175 (2018)
2. Awrejcewicz, J., Krysko, A.V., Erofeev, N.P., Dobriyan, V., Barulina, M.A., Krysko, V.A.: Quantyfying chaos by various computational methods. Part 2: Vibrations of the Bernoulli-Euler beam subjected to periodic and colored noise. Entropy **20**(3) 170 (2018)
3. Kolmogorov A.N.: General theory of dynamical systems and classical mechanics. In: International Mathematical Congress in Amsterdam, pp. 185–208. Fizmatgiz, Moscow (1961). (in Russian)
4. Benettin, G., Galgani, L., Strelcyn, J.M.: Kolmogorov entropy and numerical experiments. Phys. Rev. A **14**, 2338–2345 (1976)
5. Beklemishev, D.V.: Course of Analytical Geometry and Linear Algebra. Nauka, Moscow
6. Wolf, A., Swift, J.B., Swinney, H.L., Vastano, J.A.: Determining Lyapunov exponents from a time series. Phys. D **16**, 285–317 (1985)
7. Mackey, M.C., Glass, L.: Oscillation and chaos in physiological control systems. Science **197**, 287–289 (1977)
8. Hudson, J.L., Mankin, J.C.: Chaos in the Belousov-Zhabotinskii reaction. J. Chem. Phys. **74**, 6171 (1981)
9. Taylor, G.I.: Stability of a viscous liquid contained between two rotating cylinders. Phil. Trans. Roy. Soc. A **223**(605–615), 289–343 (1923)

10. Kantz, H.: A robust method to estimate the maximal Lyapunov exponent of a time series. Phys. Lett. A **185**, 77–87 (1994)
11. Sato, S., Sano, M., Sawada, Y.: Practical methods of measuring the generalized dimension and the largest Lyapunov exponent in high dimensional chaotic systems. Prog. Theor. Phys. **77**, 1–7 (1987)
12. Eckmann, J.-P., Ruelle, D.: Ergodic theory of chaos and strange attractors. Rev. Mod. Phys. **57**, 617–656 (1985)
13. Rosenstein, M.T., Collins, J.J., De Luca, C.J.: A practical method for calculating largest Lyapunov exponents from small data sets. Phys. D **65**, 117–134 (1993)
14. Lim, C.W., Wang, C.M.: Exact variational nonlocal stress modelling with asymptotic higher-order strain gradients for nanobeams. J. Appl. Phys. **101**, 054312 (2007)
15. Wang, C.M., Zhang, Y.Y., Kitipornchai, S.: Vibration of initially stressed micro- and nano-beams. Int. J. Struct. Stab. Dyn. **7**, 555–570 (2007)

Chapter 4
Reliability of Chaotic Vibrations of Euler-Bernoulli Beams with Clearance

4.1 Introduction

The methodology for detecting true chaos (in terms of nonlinear dynamics) developed on the example of a structure composed of two beams with a small clearance is outlined. Euler-Bernoulli hypothesis is employed, and the contact interaction between beams follows Kantor model. The complex nonlinearity results from von Kármán geometric nonlinearity as well as the nonlinearity implied by the contact interaction. The governing PDEs are reduced to ODEs by the second-order finite difference method (FDM). The obtained system of equations is solved by Runge-Kutta method of different accuracies. To purify the signal from errors introduced by numerical methods, the principal component analysis is employed and the sign of the first Lyapunov exponent is estimated by Kantz, Wolf and Rosenstein methods [1].

The proposed methodology for the detection of true chaos is based on the complex investigation of a signal/time history from the point of view of nonlinear dynamics, focusing on construction and analysis of the signals, power frequency spectra, Poincaré pseudo-maps, wavelet spectra, Lyapunov exponents as well as 2D and 3D phase portraits.

Owing to the introduced methodology, the convergence of FDM has been achieved while solving the system of nonlinear PDEs/ODEs with an account of the geometric von Kármán nonlinearity and the design nonlinearity for the mathematical model established in the frame of the kinematic first-order hypotheses.

The employed methodology has proved that the chaotic vibrations of two-layer beams can be detected based on the used methods and computational algorithms. Although the magnitude of clearance between the beams is small, it has been found that chaotic vibrations of beams appear even for small amplitudes. However, we have illustrated that in spite of small amplitudes, there is a need to take the geometric nonlinearity into account while constructing a feasible mathematical model. We have detected, the occurrence of the phase synchronization of beam vibrations for

J. Awrejcewicz et al., *Mathematical Modelling and Numerical Analysis of Size-Dependent Structural Members in Temperature Fields*, Advanced Structured Materials 142,
https://doi.org/10.1007/978-3-030-55993-9_4

the investigated system, among others. In addition, all frequencies exhibited by beam vibrations are in resonance relation with the excitation frequency.

In order to confirm the occurrence of true chaotic vibration, we have demonstrated convergence of the numerical results regarding the signal, not only with respect to the Fourier frequency power spectrum, which is recognized and often employed. Furthermore, we obtained convergence of the numerical results when the computation step (with respect to the spatial coordinate) was increased twice.

We have detected the set of parameters and we have used the methods which guarantee reliability and truth of the obtained solutions. In particular, we have demonstrated that the increase in the number of beam partitions (nodes) implies regularization of the obtained vibrations. After comparing the results obtained within linear and nonlinear problems and taking into account the contact between beams, we have also illustrated that the account of the spatial coordinate x in $u(x, t)$, i.e. the increase in the number of equations, decreases the chaotic components in the analysed signals.

Furthermore, we have presented, illustrated and discussed that the continuous mechanical system cannot be truncated to the system with a finite number of degrees of freedom, but the problem is, indeed, of an infinite dimension.

Finally, we have shown how applications of the PCA for the purification of chaotic signal of the beam allows localizing the fundamental frequency of vibrations of the studied structure composed of two beams.

4.2 Literature Review

Beams and beam constructions are widely employed as elements of numerous devices in today's industry, machine construction, rocket industry and geology. In many cases, the above-mentioned structural elements are subjected to complex external dynamical excitations. Investigations of nonlinear dynamics and contact interactions of the beam structures belong to important (but unsolved) challenging problems in the field of fabricating various sensors and amplifiers.

Owing to the complexity of equations governing the nonlinear dynamics of two geometrically nonlinear beams with a contact interaction, it is impossible to find an exact analytical solution. In general, the problem can be solved using numerical methods. However, in this case, the problem regarding reliability of the obtained results is generated [2], in particular in the case of chaos detection and monitoring. Chaotic vibrations are dangerous since they usually exhibit large amplitudes causing large-scale reactions of the system, either leading directly to the system damage or causing various harmful effects. On the other hand, in many cases, errors introduced by numerical computations are identified with chaotic oscillations. In this work, we define the truth of chaos by means of dealing with the problems of contact interaction of two beams by the employed numerical method.

It is known that the fundamental characteristics of chaos is associated with sensitivity to the initial conditions. Definition of chaos introduced in 1989 by Devaney [3] consists of three parts. In addition to the condition of existence of the dependence on the initial conditions, it includes the intermittency/transitivity condition as well as the regularity condition understood as the density exhibited by periodic points or periodicity. In 1992, Banks et al. [4] proved that sensitivity to the initial conditions can be removed, which means that transitivity and periodicity are sufficient for the occurrence of chaos.

One should also mention a definition of chaos proposed by Kundsen [5], according to which a function defined on the bounded metric space can be understood as a chaotic one if it is characterized by an essential dependence on initial conditions.

Owing to the definition of chaos given by Gulick [6], chaotic orbits exist if there is either essential sensitivity to the initial conditions or at least one of the Lyapunov exponents is positive in each point of the considered chaotic domain. In other words, the studied orbit tends finally neither to a periodic nor a chaotic one (we refer to Gulick definition while checking/validating the existence of true chaos).

The further obtained solutions depend on a chosen kinematic hypothesis, boundary and initial conditions, number of employed intervals of integration regarding beams with respect to FDM and methods used for solving Cauchy problem.

We are aimed at studying nonlinear dynamical features of a structural system composed of two beams with a small clearance. The beams are modelled using the first-order kinematic hypotheses. The lower beam (beam 2) can be viewed as an elastic foundation for the upper beam (beam 1), while the latter beam is subjected to the transverse uniformly distributed harmonic load.

In contrast to numerous other works dealing with truncation of the problem of an infinite dimension to that of a few degrees of freedom (usually two or three) or those employing the so-called reduced-order approach, we are aimed at studying the so far defined problem as that having an infinite number of degrees of freedom.

In general, the solution to the problem can essentially depend on the employed method of reduction from PDEs to ODEs, the method used to solve Cauchy problem as well as on boundary and initial conditions. The solution given by FDM of the approximation $O(c^2)$ will depend on the number of partitions of the integration interval for solving Cauchy problem with respect to time. The analysis of the obtained results is carried out by the methods of nonlinear dynamics and the theory regarding the geometrically nonlinear beams with the account of the contact interaction.

4.3 Mathematical Model

The following main steps of investigations are carried out to numerically solve nonlinear problems of structures composed of two beams with a clearance and subjected to the harmonic load, and thus to detect chaotic vibrations:

1. Since nonlinear PDEs are reduced to ODEs due to the employed FDM of the second-order accuracy, the obtained solutions essentially depend on the number of partitions of the beam. In other words, we need to find the number of n beam partitions, for which the solutions obtained for $2n$ and n partitions coincide. We

require convergence with respect to the signal/time history also in the case of chaotic vibrations. It should be mentioned that in the previous investigations [7], the signal convergence was required rather only in the case of periodic vibrations, whereas for chaos, the integral convergence was accepted.

2. Cauchy problem is also solved numerically, and hence solutions essentially depend on both the chosen method and the time step integration. Therefore, in order to achieve reliable results, Cauchy problem is solved using Runge-Kutta of the fourth (RK4) and the second (RK2) orders [8], Runge-Kutta-Fehlberg of the fourth order (RKF4) [9, 10], the Cash-Karp of the fourth order (RKCK) [11], Runge-Kutta-Prince-Dormand of the eighth order (RKPD8) [12] as well as the implicit Runge-Kutta methods of the second (IRK2) and the fourth (IRK4) orders. The explicit method is characterized by the triangular form of the matrix coefficients (including zeroth main diagonal). On the contrary, in the implicit method, the governing matrix has an arbitrary shape. Moreover, control of integration errors can be implemented in a relatively easy manner.
3. For each spatial coordinate partition as well as for the chosen method, the signals/time histories, 2D/3D phase portraits, Fourier power spectra, the Morlet wavelets, snapshots of beam deflections, the Poincaré sections and the 2D wavelet spectra are constructed while solving Cauchy problem. Haar [13], Shannon-Kotelnikov, Meyer [14] and Daubechies (from db2 to db16) [15] wavelets, coiflets and simlets, Morlet wavelet as well as the complex arbitrary Gauss function of the order higher than 8 (see also [16–18]) are implemented. The Haar and Shannon-Kotelnikov wavelets are rather not feasible for investigation of the beam constructions, i.e. the first one is badly localized in time. The carried out analysis of the wavelet spectra obtained with the help of the Daubechies wavelets, coiflets and simlets yielded a conclusion that an increase in the used wavelets order implies the improvement of the distinction of the frequency localization. However, no matter how different wavelets and different filters are employed, the wavelet spectra obtained by the Daubechies wavelets, simlets and coiflets are practically the same, but still not enough accurate for investigation of the vibration characteristics of the studied continuous mechanical systems. On the other hand, employment of the arbitrary Gauss function shows that an increase in the derivative order implies better resolution of the detected frequencies. It should be emphasized that the spectra obtained on a basis of the Meyer wavelets [17, 18] (smoothened variant of the Shannon-Kotelnikov wavelets) are localized better in the case of the low-frequency band in comparison to the Morlet wavelets. However, the higher spectrum part is better identified using the Morlet wavelets. In what follows, we report results obtained based on the Morlet wavelets. Observe that real wavelets exhibit a lack of the scaling function φ, whereas the function ψ does not have a compact carrier and is given explicitly. The complex Morlet and Gauss wavelets exhibit better localization with respect to frequency than their real counterparts. Therefore, in order to analyse complex vibrations of continuous systems composed of beams, one may employ either complex or real Morlet wavelets as well as wavelets based on the derivatives of the Gauss function of the order higher than 16.

4. Since we would like to use chaos definition given by Gulick [6], we need to compute and estimate the sign of the spectrum of Lyapunov exponents. In this work, the spectrum of Lyapunov exponents for all beam partitions has been estimated using three methods based on Kantz [19], Wolf [20] and Rosenstein et al. [21] algorithms. The final choice was made after obtaining a four-digit accuracy.

As it has been already mentioned, this paper is focused on the nonlinear dynamics and the contact interaction of two beams with a small clearance, for the case when one of the beams (beam 1) is subjected to the action of the transverse harmonic load, causing movement of the second beam. The contact interaction between beams follows Kantor and Bohatyrenko model [22].

In the literature, one can find numerous works devoted to the investigation of beams, plates and shells in the frame of Euler-Bernoulli [23] and Kirchhoff-Love [24, 25] models. The review of the state of the art in this field has been given by Alijani and Amabili [26]. In the majority of considered cases, the solutions were found using simple models including a few degrees of freedom and eventually yielded unreliable results [27]. This has been pointed out in the recent monograph [7], where the problem of chaotic vibrations of beams, plates and shells has been studied treating the mentioned structural members as those of an infinite number of degrees of freedom.

The considered structure composed of two beams occupies a 2D space within the R^2 space with the rectangular system of coordinates given in the following way: a reference line, further called the middle line $z = 0$, is fixed in the beam 2, the axis OX is directed from the left to the right of the middle line, and the axis OZ is directed downwards ($OZ \perp OX$). In the given system of coordinates, the space Ω is defined in the following way (see Fig. 4.1): $\Omega = \{x \in [0, a]; \ -h \leq z \leq h_k + 3h\}$, $0 \leq t \leq \infty$.

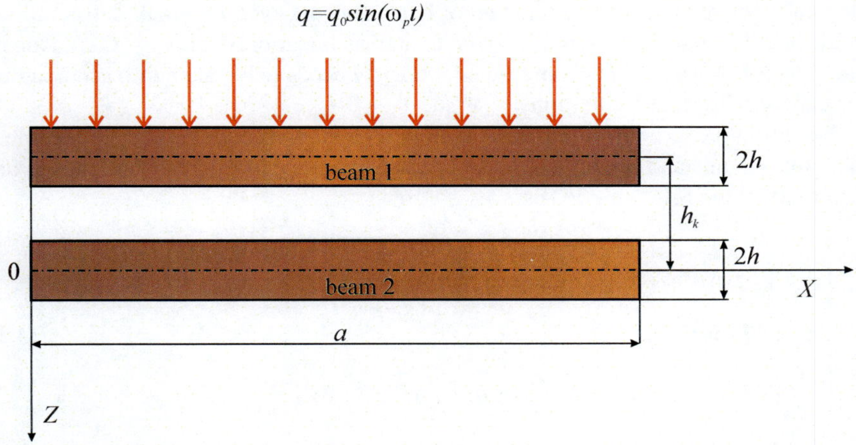

Fig. 4.1 The computation scheme of two beams with a clearance [reprinted with permission from International Journal of Non-Linear Mechanics publishers]

In order to construct the mathematical model of contact interaction of two beams, we introduce the following assumptions and hypotheses:

(i) each beam is composed of one layer;
(ii) the beams are isotropic, elastic and obey Hook's law;
(iii) the longitudinal dimensions of beams are larger than their transverse dimensions and the beams have the unit thickness;
(iv) the axis of each of the beams is a straight line;
(v) the load acts in the direction of the OZ axis and the external forces do not change their direction during the beam deformation;
(vi) contact pressure is estimated within the Kantor model [22];
(vii) normal stresses regarding the zones being parallel to the axis are small and negligible;
(viii) geometric nonlinearity is taken in the von Kármán form [28].

In order to model the contact interaction of the beam within Kantor model, we introduce the term $(-1)^i K(w_1 - w_2 - h_k)\Psi$ into the equation governing the beams vibrations. In the above term, "i" stands for the beam number and the function Ψ is defined by the formula $\Psi = \frac{1}{2}[1 + sign(w_1 - h_k - w_2)]$, i.e. if $\Psi = 1$, then the beams are in contact $w_1 > w_2 + h_k$, otherwise, there is no contact between beams [22]. The coefficient K describes the beam transverse stiffness in a contact zone, whereas h_k represents the clearance between beams (see Fig. 4.1).

In order to keep clarity, in the further parts of the paper, by "beam 1", we understand the external beam loaded, whereas "beam 2" stands for the unloaded beam.

Equations of motion of beams with the account of hypotheses of the first order as well as boundary and initial conditions are yielded from Hamilton energetic principle.

It should be noted that neglecting of von Kármán geometric nonlinearity does not reduce the investigated problem to the linear one, since the design nonlinearity implied by contact of two beams is also taken into account. We consider a small clearance between beams, i.e. contact of beams occurs even for small deformations of the beam 1, and the vibrations can be treated as linear only for $w_1 \leq 0.25$. One of our aims is to verify if the geometric nonlinearity needs to be taken into account due to small amplitudes of vibrations.

Equations governing the dynamics of two Euler-Bernoulli beams with respect to displacements and taking into account the frictional energy loss (dissipation) are governed by the following PDEs:

$$\begin{cases} \frac{1}{\lambda^2}\left\{L_2(w_i, w_i) + L_1(u_i, w_i) - \frac{1}{12}\frac{\partial^4 w_i}{\partial x^4} + q_i(t)\right\} - \frac{\partial^2 w_i}{\partial t^2} - \varepsilon_1 \frac{\partial w_i}{\partial t} + \\ (-1)^i K(w_1 - w_2 - h_k)\Psi = 0, \\ \frac{\partial^2 u_i}{\partial x^2} + L_3(w_i, w_i) - \frac{\partial^2 u_i}{\partial t^2} = 0, \qquad (i = 1,\ 2). \end{cases} \tag{4.1}$$

where $L_1(u_i, w_i) = \frac{\partial^2 u_i}{\partial x^2}\frac{\partial w_i}{\partial x} + \frac{\partial u_i}{\partial x}\frac{\partial^2 w_i}{\partial x^2}$, $L_2(w_i, w_i) = \frac{3}{2}\frac{\partial^2 w_i}{\partial x^2}\left(\frac{\partial w_i}{\partial x}\right)^2$, $L_3(w_i, w_i) = \frac{\partial^2 w_i}{\partial x^2}\frac{\partial w_i}{\partial x}$ are nonlinear operators; w_i, u_i are functions describing deflections and displacements of the upper and the lower beam, respectively; K stands for the coefficient of transverse stiffness of the contact zone, and h_k is the clearance between the beams.

We need to supplement the Eq. (4.1) by the boundary and initial conditions. Furthermore, in the case when the geometric nonlinearity is neglected, we need to take $L_1 = 0, L_2 = 0, L_3 = 0$.

The system of governing PDEs supplemented by boundary and initial conditions is reduced to the counterpart dimensionless form using the following variables:

$$\bar{w} = \frac{w}{2h}, \quad \bar{u} = \frac{ua}{(2h)^2}, \quad \bar{x} = \frac{x}{a}, \quad \lambda = \frac{a}{(2h)}, \quad \bar{q} = q\frac{a^4(1-\nu^2)}{(2h)^4 E},$$
$$\bar{t} = \frac{t}{\tau}, \quad \tau = \frac{a}{c}, \quad c = \sqrt{\frac{Eg}{(1-\nu^2)\rho}}, \quad \bar{\varepsilon}_1 = \varepsilon_1\frac{a}{c}, \tag{4.2}$$

where E—Young's modulus; g—gravity of Earth; ν—Poisson's ratio; ϱ—density of the beam material.

The system of nonlinear PDEs is reduced to the one of nonlinear ODEs by means of the FDM of the second-order approximation $O(c^2)$, where c is a step regarding the spatial coordinate.

In each node of the mesh, one obtains the following set of ODEs:

$$\begin{cases} L_{1,c}(w_{i,j}(t), u_{i,j}(t)) + q_{i,j}(t) = \varepsilon_1 \dot{w}_{i,j}(t) + \ddot{w}_{i,j}(t); \\ L_{2,c}(w_{i,j}(t), u_{i,j}(t)) = \varepsilon_2 \dot{u}_{i,j}(t) + \ddot{u}_{i,j}(t); \quad (i = 1, 2; \; j = 0, \ldots, n), \end{cases} \tag{4.3}$$

where i—beam number; j—number of beam partitions; c—step regarding the spatial coordinate; $L_{1,c}(w_{i,j}(t), u_{i,j}(t))$, $L_{2,c}(w_j(t), u_j(t))$—finite difference operators of the following form

$$L_{1,c}(w_{i,j}, u_{i,j}) = \frac{1}{\lambda^2}\left(-\frac{1}{12}\frac{1}{c^4}\left(w_{i,j+2} - 4w_{i,j+1} + 6w_{i,j} - 4w_{i,j-1} + w_{i,j-2}\right) + \right.$$
$$+ \frac{1}{2c}(w_{i,j+1} - w_{i,j-1})\frac{1}{2}\left(u_{i,j+1} - 2u_{i,j} + u_{i,j-1}\right) +$$

$$+ \frac{1}{2c}(w_{i,j+1} - w_{i,j-1})\frac{1}{2}\left(u_{i,j+1} - 2u_{i,j} + u_{i,j-1}\right) +$$
$$+ \left(\frac{1}{2c}(w_{i,j+1} - w_{i,j-1})\right)^2 \frac{1}{c^2}(w_{i,j+1} - 2w_{i,j} + w_{i,j-1}) +$$
$$+ \frac{1}{c^2}(w_{i,j+1} - 2w_{i,j} + w_{i,j-1})\left(\frac{1}{2}\left(u_{i,j+1} - u_{i,j-1}\right) + \right.$$
$$\left.\left. + \frac{1}{8^2}(w_{i,j+1} - w_{i,j-1})(w_{i,j+1} - w_{i,j-1})\right) + q_{i,j}\right) +$$
$$+ (-1)^i K(w_{1,j} - w_{2,j} - h_k)\Psi_j,$$

$$L_{2,c}(w_{i,j}, u_{i,j}) = \frac{1}{2}\left(u_{i,j+1} - 2u_{i,j} + u_{i,j-1}\right) +$$
$$+ \frac{1}{2c}(w_{i,j+1} - w_{i,j-1})\frac{1}{c^2}(w_{i,j+1} - 2w_{i,j} + w_{i,j-1}); \quad j = 0, \ldots, n.$$

In Eq. (4.3), for $j = 1, \; j = n - 1$, we need to employ the values of the contour points, which are defined by the boundary conditions.

The obtained Cauchy problem is solved by the fourth- and the second-order Runge-Kutta, the fourth-order Runge-Kutta-Fehlberg, the fourth-order Cash-Karp, the eighth-order Runge-Kutta-Prince-Dormand and the implicit second- and fourth-order Runge-Kutta methods. On the basis of the described algorithms, the program package has been developed, which allows one to solve the given problem with respect to the control parameters $\{q_0, \omega_p\}$. The main attention has been paid to control and avoid the occurrence of penetration of the structural elements. As it has been already pointed out, the studied problems are strongly nonlinear, and hence an important question regarding the reliability of the obtained results arises.

While studying the problems of contact interaction, one needs to detect the fundamental frequencies of the vibrational processes first, because, as it will be shown later, a transition into chaotic vibrations takes place after the beam contact. One of the methods devoted to studying the localization of the fundamental frequencies is the principal component analysis.

4.4 Principal Component Analysis (PCA)

It should be emphasized that, formally, the principal component analysis (PCA), aimed at the estimation of principal components with respect to the dissipative problems, can be always employed. However, in the case when the "signal" cannot be distinguished from "noise", any earlier given accuracy is helpful. Typically, the "signal" exhibits a relatively small dimension while keeping the relatively large amplitude. On the other hand, "noise" exhibits a large dimension for a relatively small amplitude. This observations allow us to understand the main features of PCA. Namely, the method works as a filter, i.e. the signal is mainly kept in projections onto the first principal components, whereas the remaining components include mainly noise (for more details see [3–32]).

As a result of solving PDEs by FDM, a matrix W composed of the values of the time-dependent beam deflections measured in the nodes are obtained. Let us present a matrix $\tilde{W}$ in the form of its linear splitting $\tilde{W} = TP^t + E$. The matrix W contains all elements, whereas the matrix $\tilde{W}$ is yielded by the series with the account of k principal components. In order to compute the score matrix T and the matrix of loading P, the singular series development of the matrix W is carried out through the so-called auto-scaling process. The auto-scaled matrix $\bar{W} = USV^t$, where $T = US$ and $P = Vt$, is obtained. The main diagonal of the matrix S is composed of eigenvalues, whereas the remaining matrix elements are equal to zero. The values of the principal components k are chosen in a way to keep the eigenvalues of the matrix S larger than 1. Equivalently, the dependence of dispersion on the number of principal components k $TRV(k)$ and the dependence of the dispersion on the number of principal components k $ERV(k)$ should rapidly change their behaviour, i.e. the truncated singular series are employed. After definition of the required number of principal components, the matrix $T = [t_{ij}]$; $t_{ij} = (x_i, a_j)$ has a dimension $m \times k$, and its each row presents a projection of the data vector onto k principal components (the number of columns k corresponds to the number of vectors of the principal components chosen for the purpose of projection). The loading matrix $P = \{a_1, ..., a_k\}$ and each of its columns correspond to the vector of principal components, whereas the number of rows n corresponds to the dimension of the data space chosen for the purpose of projection. Each row of computation matrix stands for the projection of the data vector onto k principal components. The matrices U and V are orthogonal. The characteristics of dispersion TRV and of dispersion ERV show the percentage of noise remaining after projection onto the multidimensional space PC1-PCk. In other words, the whole characteristics show what number of principal components is necessary for the purification of the signal from noise.

4.5 Numerical Experiment

It is assumed that two beams satisfy the hypotheses of the first-order Euler-Bernoulli approximation. The system (4.1) is solved taking into account the boundary (4.4) and initial conditions (4.5).

Both ends of the beams are rigidly clamped:

$$w_i(0,t) = w_i(1,t) = u_i(0,t) = u_i(1,t) = \\ = \left.\frac{\partial w_i(x,t)}{\partial x}\right|_{x=0} = \left.\frac{\partial w_i(x,t)}{\partial x}\right|_{x=1} = 0, \;\; i = 1,\; 2, \tag{4.4}$$

and the initial conditions are taken in the following form:

$$w_i(x,0) = 0, \;\; u_i(x,0) = 0, \;\; \left.\frac{\partial w_i(x,t)}{\partial t}\right|_{t=0} = 0, \;\; \left.\frac{\partial u_i(x,t)}{\partial t}\right|_{t=0} = 0, \;\; i = 1,\; 2. \tag{4.5}$$

The beam 1 is subjected to the uniformly distributed transverse harmonic excitation of the following form

$$q = q_0 \sin(\omega_p t), \tag{4.6}$$

where q_0 stands for the amplitude and ω_p for the frequency of excitation.

The obtained system of nonlinear PDEs is reduced to ODEs using the FDM with the approximation $O(c^2)$. The beam clearance equals $h_k = 0.1$. First, estimation of convergence of the FDM is conducted. For this purpose, the signals of both beams were compared employing different kinds of Runge-Kutta methods: fourth- and second-order Runge-Kutta methods, Runge-Kutta-Fehlberg method, the fourth-order Cash-Karp method, the eighth-order Runge-Kutta Prince-Dormand method as well as the implicit fourth- and second-order Runge-Kutta methods. According to the obtained results, all the above-mentioned methods ensure good convergence. In further computations, the eighth-order Runge-Kutta method keeping the Prince-Dormand accuracy is used.

Table 4.1 reports signals/time histories $w(0.5, t)$ obtained for different number of beam length partitions: $n = 40; 80; 100; 120; 140; 160$.

For $n = 40; 80$, beam deflections strongly differ from each other. Beginning from $n = 100; 120; 140; 160$, deflections of the first beam coincide. In the case of the second beam, the convergence with respect to the number of beam partitions (nodes) is significantly worse compared to the previous case and occurs only for $n = 120; 140; 160$.

Owing to the already introduced methodology, Table 4.2 represents the results of convergence of the FDM regarding the time series $w(0.5, t)$ for $500 \le t \le 506$. In the case of the test/reference signal, the calculations were made for $n = 160$. It

Table 4.1 Time histories for different number of nodes $n = 40; 80; 100; 120; 140; 160$ [reprinted with permission from International Journal of Non-Linear Mechanics publishers]

Beam 1	Beam 2
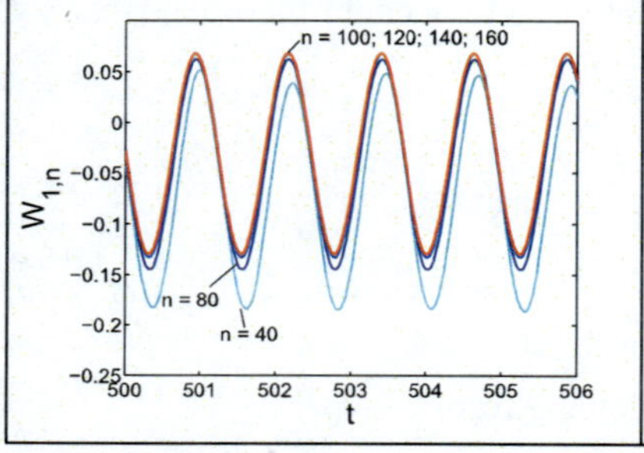	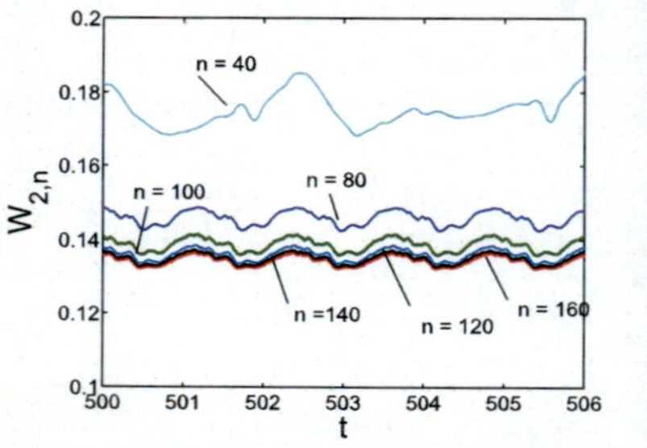

Table 4.2 Convergence of the signals $w_(i, n)(0.5, t)$ for $n = 40; 80; 120$

n	Beam 1 (%)	Beam 2 (%)
40	20.447	24.882
80	7.221	9.071
120	0.00166	0.00521

Table 4.3 Snapshots of the beam shape (for $w_{i,n}(x)$, $x \in [0;\ 1]$), computed for different $n = 40; 80; 100; 120; 140; 160$ and $t = 501.3$ [reprinted with permission from International Journal of Non-Linear Mechanics publishers]

Beam 1	Beam 2
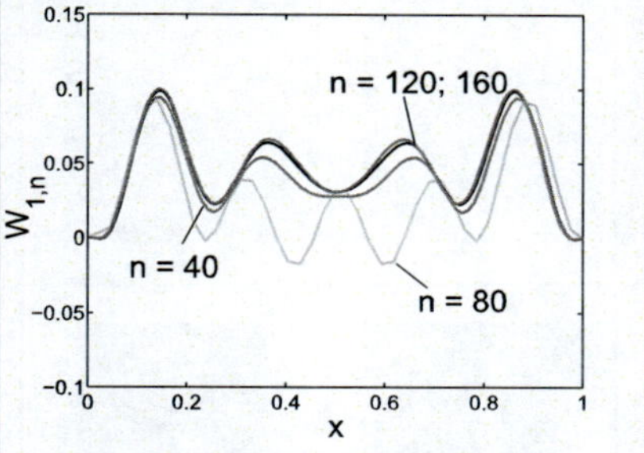	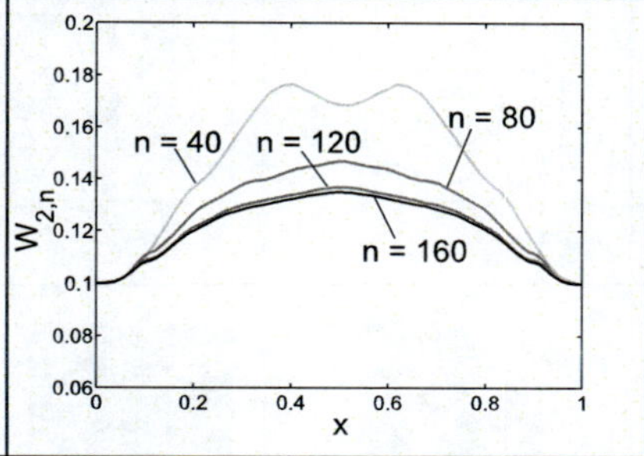

can be observed in Table 4.2 that an increase in the number of beam partitions for $x \in [0;\ 1]$ implies an increase in the convergence.

In what follows, we investigate snapshots presenting changes in the shape of beams for $x \in [0;\ 1]$, $n = 40; 80; 120; 160$ at the fixed time instants $t = 501; 501.3; 501.6$. For $t = 501$, the beam deflection $w(x = 0.5, t = 501)$ is located in its counterpart maximum. For $t = 501.3$, $w(x = 0.5, t = 501.3)$ deflection is around zero, whereas for $t = 501.6$, it is situated in the beam local minimum. For $t = 501$, one can observe five local maxima and four minima on the snapshots of the beam 1. For $t = 501.3$, (the beam deflection is around zero) the number of maxima and minima for $n = 40$ is fixed, whereas an increase in $n = 80; 120; 160$ implies a decrease down to four and five, respectively. An increase in $n = 120; 160$ causes a dome-like shape of the snapshot, i.e. the maximum is achieved only for $x = 0.5$. The deflection snapshots of the first and second beams coincide for $n = 120; 160$. Snapshots of beams configurations for $t = 501.3$ and for $n = 40; 80; 120; 160$ are reported in Table 4.3.

In Table 4.4, the frequency power spectra, the 2D and 3D phase portraits, the wavelet spectra for $\omega_p \in [0.5; 5.1]$ and $\omega_p \in [0.5; 2]$, and the Poincaré pseudo-maps are reported for different numbers of the beam spatial coordinates.

In Tables 4.4 and 4.7, the number 1 corresponds to the number of spatial coordinate partition, while 2 corresponds to the beam number.

We illustrate and discuss the dynamics of beams for $n = 40; 80; 120; 160$. For $n = 40$, linearly dependent frequencies $\frac{\omega_p}{2}, \frac{2\omega_p}{5}, \frac{3\omega_p}{5}, \frac{23\omega_p}{30}$ as well as a chaotic component are detected in the beam spectrum. The frequency spectra for the first and the second beams have the same frequencies. The 2D and 3D phase portraits exhibit a "thick ring" for the first beam, whereas "washed out cloud" is observed for the second beam. The wavelet spectrum $\omega \in [0.2; 5.1]$ exhibits two frequencies ω_p and $\frac{\omega_p}{2}$, whereas the wavelet spectrum associated with $\omega \in [0.2; 2]$ exhibits three different frequencies $\frac{\omega_p}{15}, \frac{4\omega_p}{15}, \frac{\omega_p}{3}$. The Poincaré pseudo-map presents a limit cycle for the first

Table 4.4 Dynamical characteristics of the beams 1 and 2 for $n = 40; 80; 120; 160$ [reprinted with permission from International Journal of Non-Linear Mechanics publishers]

1	2	Power spectrum	Phase portrait 3D $w(\dot{w}; \ddot{w})$	Phase portrait 2D $w(\dot{w})$	Wawelet spectrum $\omega(t)$ $\omega \in [0.2; 2]$	Poincaré pseudomap
40,80	1	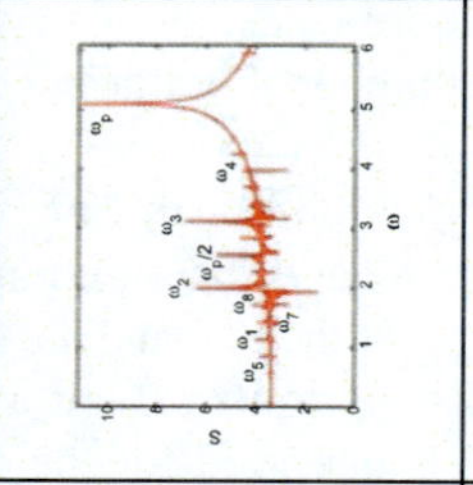	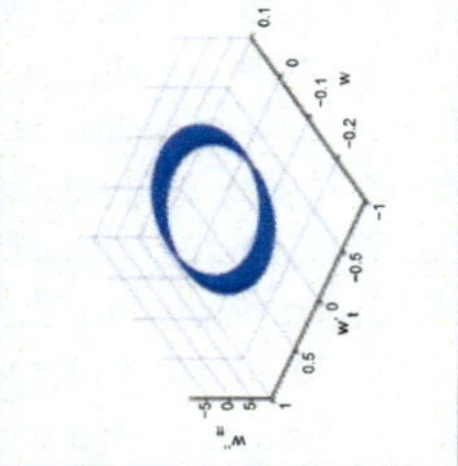	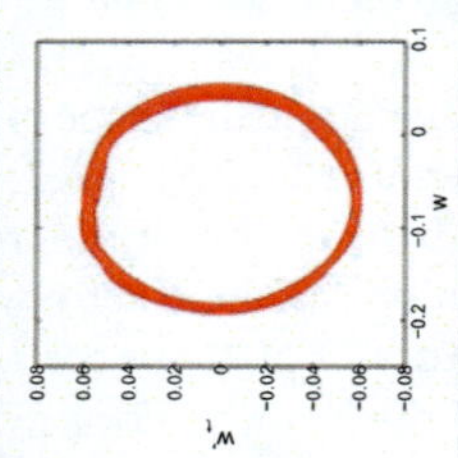	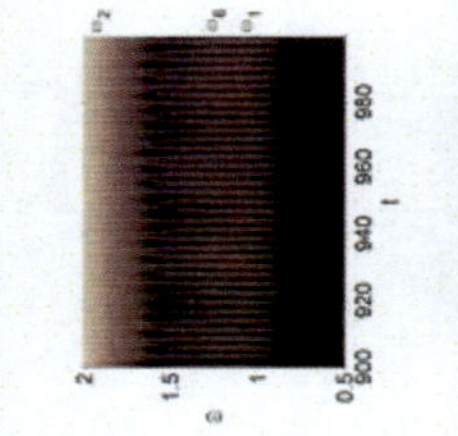	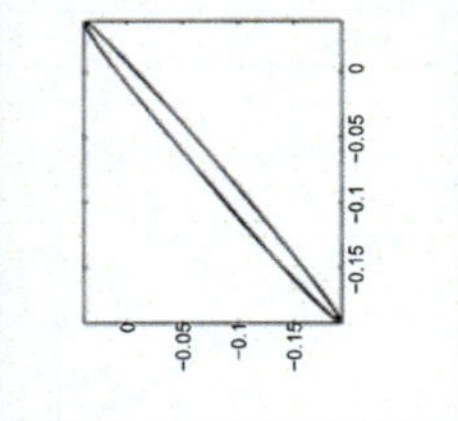
	2	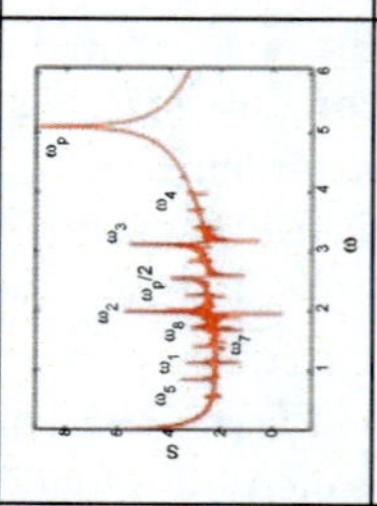	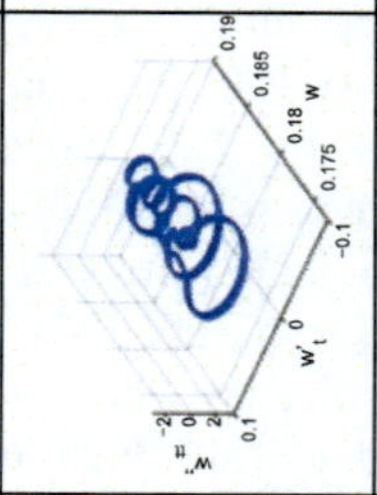	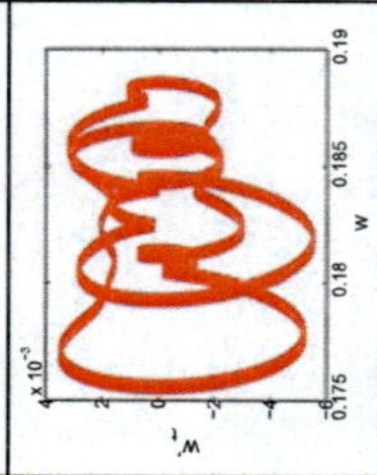	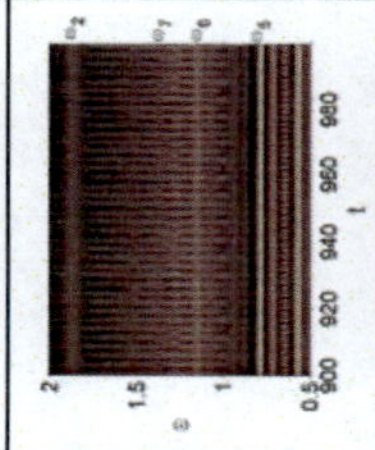	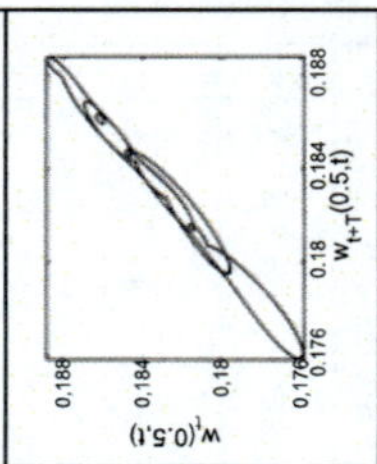
120,160	1	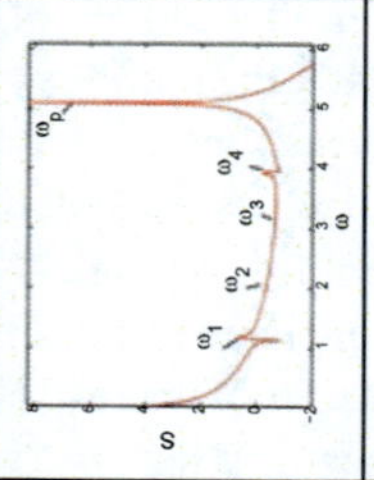	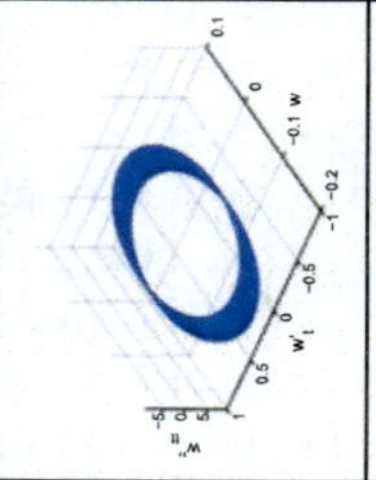	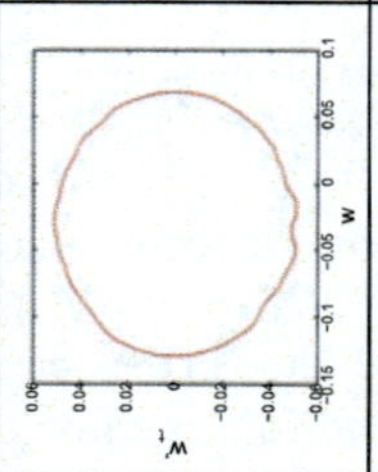	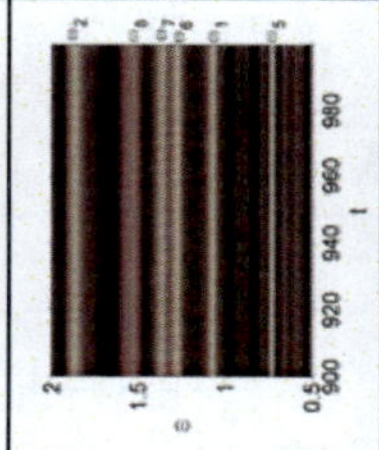	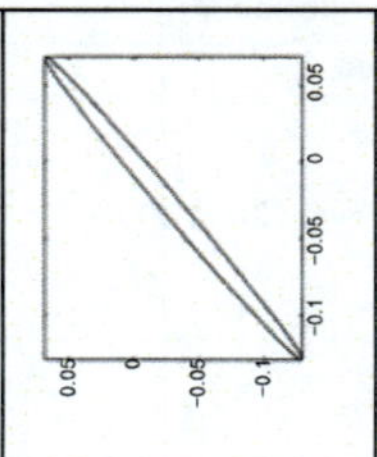
	2	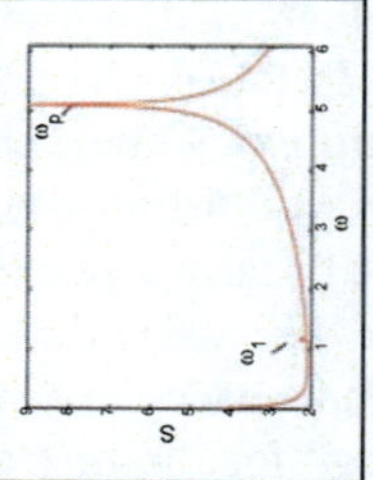	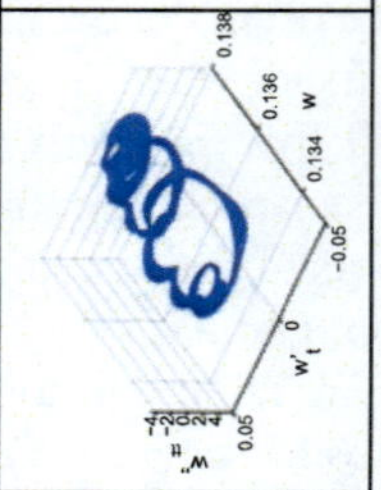	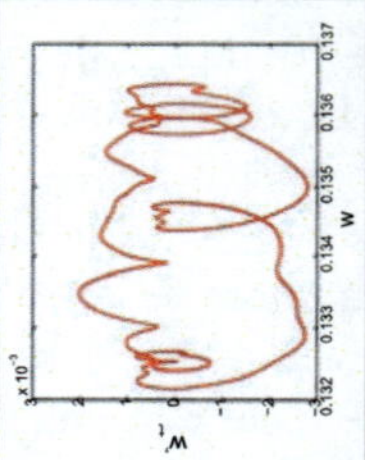	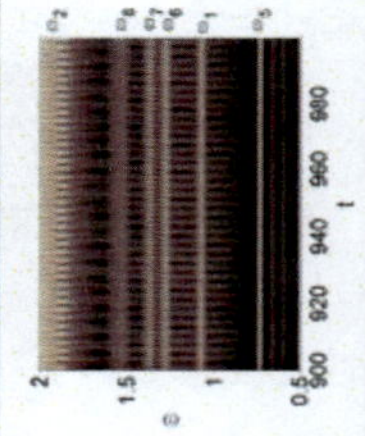	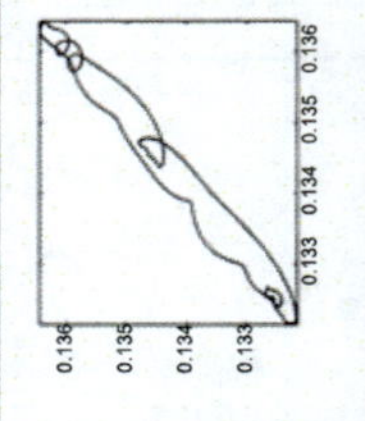

beam and two "thick rings" for the second beam. This implies the occurrence of the period doubling bifurcation.

For $n = 80$, the frequency power spectra of both beams contain the following frequencies: $\frac{\omega_p}{2}$, $\frac{\omega_p}{5}$, $\frac{4\omega_p}{5}$. The 2D phase portrait of the first beam presents a ring, whereas the 3D phase portrait exhibits a more complex structure, i.e. two rings (the structure of the phase portrait is more unique for the second beam). The wavelet spectrum for $\omega \in [0.2; 5.1]$ exhibits only the frequency ω_p, whereas the wavelet spectrum $\omega \in [0.2; 2]$ shows a set of frequencies, which are not visible in the power spectrum, whereas the Poincaré pseudo-map presents a thick ring for both beams.

For $n = 120$ and $n = 160$, the mentioned characteristics fully coincide in the frequency power spectrum for both the first and the second beams, and the following frequencies are distinguished $\omega_p, \omega_1, \omega_2, \omega_3, \omega_4$. The 2D phase portrait of the first beam presents a ring, whereas the second beam phase portrait is composed of five rings. The wavelet spectrum $\omega \in [0.2; 5.1]$ exhibits only the frequency ω_p, whereas the wavelet spectrum for $\omega \in [0.2; 2]$ possesses frequencies which are not reported by the power spectrum.

The Poincaré pseudo-map presents a thick ring for both the first and the second beams.

Since the wavelet spectrum computed for $\omega \in [0.2; 5.1]$ and for $t \in [0; 1000]$ does not exhibit any other frequencies than ω_p (frequency of excitation), it has not been presented. An increase in the number of nodes, i.e. of beams partitions, yields a decrease in the frequencies number to 5, i.e. the so-called regularization of beam vibrations takes place. Finally, let us emphasize the occurrence of the chaotic phase synchronization of vibrations of beams.

In Table 4.5, we report surfaces of the beams deflections for $t \in [95;\ 100]$ and for $n = 40; 80; 120; 160$. The horizontal axis denotes the beam length, whereas the vertical axis stands for the beam deflection, and the remaining axis denotes time.

We have achieved convergence of the deflection surfaces of the middle lines of the beams with respect to time for $n = 120$ and $n = 160$. Therefore, the convergence of the numerical results is guaranteed for the whole time and space intervals.

Figure 4.2a, b presents the curves obtained based on an estimation of the largest Lyapunov exponents by Wolf, Rosenstein and Kantz methods versus n. The obtained values result from the computation using the eighth-order Runge-Kutta method. It should be emphasized that different methods of computation of Lyapunov exponent are needed to obtain reliable/true value and, consequently, reliable estimation of chaos. When using the FDM, for the number of beams partitions $n = 40, 80, 120; 160$ for any mentioned method, Lyapunov exponents coincide with an accuracy up to the third decimal digit. Furthermore, all of the largest Lyapunov exponents (LLEs) are positive.

In what follows, we define how the LLEs depend on the method employed for the solution of Cauchy problem. For this purpose, we have computed Lyapunov exponents using Wolf, Kantz and Rosenstein algorithms for $n = 160$. On the contrary to the dynamic characteristics, we have not observed full coincidence of the results. However, the difference between the minimum and the maximum of the exponents

Table 4.5 Surfaces of the middle beam line deflections [reprinted with permission from International Journal of Non-Linear Mechanics publishers]

$n = 160$	Beam 1	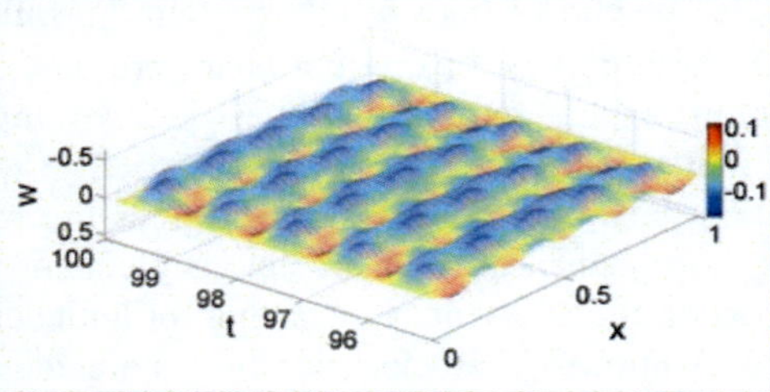
$n = 120$	Beam 2	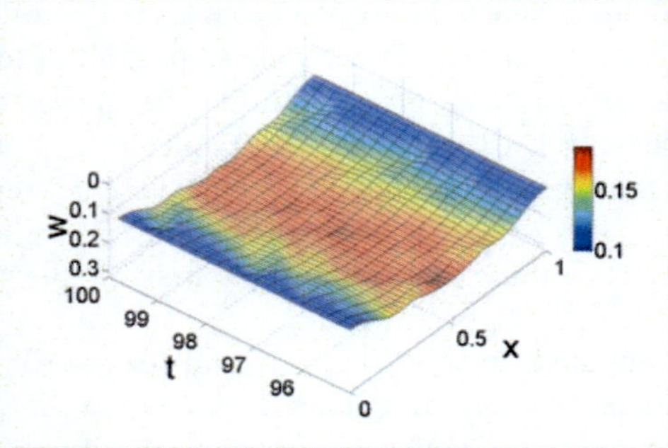
$n = 80$	Beam 1	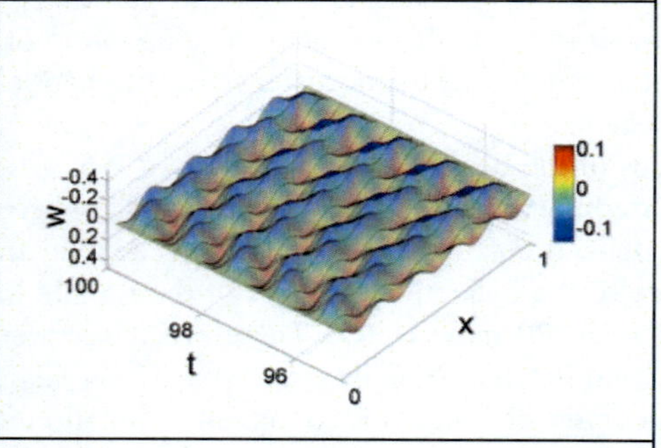
$n = 40$	Beam 2	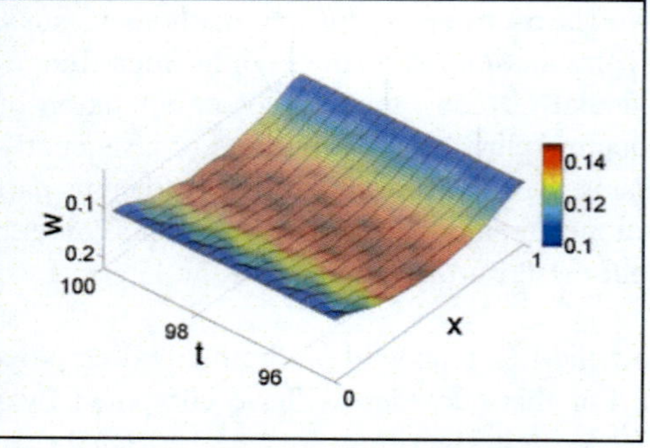

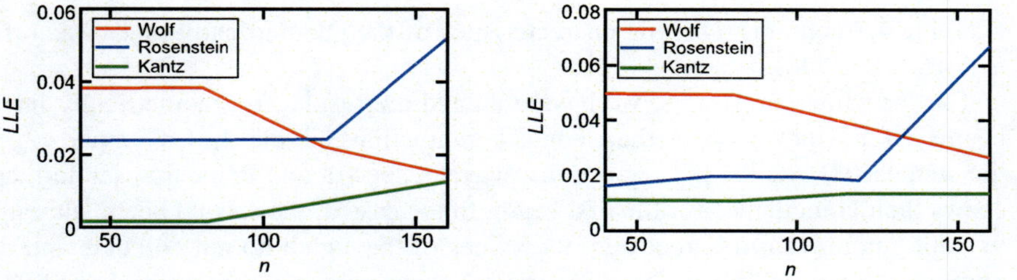

a) Beam 1 b) Beam 2

Fig. 4.2 Largest Lyapunov exponent (LLE) versus n, obtained using Wolf, Rosenstein and Kantz methods [reprinted with permission from International Journal of Non-Linear Mechanics publishers]

computed for different Runge-Kutta methods using Wolf algorithm for the beam 1 is about 0.07, whereas the same done by Rosenstein and Kantz methods yields the difference of 0.008. Convergence up to the second decimal digit has been observed for each of the computational methods of the LEs computation.

It should be noted that all values of the LLE, independently of the employed method of the solution of Cauchy problem, the beam partition number and the employed LLE method, are positive. Since the amplitude of the vibrations of the beam is small, one can assume that it is sufficient to use a linear theory of beam vibrations.

In order to validate this remark, let us compare signals of other beam characteristics, taking into account the geometric nonlinearity.

A comparison of the beams time histories with and without the account of the geometric nonlinearity are reported in Table 4.6.

Table 4.6 Comparison of the beams signals with/without the geometric nonlinearity [reprinted with permission from International Journal of Non-Linear Mechanics publishers]

Signal $w(t)$ of beam 1	Signal $w(t)$ beam 2
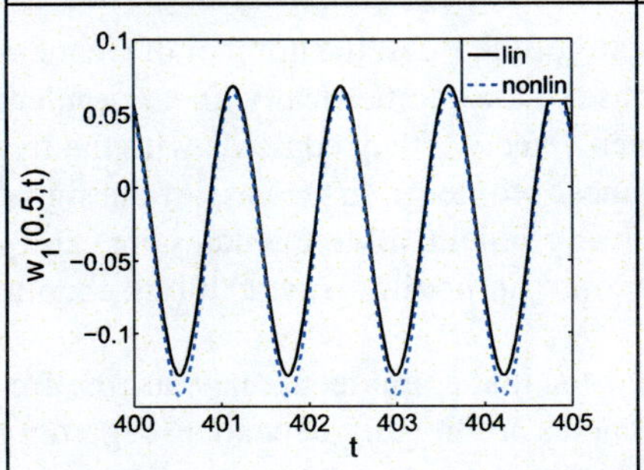	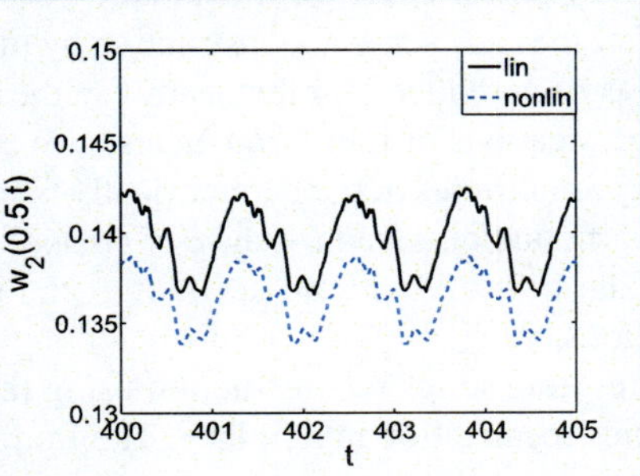

Table 4.7 contains dynamic characteristics of the geometrically linear beam for $n = 160$ (RKPD8).

For the same parameters, we have obtained the results for geometrically linear beams taking into account their contact interactions (Table 4.6). Comparison of the signals, 3D and 2D phase portraits, wavelet spectra and Poincaré pseudo-maps shows their coincidence for the first beam. In the case of the second beam, although its configuration form is repeated, the values of the signal do not coincide and the number of loops exhibited by the 2D and 3D phase portraits is different. In addition, the frequency power spectra differ from each other in the linear and nonlinear case. Namely, in the linear case, the frequencies ω_1 and ω_2 are not exhibited. This can be caused by their small power. Lyapunov exponents for both linearly geometric beams are positive.

Therefore, carrying out the complex analysis of solutions to the problem, one can conclude that vibrations of the studied two-layer beam with a small clearance between the layers are chaotic. The efficient beam partition while using the FDM is $n = 160$. Since the solutions for Cauchy problem coincide for all employed modifications of the Runge-Kutta methods, the second-order Runge-Kutta method can be used. The comparison of the geometrically linear and nonlinear problems implies the occurrence of nonlinear terms in the solution, despite the amplitude of beam vibrations coincides with the linear approximation.

4.6 Application of the Principal Component Analysis (PCA)

The PCA will be employed to find the fundamental frequencies. As it can be seen from the reported characteristics (Table 4.8), for the beam partition number $n = 160$ and for the design nonlinear problem, the frequency spectra for both beams exhibit five linearly dependent frequencies $\omega_p, \omega_1, \omega_2, \omega_3$, and ω_4, whereas for both geometrically and design nonlinear problem of the contacting beams, the noisy components of the frequency power spectra are exhibited.

After the time series reconstruction by means of the PCA, the number of frequencies has been reduced for the beam 1 in the case of the geometrically linear problem and the locations of four frequencies have been detected (they coincide with the frequency amplitudes of the geometrically nonlinear problem). In the case of the signal reconstruction obtained for the design nonlinear problem, all frequencies are exhibited. The beam 2, with no account of the geometric nonlinearity, exhibits chaotic vibrations.

After time series reconstruction using the principal component analysis, the frequencies localization takes place. The frequencies are linearly dependent regarding the geometrically nonlinear problem. As a result, the computation time has been reduced eight times due to the decrease in the number of nodes while employing the FDM, and the step of Runge-Kutta method has been increased. However, this method

Table 4.7 Dynamical characteristics of the geometrically linear beams 1 and 2 for $n = 160$ (RKPD8 method) [reprinted with permission from International Journal of Non-Linear Mechanics publishers]

1	2	Power spectrum	Phase portrait 3D $w(\dot{w}, \ddot{w})$	Phase portrait 2D $w(\dot{w})$	Wavelet spectrum $\omega(t)$, $\omega \in [0.2; 2]$	Poincaré map
160	1	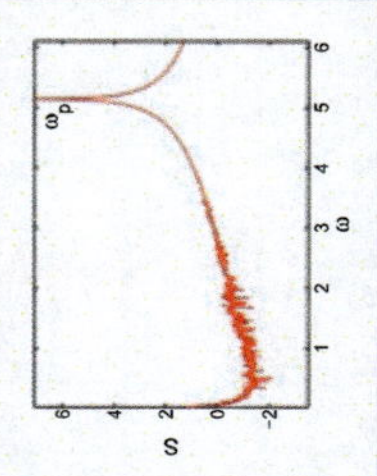	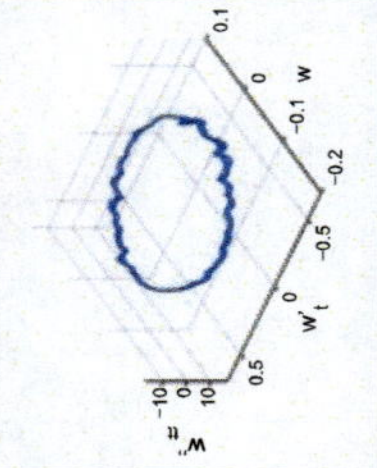	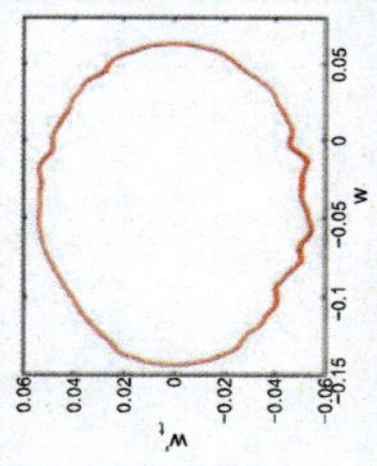	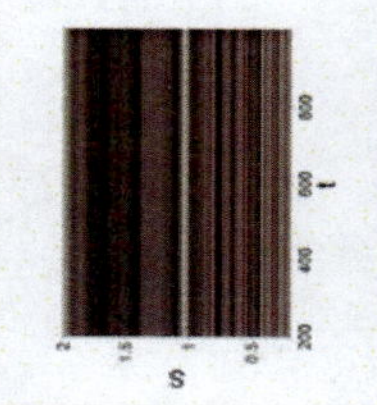	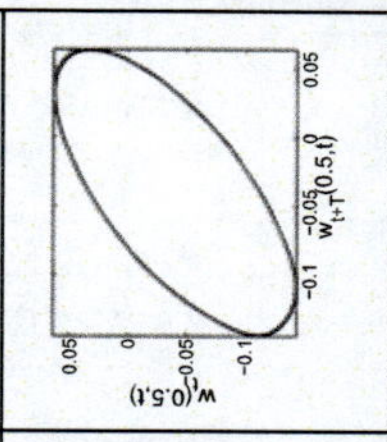
	2	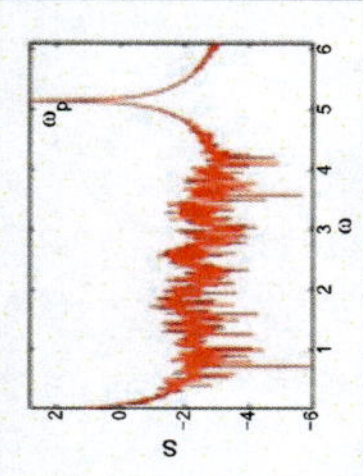	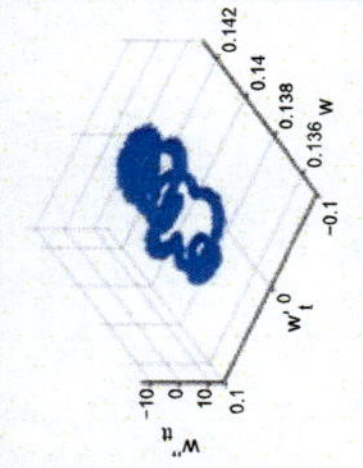	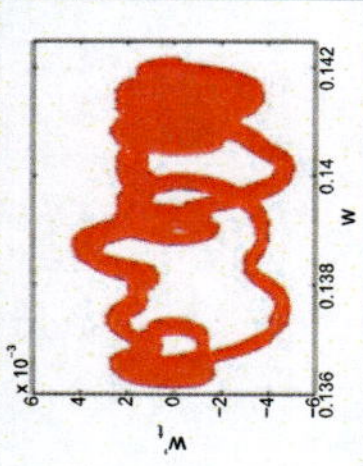	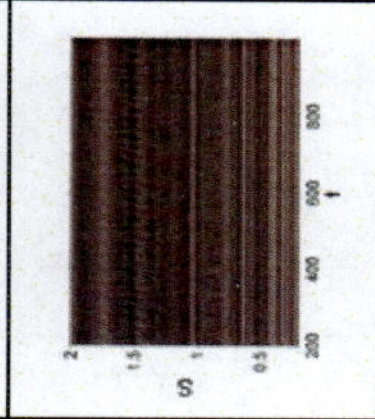	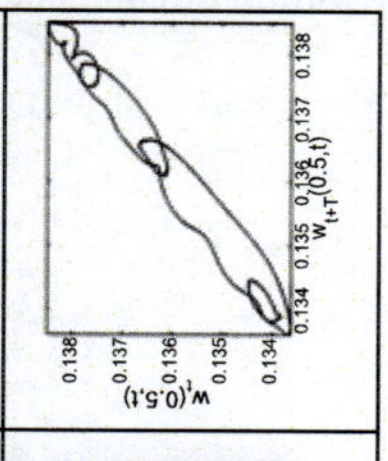

Table 4.8 Comparison of the power spectra $S(\omega)$ for the geometrically linear and nonlinear beams [reprinted with permission from International Journal of Non-Linear Mechanics publishers]

Beam No.		Geometric and design non-linearity	Geometric linearity and design non-linearity
		n =160	n =160
1	After PCA	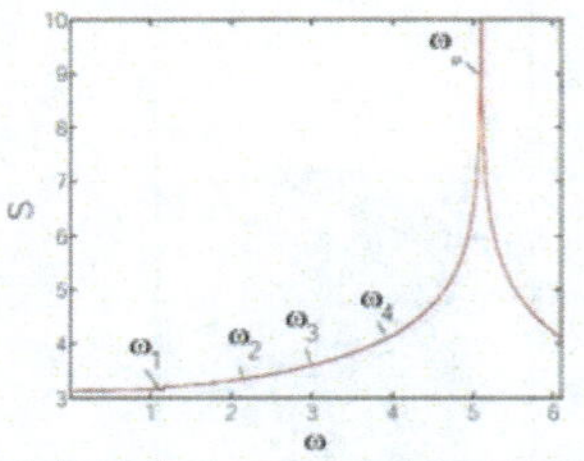	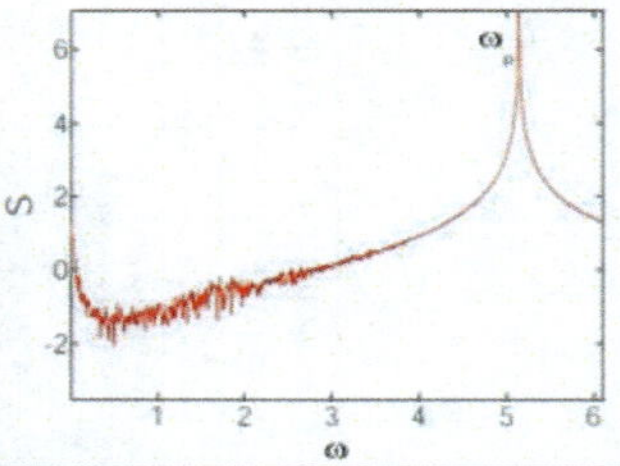
	Before PCA	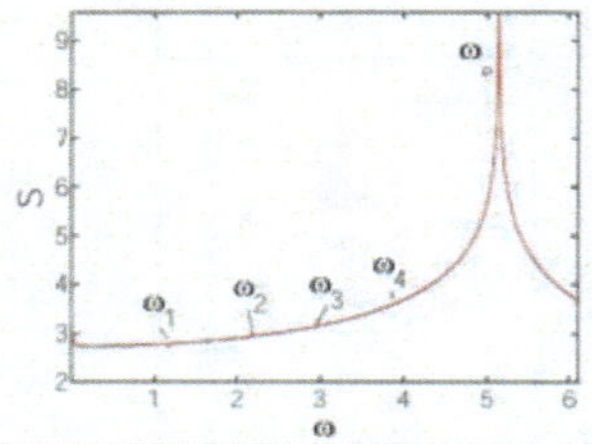	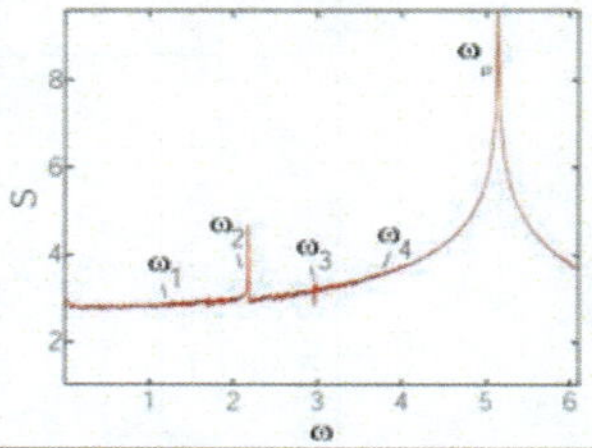
2	After PCA	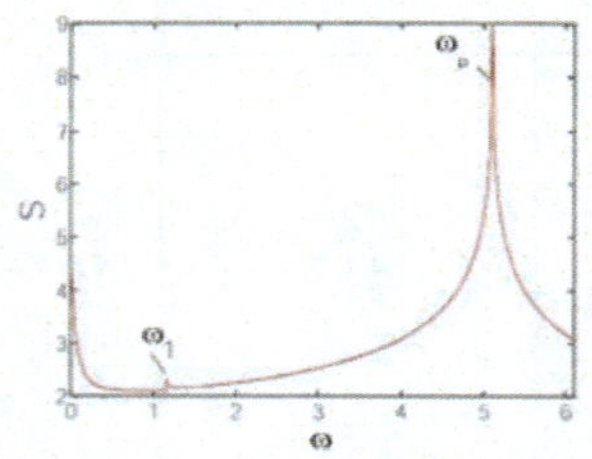	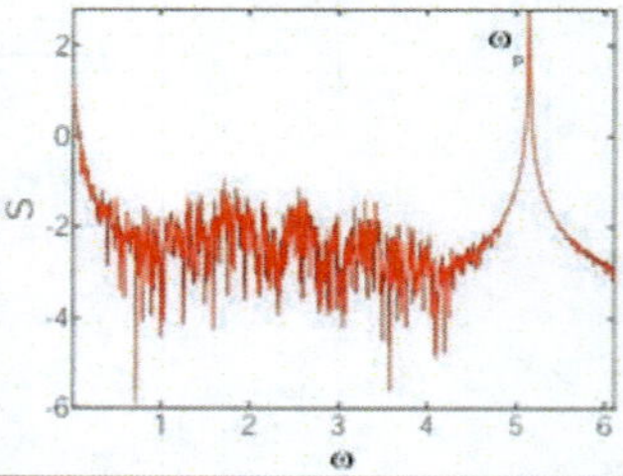
	Before PCA	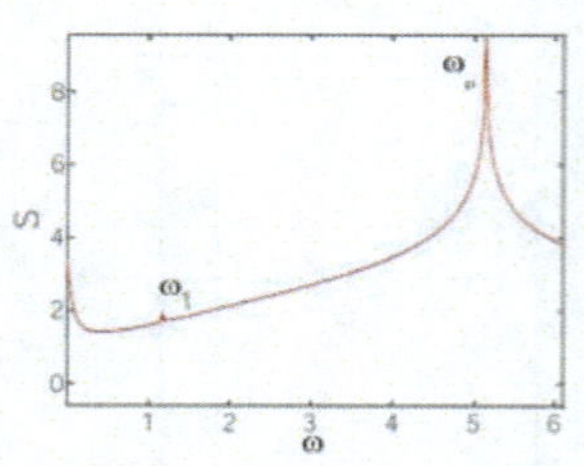	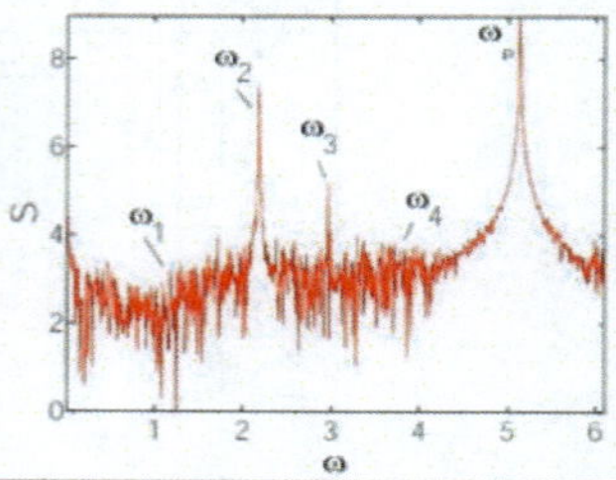

can be used only in the case of small beam deflections ($w \leq 0.25(2h)$). Comparison of the geometrically linear and nonlinear problem, taking into account the beams contact interaction, yields the conclusion that the function of displacements $u(x, t)$ implies the increase in the chaotic components of the signal.

References

1. Krysko, A.V., Awrejcewicz, J., Papkova, I.V., Saltykova, O.A., Krysko, A.V.: On reliability of chaotic dynamics of two Euler-Bernoulli beams with a small clearance. Int. J. Non-Linear. Mech. **104**, 8–18 (2018)
2. Lozi, R.: Can we trust in numerical computations of chaotic solutions of dynamical systems. In: Letellier, Ch., Gilmore, R. (eds.) Topology and Dynamics of Chaos in Celebration of Robert Gilmore's 70th Birthday, pp. 63–98. World Scientific, Singapore (2013)
3. Devaney, R.L.: An Introduction to Chaotic Dynamical Systems. Addison-Wesley, Reading (1989)
4. Banks, J., Brooks, J., Davis, G., Stacey, P.: On Devaney's definition of chaos. Am. Math. Month. **99**(4), 332–334 (1992)
5. Knudsen, C.: Chaos without periodicity. Am. Math. Mon. **101**, 563–565 (1994)
6. Gulick, D.: Encounters with Chaos. McGraw-Hill, New York (1992)
7. Awrejcewicz, J., Krysko, V.A., Papkova, I.V., Krysko, A.V.: Deterministic Chaos in One-Dimensional Continuous Systems. World Scientific, Singapore (2016)
8. Süli, E., Mayers, D.: An Introduction to Numerical Analysis. Cambridge University Press, Cambridge (2003)
9. Fehlberg, E.: Low-order classical Runge-Kutta formulas with step size control and their application to some heat transfer problems. NASA Tech. Rep. **315** (1969)
10. Fehlberg, E.: Klassische Runge-Kutta-Formeln vierter und niedrigerer Ordnung mit Schrittweiten-Kontrolle und ihre Anwendung auf Wärmeleitungsprobleme. Computing **6**(1–2), 61–71 (1970)
11. Cash, J.R., Karp, A.H.: A variable order Runge-Kutta method for initial value problems with rapidly varying right-hand sides. ACM Trans. Math. Soft. **16**, 201–222 (1990)
12. Dormand, J.R., Prince, P.J.: A family of embedded Runge-Kutta formulae. J. Comput. Appl. Math. **6**(1), 19–26 (1980)
13. Haar, A.: Zur Theorie der orthogonalen Funktionensysteme. Math. Ann. **69**(3), 331–371 (1910)
14. Meyer, Y.: Ondelettes, fonctions splines et analyses graduees. Lecture Notes for University, Torino (1986)
15. Daubechies, I., Grossmann, A., Meyer, Y.: Painless nonorthogonal expansions. J. Math. Phys. **27**, 1271–1283 (1986)
16. Grossman, A., Morlet, S.: Decomposition of Hardy functions into square separable wavelets of constant shape. SIAM J. Math. Anal. **15**(4), 723–736 (1984)
17. Meyer, Y.: Ondelettes et functions splines. Ecole Polytech. Paris, Tech. Rep. Semin. EDP (1986)
18. Meyer, Y.: Wavelets: Algorithms and Applications. SIAM, Philadelphia (1993)
19. Kantz, H.: A robust method to estimate the maximal Lyapunov exponent of a time series. Phys. Lett. A **185**, 77–87 (1994)
20. Wolf, A., Swift, J.B., Swinney, H.L., Vastano, J.A.: Determining Lyapunov exponents from a time series. Phys. D **16**, 285–317 (1985)
21. Rosenstein, M.T., Collins, J.J., De Luca, C.J.: A practical method for calculating largest Lyapunov exponents from small data sets. Phys. D **65**, 117–134 (1993)
22. Kantor, B.Y., Bohatyrenko, T.L.: The method for solving contact problems in the nonlinear theory of shells. Rep. Acad. Sci. Ukr. SSR **A**(1), 18–21 (1986)

23. Euler, L.: Sur la force des colones. Memories de L'Academie de Berlin **13**, 252–282 (1757)
24. Krysko, A.V., Awrejcewicz, J., Kutepov, I.E., Zagniboroda, N.A., Dobriyan, V., Krysko, V.A.: Chaotic dynamics of flexible Euler-Bernoulli beams. Chaos **23**(4) 043130-1 – 043130-25 (2013)
25. Krysko, A.V., Awrejcewicz, J., Saltykova, O.A., Zhigalov, M.V., Krysko, V.A.: Investigations of chaotic dynamics of multi-layer beams using taking into account rotational inertial effects. Commun. Nonlinear. Sci. Num. Simul. **19**(8), 2568–2589 (2014)
26. Alijani, F., Amabili, M.: Non-linear vibrations of shells: a literature review from 2003 to 2013. Int. J. Non-Linear. Mech. **58**(2014), 233–257 (2003)
27. Sedighi, H.M., Shirazi, K.H., Zare, J.: Novel equivalent function for deadzone nonlinearity: applied to analytical solution of beam vibration using He's parameter expanding method. Lat. Am. J. Sol. Struct. **9**(4), 443–452 (2012)
28. Kármán, T.: Festigkeitsprobleme in Maschinenbau. Encykle D Math. Wiss. **4**(4), 311–385 (1910)
29. Sylvester, J.J.: On the reduction of a bilinear quantic of the nth order to the form of a sum of n products by a double orthogonal substitution. Mess. Math. **19**, 42–46 (1889)
30. Liang, Y.C., Lee, H.P., Lim, S.P., Lin, W.Z., Lee, K.H., Wu, C.G.: Proper orthogonal decomposition and its applications. Part I: Theory. J. Sound. Vib. **252**, 527–544 (2002)
31. Hotelling, H.: Analysis of a complex of statistical variables into principal components. J. Educ. Phychol. **24**, 417–441 (1933)
32. Krysko, A.V., Awrejcewicz, J., Papkova, I.V., Szymanowska, O., Krysko, V.A.: Principal component analysis in the nonlinear dynamics of beams: purification of the signal from noise induced by the nonlinearity of beam vibrations. Adv. Math. Phys. **2017**, 1–9 (2017)

Chapter 5
Analysis of Simple Nonlinear Dynamical Systems

5.1 Introduction

Analysis of nonlinear oscillations of classical systems is carried out. It is devoted to feasible methods for the computation of Lyapunov exponents since there is no universal, verified and general method to compute their exact (in numerical sense) values. This observation leads to the conclusion that there is a need to employ qualitatively different methods while checking the reliability of "true chaotic results". Furthermore, the analysis carried out in this chapter is a helpful tool for studying systems with infinite dimensions. We show that the most perspective and useful is the modified method of neural networks. It gives excellent convergence to the original results and, as the only one (besides Benettin method), allows us to compute the spectrum of all Lyapunov exponents. In addition, very good results were obtained by Rosenstein method for all studied systems. However, the latter approach can be used to estimate only the largest Lyapunov exponents.

5.2 Gauss Wavelets [1]

In some of the engineering problems, Fourier analysis is insufficient, since it deals with the averaged spectrum of the whole studied vibration signal and presents only a general picture of the signal. On the contrary, wavelets play the role of a microscope, which allows one to observe the spectrum at each time instant, and hence to detect a birth/death of the frequencies in time.

A wavelet transform of a 1D signal consists of its development with respect to a basis being usually a soliton-like function with given properties. The basis is obtained by displacement and tension/compression of a function called a wavelet.

In the present chapter, Gauss wavelets, defined as derivatives of Gauss function, are used. Higher order derivatives have many zero moments, and hence they allow

J. Awrejcewicz et al., *Mathematical Modelling and Numerical Analysis of Size-Dependent Structural Members in Temperature Fields*, Advanced Structured Materials 142,
https://doi.org/10.1007/978-3-030-55993-9_5

one to obtain information about higher order features hidden in the investigated signal.

The eighth-order Gauss wavelets are defined in the following form:

$$g_8(x) = -(105 - 420x^2 + 210x^4 - 28x^6 + x^8)\exp^{\frac{-x^2}{2}}. \tag{5.1}$$

5.3 Logistic Map [2]

In this section, we study simple classical systems (Tables 5.1, 5.4, 5.7, 5.10, 5.13) with emphasis put on a comparison of the LEs (Tables 5.2, 5.5, 5.8, 5.11, 5.14) obtained using Wolf, Rosenstein, Kantz and neural network methods. The convergence of the mentioned methods, depending on the number of iteration steps, is illustrated and discussed (Tables 5.3, 5.6, 5.9, 5.12, 5.15).

A logistic map describes how the population changes with respect to time:

$$X_{n+1} = RX_n(1 - X_n). \tag{5.2}$$

Here, X_n takes the values from 0 to 1 and presents the population in the n-th year, whereas X_0 denotes the initial population (in the year 0); R is a positive parameter characterizing an increase in the population (computations were carried out for $R = 4$).

The first Lyapunov exponent and Kaplan-Yorke dimension were estimated by Sprott [3]. He obtained: $\lambda_1 = 0.693147181$, and Kaplan-Yorke dimension: 1.0.

Tables 5.1, 5.4, 5.7, 5.10, 5.13 report the following results: (a) signal; (b) signal window; (c) Poincaré pseudo-map; (d) Fourier power spectrum; (e) Gauss 8 wavelet; (f) bifurcation diagram with LLE; (g) graphs of LEs on the control parameters plane.

The power spectrum is noisy and it is not possible to distinguish the dominating frequency. A similar situation is exhibited by Gauss wavelet, where a large set of frequencies is visible. They are varied with respect to power, the whole interval of the signal changes and the estimated LLEs correlate with the bifurcation diagram for the same interval of the control parameter r.

As can be seen in Table 5.2, all computational methods were compared with Benettin's original results. A good coincidence was exhibited by the neural network method, Rosenstein method and the method of synchronization. Kantz/Wolf method gave decreased/increased value of LLE in comparison to the original value.

5.4 Hénon Map [4]

Hénon map takes a point (X_n, Y_n) and maps it into another point by the following formulas:

Table 5.1 Nonlinear characteristics of the oscillation signal: (a) time histories; (b) time window; (c) Poincaré pseudo-map; (d) Fourier frequency spectrum; (e) wavelet spectrum; (f) bifurcation diagram and LLE; (g) no graph of Lyapunov exponents (logistic map)

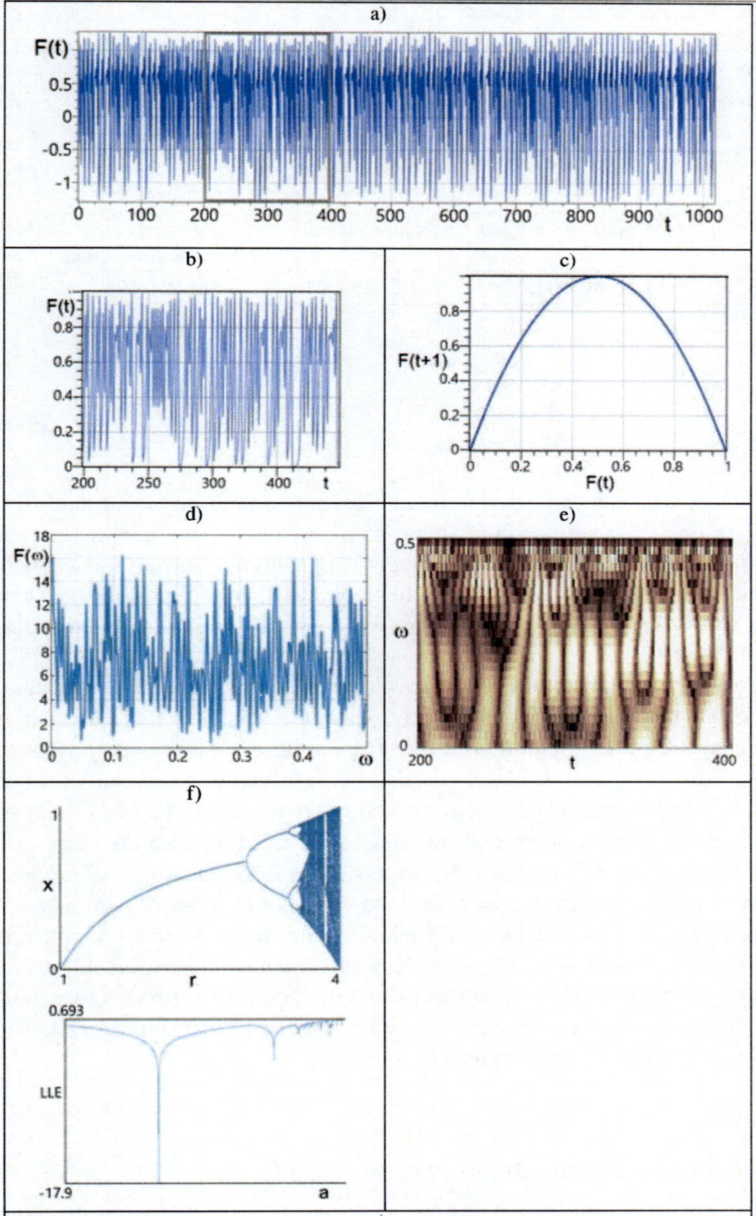

g)

the system consists of one control parameter, and hence the graph of Lyapunov exponents cannot be constructed

Table 5.2 Spectrum of Lyapunov exponents and LLEs computed by different methods (logistic map)

LE spectrum			
Benettin method		Neural network	
(LEs): 0.69315		LEs: 0.69290	
Dimension Kaplan–Yorke (DKY): 1		DKY: 1	
Kolmogorov–Sinai entropy (KSE): 0.69315		EKS: 0.69290	
Phase volume compression (PVC): 0.69315		PVC: 0.69290	
LLE			
Wolf method	Rosenstein method	Kantz method	Method of synchronization
LLE: 0.99683	LLE: 0.690553	LLE: 0.31321	LLE: 0.696

$$
\begin{aligned}
X_{n+1} &= 1 - aX_n^2 + Y_n, \\
Y_{n+1} &= bX_n.
\end{aligned}
\tag{5.3}
$$

The following parameters are fixed for numerical experiments: $a = 1.4$, $b = 0.3$. Since the equations (5.3) do not correspond to a real object, the parameters are replaced with fixed values. Sprott et al. [5] computed Lyapunov spectrum and Kaplan-Yorke dimension of the map using Benettin method [6] by solving (5.3). They obtained the following LEs: $\lambda_1 = 0.419217$, $\lambda_2 = -1.623190$ and Kaplan-Yorke dimension: 1.258267.

Similarly to the logistic map, the power spectrum exhibits a uniform noisy shape. However, one can distinguish a dominating frequency ($\omega_1 \approx 0.45$). It is also visible on the wavelet spectrum as a region of the largest amplitudes along with the whole signal. Plots of the change in the LLE correlate with bifurcation diagrams for the same interval of changes in the parameters a and b. Dynamics of the LLE changes increases with the increase in both control parameters. Starting with the graphs of LEs for a given set of control parameters, the system mainly remains in a periodic regime, but it exhibits chaotic dynamics for large values of the control parameters.

Beginning from the results shown in Table 5.5, the majority of the employed computational methods yielded good results. However, the most accurate results were obtained by the neural network method (for the whole spectrum of LEs), Rosenstein method and the method of synchronization (in the case of LLEs). Wolf and Kantz methods gave decreased estimated values of the LLEs.

5.5 Hyperchaotic Generalized Hénon Map [7]

To obtain the hyperchaotic Hénon map, one needs to take a point (X_n, Y_n, Z_n) and map it into the following one:

Table 5.3 Fourier power spectra and Gauss wavelet obtained for $\Delta t = 1, 2$ and the LLEs computed by different methods (logistic map)

$\Delta t = 1$	$\Delta t = 2$
Fourier power spectrum	
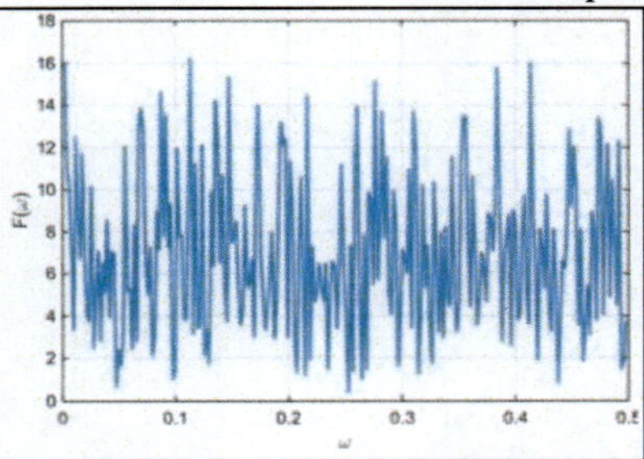	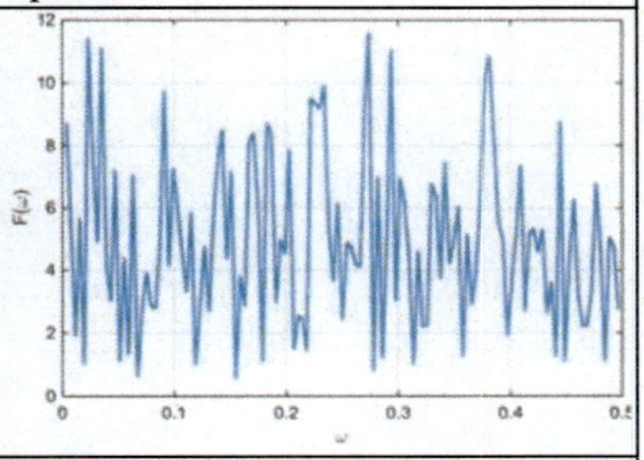
Gauss wavelet	
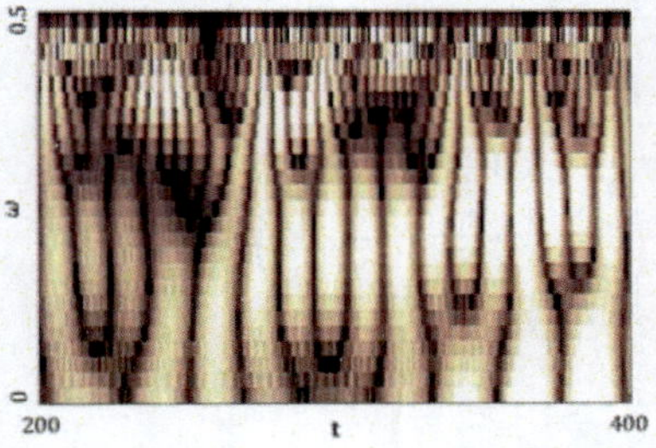	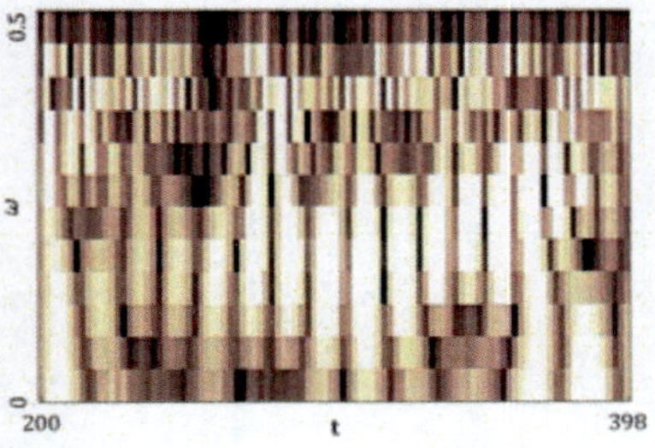
LLE (Wolf)	
0.99961	1.00014
LLE (Rosenstein)	
0.69231	0.69065
LLE (Kantz)	
0.31321	
LLE (Synchronization)	
0.69400	0.69330
LEs (Benettin)	
LES: 0.69318 DKY: 1.00000 KSE: 0.69318 PVC: 0.69318	LES: 0.69400 DKY: 1.00000 KSE: 0.69400 PVC: 0.69400
LEs (neural network)	
LES: 0.69290 DKY: 1 KSE: 0.69290 PVC: 0.69290	LES: 0.69107 DKY: 1.00000 KSE: 0.69107 PVC: 0.69107

Table 5.4 Characteristics of Hénon map: (a) time history; (b) time window; (c) Poincaré pseudo-map; (d) Fourier frequency spectrum; (e) wavelet spectrum; (f) bifurcation diagrams and LLEs; (g) graph of Lyapunov exponents

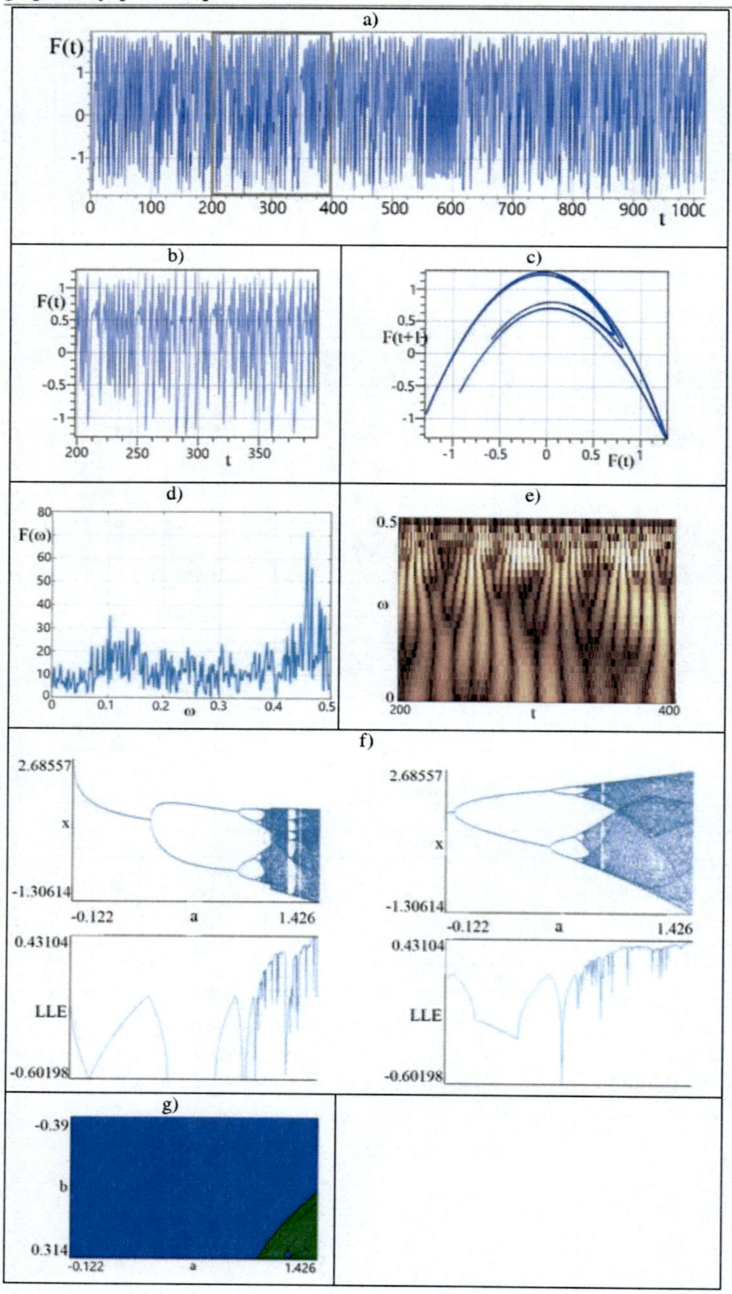

Table 5.5 Lyapunov exponents spectrum and LLEs computed by different methods (Hénon map)

Spectrum of LEs			
Benettin method		Neural network	
LEs: 0.41919–1.62316		LEs: 0.41919–1.62316	
DKY: 1.25826		DKY: 1.25826	
EKS: 0.41919		EKS: 0.41919	
PVC: –1.20397		PVC: –1.20397	
LLEs			
Wolf method	Rosenstein method	Kantz method	Method of synchronization
LLE: 0.38788	LLE: 0.414218	LLE: 0.17759	LLE: 0.40608

$$\begin{aligned} X_{n+1} &= a - aY_n^2 - bZ_n, \\ Y_{n+1} &= X_n, \\ Z_{n+1} &= Y_n. \end{aligned} \tag{5.4}$$

The computations were carried out for the following fixed parameters: $a = 3.4$, $b = 0.1$. Lyapunov spectrum reported in reference [7] is: 0.276; 0.257; 4.040.

One can distinguish a large number of frequencies in the power spectrum. Frequencies with the largest amplitude are located in the interval [0.15; 0.3] (frequencies $\omega_1 - \omega_4$), but the remaining part of the spectrum is noisy. This interval corresponds to the brightest region on Gauss wavelet, which is correlated with the values of the power spectrum. Changes in LLEs coincide with the bifurcation diagrams constructed for the same intervals of changes in the control parameters a and b. Dynamics of LLEs increases with the increase in control parameters. As in the case of Hénon map, the chart of LEs for the selected control parameters exhibits, for a majority of studied parameters, periodic dynamics. It transits into chaos for $a \approx 1.4$, and is almost suddenly shifted into hyper-chaos (2 positive LEs).

Good results are obtained by Benettin, Rosenstein and synchronization methods (divergence from the third decimal place). The neural network yielded slightly increased estimates of two first LEs, whereas the third LE is estimated almost exactly. Kantz method gave a decreased result in comparison to reference data. Wolf method resulted in the largest error.

Table 5.6 Fourier power spectra and Gauss wavelet spectra obtained for $\Delta t = 1, 2$ and the computed LLEs by different methods (Hénon map)

$\Delta t = 1$	$\Delta t = 2$
Fourier power spectrum	
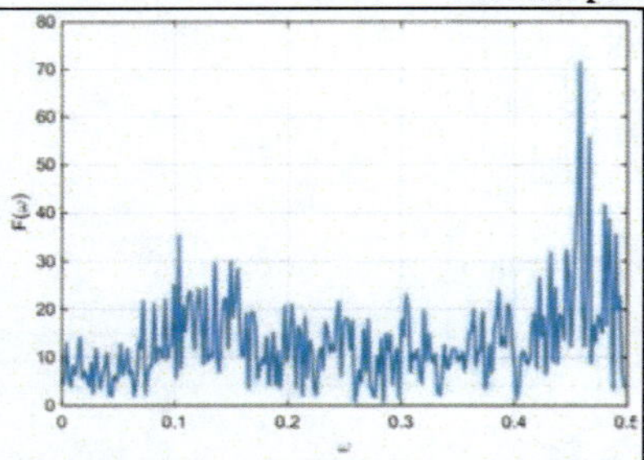	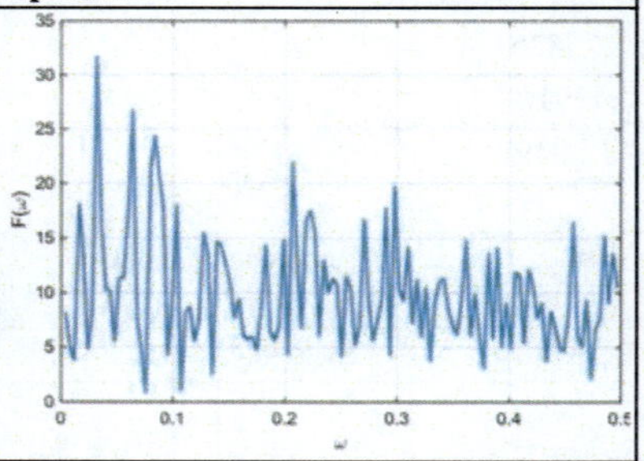
Gauss wavelet	
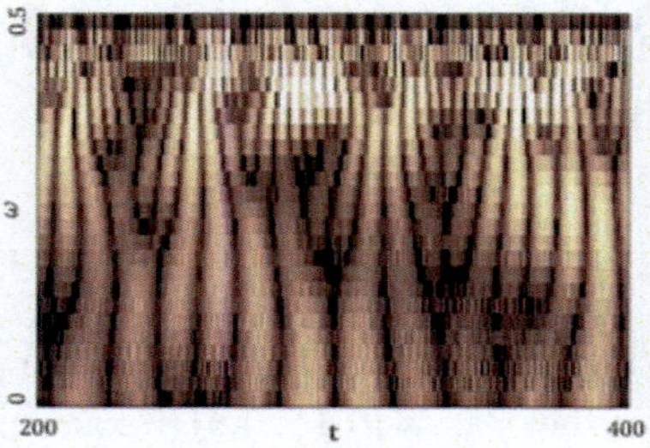	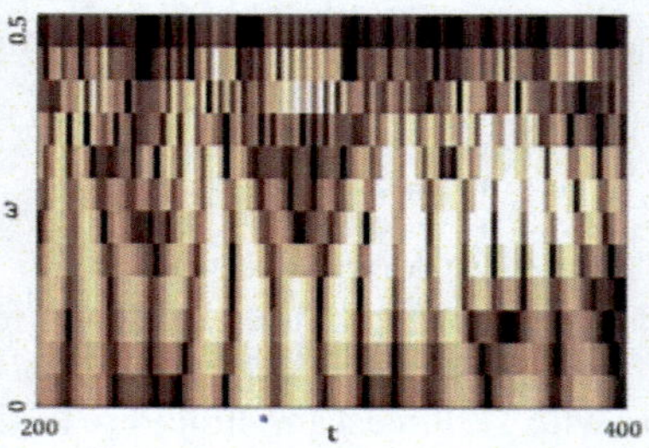
LLE (Wolf)	
0.4158	0.39734
LLE (Rosenstein)	
0.41637	0.400635
LLE (Kantz)	
0.17759	0.105365
LLE (Synchronization)	
0.40608	0.40510
All LEs (Benettin)	
LEs: 0.41919 - 1.62316 DKY: 1.25826 EKs: 0.41919 PVC: -1.20397	LEs: 0.41917 - 1.62315 DKY: 1.25825 EKs: 0.41917 PVC: -1.20397
All LEs (neural network)	
LEs: 0.41919 - 1.62316 DKY: 1.25826 KSE: 0.41919 PVC: -1.20397	LES: 0.40924 - 1.61321 DKY: 1.25368 KSE: 0.40924 PVC: -1.20397

Table 5.7 Signal characteristics: (a) time history; (b) time window; (c) Poincaré pseudo-map; (d) Fourier frequency spectrum; (e) wavelet spectrum; (f) bifurcation diagram and LLE; (g) graph of Lyapunov exponents (generalized Hénon map)

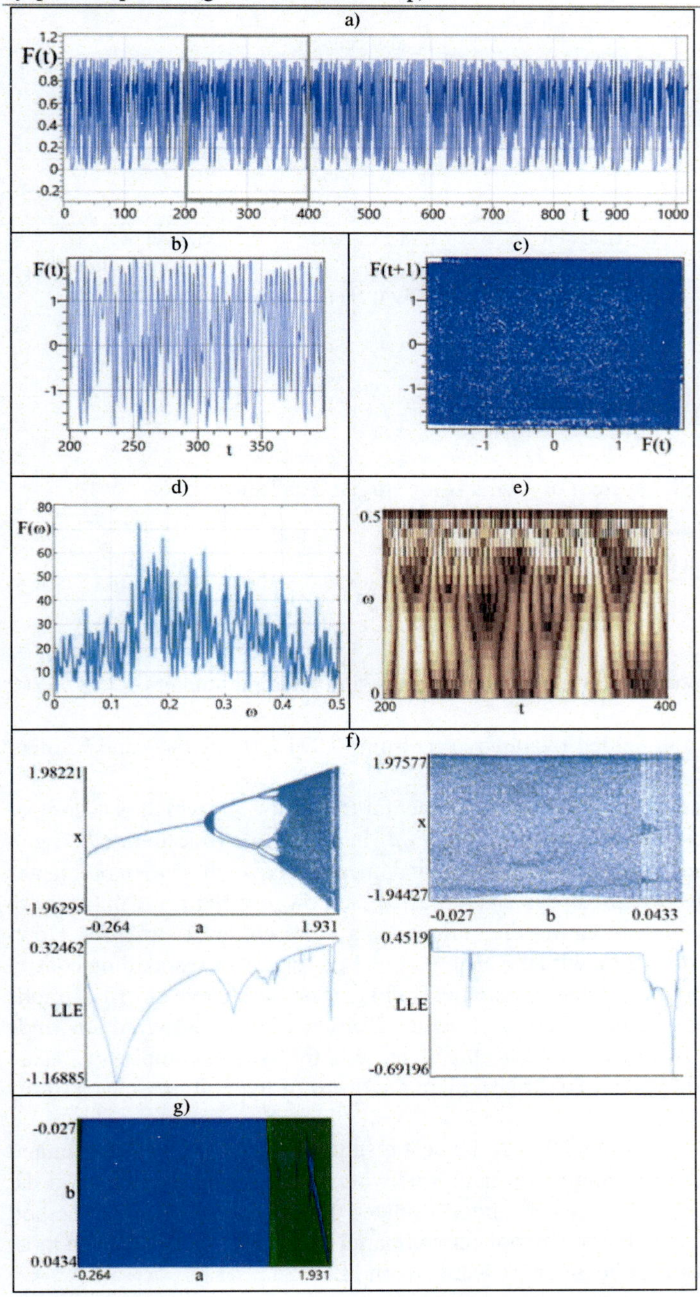

Table 5.8 Lyapunov exponents spectrum and LLEs computed by different methods (generalized Hénon map)

Spectrum of LEs			
Benettin method		Neural network	
LEs: 0.27628 0.25770–4.04053		LEs: 0.29251 0.27104–4.04583	
DKY: 2.13215		DKY: 2.13929	
EKS: 0.53397		EKS: 0.56355	
PVC: –3.50656		PVC: –3.48227	
LLEs			
Wolf method	Rosenstein method	Kantz method	Method of synchronization
LLE: 0.45214	LLE: 0.27930	LLE: 0.26601	LLE: 0.27250

5.6 Rössler Attractor [8]

The following Rössler system of ODEs is investigated as follows:

$$\begin{cases} \dot{x} = -y - z, \\ \dot{y} = x + ay, \\ \dot{z} = b + z(x - c), \end{cases} \tag{5.5}$$

and the computations are carried out for the following fixed parameters $a = b = 0.2$ and $c = 5.7$.

The original study yielded Lyapunov spectrum: 0.0714, 0, –5:3943 and Kaplan-Yorke dimension equals to 2.0132.

The power spectrum contains the fundamental frequency ω_1, which is accompanied by damped bursts (frequencies $\omega_2 - \omega_{10}$). In the whole time interval, Gauss wavelet exhibits the brightest region of the fundamental frequency with darker peaks going to zero. Thus, the picture is analogous to the power spectrum. Contrarily to the studied maps, the bifurcation diagrams have a more complex structure. However, there is still correlation with the changes in LLEs for the corresponding control parameters. The parameter b has the smallest influence on the change in LLE. Graphs of LLEs also exhibit a more complex structure. Borders of different vibration kinds have complex forms, which illustrates the increase in the system complexity. Aside from the chaos and hyper-chaos zones, there are drops of hyper hyper-chaos (three positive LEs).

As far as Table 5.11 is considered, the best results were yielded by Benettin and Rosenstein methods. The method of neural networks gave very good results in the case of estimates of two first LEs but underestimated the third exponent. Wolf method yielded smaller value of the first exponent compared to the reference data. The most underestimated results were given by Kantz method.

Table 5.9 Fourier power spectra and Gauss wavelet spectra obtained for $\Delta t = 1, 2$ and the computed LLEs by different methods (generalized Hénon map)

$\Delta t = 1$	$\Delta t = 2$
Fourier power spectrum	
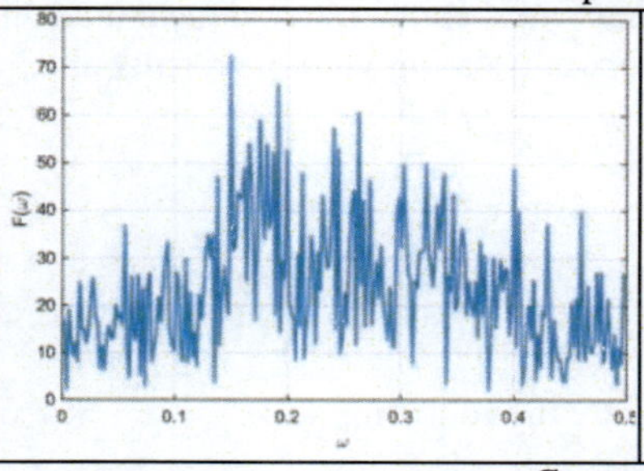	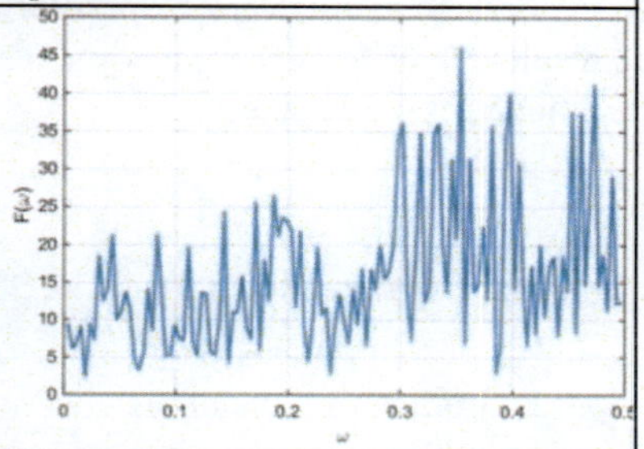
Gauss wavelet	
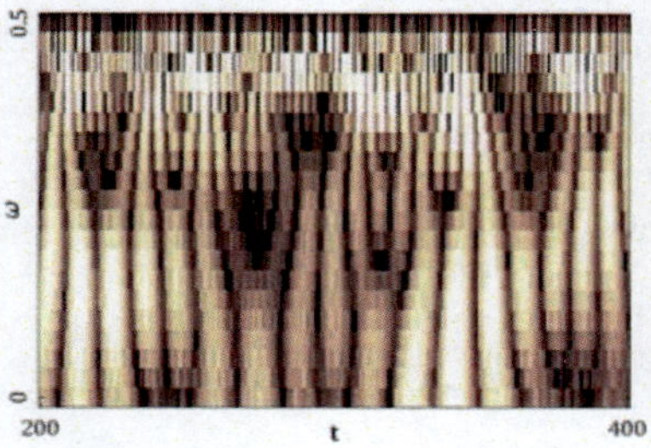	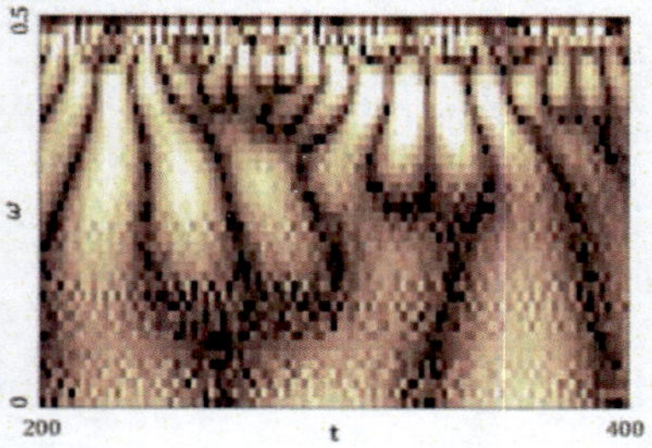
LLE (Wolf)	
0.45214	0.46706
LLE (Rosenstein)	
0.27930	0.27459 (0.62515)
LLE (Kantz)	
0.26601	
LLE (Synchronization)	
0.27250	0.27200
All LEs (Benettin)	
LEs: 0.27628 0.25770 - 4.04053 DKY: 2.13215 KSE: 0.53397 PVC: -3.50656	LEs: 0.27487 0.25631 - 4.03774 DKY: 2.13155 EKS: 0.53118 PVC: -3.50656
All LEs (neural network)	
LEs: 0.29251 0.27104 - 4.04583 DKY: 2.13929 KSE: 0.56355 PVC: -3.48227	LES: 0.26304 0.24387 - 4.14321 DKY: 2.12235 KSE: 0.50691 PVC: -3.63630

The carried out numerical experiments showed that using the different sampling frequency, the power spectrum and wavelet spectrum were not changed. This was also validated by results obtained by Benettin, neural networks and Rosenstein methods which yielded the results very close to original ones. Kantz method gave underestimated results for different frequency selection, correlating with the results obtained for the standard sample size.

5.7 Lorenz Attractor [9]

The input hydrodynamic system is governed by the following ODEs:

$$\begin{cases} \dot{x} = \sigma(y - x), \\ \dot{y} = x(r - z) - y, \\ \dot{z} = xy - bz, \end{cases} \tag{5.6}$$

where r stands for the normalized Rayleigh number (non-dimensional number defining fluid behaviour under gradient):

$$r = \frac{g\beta\Delta T L^3}{\nu\chi}. \tag{5.7}$$

In the above equation, the following notation is used: g—gravity of Earth; L—characteristic dimension of the fluid space; ΔT—temperature difference between fluid walls; ν—kinematic fluid viscosity, χ—thermal conductivity of the fluid; β—coefficient of heat fluid extension; σ—Prandtl number (takes into account heat source property) governed by the following equation

$$\sigma = \frac{\nu}{\alpha} = \frac{\eta C_p}{\aleph}, \tag{5.8}$$

where $\nu = \eta/\rho$—kinematic viscosity, η—dynamic viscosity, ρ—density, $\alpha = \frac{\aleph}{\rho C_p}$—temperature transfer coefficient, $\aleph$—heat transfer coefficient, C_p—specific heat capacity under constant pressure; and ρ—information about the geometry of the convective cell.

Moreover, for fixed: $\sigma = 10.0$, $r = 28.0$, $b = 8/3$. The original results follow: LEs: 0.9056, 0, -14.5723; the Kaplan-York dimension: 2.06215.

The power spectrum of the attractor uniformly decreases when approaching a finite frequency, and there is a lack of frequencies with a strongly dominating amplitude. The latter observation is also verified by Gauss wavelet spectrum. The bifurcation diagrams, similar to those for Rössler system, exhibit a complex structure, but the correlation to the LLEs change is conserved. The richest/lowest dynamics of LLE is obtained for changing parameter r/σ. Based on the reported graphs of LEs, one can

Table 5.10 Signal characteristics: (a) time history; (b) time window; (c) Poincaré pseudo-map; (d) Fourier frequency spectrum; (e) wavelet spectrum; (f) bifurcation diagrams and LLEs; (g) graphs of Lyapunov exponents (Rössler attractor)

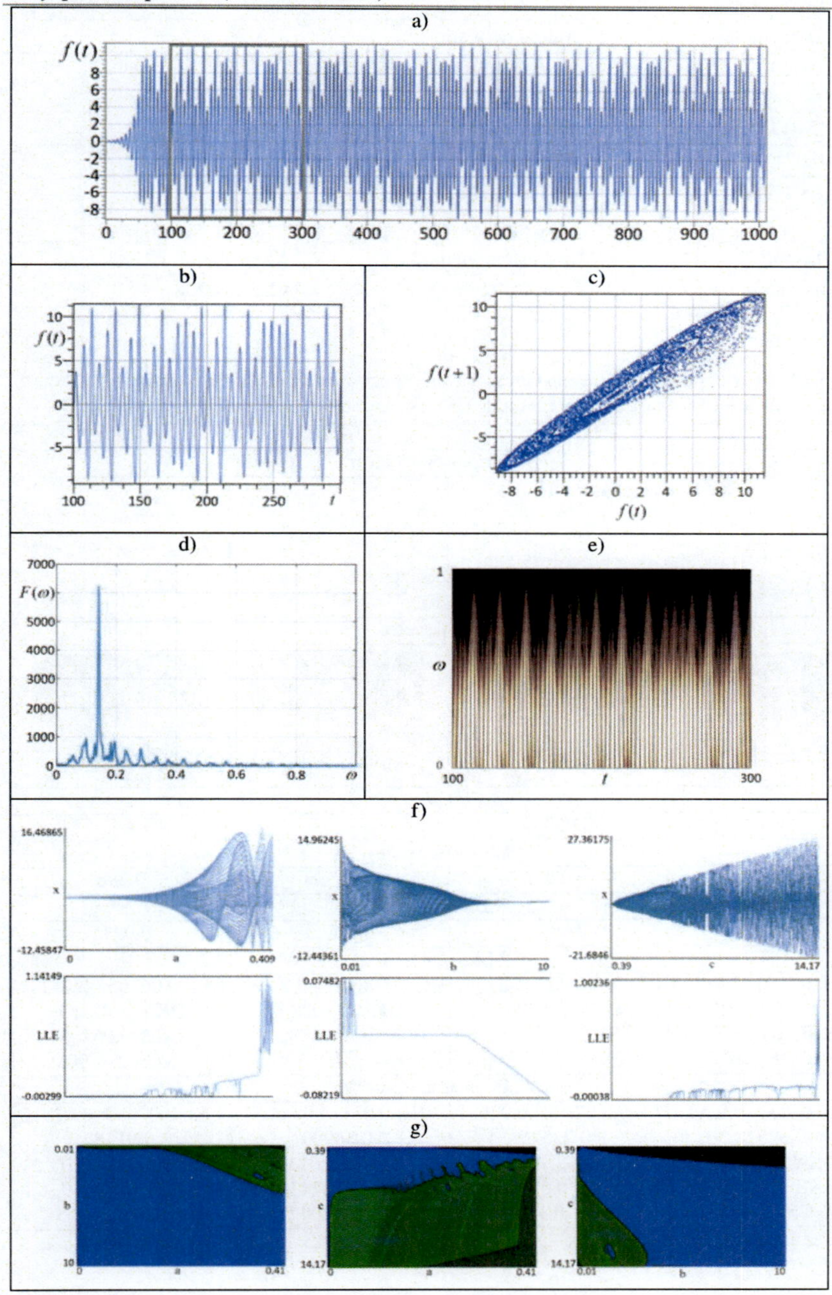

Table 5.11 Lyapunov exponents spectrum and LLEs computed by different methods (Rössler attractor)

Spectrum of LEs		
Benettin method	Neural network	
LEs: 0.07135 0.00000–5.39420	LEs: 0.07593 -0.00060–0.78178	
DKY: 2.01323	DKY: 2.09635	
EKS: 0.07135	EKS: 0.07593	
PVC: –5.32285	PVC: –0.70646	
LLEs		
Wolf method	Rosenstein method	Kantz method
LLE: 0.05855	LLE: 0.0726	LLE: 0.0208

Table 5.12 Fourier power spectra and Gauss wavelet spectra obtained for $\Delta t = 0.05, 0.1, 0.15, 0.2$ and the computed LLEs by different methods (Rössler attractor)

$\Delta t = 0.05$	$\Delta t = 0.1$	$\Delta t = 0.15$	$\Delta t = 0.2$
Fourier power spectrum			
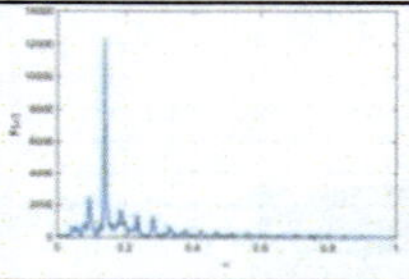	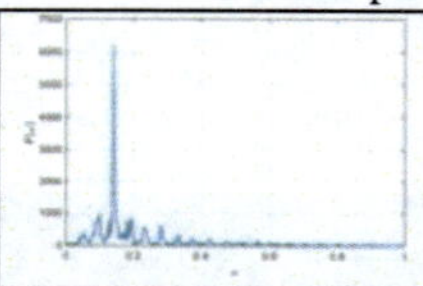	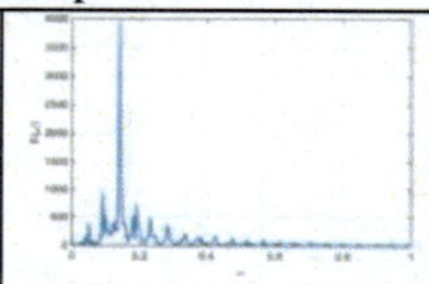	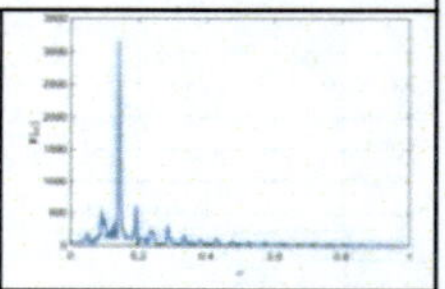
Gauss wavelet			
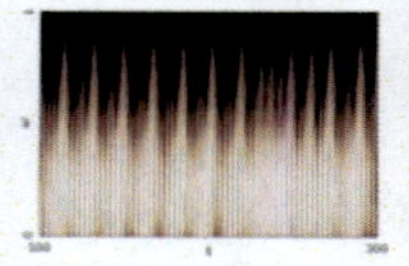	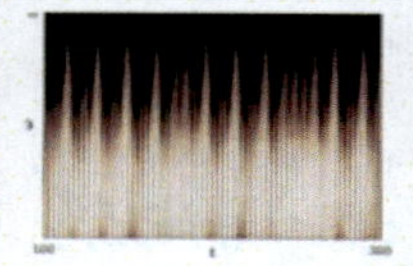	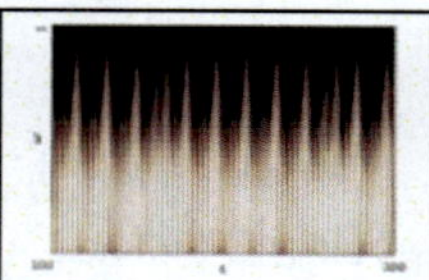	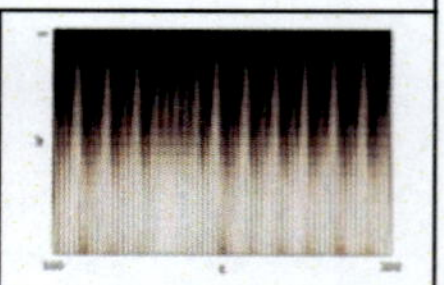
LLE (Wolf)			
	0.05855		
LLE (Rosenstein)			
0.083	0.0726	0.06553	0.606
LLE (Kantz)			
0.0234	0.0208	0.02133	0.0215
All LEs (Benettin)			
LEs: 0.07156 -5.38768 DKY: 2.01328 KSE: 0.07156 PVC: -5.31612	LEs: 0.06959 - 5.21949 DKY: 2.01333 EKS: 0.06959 PVC: -5.14990	LEs: 0.06789 - 4.34385 DKY: 2.01563 KSE: 0.06789 PVC: -4.27596	LEs: 0.06205 - 2.84111 DKY: 2.02184 KSE: 0.06205 PVC: -2.77906
All LEs (neural network)			
LEs: 0.06259 -0.07984 -0.32528 DKY: 1.78396 KSE: 0.06259 PVC: -0.34253	LEs: 0.07340 -0.02681 -0.23525 DKY: 2.19807 KSE: 0.07340 PVC: -0.18865	LEs: 0.07374 0.00057 -0.36909 DKY: 2.20135 KSE: 0.07432 PVC:-0.29477	LEs: 0.07983 -0.02816 -0.91182 DKY: 2.05667 KSE: 0.07983 PVC: -0.86015

Table 5.13 Signal characteristics: (a) time history; (b) time window; (c) Poincaré pseudo-map; (d) Fourier frequency spectrum; (e) wavelet spectrum; (f) bifurcation diagrams and LLEs; (g) graphs of Lyapunov exponents (Lorenz attractor)

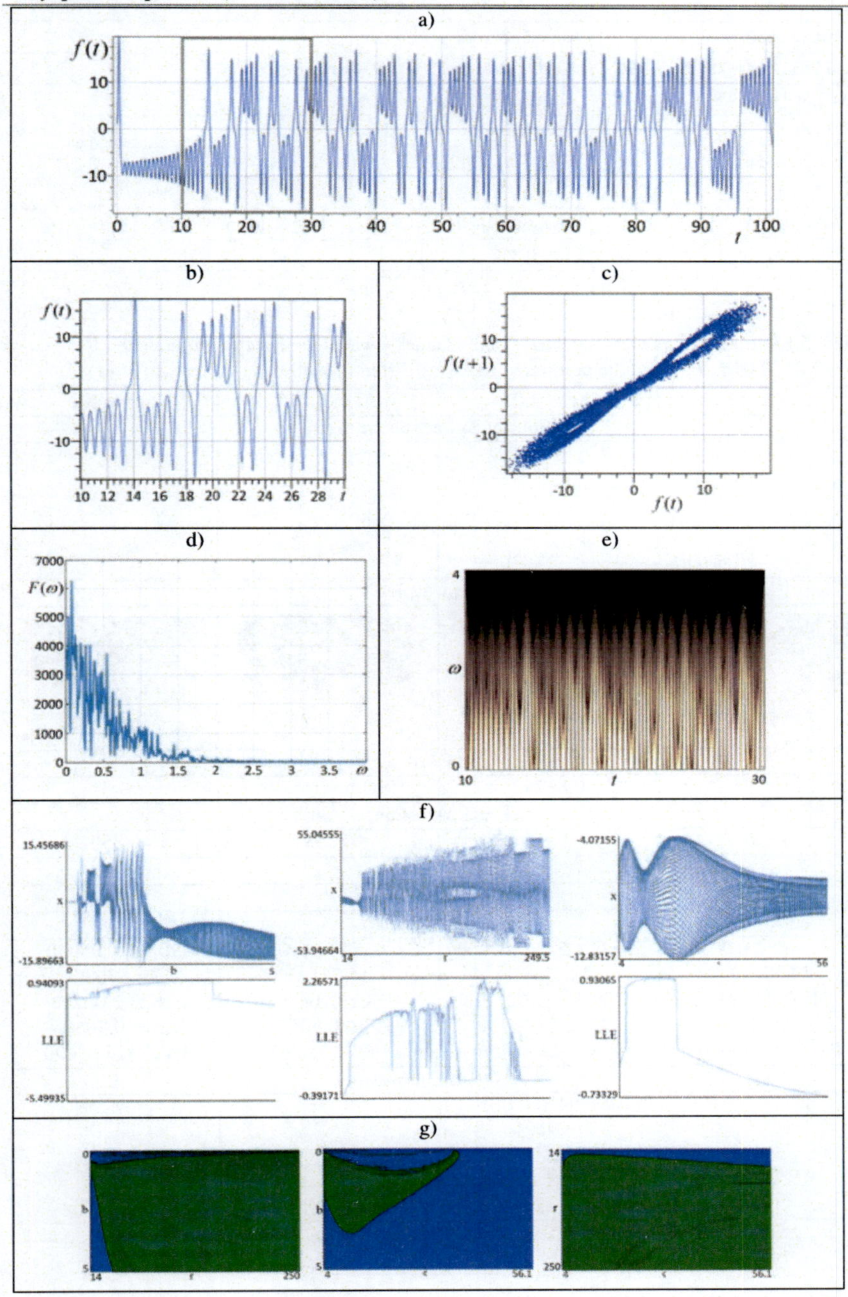

Table 5.14 Lyapunov exponents spectrum and LLEs computed by different methods (Lorenz attractor)

Spectrum of LEs		
Benettin method	Neural network	
LEs: 0.90557 0.00000–14.57214	LEs: 0.9490 0.0610–13.9101	
DKY: 2.06214	DKY: 2.07261	
EKS: 0.90557	EKS: 1.0101	
PVC: –13.66656	PVC: –12.9000	
LLEs		
Wolf method	Rosenstein method	Kantz method
LLE: 0.81704	LLE: 0.836	LLE: 0.807185

Table 5.15 Fourier power spectra and Gauss wavelet spectra obtained for $\Delta t = 0.005, 0.01, 0.015, 0.02$ and the computed LLEs by different methods (Lorenz attractor)

$\Delta t = 0.005$	$\Delta t = 0.01$	$\Delta t = 0.015$	$\Delta t = 0.02$
Fourier power spectrum			
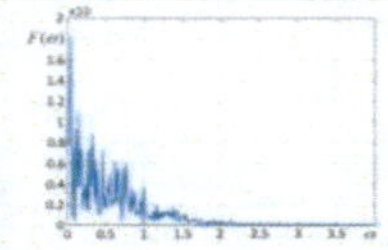	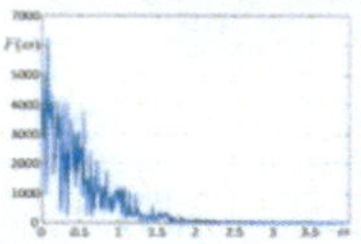	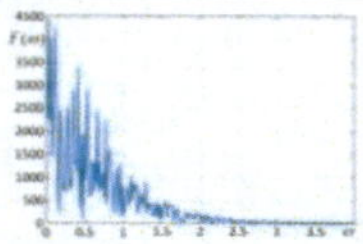	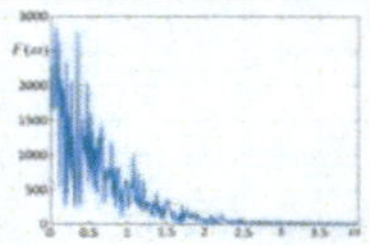
Gauss wavelet			
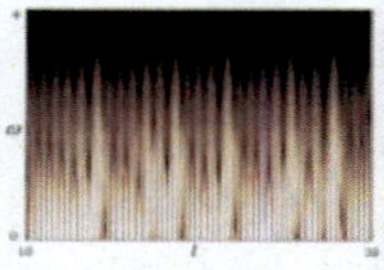	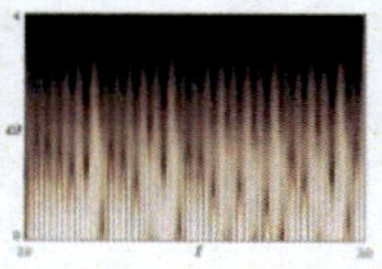	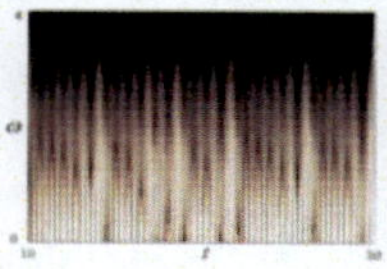	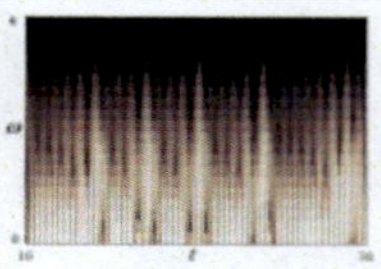
LLE (Wolf)			
	0.81704		
LLE (Rosenstein)			
0.876	0.836	0.858	0.859
LLE (Kantz)			
All LEs (Benettin)			
LEs: 0.90632 0.00000 -14.57297 DKY: 2.06219 KSE: 0.90632 PVC: -13.66666	LEs: 0.90523 0.00000 -14.57179 DKY: 2.06212 KSE: 0.90523 PVC: -13.66656	LEs: 0.90551 0.00000 -14.57163 DKY: 2.06214 KSE: 0.90551 PVC: -13.66613	0.90596 0.00000 -14.57086 DKY: 2.06218 KSE: 0.90596 PVC: -13.66490
LEs (neural network)			
	LEs: 0.9490 0.0610 -13.9101 DKY: 2.07261 KSE: 1.0101 PVC: -12.9000		

conclude that the system dynamics is fully chaotic. There are also narrow windows of hyperchaotic dynamics.

A comparison of the results reported in Table 5.14 with the original results exhibit an excellent coincidence of Benettin method (original results) and the neural network method (+4.79%). Wolf and Rosenstein methods yielded the underestimated results of the LLE value. The worst estimation is obtained by Kantz method.

Employing different sampling frequency does not change a picture of Fourier and wavelet power spectra. This is also validated by Benettin and Rosenstein methods, which yield the results very close to the original values in spite of the arbitrary choice of the sampling frequency.

References

1. Astafeva, N.M.: Wavelet-analysis: basic theory and examples of applications. Succ. Phys. Sci. **166**(11), 1145–1170 (1996)
2. May, R.: Simple mathematical model with very complicated dynamics. Nature **261**, 45–67 (1976)
3. Sprott, J.C.: Elegant Chaos. Algebraically Simple Chaotic Flows. World Scientific, Singapore (2010)
4. Hénon, M.: A two-dimensional mapping with a strange attractor. Commun. Math. Phys. **50**(1), 69–77 (1976)
5. Sprott, J.C.: Chaos and Time Series Analysis. Oxford University Press, Oxford (2003)
6. Benettin, G., Galgani, L., Strelcyn, J.M.: Kolmogorov entropy and numerical experiments. Phys. Rev. A **14**, 2338–2345 (1976)
7. Baier, G., Klein, M.: Maximum hyperchaos in generalized Henon maps. Phys. Lett. A **151**(6–7), 281–284 (1990)
8. Peitgen, H.-O., Jürgens, H., Saupe, D.: The Rössler attractor. In: Chaos and Fractals: New Frontiers of Science, pp. 636–646. Springer, Berlin (2004)
9. Lorenz, E.N.: Deterministic nonperiodic flow. J. Atm. Sci. **20**(2), 130–141 (1963)

Chapter 6
Mathematical Models of Micro- and Nano-cylindrical Panels in Temperature Field

6.1 Introduction

Mathematical models of nonlinear micro- and nano-cylindrical panels in temperature fields are introduced and studied. First, the application of the modified couple stress theory of thermoelastic curvilinear panels based on the third-order hypotheses has been described. Then, a technical theory of the Sheremetev–Pelekh, Timoshenko and Bernoulli–Euler models is presented. The method of solving static problems is outlined in Sect. 6.4. Chaotic dynamics of the size-dependent flexible Bernoulli–Euler, Timoshenko and Sheremetev–Pelekh beams based on the modified couple stress theory of elasticity is investigated in Sect. 6.5. The two last sections are devoted to the construction of so-called charts of vibration character with regard to amplitude and frequency of the external excitation and their study with the use of the first, second and third kinematic hypotheses.

6.2 Literature Review

Occurrence of new polymer composite material (reinforced plastics, etc.) in engineering practice essentially increased motivation for the development and generalization of the so far used classical theories of beams, plates and shells. Construction of the theories focused on reliable computations of the structural members made from the new materials with an account of specific features in their behaviour, grading of material in one or two directions as well as the size-dependent behaviour under dimensions of micro- and nano-meter orders requires novel challenging modelling and mathematical/numerical approaches.

One of the ways to improve the accuracy of classical theory of shells is associated with the employment of higher order models. Two Ukrainian scientists Sheremetev

J. Awrejcewicz et al., *Mathematical Modelling and Numerical Analysis of Size-Dependent Structural Members in Temperature Fields*, Advanced Structured Materials 142,
https://doi.org/10.1007/978-3-030-55993-9_6

and Pelekh [1] proposed in 1964 development of the Timoshenko theory aimed at account not only of rotation of beam middle curve but also its curving. The transversal tangential stresses are equal to zero on the low and upper panel surfaces. The latter model can be recognized as third-order model. In the series of publications [2, 3], the latter theory was used to compute beams, plates and shells. After 27 years, the problem was revisited by Levinson [4] and Reddy [5]. This is why in this book, and beyond, we refer to the name as the Sheremetev–Pelekh (SP) model.

Micro- and nano-sized beams, plates and shells are widely employed in the micro- and nano-electromechanical systems such as the vibration sensors [6], micro-conductors [7] and micro-switches [8].

The dependence of the elastic behaviour on the body size in the microscale has been observed experimentally in metals [9, 10], alloys [11], polymers [12] and crystals [13]. In spite of the numerous works on the mentioned topic, in general, the carried out numerical analysis is based on the linear models though the experimental results point out a need of account of nonlinear behaviour of micro- and nano-mechanical systems [14].

The classical mechanics of a rigid body does not allow for a proper interpretation and forecast of the size-dependent behaviour exhibited in structures of micro- and submicroscales due to lack of a parameter responsible for the scale effects. In the recent years, a big effort has been observed focused on developing various theories for modelling scale effects occurred in continuum including coupled stress theory of elasticity [15, 16], nonlocal theory of elasticity [17], gradient theory of elasticity [18] and surface elasticity [19].

We focus on the works devoted investigations of the problems of theory of elasticity based on the couple stress-based gradient theory. The fundamental theoretical background of the couple stress theory was presented by Yang [20]. He introduced, in spite of two classical material Lamé constants, an additional material constant of a higher order. Fleck and Hutchinson [21] employed the modified couple stress theory of elasticity in order to account and explanation of the size-dependent behaviour. In the recent years, the latter theory has been used by numerous researchers for a proper interpretation of the size-dependent dynamic behaviour of microstructures [22–26].

One of the important aspects of the use of the couple stress theory of elasticity relies on its application to the static and dynamic problems of beams. The latter structural members are widely used in fabrication of nano sensor, nano conductors and nano switchers. In order to construct mathematical models of beams, the hypotheses of zero-order approximation (Bernoulli–Euler), first-order approximation (Timoshenko) and third-order approximation (Sheremetev–Pelekh) are used. Each of the mentioned hypotheses can be viewed as an approximation of the beam treated as 2D and 3D body. Recall that the first approximation being based on the Bernoulli–Euler hypotheses does not take into account curvature of a normal to the beam axis.

In references [27, 28] with the help of modified couple stress theory of elasticity, the governing linear equations, the initial and boundary conditions of the size-dependent Euler–Bernoulli model are derived. Influence of the size-dependent length parameter on the static deformation and magnitude of the natural frequencies is investigated. In the case of static problem, the linear fourth-order equation with regard to the longitudinal displacement has been analysed. Investigation of the natural frequencies behaviour for small deflection has been carried out based on the sixth-order linear equation with regard to the deflection. The Bubnov–Galerkin method in the first approximation has been used which allowed to transfer the studied PDE to ODE. Rajabi and Ramezani [29] derived the governing equations for the geometrically nonlinear Euler–Bernoulli beam based on the von Kármán relations. In order to obtain a numerical solution, the Bubnov–Galerkin method in the first-order approximation has been employed. The influence of the length scale parameter on the natural frequency of nonlinear vibrations has been analysed.

Mathematical model of the Timoshenko beam and, on contrary to the mathematical model of the Bernoulli–Euler beam, a rotation of a normal to the beam axis and deformation of the transversal shear effects are taken into account. It stands for the next step of approximation to the real behaviour of a 2D and 3D body. Though the order of the system of differential equations remains unchanged, there is added a function responsible for the normal rotation. Investigation of the size-dependent models of Timoshenko beam is realized via various approaches including the modified couple stress theory of elasticity [30], while Asghari [31] derived the geometrically nonlinear governing PDEs and boundary conditions based on the strain gradient theory.

A numerical study has been carried by considering simple cases without taking into account the longitudinal inertia and with the help of first-order approximation of the Bubnov–Galerkin method. Moreover, a series of simplifications are used to reduce the problem to that of solving a linear ODE to get the vibration frequencies. In addition, nonlinear problems of static deformations have been also considered. Ansan et al. [32] employed strain gradient theory to analyse free vibration of the Timoshenko beam. The derived governing PDEs have been reduced to a system composed of two ODEs through the Bubnov–Galerkin method. Numerical results have been compared with the earlier reported results based on the models of linear deformation gradient, nonlinear and linear modified couple stress theory of elasticity, as well as linear and nonlinear classical models. Deflections obtained based on the mentioned two models differ essentially when the beam is relatively short. This effect is due to the neglection of the shear deformation in the Bernoulli–Euler model.

Ma et al. [26] employed the SP linear model to solve the static problems and to detect the natural frequencies. The obtained results have been compared with the results based on other models with an account (no account) of the size-dependent behaviour.

The so far carried out analysis of the state of the art of literature devoted to investigation of the size-dependent beams based on the models of Bernoulli–Euler, Timoshenko and Sheremetev–Pelekh is mainly reduced to an ordinary differential equation of the Duffing type using the Bubnov–Galerkin method in its first approximation to the input governing PDEs. Linear problems focused on the frequencies estimation and static problems aimed at investigation of the influence of parameter responsible for the size dependence have been considered. However, nonlinear problems of statics and dynamics, in particular, problems related to chaos have not been analysed.

This stands as motivation for us to study more deeply nonlinear deformations of the size-dependent beams under static and dynamic loads. In order to study dynamics of the size-dependent beams, there is a must to employ apparatus of nonlinear dynamics based on the Fourier and wavelet spectra, phase portraits, Poincré maps, LLEs, autocorrelations function, etc. That apparatus has been earlier applied to investigate nonlinear dynamics of structural members including beams, plates and shells [33–39]. The types of nonlinearities have been considered physical, geometric and design (contact interaction in time). However, the results have been obtained based on the classical theory of elasticity, without account of the size-dependent, behaviour of the nano/microconstructions, and the construction material has been assumed to be homogeneous.

This chapter is devoted to successive presentation of the fundamental steps of theory of elastic curvilinear beams and cylindrical panels on the basis of the model of third order (SP) which account of material gradients in two directions, the thermal dependence of the material properties, and the modified theory of couple stresses and nonlinear von Kármán-type deformations. In particular, we extend the modified couple stress theory [40] into the case of functionally gradient cylindrical panels [41] with the use of the third-order approximations and von Karman nonlinearity. The governing PDEs are yielded by the Hamilton principle [40]. Since the majority of nanosized devices include beam-like elements which can be functionally gradient and exhibit rotations, then the Sheremetev–Pelekh theory further extended by Levinson-Reddy can be used to explain the size-dependent effects for the functionally gradient curvilinear microbeams and micropanels.

6.3 Modified Couple Stress Theory of Thermoelastic Curvilinear Panels Based on the Third-Order Sheremetev–Pelekh, Timoshenko and Bernoulli–Euler Hypotheses

Owing to references [20, 42, 43], the governing equations of thermoelasticity of isotropic linear elastic material have the following form:

$$\sigma_{ij} = 2G\left(\varepsilon_{ij} + \frac{\nu\varepsilon_{kk}}{1-2\nu}\delta_{ij} - \frac{1+\nu}{1-2\nu}\gamma\, T\delta_{ij}\right), \tag{6.1}$$

where σ_{ij} stands for components of stress tensor, G—the shear modulus, ε_{ij}—component of deformation tensor, ν—Poisson's coefficient, δ_{ij}—Kronecker's symbol, γ—temperature coefficient of linear expansion and T— temperature increase with regard to the input temperature. The increased relations have the following form:

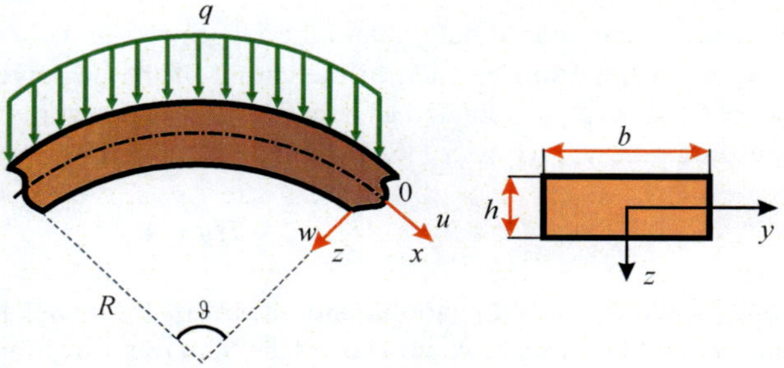

Fig. 6.1 Scheme of non-homogeneous curvilinear panel [reprinted with permission from International Journal of Non-Linear Mechanics publishers]

$$\varepsilon_{ij} = \frac{1+\nu}{E}\sigma_{ij} + \left(\gamma T - \frac{\nu}{E}\sigma_{kk}\right)\delta_{ij}, \tag{6.2}$$

where E is Young's modulus.

In the modified couple stress theory of elasticity, besides Eq. (6.1), the following additional equation is introduced:

$$m_{ij} = 2Gl^2\chi_{ij}. \tag{6.3}$$

Here, m_{ij} are component of deviatory part of the higher order couple stress tensor and l stands for the scale length parameter taking into account of couple stress effects of a higher order [16]. By χ_{ij}, we denote the components of symmetric curvature tensor [20].

Moreover,

$$\varepsilon_{ij} = \frac{1}{2}\left(u_{i,j} + u_{j,i} + u_{m,i}u_{m,j}\right), \tag{6.4}$$

$$\chi_{ij} = \frac{1}{2}\left(\theta_{i,j} + \theta_{j,i}\right), \tag{6.5}$$

where u_i are components of the vector of displacements $\mathbf{u}$, and $\boldsymbol{\Theta}$ stands for the infinitely small vector of rotation with components θ_i.

Theory of deformations of thin curvilinear panels can be constructed in an analogous way to that of theory of plates deformation. Since middle surface of a curvilinear panel in the non-deformed state is cylindrical, the location of middle surface points of the panel is defined by a Gauss curvilinear coordinate ϑ of that surface, i.e. the curvilinear coordinate coincides with the main curvature of the middle surface (Fig. 6.1). That curvilinear coordinate corresponds to the Lame coefficient H_1, whereas the main curvature radius is denoted by R (curvature of the middle surface $k = 1/R$).

Based on the introduced 1D curvilinear coordinate we construct a 2D system. For this purpose, we take an arbitrary point not lying on the middle curve and make a perpendicular projection onto that line, then localization of the given point in plane is fixed by two parameters: coordinate ϑ and the length z (Fig. 6.1). Coordinate z is positive if the point lies from the side of the centre of the negative curvature of the middle curve. Otherwise, z is treated as negative. Observe that the 3D system of curvilinear coordinate ϑ, y, z (Fig. 6.1) is orthogonal. Its Lamé coefficients are as follows:

$$H_1 = R\,(1 + kz)\,, \;\; H_2 = 1\,, \;\; H_3 = 1. \tag{6.6}$$

In what follows we define elongations and shears in the panel layer lying in distance z from the middle surface. Assume displacement of an arbitrary point of the equidistant surface is the following way:

$$\mathbf{u} = u^z \mathbf{e}_\vartheta + w\,\mathbf{e}_n. \tag{6.7}$$

We start with the general relations for the deformation components in the introduced curvilinear system of coordinates ϑ, y, z, then the corresponding component of the nonlinear, in the von Kármán sense, deformation takes the following form $dx = Rd\vartheta$) [44, 45]:

$$\varepsilon_{11} = \frac{1}{1+kz}\left(\frac{\partial u^z}{\partial x} + kw\right) + \frac{1}{2}\frac{1}{(1+kz)^2}\left(\frac{\partial w}{\partial x} - \beta\,ku^z\right)^2, \tag{6.8}$$

$$\varepsilon_{13} = \frac{1}{(1+kz)}\left[\frac{\partial w}{\partial x} - \beta ku^z + (1+kz)\frac{\partial u^z}{\partial z}\right], \tag{6.9}$$

where β stands for the parameter which takes the value of 1 or 0 (1—full theory, 0—simplified theory). It is assumed that the panel thickness h is small in comparison with the curvature radius R, i.e. $h/R \;<<1$ and $|z/R\,| << 1$. We employ the following rule of the change of displacement u^z along panel thickness:

$$u^z = u + z\varphi + a\,z^2 + b\,z^3\frac{\partial w}{\partial x}, \tag{6.10}$$

where u and w are components of the displacement vector of the points of the middle surface along x and z, respectively; φ stands as an angle of rotation (about axis y) of the transversal cross section with respect to the vertical direction (here axis z, see Fig. 6.2), and a, b are coefficients not yet defined.

We define the parameters a and b in (6.10) into a way to make shear deformation equal to zero on both upper and lower surfaces, i.e. for $z = \pm h/2$, where h denotes panel thickness. Formula (6.9) implies that the shape of the function u^z is governed by the following relation:

$$u^z = u + z\varphi - \alpha\,z^3\left(\varphi + \frac{\partial w}{\partial x} - \beta ku\right), \tag{6.11}$$

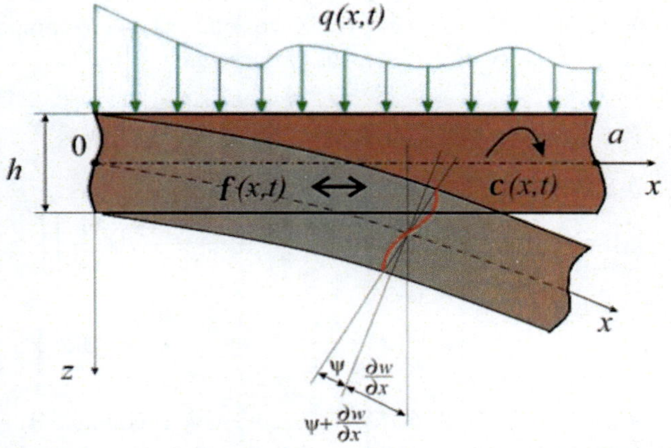

Fig. 6.2 Kinematic parameters and loads of the Sheremetev–Pelekh beam [reprinted with permission from International Journal of Non-Linear Mechanics publishers]

where $\alpha = \frac{4}{3h^2}$. Observe that the field of displacement defined by Eq. (6.11) takes into account the panel shear deformation of the third order and differs from the deformation of the Timoshenko first-order model through an occurrence of the additional term of third order in u^z. However, the latter field of displacement can be reduced to the field of displacement of the Timoshenko model when we take $\alpha = 0$ in Eq. (6.11). It can be also reduced to the field of displacements of the Bernoulli–Euler model when in Eq. (6.11) we assume that the transversal cross section are normal, i.e. $\varphi = -\partial w / \partial x$.

Employing approximation $(1 + z/R) \approx 1$, displacement (6.11) can be written in the following simplified form:

$$u^z \approx u + z\varphi + \alpha z^3 \left(\varphi + \frac{\partial w}{\partial x} \right). \tag{6.12}$$

It should be emphasized that (6.12) cannot be used for the relations containing partial differentiation with regard to z.

Substituting relations (6.11) into (6.8) and (6.9) yields the estimation of deformation tensor components in an arbitrary point

$$\varepsilon_{11} = \frac{\partial u}{\partial x} + z \frac{\partial \varphi}{\partial x} - \alpha z^3 \left(\frac{\partial \varphi}{\partial x} + \frac{\partial^2 w}{\partial x^2} \right) + kw + \frac{1}{2} \left(\frac{\partial w}{\partial x} - \beta k u \right)^2, \tag{6.13}$$

$$\varepsilon_{13} = \frac{1}{2} (1 - 3\alpha z^2) \left[\varphi + \frac{\partial w}{\partial x} - \beta k u \right]. \tag{6.14}$$

Analogously, substituting (6.11) into (6.5), we define non-zero components of the rotation vector and of the symmetric tensor of curvature

$$\theta_2 = \frac{1}{2}\left[\frac{\partial u^z}{\partial z} - (1+kz)^{-1}\left(\frac{\partial w}{\partial x} - \beta k u^z\right)\right], \tag{6.15}$$

$$\chi_{12} = \frac{1}{4}\left\{2\frac{\partial \varphi}{\partial x} - (1+3\alpha z^2)\left[\frac{\partial \varphi}{\partial x} + \frac{\partial^2 w}{\partial x^2} - \beta k \frac{\partial u}{\partial x}\right]\right\}, \tag{6.16}$$

$$\chi_{23} = \frac{1}{4}\left\{(k - 6\alpha z)\left(\varphi + \frac{\partial w}{\partial x} - \beta k u\right) - (1-\beta)k\varphi\right\}. \tag{6.17}$$

In relations (6.13), (6.14) and (6.16), (6.17), we have accounted $(1 + z/R\ ) \approx 1$.

Let us transform (6.1) to fit 2D problems. For this purpose, we consider two types of the problems, general plane stress state and plane deformation state.

In the case of the phase stress state, we take $\sigma_{22} = \sigma_{33} = \sigma_{12} = \sigma_{23} = 0$, then $\sigma_{kk} = \sigma_{11}$ and (6.1) yields

$$\sigma_{11} = E\,\varepsilon_{11} - E\gamma\,T, \quad \sigma_{13} = 2G\,\varepsilon_{13}, \quad \varepsilon_{22} = \varepsilon_{33} = -\nu\,\varepsilon_{11} + (1+\nu)\gamma\,T. \tag{6.18}$$

In the case of the plane deformable state, we take stresses $\sigma_{33} = \sigma_{23} = 0$ and deformation $\varepsilon_{22} = \varepsilon_{12} = \varepsilon_{23} = 0$, and hence (6.1) yields

$$\sigma_{11} = \frac{E}{1-\nu^2}\varepsilon_{11} - \frac{E}{1-\nu}\gamma T, \qquad \sigma_{13} = 2G\,\varepsilon_{13}, \tag{6.19}$$

$$\sigma_{22} = \frac{\nu E}{1-\nu^2}\varepsilon_{11} - \frac{E}{1-\nu}\gamma T, \quad \varepsilon_{33} = \frac{-\nu}{1-\nu}\varepsilon_{11} - \frac{1+\nu}{1-\nu}\gamma T. \tag{6.20}$$

Observe that, if we take

$$\tilde{G} = G, \quad \tilde{E} = \frac{E}{1-\nu^2}, \quad \tilde{\gamma} = (1+\nu)\,\gamma, \quad \tilde{\nu} = \frac{\nu}{1-\nu} \tag{6.21}$$

then relations σ_{11}, σ_{13}, ε_{33} in (6.18) and (6.19), (6.20) take the same form. Owing to this observation, the next step we consider is the only relations for the plane stress state (6.18) for beams. The mentioned relations with an account of (6.21) are valid also for panels.

Therefore, for the stresses and couple of stresses of higher order the following relations hold (in displacements):

$$\sigma_{11} = E\left(\frac{\partial u}{\partial x} + z\frac{\partial \varphi}{\partial x} - \alpha\,z^3\left(\frac{\partial \varphi}{\partial x} + \frac{\partial^2 w}{\partial x^2}\right) + kw + \frac{1}{2}\left(\frac{\partial w}{\partial x} - \beta k u\right)^2\right) - E\gamma\,T, \tag{6.22}$$

$$\sigma_{13} = G\left(1 - 3\alpha z^2\right)\left[\varphi + \frac{\partial w}{\partial x} - \beta k u\right], \quad \varepsilon_{22} = \varepsilon_{33} = -\nu\,\varepsilon_{11} + (1+\nu)\,\gamma\,T, \tag{6.23}$$

$$m_{12} = \frac{1}{2}Gl^2\left\{2\frac{\partial\varphi}{\partial x} - (1 + 3\alpha z^2)\left[\frac{\partial\varphi}{\partial x} + \frac{\partial^2 w}{\partial x^2} - \beta k\frac{\partial u}{\partial x}\right]\right\}, \tag{6.24}$$

$$m_{23} = \frac{1}{2}Gl^2\left\{(k - 6\alpha z)\left(\varphi + \frac{\partial w}{\partial x} - \beta k u\right) - (1-\beta)k\varphi\right\}. \tag{6.25}$$

General energy of an elastic beam deformation is as follows:

$$U = \frac{1}{2}\int_V (\sigma_{ij}\delta\varepsilon_{ij} + m_{ij}\delta\chi_{ij})dw = \\ = \frac{1}{2}\int_0^L\int_A (\sigma_{11}\varepsilon_{11} + 2\sigma_{13}\varepsilon_{13} + 2m_{12}\chi_{12} + 2m_{23}\chi_{23})dAdx. \tag{6.26}$$

Variation of elastic deformations energy for the kinematically allowed displacement fields takes the following form:

$$\delta U = \int_0^L\int_A \sigma_{11}\delta\,\varepsilon_{11} + \sigma_{13}\delta\,\varepsilon_{13} + m_{12}\delta\chi_{12} + m_{23}\delta\chi_{23})dAdx, \tag{6.27}$$

where A is the transversal beam cross section.

Substituting the variations $\delta\,\varepsilon_{11}$, $\delta\,\varepsilon_{13}$, $\delta\chi_{12}$, $\delta\chi_{23}$ from (6.13), (6.14), (6.16), (6.17) into (6.26) yields

$$\delta U = \int_0^L\int_A \sigma_{11}\delta\left[\frac{\partial u}{\partial x} + kw + \frac{1}{2}\left(\frac{\partial w}{\partial x} - \beta k u\right)^2 + z\frac{\partial\varphi}{\partial x} - \alpha z^3\frac{\partial\varphi}{\partial x} - \alpha z^3\frac{\partial^2 w}{\partial x^2}\right] + \\ + \sigma_{13}\delta\left[(1 - 3\alpha z^2)\left(\varphi + \frac{\partial w}{\partial x} - \beta k u\right)\right] + \\ + \frac{m_{12}}{2}\delta\left[2\frac{\partial\varphi}{\partial x} - (1 + 3\alpha z^2)\left(\frac{\partial\varphi}{\partial x} + \frac{\partial^2 w}{\partial x^2} - \beta k\frac{\partial u}{\partial x}\right)\right] + \\ + \frac{m_{23}}{2}\delta\left[(k - 6\alpha z)\left(\varphi + \frac{\partial w}{\partial x} - \beta k u\right) - (1-\beta)k\varphi\right]\Bigg\}. \tag{6.28}$$

We define classical versus higher order resultants with regard to the transversal beam cross section as follows:

$$(N, M, P) = \int_A \sigma_{11}(1, z, z^3)dA, \quad (Q, R) = \int_A \sigma_{13}(1, z^2)dA, \tag{6.29}$$

where N, M, Q stand for classical axial force, moment and cutting force, whereas P, R are higher order resultants. Besides, we introduce couples of stresses of higher orders

$$(Y_{12}, T_{12}) = \int_A m_{12}(1, z^2)dA, \quad (Y_{23}, T_{23}) = \int_A m_{23}(1, z)dA. \tag{6.30}$$

Formula (6.28) with an account of (6.29), (6.30) takes the following form:

$$\begin{gathered}
\delta U = \int_0^L \left\{ N \left[\frac{\partial \delta u}{\partial x} + k\delta w + \left(\frac{\partial w}{\partial x} - \beta k u \right) \left(\frac{\partial \delta w}{\partial x} - \beta k \delta u \right) \right] + \right. \\
+ M \frac{\partial \delta \varphi}{\partial x} - \alpha P \left(\frac{\partial \delta \varphi}{\partial x} + \frac{\partial^2 \delta w}{\partial x^2} \right) + \\
+ Q \left(\delta \varphi + \frac{\partial \delta w}{\partial x} - \beta k \delta u \right) - 3\alpha R \left(\delta \varphi + \frac{\partial \delta w}{\partial x} - \beta k \delta u \right) + \\
+ \frac{Y_{12}}{2} \left(\frac{\partial \delta \varphi}{\partial x} - \frac{\partial^2 \delta w}{\partial x^2} + \beta k \frac{\partial \delta u}{\partial x} \right) - \frac{3}{2} \alpha T_{12} \left(\frac{\partial \delta \varphi}{\partial x} + \frac{\partial^2 \delta w}{\partial x^2} - \beta k \frac{\partial \delta u}{\partial x} \right) + \\
\left. + \frac{Y_{23}}{2} \left(\beta k \delta \varphi + k \frac{\partial \delta w}{\partial x} - \beta k^2 \delta u \right) - 3\alpha T_{23} \left(\delta \varphi + \frac{\partial \delta w}{\partial x} - \beta k \delta u \right) \right\} dx.
\end{gathered} \tag{6.31}$$

After integration by parts and algebraic transformations, we get

$$\begin{gathered}
\delta U = \int_0^L \left\{ - \left[\frac{\partial N}{\partial x} + \beta k N \left(\frac{\partial w}{\partial x} - \beta k u \right) + \beta k Q + \frac{\beta k}{2} \left(\frac{\partial Y_{12}}{\partial x} + k Y_{23} \right) - \right. \right. \\
\left. - 3\alpha \beta k \left(R + T_{23} - \frac{1}{2} \frac{\partial T_{12}}{\partial x} \right) \right] \delta u - \left[\frac{\partial M}{\partial x} - Q + \frac{1}{2} \left(\frac{\partial Y_{12}}{\partial x} - \beta k Y_{23} \right) - \right. \\
\left. - \alpha \left(\frac{\partial P}{\partial x} + \frac{3}{2} \frac{\partial T_{12}}{\partial x} \right) + 3\alpha \left(R + T_{23} \right) \right] \delta \varphi - \left[\frac{\partial Q}{\partial x} - kN + \right. \\
+ \frac{\partial}{\partial x} \left(N \left(\frac{\partial w}{\partial x} - \beta k u \right) \right) + \frac{1}{2} \left(\frac{\partial^2 Y_{12}}{\partial x^2} + k \frac{\partial Y_{23}}{\partial x} \right) +
\end{gathered} \tag{6.32}$$

$$+\alpha\frac{\partial^2 P}{\partial x^2}-3\alpha\left(\frac{\partial R}{\partial x}+\frac{\partial T_{23}}{\partial x}-\frac{1}{2}\frac{\partial^2 T_{12}}{\partial x^2}\right)\Big]\delta w\Big\}dx\ +$$

$$+\left[N+\frac{3}{2}\alpha\beta kT_{12}+\beta k\frac{Y_{12}}{2}\right]\delta u\Big|_{x=0}^{x=L}+\left[M-\alpha\left(P+\frac{3}{2}T_{12}\right)+\frac{Y_{12}}{2}\right]\delta\varphi\Big|_{x=0}^{x=L}+$$

$$+\left[Q+\frac{1}{2}kY_{23}+\frac{1}{2}\frac{\partial Y_{12}}{\partial x}+N\left(\frac{\partial w}{\partial x}-\beta ku\right)+\alpha\frac{\partial P}{\partial x}-\right.$$

$$\left.-3\alpha\left(R+T_{23}-\frac{1}{2}\frac{\partial T_{12}}{\partial x}\right)\right]\delta w\Big|_{x=0}^{x=L}+\left[\alpha P+\frac{3}{2}\alpha T_{12}+\frac{Y_{12}}{2}\right]\frac{\partial\delta w}{\partial x}\Big|_{x=0}^{x=L}.$$

Let us find variation of the kinetic beam energy. Let $\rho\,(x,\ z,\ T)$ denote the density of beam material. Its kinetic energy is as follows:

$$K=\frac{1}{2}\int_0^L\int_A\rho\,(x,\ z,\ T)\left[\left(\frac{\partial\,u^z}{\partial\,t}\right)^2+\left(\frac{\partial\,w}{\partial\,t}\right)^2\right]dAdx\ . \tag{6.33}$$

Let us introduce the following resultants with regard to the beam cross section

$$(I_0,\,I_1,\,I_2,\,I_3,\,I_4,\,I_5,\,I_6)=\int_A\rho(x,z,\ T)(1,z,z^2,z^3,z^4,z^5,z^6)dz. \tag{6.34}$$

Now, the first variation of the kinetic energy with an account of (6.12) takes the following form:

$$\delta K=-\int_0^L\left\{\left[I_0\frac{\partial^2 u}{\partial t^2}+(I_1-\alpha I_3)\frac{\partial^2\varphi}{\partial t^2}-\alpha I_3\frac{\partial^3 w}{\partial x\partial t^2}\right]\delta u+\right.$$

$$+\left[(I_1-\alpha I_3)\frac{\partial^2 u}{\partial t^2}+(I_2-\alpha I_4)\frac{\partial^2\varphi}{\partial t^2}-\alpha\,(I_4-\alpha I_6)\left(\frac{\partial^2\varphi}{\partial t^2}+\frac{\partial^3 w}{\partial x\partial t^2}\right)\right]\delta\varphi+$$

$$\left.+\left[I_0\frac{\partial^2 w}{\partial t^2}+\alpha I_3\frac{\partial^3 u}{\partial x\,\partial t^2}+\alpha I_4\frac{\partial^3\varphi}{\partial x\partial t^2}-\alpha^2 I_6\left(\frac{\partial^4 w}{\partial x^2\partial t^2}+\frac{\partial^3\varphi}{\partial x\partial t^2}\right)\right]\delta w\right\}dx+$$

$$+\left[\alpha I_3\frac{\partial^2 u}{\partial t^2}+\alpha\,(I_4-\alpha I_6)\frac{\partial^2\varphi}{\partial t^2}-\alpha^2 I_6\frac{\partial^3\,w}{\partial x\partial t^2}\right]\delta w\Big|_{x=0}^{x=L}. \tag{6.35}$$

The work of external forces carried out in the deformed elastic body under hypotheses of the modified couple stress theory is defined by the following formula: [46]

$$W = \int_0^L (fu + qw + c\theta_y)\, dx + \left(\bar{N}u + \bar{Q}w + \bar{M}\varphi + \alpha \bar{P}\frac{\partial w}{\partial x} \right)\Bigg|_{x=0}^{x=L}, \tag{6.36}$$

where $f(x, t)$ denotes the force per unit length acting on the axial cross section along the body axis OX. The quantity $q(x, t)$ presents resultant of the transverse stress on the beam's upper surface as well as the transverse volume forces per unit length of beam. By $c\,(x,\ t)$ we denote y-th component of volume couple stresses of higher order acting on the beam points, and $\bar{N}$, $\bar{Q}$, $\bar{M}$ and $\bar{P}$ stand for the axial stress, cutting force, bending moments of the first and second order appeared on the beam ends, respectively. We further employ the following notation:

$$(C_0, C_2) = \int_A c(1, z^2) dA$$

implied by an action of the volume moment of higher order along the beam cross section.

Then, variation of the work carried out by beam forces and moments can be recast to the following form:

$$\delta W = \int_0^L \Bigg\{ \left[f + \frac{\beta k}{2} (C_0 + 3\alpha C_2) \right] \delta u + \left[q + \frac{1}{2}\frac{\partial}{\partial x} (C_0 + 3\alpha C_2) \right] \delta w +$$
$$+ \frac{1}{2} (C_0 - 3\alpha C_2)\, \delta\varphi \Bigg\}\, dx + \left[\bar{N}\delta u + \left(\bar{Q} - \frac{C_0 + 3\alpha C_2}{2} \right) \delta w + \right. \tag{6.37}$$
$$\left. + \bar{M}\delta\varphi + \alpha \bar{P}\delta \left(\frac{\partial w}{\partial x} \right) \right]\Bigg|_{x=0}^{x=L}.$$

The Hamilton principle yields

$$\int_{t_2}^{t_1} \delta\,[K - U + W]\, dt = 0, \tag{6.38}$$

where t_1 and t_2 stands for the initial and terminal motion, δK is variation of the kinetic energy, δU stands for variation of the deformation energy and δW is variation of the works of the external loads.

Substituting (6.32), (6.35) and (6.37) into (6.38) yields

$$\int_{t_1}^{t_2} \int_0^L \Bigg\{ \Bigg[- \left(I_0 \frac{\partial^2 u}{\partial t^2} + (I_1 - \alpha I_3) \frac{\partial^2 \varphi}{\partial t^2} - \alpha I_3 \frac{\partial^3 w}{\partial x t^2} \right) + f +$$

$$+\frac{\beta k}{2}(C_0+3\alpha C_2)+\frac{\partial N}{\partial x}+\beta kN\left(\frac{\partial w}{\partial x}-\beta ku\right)+\beta kQ+$$

$$+\frac{\beta k}{2}\left(\frac{\partial Y_{12}}{\partial x}+kY_{23}\right)-3\alpha\beta k\left(R+T_{23}-\frac{1}{2}\frac{\partial T_{12}}{\partial x}\right)\Bigg]\delta u+$$

$$\Bigg[-\left((I_1-\alpha I_3)\frac{\partial^2 u}{\partial t^2}+(I_2-\alpha I_4)\frac{\partial^2\varphi}{\partial t^2}-\alpha\,(I_4-\alpha I_6)\left(\frac{\partial^2\varphi}{\partial t^2}+\frac{\partial^3 w}{\partial x\partial t^2}\right)\right)+$$

$$+\frac{1}{2}(C_0-3\alpha\,C_2)+\frac{\partial M}{\partial x}-Q+\frac{1}{2}\left(\frac{\partial Y_{12}}{\partial x}-\beta kY_{23}\right)-\alpha\left(\frac{\partial P}{\partial x}+\frac{3}{2}\frac{\partial T_{12}}{\partial x}\right)+$$

$$+3\alpha\,(R+T_{23})]\,\delta\varphi+\Bigg[-\left(I_0\frac{\partial^2 w}{\partial t^2}+\alpha\,I_3\frac{\partial^3 u}{\partial x\,\partial t^2}+\alpha I_4\frac{\partial^3\varphi}{\partial x\,\partial t^2}-\right.$$

$$\left.-\alpha^2 I_6\left(\frac{\partial^4 w}{\partial x^2\,\partial t^2}+\frac{\partial^3\varphi}{\partial x\partial t^2}\right)\right)+\frac{1}{2}\frac{\partial}{\partial x}(C_0+3\alpha C_2)+\tag{6.39}$$

$$\frac{\partial Q}{\partial x}-kN+\frac{\partial}{\partial x}\left(N\left(\frac{\partial w}{\partial x}-\beta ku\right)\right)+\frac{1}{2}\left(\frac{\partial^2 Y_{12}}{\partial x^2}+k\frac{\partial Y_{23}}{\partial x}\right)+\alpha\frac{\partial^2 P}{\partial x^2}-$$

$$-3\alpha\left(\frac{\partial R}{\partial x}+\frac{\partial T_{23}}{\partial x}-\frac{3}{2}\frac{\partial^2 T_{12}}{\partial x^2}\right)+q\Bigg]\delta w\Bigg\}\,dxdt+$$

$$+\int_{t_1}^{t_2}\Bigg\{\left(\bar N-N-\frac{3}{2}\alpha\beta kT_{12}-\beta k\frac{Y_{12}}{2}\right)\delta u+\left(\bar M-M+\alpha\left(P+\frac{3}{2}T_{12}\right)-\frac{Y_{12}}{2}\right)\delta\varphi+$$

$$+\Bigg[\bar Q+\left(\alpha\,I_3\frac{\partial^2 u}{\partial t^2}+\alpha\,(I_4-\alpha I_6)\frac{\partial^2\varphi}{\partial t^2}-\alpha^2 I_6\frac{\partial^3 w}{\partial x\partial t^2}\right)-\frac{C_0+3\alpha C_2}{2}-$$

$$\left(Q+\frac{1}{2}kY_{23}+\frac{1}{2}\frac{\partial Y_{12}}{\partial x}+N\left(\frac{\partial w}{\partial x}-\beta ku\right)+\alpha\frac{\partial P}{\partial x}-3\alpha\left(R+T_{23}-\frac{1}{2}\frac{\partial T_{12}}{\partial x}\right)\right)\Bigg]\delta w+$$

$$+\left(\alpha\bar P-\alpha P-\frac{Y_{12}+3\alpha T_{12}}{2}\right)\frac{\partial\delta w}{\partial x}\Bigg\}\Bigg|_{x=0}^{x=L}dt=0.$$

Employing the fundamental lemma of the variational computations into equation (6.39) with an account of arbitrariness of kinematically allowed variable $\delta u\delta\varphi$ and δw for $x\in[0,L]$ and $t\in(t_1,t_2)$, the following system of the Euler–Lagrange equations is obtained for the beams:

$$\frac{\partial N}{\partial x}+\beta kN\left(\frac{\partial w}{\partial x}-\beta ku\right)+\beta kQ+\frac{\beta k}{2}\left(\frac{\partial Y_{12}}{\partial x}+kY_{23}\right)-$$

$$- 3\alpha\beta k \left(R + T_{23} - \frac{1}{2}\frac{\partial T_{12}}{\partial x} \right) + \frac{\beta k}{2} \left(C_0 + 3\alpha C_2 \right) + f = \tag{6.40}$$

$$= I_0 \frac{\partial^2 u}{\partial t^2} + \left(I_1 - \alpha I_3 \right) \frac{\partial^2 \varphi}{\partial t^2} - \alpha I_3 \frac{\partial^3 w}{\partial x \partial t^2},$$

$$\frac{\partial M}{\partial x} - Q + 3\alpha \left(R + T_{23} \right) + \frac{1}{2} \left(\frac{\partial Y_{12}}{\partial x} - \beta k Y_{23} \right) -$$

$$- \alpha \left(\frac{\partial P}{\partial x} + \frac{3}{2}\frac{\partial T_{12}}{\partial x} \right) + \frac{1}{2} \left(C_0 - 3\alpha C_2 \right) = \tag{6.41}$$

$$= \left(I_1 - \alpha I_3 \right) \frac{\partial^2 u}{\partial t^2} + \left(I_2 - \alpha I_4 \right) \frac{\partial^2 \varphi}{\partial t^2} - \alpha \left(I_4 - \alpha I_6 \right) \left(\frac{\partial^2 \varphi}{\partial t^2} + \frac{\partial^3 w}{\partial x \partial t^2} \right),$$

$$\frac{\partial Q}{\partial x} - kN + \frac{\partial}{\partial x} \left(N \left(\frac{\partial w}{\partial x} - \beta k u \right) \right) + \frac{1}{2} \left(\frac{\partial^2 Y_{12}}{\partial x^2} + k \frac{\partial Y_{23}}{\partial x} \right) + \alpha \frac{\partial^2 P}{\partial x^2} -$$

$$- 3\alpha \left(\frac{\partial R}{\partial x} + \frac{\partial T_{23}}{\partial x} - \frac{3}{2}\frac{\partial^2 T_{12}}{\partial x^2} \right) + \frac{1}{2}\frac{\partial}{\partial x} \left(C_0 + 3\alpha C_2 \right) + q = \tag{6.42}$$

$$= I_0 \frac{\partial^2 w}{\partial t^2} + \alpha I_3 \frac{\partial^3 u}{\partial x \partial t^2} + \alpha I_4 \frac{\partial^3 \varphi}{\partial x \partial t^2} - \alpha^2 I_6 \left(\frac{\partial^4 w}{\partial x^2 \partial t^2} + \frac{\partial^3 \varphi}{\partial x \partial t^2} \right),$$

and the corresponding boundary conditions

$$N + \beta k \left(\frac{Y_{12} + 3\alpha T_{12}}{2} \right) \Bigg|_{x=0}^{x=L} = \bar{N} \quad \text{or} \quad u|_{x=0}^{x=L} = \bar{u}, \tag{6.43}$$

$$M - \alpha P + \frac{Y_{12} - 3\alpha T_{12}}{2} \Bigg|_{x=0}^{x=L} = \bar{M} \quad \text{or} \quad \varphi|_{x=0}^{x=L} = \bar{\varphi}, \tag{6.44}$$

$$\left[\alpha^2 I_6 \frac{\partial^3 w}{\partial x \partial t^2} - \alpha I_3 \frac{\partial^2 u}{\partial t^2} - \alpha \left(I_4 - \alpha I_6 \right) \frac{\partial^2 \varphi}{\partial t^2} + \frac{C_0 + 3\alpha C_2}{2} + \right.$$

$$\left. + \left(Q + \frac{1}{2} k Y_{23} + \frac{1}{2}\frac{\partial Y_{12}}{\partial x} + N \left(\frac{\partial w}{\partial x} - \beta k u \right) + \alpha \frac{\partial P}{\partial x} - 3\alpha \left(R + T_{23} - \frac{1}{2}\frac{\partial T_{12}}{\partial x} \right) \right) \right] \Bigg|_{x=0}^{x=L} = \bar{Q} \tag{6.45}$$

or

$$w|_{x=0}^{x=L} = \bar{w},$$

$$\alpha P + \left.\frac{Y_{12} + 3\alpha T_{12}}{2}\right|_{x=0}^{x=L} = \alpha \bar{P} \quad \text{or} \quad \left.\frac{\partial w}{\partial x}\right|_{x=0}^{x=L} = \overline{\frac{\partial w}{\partial x}}. \tag{6.46}$$

In relations (6.22)–(6.25) and (6.29), (6.30), we describe stresses and moments by displacements u, φ and w and we obtain

$$N = \int\limits_A (1 + kz)\sigma_{11} dA \cong \int\limits_A \left[E(x, z, T(x, z))\varepsilon_{11} - \gamma E(x, z, T(x, z))T(x, z)\right] dA =$$

$$= \int\limits_A \left\{ E\left(\frac{\partial u}{\partial x} + kw + \frac{1}{2}\left(\frac{\partial w}{\partial x} - \beta ku\right)^2 + Ez\frac{\partial \varphi}{\partial x} - \right.\right. \tag{6.47}$$

$$\left.\left. -\alpha E z^3 \left(\frac{\partial \varphi}{\partial x} + \frac{\partial^2 w}{\partial x^2}\right)\right) - E\gamma T\right\} dA.$$

Introducing the notation

$$(J_0, J_1, J_2, J_3, J_4, J_6) = \int\limits_A E(x, z, T(x, z))(1, z, z^2, z^3, z^4, z^6) dA, \tag{6.48}$$

$$N^T = \int\limits_A E\gamma T dA, \quad M^T = \int\limits_A E\gamma T z dA, \quad P^T = \int\limits_A E\gamma T z^3 dA. \tag{6.49}$$

yields $N = N^e - N^T$, where

$$N^e = J_0 \left(\frac{\partial u}{\partial x} + kw + \frac{1}{2}\left(\frac{\partial w}{\partial x} - \beta ku\right)^2\right) + (J_1 - \alpha J_3)\frac{\partial \varphi}{\partial x} - \alpha J_3 \frac{\partial^2 w}{\partial x^2}. \tag{6.50}$$

We find equations governing M and P in the way analogous to N, i.e. we have $M = M^e - M^T$, where

$$M^e = J_1 \left(\frac{\partial u}{\partial x} + kw + \frac{1}{2}\left(\frac{\partial w}{\partial x} - \beta ku\right)^2\right) + (J_2 - \alpha J_4)\frac{\partial \varphi}{\partial x} - \alpha J_4 \frac{\partial^2 w}{\partial x^2} \tag{6.51}$$

and $P = P^e - P^T$, where

$$P^e = J_3 \left(\frac{\partial u}{\partial x} + kw + \frac{1}{2}\left(\frac{\partial w}{\partial x} - \beta ku\right)^2\right) + (J_4 - \alpha J_6)\frac{\partial \varphi}{\partial x} - \alpha J_6 \frac{\partial^2 w}{\partial x^2}. \tag{6.52}$$

Next, to get the resultants Q, R, expressed in terms of displacements.
We denote

$$(A_0, A_2, A_4) = \int_A G(x, z, T(x, z))(1, z^2, z^4)dA. \tag{6.53}$$

Hence, we have

$$Q = (A_0 - 3\alpha A_2)\left[\varphi + \frac{\partial w}{\partial x} - \beta k u\right], \tag{6.54}$$

$$R = (A_2 - 3\alpha A_4)\left[\varphi + \frac{\partial w}{\partial x} - \beta k u\right]. \tag{6.55}$$

Proceeding in the analogous way, to find the higher order resultants, we employ the following notation:

$$(B_0, B_1, B_2, B_4) = \int_A G(x, z, T(x, z))\, l^2(x, z, T(x, z))(1, z, z^2, z^4)dA, \tag{6.56}$$

and hence

$$Y_{12} = \frac{1}{2}\left[B_0\left(\frac{\partial \varphi}{\partial x} - \frac{\partial^2 w}{\partial x^2} + \beta k \frac{\partial u}{\partial x}\right) - 3\alpha B_2\left(\frac{\partial \varphi}{\partial x} + \frac{\partial^2 w}{\partial x^2} - \beta k \frac{\partial u}{\partial x}\right)\right], \tag{6.57}$$

$$T_{12} = \frac{1}{2}\left[B_2\left(\frac{\partial \varphi}{\partial x} - \frac{\partial^2 w}{\partial x^2} + \beta k \frac{\partial u}{\partial x}\right) - 3\alpha B_4\left(\frac{\partial \varphi}{\partial x} + \frac{\partial^2 w}{\partial x^2} - \beta k \frac{\partial u}{\partial x}\right)\right], \tag{6.58}$$

$$Y_{23} = \frac{1}{2}\left[B_0 k\left(\beta \varphi + \frac{\partial w}{\partial x} - \beta k u\right) - 6\alpha B_1\left(\varphi + \frac{\partial w}{\partial x} - \beta k u\right)\right], \tag{6.59}$$

$$T_{23} = \frac{1}{2}\left[B_1 k\left(\beta \varphi + \frac{\partial w}{\partial x} - \beta k u\right) - 6\alpha B_2\left(\varphi + \frac{\partial w}{\partial x} - \beta k u\right)\right]. \tag{6.60}$$

Occurrence of $Y_{12}, Y_{23}, T_{12}, T_{23}$ in the consider model (see (6.40)–(6.46) is implied by direct contribution of the couples of stresses of higher orders to the beam deformation energy (see (6.26), (6.30). The derived four results explicitly depend on the length scale material parameter l, which exhibits the influence of the higher order couples of stresses.

Finally, substitution of relations (6.50)–(6.52), (6.54), (6.55), and (6.57)–(6.60) into equations (6.40)–(6.46) yields equation of motion with third-order shear deformation of the panel with regard to displacements u, φ, w.

6.4 Technical Theory for the Sheremetev–Pelekh, Timoshenko and Bernoulli–Euler Models

The structure of equations (6.40)–(6.42) of the general theory of curvilinear panels is rather complex, and hence getting a reliable solution to those equations for a series of problems requires to overcome of serious mathematical difficulties. Therefore, there is a need to simplify those fundamental equations and choose their variants being suitable for account of features of the most important problems that occurred in practice. On a basis of numerous theoretical investigations [45, 47, 48], it has been shown that in majority of the engineering problems it is sufficient to consider the middle panel surface as shallow one. In the following, we will present the simplified variant of the fundamental governing equations of theory of curvilinear panels being based on that shallowly panel hypothesis. Recall that a panel or a beam is defined as shallow if owing to the Vlasov statement $kh \leq 1/30$. The theory with the introduced assumptions will be further called technical theory of curvilinear panels.

Besides the fundamental hypothesis, the so far introduced technical theory employs the following hypotheses in the geometric relations (6.8), and the deformations caused by tangential displacement u^z can be neglected, i.e. we take $\beta = 0$; the curvilinear coordinate ϑ is chosen to satisfy the inequality $kh << 1$. Validation of the latter assumption into theory of shell has been carried out by Mushtari [47, 48]. He showed that neglection of the tangential displacements will be more strongly realized by decreasing stresses caused by moments in comparison to stresses originated from forces.

An account of the given so far assumptions of the theory of isotropic curvilinear panels, as it follows from Eqs. (6.40)–(6.42), yields the following relations:

$$\frac{\partial N}{\partial x} + f = I_0 \frac{\partial^2 u}{\partial t^2} + (I_1 - \alpha I_3)\frac{\partial^2 \varphi}{\partial t^2} - \alpha I_3 \frac{\partial^3 w}{\partial x \partial t^2}, \tag{6.61}$$

$$\begin{aligned}&\frac{\partial M}{\partial x} - Q + 3\alpha\,(R + T_{23}) + \frac{1}{2}\frac{\partial Y_{12}}{\partial x} - \alpha\left(\frac{\partial P}{\partial x} + \frac{3}{2}\frac{\partial T_{12}}{\partial x}\right) + \frac{1}{2}\,(C_0 - 3\alpha C_2) = \\ &= (I_1 - \alpha\, I_3)\,\frac{\partial^2 u}{\partial t^2} + (I_2 - \alpha\, I_4)\,\frac{\partial^2 \varphi}{\partial t^2} - \alpha\,(I_4 - \alpha I_6)\left(\frac{\partial^2 \varphi}{\partial t^2} + \frac{\partial^3 w}{\partial x \partial t^2}\right),\end{aligned} \tag{6.62}$$

$$\begin{aligned}&\frac{\partial Q}{\partial x} - kN + \frac{\partial}{\partial x}\left(N\frac{\partial w}{\partial x}\right) + \frac{1}{2}\left(\frac{\partial^2 Y_{12}}{\partial x^2} + k\frac{\partial Y_{23}}{\partial x}\right) + \alpha\frac{\partial^2 P}{\partial x^2} - \\ &- 3\alpha\left(\frac{\partial R}{\partial x} + \frac{\partial T_{23}}{\partial x} - \frac{3}{2}\frac{\partial^2 T_{12}}{\partial x^2}\right) + \frac{1}{2}\frac{\partial}{\partial x}\,(C_0 + 3\alpha C_2) + q \\ &= I_0\frac{\partial^2 w}{\partial t^2} + \alpha I_3\frac{\partial^3 u}{\partial x \partial t^2} + \alpha I_4\frac{\partial^3 \varphi}{\partial x \partial t^2} - \alpha^2 I_6\left(\frac{\partial^4 w}{\partial x^2 \partial t^2} + \frac{\partial^3 \varphi}{\partial x \partial t^2}\right).\end{aligned} \tag{6.63}$$

Boundary conditions (6.43)–(6.46) are also simplified. This requires to take $\beta = 0$ in equations (6.50)–(6.52), (6.54), (6.55) and (6.57)–(6.60), and one gets

$$N|_{x=0}^{x=L} = \bar{N} \quad \text{or} \quad u|_{x=0}^{x=L} = \bar{u} \tag{6.64}$$

$$M - \alpha P + \left.\frac{Y_{12} - 3\alpha T_{12}}{2}\right|_{x=0}^{x=L} = \bar{M} \quad \text{or} \quad \varphi|_{x=0}^{x=L} = \bar{\varphi}, \tag{6.65}$$

$$\left[\alpha^2 I_6 \frac{\partial^3 w}{\partial x \partial t^2} - \alpha I_3 \frac{\partial^2 u}{\partial t^2} - \alpha \left(I_4 - \alpha I_6\right) \frac{\partial^2 \varphi}{\partial t^2} + \frac{C_0 + 3\alpha C_2}{2} + \right.$$
$$\left.\left. + \left(Q + \frac{1}{2} k Y_{23} + \frac{1}{2}\frac{\partial Y_{12}}{\partial x} + N\frac{\partial w}{\partial x} + \alpha \frac{\partial P}{\partial x} - 3\alpha \left(R + T_{23} - \frac{1}{2}\frac{\partial T_{12}}{\partial x}\right)\right)\right]\right|_{x=0}^{x=L} = \bar{Q} \tag{6.66}$$

or

$$w|_{x=0}^{x=L} = \bar{w}$$

$$\alpha P + \left.\frac{Y_{12} + 3\alpha T_{12}}{2}\right|_{x=0}^{x=L} = \alpha \bar{P} \quad \text{or} \quad \left.\frac{\partial w}{\partial x}\right|_{x=0}^{x=L} = \overline{\frac{\partial w}{\partial x}}. \tag{6.67}$$

Let us rewrite Eqs. (6.61)–(6.63) in displacements:

$$\frac{\partial}{\partial x}\left[J_0 \left(\frac{\partial u}{\partial x} + kw + \frac{1}{2}\left(\frac{\partial w}{\partial x}\right)^2\right) + \left(J_1 - \alpha J_3\right)\frac{\partial \varphi}{\partial x} - \alpha\, J_3 \frac{\partial^2 w}{\partial x^2}\right] +$$
$$+ f - \frac{\partial N^T}{\partial x} = I_0 \frac{\partial^2 u}{\partial t^2} + \left(I_1 - \alpha I_3\right) \frac{\partial^2 \varphi}{\partial t^2} - \alpha I_3 \frac{\partial^3 w}{\partial x \partial t^2}, \tag{6.68}$$

$$\frac{\partial}{\partial x}\left[J_1 \left(\frac{\partial u}{\partial x} + kw + \frac{1}{2}\left(\frac{\partial w}{\partial x}\right)^2\right) + \left(J_2 - \alpha J_4\right)\frac{\partial \varphi}{\partial x} - \alpha J_4 \frac{\partial^2 w}{\partial x^2}\right] - \left(A_0 - 3\alpha A_2\right)\left(\varphi + \frac{\partial w}{\partial x}\right) +$$
$$+ 3\alpha \left[\left(A_2 - 3\alpha A_4\right)\left(\varphi + \frac{\partial w}{\partial x}\right) + \frac{1}{2}\left(\left(B_1 k - 6\alpha B_2\right)\frac{\partial w}{\partial x} - 6\alpha B_2 \varphi\right)\right] +$$
$$+ \frac{1}{4}\frac{\partial}{\partial x}\left(B_0 \left(\frac{\partial \varphi}{\partial x} - \frac{\partial^2 w}{\partial x^2}\right) - 3\alpha B_2 \left(\frac{\partial \varphi}{\partial x} + \frac{\partial^2 w}{\partial x^2}\right)\right) - \tag{6.69}$$
$$-\alpha \frac{\partial}{\partial x}\left\{J_3 \left(\frac{\partial u}{\partial x} + kw + \frac{1}{2}\left(\frac{\partial w}{\partial x}\right)^2\right) + \left(J_4 - \alpha J_6\right)\frac{\partial \varphi}{\partial x} - \alpha J_6 \frac{\partial^2 w}{\partial x^2} + \right.$$

$$+\frac{3}{4}\left[B_2\left(\frac{\partial\varphi}{\partial x}-\frac{\partial^2 w}{\partial x^2}\right)-3\alpha B_4\left(\frac{\partial\varphi}{\partial x}+\frac{\partial^2 w}{\partial x^2}\right)\right]\Bigg\}+$$

$$+\frac{1}{2}\left(C_0-3\alpha\, C_2\right)-\frac{\partial M^T}{\partial x}+\alpha\frac{\partial P^T}{\partial x}=$$

$$=\left(I_1-\alpha I_3\right)\frac{\partial^2 u}{\partial t^2}+\left(I_2-\alpha I_4\right)\frac{\partial^2\varphi}{\partial t^2}-\alpha\,\left(I_4-\alpha I_6\right)\left(\frac{\partial^2\varphi}{\partial t^2}+\frac{\partial^3 w}{\partial x\partial t^2}\right),$$

$$\frac{\partial}{\partial x}\left[\left(A_0-3\alpha A_2\right)\left(\varphi+\frac{\partial w}{\partial x}\right)\right]-k\left[J_0\left(\frac{\partial u}{\partial x}+kw+\frac{1}{2}\left(\frac{\partial w}{\partial x}\right)^2\right)+\right.$$

$$\left.+\left(J_1-\alpha J_3\right)\frac{\partial\varphi}{\partial x}-\alpha J_3\frac{\partial^2 w}{\partial x^2}\right]+\frac{\partial}{\partial x}\left[\left(J_0\left(\frac{\partial u}{\partial x}+kw+\frac{1}{2}\left(\frac{\partial w}{\partial x}\right)^2\right)+\right.\right.$$

$$\left.\left.+\left(J_1-\alpha J_3\right)\frac{\partial\varphi}{\partial x}-\alpha J_3\frac{\partial^2 w}{\partial x^2}\right)\left(\frac{\partial w}{\partial x}\right)\right]+\frac{1}{4}\frac{\partial}{\partial x}\left(\frac{\partial}{\partial x}\left[B_0\left(\frac{\partial\varphi}{\partial x}-\frac{\partial^2 w}{\partial x^2}\right)-\right.\right.$$

$$\left.\left.-3\alpha B_2\left(\frac{\partial\varphi}{\partial x}+\frac{\partial^2 w}{\partial x^2}\right)\right]+k\left[\left(B_0k-6\alpha B_1\right)\frac{\partial w}{\partial x}-6\alpha B_1\varphi\right]\right)+\tag{6.70}$$

$$+\alpha\frac{\partial^2}{\partial x^2}\left(J_3\left(\frac{\partial u}{\partial x}+kw+\frac{1}{2}\left(\frac{\partial w}{\partial x}\right)^2\right)+\left(J_4-\alpha J_6\right)\frac{\partial\varphi}{\partial x}-\alpha J_6\frac{\partial^2 w}{\partial x^2}\right)-$$

$$-3\alpha\frac{\partial}{\partial x}\Bigg\{\left(A_2-3\alpha A_4\right)\left(\varphi+\frac{\partial w}{\partial x}\right)+\frac{1}{2}\left[\left(B_1k-6\alpha B_2\right)\frac{\partial w}{\partial x}-6\alpha B_2\varphi\right]-$$

$$-\frac{3}{4}\frac{\partial}{\partial x}\left[B_2\left(\frac{\partial\varphi}{\partial x}-\frac{\partial^2 w}{\partial x^2}\right)-3\alpha B_4\left(\frac{\partial\varphi}{\partial x}+\frac{\partial^2 w}{\partial x^2}\right)\right]\Bigg\}+$$

$$+\frac{1}{2}\frac{\partial}{\partial x}\left(C_0+3\alpha C_2\right)+kN^T-\frac{\partial}{\partial x}\left(N^T\frac{\partial w}{\partial x}\right)-\alpha\frac{\partial^2 P^T}{\partial x^2}+q=$$

$$=I_0\frac{\partial^2 w}{\partial t^2}+\alpha I_3\frac{\partial^3 u}{\partial x\partial t^2}+\alpha I_4\frac{\partial^3\varphi}{\partial x\partial t^2}-\alpha^2 I_6\left(\frac{\partial^4 w}{\partial x^2\partial t^2}+\frac{\partial^3\varphi}{\partial x\partial t^2}\right).$$

The boundary conditions (6.64)–(6.67) are also rewritten in displacements, i.e. we have

$$J_0\left(\frac{\partial u}{\partial x}+kw+\frac{1}{2}\left(\frac{\partial w}{\partial x}\right)^2\right)+(J_1-\alpha J_3)\frac{\partial \varphi}{\partial x}- \\ -\alpha J_3\frac{\partial^2 w}{\partial x^2}-N^T\Bigg|_{x=0}^{x=L}=\bar{N}\quad \text{or}\quad u|_{x=0}^{x=L}=\bar{u}, \tag{6.71}$$

$$\Bigg\{J_1\left(\frac{\partial u}{\partial x}+kw+\frac{1}{2}\left(\frac{\partial w}{\partial x}\right)^2\right)+(J_2-\alpha\, J_4)\frac{\partial \varphi}{\partial x}-\alpha\, J_4\frac{\partial^2 w}{\partial x^2}- \\ -\alpha\left(J_3\left(\frac{\partial u}{\partial x}+kw+\frac{1}{2}\left(\frac{\partial w}{\partial x}\right)^2\right)+(J_4-\alpha\, J_6)\frac{\partial \varphi}{\partial x}-\alpha\, J_6\frac{\partial^2 w}{\partial x^2}\right)+ \\ +\frac{1}{4}\left(\left[B_0\left(\frac{\partial \varphi}{\partial x}-\frac{\partial^2 w}{\partial x^2}\right)-3\alpha B_2\left(\frac{\partial \varphi}{\partial x}+\frac{\partial^2 w}{\partial x^2}\right)\right]- \\ -3\alpha\left[B_2\left(\frac{\partial \varphi}{\partial x}-\frac{\partial^2 w}{\partial x^2}\right)-3\alpha B_4\left(\frac{\partial \varphi}{\partial x}+\frac{\partial^2 w}{\partial x^2}\right)\right]\right)+ \\ +\alpha P^T-M^T\Bigg\}\Bigg|_{x=0}^{x=L}=\bar{M}\quad \text{or}\quad \varphi|_{x=0}^{x=L}=\bar{\varphi}. \tag{6.72}$$

$$\Bigg\{\alpha^2 I_6\frac{\partial^3 w}{\partial x\partial t^2}-\alpha I_3\frac{\partial^2 u}{\partial t^2}-\alpha\,(I_4-\alpha I_6)\frac{\partial^2 \varphi}{\partial t^2}+\frac{C_0+3\alpha\, C_2}{2}+ \\ +(A_0-3\alpha A_2)\left(\varphi+\frac{\partial w}{\partial x}\right)+\frac{1}{4}k\left[B_0 k\left(\frac{\partial w}{\partial x}\right)-6\alpha B_1\left(\varphi+\frac{\partial w}{\partial x}\right)\right]+ \\ +\frac{1}{4}\frac{\partial}{\partial x}\left[B_0\left(\frac{\partial \varphi}{\partial x}-\frac{\partial^2 w}{\partial x^2}\right)-3\alpha B_2\left(\frac{\partial \varphi}{\partial x}+\frac{\partial^2 w}{\partial x^2}\right)\right]+ \\ +\left[J_0\left(\frac{\partial u}{\partial x}+kw+\frac{1}{2}\left(\frac{\partial w}{\partial x}\right)^2\right)+(J_1-\alpha J_3)\frac{\partial \varphi}{\partial x}-\alpha J_3\frac{\partial^2 w}{\partial x^2}\right]\left(\frac{\partial w}{\partial x}\right)+ \\ +\alpha\frac{\partial}{\partial x}\left[J_3\left(\frac{\partial u}{\partial x}+kw+\frac{1}{2}\left(\frac{\partial w}{\partial x}\right)^2\right)+(J_4-\alpha\, J_6)\frac{\partial \varphi}{\partial x}-\alpha\, J_6\frac{\partial^2 w}{\partial x^2}\right]- \\ -3\alpha\,(A_2-3\alpha A_4)\left(\varphi+\frac{\partial w}{\partial x}\right)-\frac{3}{2}\alpha\left[B_1 k\frac{\partial w}{\partial x}-6\alpha B_2\left(\varphi+\frac{\partial w}{\partial x}\right)\right]+ \\ +\frac{3}{4}\alpha\frac{\partial}{\partial x}\left[B_2\left(\frac{\partial \varphi}{\partial x}-\frac{\partial^2 w}{\partial x^2}\right)-3\alpha B_4\left(\frac{\partial \varphi}{\partial x}+\frac{\partial^2 w}{\partial x^2}\right)\right]- \tag{6.73}$$

$$- N^T \left(\frac{\partial w}{\partial x}\right) - \alpha \frac{\partial P^T}{\partial x} \Bigg\} \Bigg|_{x=0}^{x=L} = \bar{Q} \quad \text{or} \quad w|_{x=0}^{x=L} = \bar{w}$$

$$\left\{ \alpha \left[J_3 \left(\frac{\partial u}{\partial x} + kw + \frac{1}{2}\left(\frac{\partial w}{\partial x}\right)^2 \right) + (J_4 - \alpha J_6)\frac{\partial \varphi}{\partial x} - \alpha J_6 \frac{\partial^2 w}{\partial x^2} \right] + \right.$$

$$+ \frac{1}{2}\left[B_0 \left(\frac{\partial \varphi}{\partial x} - \frac{\partial^2 w}{\partial x^2} \right) - 3\alpha B_2 \left(\frac{\partial \varphi}{\partial x} + \frac{\partial^2 w}{\partial x^2} \right) \right] + \tag{6.74}$$

$$\left. + \frac{3\alpha}{4}\left[B_2 \left(\frac{\partial \varphi}{\partial x} - \frac{\partial^2 w}{\partial x^2} \right) - 3\alpha B_4 \left(\frac{\partial \varphi}{\partial x} + \frac{\partial^2 w}{\partial x^2} \right) \right] - \alpha P^T \right\} \Bigg|_{x=0}^{x=L} = \alpha \bar{P}$$

or

$$\frac{\partial w}{\partial x}\bigg|_{x=0}^{x=L} = \overline{\frac{\partial w}{\partial x}}.$$

Equations of Motion for the Timoshenko and Bernoulli–Euler Models

The second approximation model proposed by Timoshenko [49] does not take into account curving of a normal (see Fig. 6.2) while it accounts only of the normal rotation (see Fig. 6.3).

In the case of the curvilinear shallow Timoshenko panels, in Eqs. (6.68)–(6.74), we take $\alpha = 0$. The obtained equations take the following form:

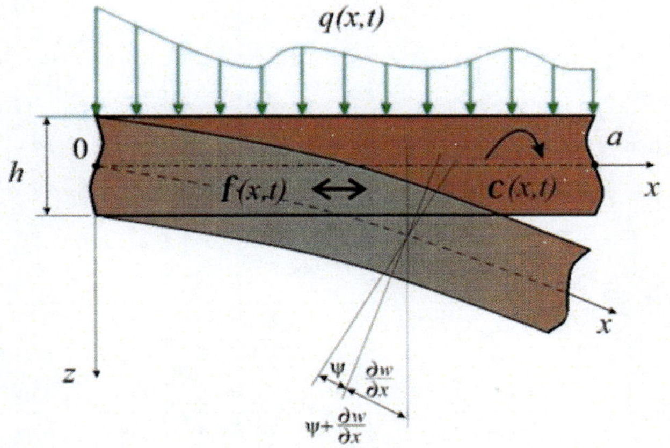

Fig. 6.3 Timoshenko beam subjected to kinematic parameters and loads [reprinted with permission from International Journal of Non-Linear Mechanics publishers]

$$\frac{\partial}{\partial x}\left[J_0\left(\frac{\partial u}{\partial x}+kw+\frac{1}{2}\left(\frac{\partial w}{\partial x}\right)^2\right)+J_1\frac{\partial \varphi}{\partial x}\right]+f-\frac{\partial N^T}{\partial x}=I_0\frac{\partial^2 u}{\partial t^2}+I_1\frac{\partial^2 \varphi}{\partial t^2}, \tag{6.75}$$

$$\begin{gathered}\frac{\partial}{\partial x}\left[J_1\left(\frac{\partial u}{\partial x}+kw+\frac{1}{2}\left(\frac{\partial w}{\partial x}\right)^2\right)+J_2\frac{\partial \varphi}{\partial x}\right]-A_0\left(\varphi+\frac{\partial w}{\partial x}\right)+\\+\frac{1}{4}\frac{\partial}{\partial x}\left[B_0\left(\frac{\partial \varphi}{\partial x}-\frac{\partial^2 w}{\partial x^2}\right)\right]+\frac{C_0}{2}-\frac{\partial M^T}{\partial x}=I_1\frac{\partial^2 u}{\partial t^2}+I_2\frac{\partial^2 \varphi}{\partial t^2},\end{gathered} \tag{6.76}$$

$$\begin{gathered}\frac{\partial}{\partial x}\left[A_0\left(\varphi+\frac{\partial w}{\partial x}\right)\right]-k\left[J_0\left(\frac{\partial u}{\partial x}+kw+\frac{1}{2}\left(\frac{\partial w}{\partial x}\right)^2\right)+J_1\frac{\partial \varphi}{\partial x}\right]+\\+\frac{\partial}{\partial x}\left\{\left[J_0\left(\frac{\partial u}{\partial x}+kw+\frac{1}{2}\left(\frac{\partial w}{\partial x}\right)^2\right)+J_1\frac{\partial \varphi}{\partial x}\right]\left(\frac{\partial w}{\partial x}\right)\right\}+\\+\frac{1}{4}\frac{\partial}{\partial x}\left\{\frac{\partial}{\partial x}\left[B_0\left(\frac{\partial \varphi}{\partial x}-\frac{\partial^2 w}{\partial x^2}\right)\right]+k^2B_0\frac{\partial w}{\partial x}\right\}+\\+\frac{1}{2}\frac{\partial C_0}{\partial x}+kN^T-\frac{\partial}{\partial x}\left(N^T\frac{\partial w}{\partial x}\right)+q=I_0\frac{\partial^2 w}{\partial t^2}\end{gathered} \tag{6.77}$$

with the corresponding boundary conditions

$$J_0\left(\frac{\partial u}{\partial x}+kw+\frac{1}{2}\left(\frac{\partial w}{\partial x}\right)^2\right)+J_1\frac{\partial \varphi}{\partial x}-N^T\Bigg|_{x=0}^{x=L}=\bar{N}\quad\text{or}\quad u|_{x=0}^{x=L}=\bar{u}, \tag{6.78}$$

$$J_1\left(\frac{\partial u}{\partial x}+kw+\frac{1}{2}\left(\frac{\partial w}{\partial x}\right)^2\right)+J_2\frac{\partial \varphi}{\partial x}+\frac{1}{4}B_0\left(\frac{\partial \varphi}{\partial x}-\frac{\partial^2 w}{\partial x^2}\right)-M^T\Bigg|_{x=0}^{x=L}=\bar{M} \tag{6.79}$$

or

$$\varphi|_{x=0}^{x=L}=\bar{\varphi},$$

$$\begin{gathered}\left\{\frac{C_0}{2}+A_0\left(\varphi+\frac{\partial w}{\partial x}\right)+\frac{1}{4}k^2B_0\left(\frac{\partial w}{\partial x}\right)+\frac{1}{4}\frac{\partial}{\partial x}\left[B_0\left(\frac{\partial \varphi}{\partial x}-\frac{\partial^2 w}{\partial x^2}\right)\right]+\right.\\\left.+\left[J_0\left(\frac{\partial u}{\partial x}+kw+\frac{1}{2}\left(\frac{\partial w}{\partial x}\right)^2\right)+J_1\frac{\partial \varphi}{\partial x}\right]\left(\frac{\partial w}{\partial x}\right)-N^T\left(\frac{\partial w}{\partial x}\right)\right\}\Bigg|_{x=0}^{x=L}=\bar{Q}\end{gathered} \tag{6.80}$$

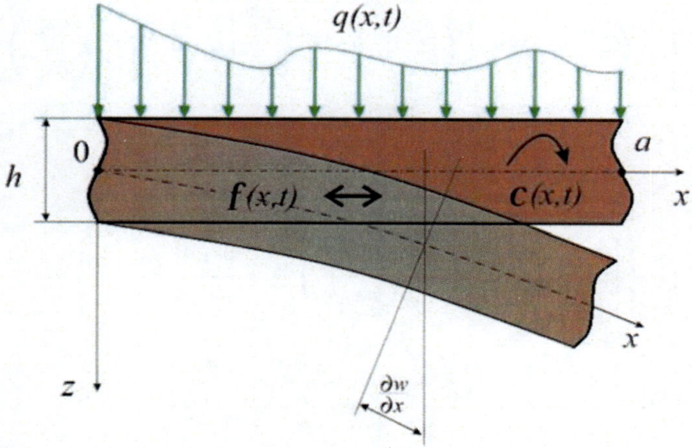

Fig. 6.4 The Bernoulli–Euler beam under kinematic parameters and loads [reprinted with permission from International Journal of Non-Linear Mechanics publishers]

or

$$w|_{x=0}^{x=L} = \bar{w},$$

$$\left(\frac{\partial \varphi}{\partial x} - \frac{\partial^2 w}{\partial x^2}\right)\Bigg|_{x=0}^{x=L} = 0 \quad \text{or} \quad \frac{\partial w}{\partial x}\Bigg|_{x=0}^{x=L} = \overline{\frac{\partial w}{\partial x}}. \tag{6.81}$$

The first-order model for beams was first proposed by Bernoulli [50] and Euler [51], whereas for plates by Kirchhoff [27], and for shells by Love [52]. Its main idea follows: curves being straight and normal to the middle curve (for beams) or surface (for plates and shells) before deformations remain normal and straight with regard to the middle curve or surface after deformations and their length is conserved. On contrary to Figs. 6.2 and 6.3, the geometric interpretation of the Bernoulli–Euler model shown in Fig. 6.4 exhibits only the beam bending without rotation and curvature of the transverse cross section.

The resolving equations based on the Bernoulli–Euler hypotheses follow from Eqs. (6.75)–(6.81), if we take $-\frac{\partial w(x,t)}{\partial x}$ instead of the function $\varphi(x, t)$. In result, the following system of governing equations is obtained:

$$\frac{\partial}{\partial x}\left[J_0\left(\frac{\partial u}{\partial x} + kw + \frac{1}{2}\left(\frac{\partial w}{\partial x}\right)^2\right) - J_1\frac{\partial^2 w}{\partial x^2}\right] + f - \frac{\partial N^T}{\partial x} = I_0\frac{\partial^2 u}{\partial t^2} - I_1\frac{\partial^3 w}{\partial t^2 \partial x}, \tag{6.82}$$

$$\frac{\partial^2}{\partial x^2}\left[J_1\left(\frac{\partial u}{\partial x}+kw+\frac{1}{2}\left(\frac{\partial w}{\partial x}\right)^2\right)-J_2\frac{\partial^2 w}{\partial x^2}\right]-\frac{\partial^2}{\partial x^2}\left[B_0\left(\frac{\partial^2 w}{\partial x^2}\right)\right]-$$

$$-k\left[J_0\left(\frac{\partial u}{\partial x}+kw+\frac{1}{2}\left(\frac{\partial w}{\partial x}\right)^2\right)-J_1\frac{\partial^2 w}{\partial x^2}\right]+$$

$$+\frac{\partial}{\partial x}\left\{\left[J_0\left(\frac{\partial u}{\partial x}+kw+\frac{1}{2}\left(\frac{\partial w}{\partial x}\right)^2\right)-J_1\frac{\partial^2 w}{\partial x^2}\right]\left(\frac{\partial w}{\partial x}\right)\right\}+ \tag{6.83}$$

$$+\frac{k^2}{4}\frac{\partial}{\partial x}\left\{B_0\frac{\partial w}{\partial x}\right\}+\frac{\partial C_0}{\partial x}+kN^T-\frac{\partial}{\partial x}\left(N^T\frac{\partial w}{\partial x}\right)-\frac{\partial^2 M^T}{\partial x^2}+q=$$

$$=I_0\frac{\partial^2 w}{\partial t^2}+I_1\frac{\partial^3 u}{\partial t^2\partial x}-I_2\frac{\partial^4 w}{\partial t^2\partial x^2}$$

with the corresponding boundary conditions

$$J_0\left(\frac{\partial u}{\partial x}+kw+\frac{1}{2}\left(\frac{\partial w}{\partial x}\right)^2\right)-J_1\frac{\partial^2 w}{\partial x^2}-N^T\Bigg|_{x=0}^{x=L}=\bar{N}\quad\text{or}\quad u|_{x=0}^{x=L}=\bar{u}, \tag{6.84}$$

$$J_1\left(\frac{\partial u}{\partial x}+kw+\frac{1}{2}\left(\frac{\partial w}{\partial x}\right)^2\right)-J_2\frac{\partial^2 w}{\partial x^2}-B_0\frac{\partial^2 w}{\partial x^2}-M^T\Bigg|_{x=0}^{x=L}=\bar{M} \tag{6.85}$$

or

$$\frac{\partial w}{\partial x}\Bigg|_{x=0}^{x=L}=\overline{\frac{\partial w}{\partial x}},$$

$$\left\{\frac{\partial}{\partial x}\left[J_1\left(\frac{\partial u}{\partial x}+kw+\frac{1}{2}\left(\frac{\partial w}{\partial x}\right)^2\right)-J_2\frac{\partial^2 w}{\partial x^2}\right]+\frac{1}{4}k^2B_0\left(\frac{\partial w}{\partial x}\right)-\frac{\partial}{\partial x}\left[B_0\frac{\partial^2 w}{\partial x^2}\right]+\right.$$
$$\left.+\left[J_0\left(\frac{\partial u}{\partial x}+kw+\frac{1}{2}\left(\frac{\partial w}{\partial x}\right)^2\right)-J_1\frac{\partial^2 w}{\partial x^2}\right]\left(\frac{\partial w}{\partial x}\right)-N^T\left(\frac{\partial w}{\partial x}\right)-\frac{\partial M^T}{\partial x}+\frac{C_0}{2}\right\}\Bigg|_{x=0}^{x=L}=\bar{Q} \tag{6.86}$$

or

$$w|_{x=0}^{x=L}=\bar{w}.$$

In Sect. 6.4, the resolving equations of motion of size-dependent Sheremetev–Pelekh beam on the basis of modified couple stress theory have been obtained. If we take the material length parameter $l=0$ in (6.56) and if we assume a homogeneous panel then $B_i = 0$ $(i = 1, \ldots, 4)$, $J_1 = J_3 = J_5 = 0$, $I_1 = I_3 = I_5 = 0$. In this case the resolving Eqs. (6.68)–(6.70) and the corresponding boundary conditions (6.71)–(6.74) coincide with the classical Sheremetev–Pelekh equations [53]. The obtained equations of motion are general, i.e. they imply the Timoshenko and Bernoulli–Euler equations with/without the size-dependent behaviour.

For our further investigations, we introduce the following non-dimensional (with bars) parameters:

$$\begin{aligned}
&\bar{x} = x/L\,,\ \ \bar{z} = z/h\,,\ \ \bar{w} = w/h\,,\ \ \bar{\varphi} = \varphi L/h\,,\ \ \bar{u} = u\,L/h^2\,,\ \ \bar{E} = E/E_0,\\
&\bar{G} = G/E_0,\ \ \bar{\rho} = \rho/\rho_0\,,\ \ \lambda = L/h\,,\ \ \bar{t} = \omega_0 t\,,\ \ \omega_0 = \sqrt{E_0/\rho_0\lambda^2 L^2}\,,\\
&\bar{k} = kL^2/h\,,\ \ \bar{T} = \gamma T,\ \ \bar{N}^T = N^T\lambda^2/AE_0\,,\ \ \bar{M}^T = M^T\lambda^2/AE_0 h\,,\\
&\bar{P}^T = P^T\lambda^2/AE_0 h^3,\ \ \bar{f} = \lambda^4 Lf/AE_0,\ \ \bar{q} = \lambda^3 Lq/AE_0\,,\\
&\bar{C}_0 = \lambda^3 C_0/AE_0,\ \ \bar{C}_2 = \lambda^3 C_2/AE_0 h^2\,,\ \ \bar{\alpha} = \alpha h^2\,.
\end{aligned} \tag{6.87}$$

First, we compute the non-dimensional values of the resultants

$$\begin{aligned}
&(\bar{J}_0,\ \bar{J}_1,\ \bar{J}_2,\ \bar{J}_3,\ \bar{J}_4,\ \bar{J}_5,\ \bar{J}_6) = \int_{-1/2}^{1/2} \bar{E}\,(1,\ \bar{z},\ \bar{z}^2,\ \bar{z}^3,\ \bar{z}^4,\ \bar{z}^5,\ \bar{z}^6)\,d\bar{z} =\\
&= \frac{1}{AE_0}\left(\frac{J_0}{1},\ \frac{J_1}{h},\ \frac{J_2}{h^2},\ \frac{J_3}{h^3},\ \frac{J_4}{h^4},\ \frac{J_5}{h^5},\ \frac{J_6}{h^6}\right),\\
&(\bar{I}_0,\ \bar{I}_1,\ \bar{I}_2,\ \bar{I}_3,\ \bar{I}_4,\ \bar{I}_5,\ \bar{I}_6) = \int_{-1/2}^{1/2} \bar{\rho}\,(1,\ \bar{z},\ \bar{z}^2,\ \bar{z}^3,\ \bar{z}^4,\ \bar{z}^5,\ \bar{z}^6)\,d\bar{z} =\\
&= \frac{1}{A\rho_0}\left(\frac{I_0}{1},\ \frac{I_1}{h},\ \frac{I_2}{h^2},\ \frac{I_3}{h^3},\ \frac{I_4}{h^4},\ \frac{I_5}{h^5},\ \frac{I_6}{h^6}\right),\\
&(\bar{A}_0,\ \bar{A}_2,\ \bar{A}_4) = \int_{-\frac{1}{2}}^{\frac{1}{2}} \overline{G}\,(1,\ \bar{z}^2,\ \bar{z}^4)\,d\bar{z} = \frac{1}{AE_0}\left(\frac{A_0}{1},\ \frac{A_2}{h^2},\ \frac{A_4}{h^4}\right),
\end{aligned} \tag{6.88}$$

$$(\bar{B}_0,\ \bar{B}_1,\ \bar{B}_2,\ \bar{B}_4) = \int_{-\frac{1}{2}}^{\frac{1}{2}} \overline{G}\left(\frac{l}{h}\right)^2 (1,\ \bar{z},\ \bar{z}^2,\ \bar{z}^4)d\bar{z} = \frac{1}{AE_0 h^2}\left(\frac{B_0}{1},\ \frac{B_1}{h},\ \frac{B_2}{h^2},\ \frac{B_3}{h^3}\right),$$

$$\bar{M}^T = \lambda^2 \int_{-\frac{1}{2}}^{\frac{1}{2}} \bar{E}\bar{T}\bar{z}d\bar{z},\ \ \bar{N}^T = \lambda^2 \int_{-\frac{1}{2}}^{\frac{1}{2}} \bar{E}\bar{T}d\bar{z}\,,\ \ \bar{P}^T = \lambda^2 \int_{-\frac{1}{2}}^{\frac{1}{2}} \bar{E}\bar{T}\bar{z}^3 d\bar{z}\,.$$

Equations (6.68)–(6.74) are recast to the following counterpart non-dimensional forms (bars over the non-dimensional parameters are omitted):

$$\lambda^2\frac{\partial}{\partial x}\left[J_0\left(\frac{\partial u}{\partial x}+kw+\frac{1}{2}\left(\frac{\partial w}{\partial x}\right)^2\right)+(J_1-\alpha J_3)\frac{\partial \varphi}{\partial x}-\alpha J_3\frac{\partial^2 w}{\partial x^2}\right]+$$
$$+f-\frac{\partial N^T}{\partial x}=I_0\frac{\partial^2 u}{\partial t^2}+(I_1-\alpha I_3)\frac{\partial^2 \varphi}{\partial t^2}-\alpha I_3\frac{\partial^3 w}{\partial x\partial t^2}, \tag{6.89}$$

$$\lambda^2\frac{\partial}{\partial x}\left[J_1\left(\frac{\partial u}{\partial x}+kw+\frac{1}{2}\left(\frac{\partial w}{\partial x}\right)^2\right)+J_2\frac{\partial \varphi}{\partial x}-\alpha J_4\left(\frac{\partial \varphi}{\partial x}+\frac{\partial^2 w}{\partial x^2}\right)\right]-$$
$$-\lambda^4(A_0-3\alpha A_2)\left(\varphi+\frac{\partial w}{\partial x}\right)+3\alpha\left[\lambda^4(A_2-3\alpha A_4)\left(\varphi+\frac{\partial w}{\partial x}\right)+\right.$$
$$\left.+\frac{1}{2}\left(\lambda^2B_1k\frac{\partial w}{\partial x}-6\alpha B_2\lambda^4\left(\varphi+\frac{\partial w}{\partial x}\right)\right)\right]+$$
$$+\frac{\lambda^2}{4}\frac{\partial}{\partial x}\left(B_0\left(\frac{\partial \varphi}{\partial x}-\frac{\partial^2 w}{\partial x^2}\right)-3\alpha B_2\left(\frac{\partial \varphi}{\partial x}+\frac{\partial^2 w}{\partial x^2}\right)\right)- \tag{6.90}$$
$$+\frac{\lambda^2}{4}\frac{\partial}{\partial x}\left(B_0\left(\frac{\partial \varphi}{\partial x}-\frac{\partial^2 w}{\partial x^2}\right)-3\alpha B_2\left(\frac{\partial \varphi}{\partial x}+\frac{\partial^2 w}{\partial x^2}\right)\right)-$$
$$-\alpha\lambda^2\frac{\partial}{\partial x}\left\{J_3\left(\frac{\partial u}{\partial x}+kw+\frac{1}{2}\left(\frac{\partial w}{\partial x}\right)^2\right)+(J_4-\alpha J_6)\frac{\partial \varphi}{\partial x}-\alpha J_6\frac{\partial^2 w}{\partial x^2}+\right.$$
$$\left.+\frac{3}{4}\left[B_2\left(\frac{\partial \varphi}{\partial x}-\frac{\partial^2 w}{\partial x^2}\right)-3\alpha B_4\left(\frac{\partial \varphi}{\partial x}+\frac{\partial^2 w}{\partial x^2}\right)\right]\right\}+\frac{\lambda^2}{2}(C_0-3\alpha C_2)-$$
$$-\lambda^2\frac{\partial M^T}{\partial x}+\alpha\lambda^2\frac{\partial P^T}{\partial x}=(I_1-\alpha I_3)\frac{\partial^2 u}{\partial t^2}+(I_2-\alpha I_4)\frac{\partial^2 \varphi}{\partial t^2}-$$
$$-\alpha\,(I_4-\alpha I_6)\left(\frac{\partial^2 \varphi}{\partial t^2}+\frac{\partial^3 w}{\partial x\partial t^2}\right),$$

$$\lambda^2\frac{\partial}{\partial x}\left[(A_0-3\alpha A_2)\left(\varphi+\frac{\partial w}{\partial x}\right)\right]-k\left[J_0\left(\frac{\partial u}{\partial x}+kw+\frac{1}{2}\left(\frac{\partial w}{\partial x}\right)^2\right)+\right.$$

$$
\begin{aligned}
&+ (J_1 - \alpha J_3)\frac{\partial \varphi}{\partial x} - \alpha J_3 \frac{\partial^2 w}{\partial x^2}\Bigg] + \frac{\partial}{\partial x}\Bigg[\Bigg(J_0\left(\frac{\partial u}{\partial x} + kw + \frac{1}{2}\left(\frac{\partial w}{\partial x}\right)^2\right) + \\
&+ (J_1 - \alpha J_3)\frac{\partial \varphi}{\partial x} - \alpha J_3 \frac{\partial^2 w}{\partial x^2}\Bigg)\left(\frac{\partial w}{\partial x}\right)\Bigg] + \frac{1}{4}\frac{\partial}{\partial x}\left(\frac{\partial}{\partial x}\left[B_0\left(\frac{\partial \varphi}{\partial x} - \frac{\partial^2 w}{\partial x^2}\right) - \right.\right. \\
&\left.\left. -3\alpha B_2\left(\frac{\partial \varphi}{\partial x} + \frac{\partial^2 w}{\partial x^2}\right)\right] + k\left[(B_0 k - 6\alpha B_1)\frac{\partial w}{\partial x} - 6\alpha B_1 \varphi\right]\right) + \qquad (6.91) \\
&+\alpha \frac{\partial^2}{\partial x^2}\left(J_3\left(\frac{\partial u}{\partial x} + kw + \frac{1}{2}\left(\frac{\partial w}{\partial x}\right)^2\right) + (J_4 - \alpha J_6)\frac{\partial \varphi}{\partial x} - \alpha J_6 \frac{\partial^2 w}{\partial x^2}\right) - \\
&-3\lambda^2\alpha \frac{\partial}{\partial x}\left\{(A_2 - 3\alpha A_4)\left(\varphi + \frac{\partial w}{\partial x}\right) + \frac{1}{2}\left[(B_1 k - 6\alpha B_2)\frac{\partial w}{\partial x} - 6\alpha B_2 \varphi\right] - \right. \\
&\left. -\frac{3}{4}\frac{\partial}{\partial x}\left[B_2\left(\frac{\partial \varphi}{\partial x} - \frac{\partial^2 w}{\partial x^2}\right) - 3\alpha B_4\left(\frac{\partial \varphi}{\partial x} + \frac{\partial^2 w}{\partial x^2}\right)\right]\right\} + \\
&+\frac{1}{2}\frac{\partial}{\partial x}(C_0 + 3\alpha C_2) + kN^T - \frac{\partial}{\partial x}\left(N^T \frac{\partial w}{\partial x}\right) - \alpha \frac{\partial^2 P^T}{\partial x^2} + q = \\
&= I_0 \frac{\partial^2 w}{\partial t^2} + \alpha I_3 \frac{\partial^3 u}{\partial x \partial t^2} + \alpha I_4 \frac{\partial^3 \varphi}{\partial x \partial t^2} - \alpha^2 I_6\left(\frac{\partial^4 w}{\partial x^2 \partial t^2} + \frac{\partial^3 \varphi}{\partial x \partial t^2}\right)
\end{aligned}
$$

and with the boundary conditions

$$
\left[J_0\left(\frac{\partial u}{\partial x} + kw + \frac{1}{2}\left(\frac{\partial w}{\partial x}\right)^2\right) + (J_1 - \alpha J_3)\frac{\partial \varphi}{\partial x} - \alpha J_3 \frac{\partial^2 w}{\partial x^2}\right] - N^T \Bigg|_{x=0}^{x=1} = \bar{N} \tag{6.92}
$$

or

$$
u|_{x=0}^{x=1} = \bar{u},
$$

$$
\begin{aligned}
&\left\{J_1\left(\frac{\partial u}{\partial x} + kw + \frac{1}{2}\left(\frac{\partial w}{\partial x}\right)^2\right) + (J_2 - \alpha J_4)\frac{\partial \varphi}{\partial x} - \alpha J_4 \frac{\partial^2 w}{\partial x^2} - \right. \\
&-\alpha\left(J_3\left(\frac{\partial u}{\partial x} + kw + \frac{1}{2}\left(\frac{\partial w}{\partial x}\right)^2\right) + (J_4 - \alpha J_6)\frac{\partial \varphi}{\partial x} - \alpha J_6 \frac{\partial^2 w}{\partial x^2}\right) + \\
&+\frac{1}{4}\left(\left[B_0\left(\frac{\partial \varphi}{\partial x} - \frac{\partial^2 w}{\partial x^2}\right) - 3\alpha B_2\left(\frac{\partial \varphi}{\partial x} + \frac{\partial^2 w}{\partial x^2}\right)\right] - \right. \qquad (6.93)
\end{aligned}
$$

$$- 3\alpha\left[B_2\left(\frac{\partial\varphi}{\partial x}-\frac{\partial^2 w}{\partial x^2}\right)-3\alpha B_4\left(\frac{\partial\varphi}{\partial x}+\frac{\partial^2 w}{\partial x^2}\right)\right]\Bigg)+$$

$$+\alpha P^T - M^T\Big\}\Big|_{x=0}^{x=1} = \bar{M} \quad \text{or} \quad \varphi|_{x=0}^{x=1} = \bar{\varphi},$$

$$\Bigg\{\alpha^2 I_6\frac{\partial^3 w}{\partial x\partial t^2} - \alpha I_3\frac{\partial^2 u}{\partial t^2} - \alpha\left(I_4-\alpha I_6\right)\frac{\partial^2\varphi}{\partial t^2} + \frac{C_0+3\alpha C_2}{2} +$$

$$+\left(A_0-3\alpha A_2\right)\left(\varphi+\frac{\partial w}{\partial x}\right)+\frac{1}{4}k\left[B_0 k\left(\frac{\partial w}{\partial x}\right)-6\alpha B_1\left(\varphi+\frac{\partial w}{\partial x}\right)\right]+$$

$$+\frac{1}{4}\frac{\partial}{\partial x}\left[B_0\left(\frac{\partial\varphi}{\partial x}-\frac{\partial^2 w}{\partial x^2}\right)-3\alpha B_2\left(\frac{\partial\varphi}{\partial x}+\frac{\partial^2 w}{\partial x^2}\right)\right]+ \tag{6.94}$$

$$+\left[J_0\left(\frac{\partial u}{\partial x}+kw+\frac{1}{2}\left(\frac{\partial w}{\partial x}\right)^2\right)+\left(J_1-\alpha J_3\right)\frac{\partial\varphi}{\partial x}-\alpha J_3\frac{\partial^2 w}{\partial x^2}\right]\left(\frac{\partial w}{\partial x}\right)+$$

$$+\alpha\frac{\partial}{\partial x}\left[J_3\left(\frac{\partial u}{\partial x}+kw+\frac{1}{2}\left(\frac{\partial w}{\partial x}\right)^2\right)+\left(J_4-\alpha J_6\right)\frac{\partial\varphi}{\partial x}-\alpha J_6\frac{\partial^2 w}{\partial x^2}\right]-$$

$$-3\alpha\left(A_2-3\alpha A_4\right)\left(\varphi+\frac{\partial w}{\partial x}\right)-\frac{3}{2}\alpha\left[B_1 k\frac{\partial w}{\partial x}-6\alpha B_2\left(\varphi+\frac{\partial w}{\partial x}\right)\right]+$$

$$+\frac{3}{4}\alpha\frac{\partial}{\partial x}\left[B_2\left(\frac{\partial\varphi}{\partial x}-\frac{\partial^2 w}{\partial x^2}\right)-3\alpha B_4\left(\frac{\partial\varphi}{\partial x}+\frac{\partial^2 w}{\partial x^2}\right)\right]-N^T\left(\frac{\partial w}{\partial x}\right)-\alpha\frac{\partial P^T}{\partial x}\Bigg\}\Bigg|_{x=0}^{x=1}=\bar{Q}$$

or

$$w|_{x=0}^{x=1} = \bar{w}$$

$$\left(\frac{\partial\varphi}{\partial x}-\frac{\partial^2 w}{\partial x^2}\right)\Bigg|_{x=0}^{x=1} = 0 \quad \text{or} \quad \frac{\partial w}{\partial x}\Bigg|_{x=0}^{x=1} = \overline{\frac{\partial w}{\partial x}}. \tag{6.95}$$

The initial conditions are as follows:

$$\begin{aligned} w(x,0) &= \psi_1(x); & \left.\frac{\partial w(x,t)}{\partial t}\right|_{t=0} &= \psi_2(x), \\ u(x,0) &= \psi_3(x); & \left.\frac{\partial u(x,t)}{\partial t}\right|_{t=0} &= \psi_4(x), \\ \varphi(x,0) &= \psi_5(x); & \left.\frac{\partial \varphi(x,t)}{\partial t}\right|_{t=0} &= \psi_6(x). \end{aligned} \tag{6.96}$$

Introduction of the dimensionless variables through formulas (6.87) (6.88), and the simplified assumptions being analogous to that employed the Sheremetev–Pelekh model and taking $\alpha = 0$ in relations (6.89)–(6.95) yields the non-dimensional form (bars are omitted) of the governing equations for the Timoshenko model:

$$\lambda^2 \frac{\partial}{\partial x}\left[J_0\left(\frac{\partial u}{\partial x} + kw + \frac{1}{2}\left(\frac{\partial w}{\partial x}\right)^2\right) + J_1\frac{\partial \varphi}{\partial x}\right] + f - \lambda^2 \frac{\partial N^T}{\partial x} = I_0 \frac{\partial^2 u}{\partial t^2} + I_1 \frac{\partial^2 \varphi}{\partial t^2}, \tag{6.97}$$

$$\begin{aligned}
&\lambda^2 \frac{\partial}{\partial x}\left[J_1\left(\frac{\partial u}{\partial x} + kw + \frac{1}{2}\left(\frac{\partial w}{\partial x}\right)^2\right) + J_2\frac{\partial \varphi}{\partial x}\right] - \\
&- \lambda^4 A_0\left(\varphi + \frac{\partial w}{\partial x}\right) + \frac{\lambda^2}{4}\frac{\partial}{\partial x}\left(B_0\left(\frac{\partial \varphi}{\partial x} - \frac{\partial^2 w}{\partial x^2}\right)\right) + \\
&+ \lambda^2 \frac{C_0}{2} - \lambda^2 \frac{\partial M^T}{\partial x} = I_1 \frac{\partial^2 u}{\partial t^2} + I_2 \frac{\partial^2 \varphi}{\partial t^2},
\end{aligned} \tag{6.98}$$

$$\begin{aligned}
&\lambda^2 \frac{\partial}{\partial x}\left[A_0\left(\varphi + \frac{\partial w}{\partial x}\right)\right] - k\left[J_0\left(\frac{\partial u}{\partial x} + kw + \frac{1}{2}\left(\frac{\partial w}{\partial x}\right)^2\right) + J_1\frac{\partial \varphi}{\partial x}\right] + \\
&+ \frac{\partial}{\partial x}\left\{\left[J_0\left(\frac{\partial u}{\partial x} + kw + \frac{1}{2}\left(\frac{\partial w}{\partial x}\right)^2\right) + J_1\frac{\partial \varphi}{\partial x}\right]\left(\frac{\partial w}{\partial x}\right)\right\} + \\
&+ \frac{1}{4}\frac{\partial}{\partial x}\left\{\frac{\partial}{\partial x}\left[B_0\left(\frac{\partial \varphi}{\partial x} - \frac{\partial^2 w}{\partial x^2}\right)\right] + \frac{k^2 B_0}{\lambda^2}\frac{\partial w}{\partial x}\right\} + \\
&+ \frac{1}{2}\frac{\partial C_0}{\partial x} + kN^T - \frac{\partial}{\partial x}\left(N^T \frac{\partial w}{\partial x}\right) + q = I_0 \frac{\partial^2 w}{\partial t^2}
\end{aligned} \tag{6.99}$$

with the corresponding boundary conditions

$$J_0\left(\frac{\partial u}{\partial x} + kw + \frac{1}{2}\left(\frac{\partial w}{\partial x}\right)^2\right) + J_1\frac{\partial \varphi}{\partial x} - N^T\Bigg|_{x=0}^{x=1} = \bar{N} \quad \text{or} \quad u|_{x=0}^{x=1} = \bar{u}, \tag{6.100}$$

$$J_1\left(\frac{\partial u}{\partial x} + kw + \frac{1}{2}\left(\frac{\partial w}{\partial x}\right)^2\right) + J_2\frac{\partial \varphi}{\partial x} + \frac{1}{4}B_0\left(\frac{\partial \varphi}{\partial x} - \frac{\partial^2 w}{\partial x^2}\right) - M^T\Bigg|_{x=0}^{x=1} = \bar{M} \tag{6.101}$$

or

$$\varphi|_{x=0}^{x=1} = \bar{\varphi},$$

$$\left\{\frac{C_0}{2}+\lambda^2 A_0\left(\varphi+\frac{\partial w}{\partial x}\right)+\frac{1}{4\lambda^2}k^2 B_0\left(\frac{\partial w}{\partial x}\right)+\frac{1}{4}\frac{\partial}{\partial x}\left[B_0\left(\frac{\partial \varphi}{\partial x}-\frac{\partial^2 w}{\partial x^2}\right)\right]+\right.$$
$$\left.+\left[J_0\left(\frac{\partial u}{\partial x}+kw+\frac{1}{2}\left(\frac{\partial w}{\partial x}\right)^2\right)+J_1\frac{\partial \varphi}{\partial x}\right]\left(\frac{\partial w}{\partial x}\right)-N^T\left(\frac{\partial w}{\partial x}\right)\right\}\Bigg|_{x=0}^{x=1}=\bar{Q} \tag{6.102}$$

or

$$w|_{x=0}^{x=1}=\bar{w},$$

$$\left(\frac{\partial \varphi}{\partial x}-\frac{\partial^2 w}{\partial x^2}\right)\Bigg|_{x=0}^{x=1}=0 \quad \text{or} \quad \frac{\partial w}{\partial x}\Bigg|_{x=0}^{x=1}=\overline{\frac{\partial w}{\partial x}}. \tag{6.103}$$

The initial conditions remain the same as in the case of the Sheremetev–Pelekh beam (6.96).

Equations of beam motion based on the Bernoulli–Euler hypotheses are yielded by relations (6.97)–(6.103), and they have the following form:

$$\lambda^2\frac{\partial}{\partial x}\left[J_0\left(\frac{\partial u}{\partial x}+kw+\frac{1}{2}\left(\frac{\partial w}{\partial x}\right)^2\right)-J_1\frac{\partial^2 w}{\partial x^2}\right]+$$
$$+f-\lambda^2\frac{\partial N^T}{\partial x}=I_0\frac{\partial^2 u}{\partial t^2}-I_1\frac{\partial^3 w}{\partial t^2\partial x}, \tag{6.104}$$

$$\frac{\partial^2}{\partial x^2}\left[J_1\left(\frac{\partial u}{\partial x}+kw+\frac{1}{2}\left(\frac{\partial w}{\partial x}\right)^2\right)-J_2\frac{\partial^2 w}{\partial x^2}\right]-\frac{\partial^2}{\partial x^2}\left[B_0\left(\frac{\partial^2 w}{\partial x^2}\right)\right]-$$
$$-k\left[J_0\left(\frac{\partial u}{\partial x}+kw+\frac{1}{2}\left(\frac{\partial w}{\partial x}\right)^2\right)-J_1\frac{\partial^2 w}{\partial x^2}\right]+$$
$$+\frac{\partial}{\partial x}\left\{\left[J_0\left(\frac{\partial u}{\partial x}+kw+\frac{1}{2}\left(\frac{\partial w}{\partial x}\right)^2\right)-J_1\frac{\partial^2 w}{\partial x^2}\right]\left(\frac{\partial w}{\partial x}\right)\right\}+ \tag{6.105}$$
$$+\frac{1}{\lambda^2}\frac{k^2}{4}\frac{\partial}{\partial x}\left\{B_0\frac{\partial w}{\partial x}\right\}+\frac{\partial C_0}{\partial x}+kN^T-\frac{\partial}{\partial x}\left(N^T\frac{\partial w}{\partial x}\right)-\frac{\partial^2 M^T}{\partial x^2}+q=$$
$$=I_0\frac{\partial^2 w}{\partial t^2}+\frac{I_1}{\lambda^2}\frac{\partial^3 u}{\partial t^2\partial x}-\frac{I_2}{\lambda^2}\frac{\partial^4 w}{\partial t^2\partial x^2}.$$

The corresponding boundary conditions are as follows:

$$J_0\left(\frac{\partial u}{\partial x}+kw+\frac{1}{2}\left(\frac{\partial w}{\partial x}\right)^2\right)-J_1\frac{\partial^2 w}{\partial x^2}-N^T\Bigg|_{x=0}^{x=L}=\bar{N}\quad\text{or}\quad u|_{x=0}^{x=1}=0, \tag{6.106}$$

$$\begin{aligned}&\left\{\frac{\partial}{\partial x}\left[J_1\left(\frac{\partial u}{\partial x}+kw+\frac{1}{2}\left(\frac{\partial w}{\partial x}\right)^2\right)-J_2\frac{\partial^2 w}{\partial x^2}\right]+\frac{1}{\lambda^2}\frac{1}{4}k^2B_0\left(\frac{\partial w}{\partial x}\right)-\right.\\&-\frac{\partial}{\partial x}\left[B_0\frac{\partial^2 w}{\partial x^2}\right]+\left[J_0\left(\frac{\partial u}{\partial x}+kw+\frac{1}{2}\left(\frac{\partial w}{\partial x}\right)^2\right)-J_1\frac{\partial^2 w}{\partial x^2}\right]\left(\frac{\partial w}{\partial x}\right)-\\&\left.-N^T\left(\frac{\partial w}{\partial x}\right)-\frac{\partial M^T}{\partial x}+\frac{C_0}{2}\right\}\Bigg|_{x=0}^{x=1}=\bar{Q}\quad\text{or}\quad w|_{x=0}^{x=1}=\bar{w},\end{aligned} \tag{6.107}$$

$$J_1\left(\frac{\partial u}{\partial x}+kw+\frac{1}{2}\left(\frac{\partial w}{\partial x}\right)^2\right)-J_2\frac{\partial^2 w}{\partial x^2}-B_0\frac{\partial^2 w}{\partial x^2}-M^T\Bigg|_{x=0}^{x=1}=\bar{M} \tag{6.108}$$

$$\text{or}\quad \frac{\partial w}{\partial x}\Bigg|_{x=0}^{x=1}=0.$$

The initial conditions for the Bernoulli–Euler beam are introduced only for u and w:

$$\begin{aligned}&w(x,0)=\psi_1(x),\qquad \frac{\partial w(x,t)}{\partial t}\Bigg|_{t=0}=\psi_2(x),\\&u(x,0)=\psi_3(x)\qquad \frac{\partial u(x,t)}{\partial t}\Bigg|_{t=0}=\psi_4(x).\end{aligned} \tag{6.109}$$

6.5 Static Solutions

For the purpose of further numerical static investigations, the following simplifications are introduced: the beam is homogeneous, straight line and $f(x,t)=0$, $C(x,t)=0$, $T=0$, and the beam material is linearly elastic and isotropic. Since the beam is homogeneous and straight line, then we take $k=0$, $B_1=0$, $J_1=J_3=J_5=0$, $I_1=I_3=I_5=0$ in Eqs. (6.89)–(6.95). In result, for the Sheremetev–Pelekh model, we obtain the following system of nonlinear PDEs:

$$\lambda^2\frac{\partial}{\partial x}\left[J_0\left(\frac{\partial u}{\partial x}+\frac{1}{2}\left(\frac{\partial w}{\partial x}\right)^2\right)\right]=I_0\frac{\partial^2 u}{\partial t^2}, \tag{6.110}$$

$$\lambda^2\frac{\partial}{\partial x}\left[J_2\frac{\partial\varphi}{\partial x}-\alpha J_4\left(\frac{\partial\varphi}{\partial x}+\frac{\partial^2 w}{\partial x^2}\right)\right]-\lambda^4(A_0-3\alpha A_2)\left(\varphi+\frac{\partial w}{\partial x}\right)+$$
$$+3\alpha\left[\lambda^4(A_2-3\alpha A_4)\left(\varphi+\frac{\partial w}{\partial x}\right)-3\alpha B_2\lambda^4\left(\varphi+\frac{\partial w}{\partial x}\right)\right]+$$
$$+\frac{\lambda^2}{4}\frac{\partial}{\partial x}\left(B_0\left(\frac{\partial\varphi}{\partial x}-\frac{\partial^2 w}{\partial x^2}\right)-3\alpha B_2\left(\frac{\partial\varphi}{\partial x}+\frac{\partial^2 w}{\partial x^2}\right)\right)- \tag{6.111}$$
$$-\alpha\lambda^2\frac{\partial}{\partial x}\left\{(J_4-\alpha J_6)\frac{\partial\varphi}{\partial x}-\alpha J_6\frac{\partial^2 w}{\partial x^2}+\frac{3}{4}\left[B_2\left(\frac{\partial\varphi}{\partial x}-\frac{\partial^2 w}{\partial x^2}\right)-3\alpha B_4\left(\frac{\partial\varphi}{\partial x}+\frac{\partial^2 w}{\partial x^2}\right)\right]\right\}$$
$$+\frac{\lambda^2}{2}(C_0-3\alpha C_2)=\frac{\partial^2 u}{\partial t^2}+(I_2-\alpha I_4)\frac{\partial^2\varphi}{\partial t^2}-\alpha\,(I_4-\alpha I_6)\left(\frac{\partial^2\varphi}{\partial t^2}+\frac{\partial^3 w}{\partial x\partial t^2}\right),$$

$$\lambda^2\frac{\partial}{\partial x}\left[(A_0-3\alpha_2)\left(\varphi+\frac{w}{\partial x}\right)\right]+\frac{\partial}{\partial x}\left[\left(J_0\left(\frac{\partial u}{\partial x}+\frac{1}{2}\left(\frac{\partial w}{\partial x}\right)^2\right)\right)\left(\frac{\partial w}{\partial x}\right)\right]+$$
$$+\frac{1}{4}\frac{\partial}{\partial x}\left(\frac{\partial}{\partial x}\left[B_0\left(\frac{\partial\varphi}{\partial x}-\frac{\partial^2 w}{\partial x^2}\right)-3\alpha B_2\left(\varphi+\frac{\partial^2 w}{\partial x^2}\right)\right]\right)+$$
$$+\alpha\frac{\partial^2}{\partial x^2}\left((J_4-\alpha J_6)\frac{\partial\varphi}{\partial x}-\alpha J_6\frac{\partial^2 w}{\partial x^2}\right)-$$
$$-3\lambda^2\alpha\frac{\partial}{\partial x}\left\{(A_2-3\alpha A_4)\left(\varphi+\frac{\partial w}{\partial x}\right)-3\alpha\left[B_2\left(\frac{\partial w}{\partial x}+\varphi\right)\right]-\right.$$
$$\left.-\frac{3}{4}\frac{\partial}{\partial x}\left[B_2\left(\frac{\partial\varphi}{\partial x}-\frac{\partial^2 w}{\partial x^2}\right)-3\alpha B_4\left(\frac{\partial\varphi}{\partial x}+\frac{\partial^2 w}{\partial x^2}\right)\right]\right\}+ \tag{6.112}$$
$$+\frac{1}{2}\frac{\partial}{\partial x}(C_0+3\alpha C_2)+q=\varepsilon\frac{\partial w}{\partial t}+I_0\frac{\partial^2 w}{\partial t^2}+\alpha I_4\frac{\partial^3\varphi}{\partial x\partial t^2}-\alpha^2 I_6\left(\frac{\partial^4 w}{\partial x^2\partial t^2}+\frac{\partial^3\varphi}{\partial x\partial t^2}\right),$$

where in Eq. (6.112) the term $\varepsilon\partial w/\partial t$ is introduced which follows idea of the set-up method and guarantees damping of vibrations.

The boundary conditions for the homogeneous beam take the following form:

$$J_0\left(\frac{\partial u}{\partial x}+\frac{1}{2}\left(\frac{\partial w}{\partial x}\right)^2\right)-N^T\Bigg|_{x=0}^{x=1}=\bar{N}\quad\text{or}\quad u|_{x=0}^{x=1}=\bar{u}, \tag{6.113}$$

$$\left\{(J_2-\alpha J_4)\frac{\partial\varphi}{\partial x}-\alpha J_4\frac{\partial^2 w}{\partial x^2}-\alpha\left((J_4-\alpha J_6)\frac{\partial\varphi}{\partial x}-\alpha J_6\frac{\partial^2 w}{\partial x^2}\right)+\right.$$

$$+\frac{1}{4}\left(\left[(B_0-3\alpha B_2)\left(\frac{\partial\varphi}{\partial x}-\frac{\partial^2 w}{\partial x^2}\right)-3\alpha B_2\left(\frac{\partial\varphi}{\partial x}+\frac{\partial^2 w}{\partial x^2}\right)\right]+\right. \tag{6.114}$$

$$\left.\left.+9\alpha^2 B_4\left(\frac{\partial\varphi}{\partial x}+\frac{\partial^2 w}{\partial x^2}\right)\right)\right\}\Bigg|_{x=0}^{x=1}=\bar{M}\quad\text{or}\quad\varphi|_{x=0}^{x=1}=\bar{\varphi},$$

$$\left\{\alpha^2 I_6\frac{\partial^3 w}{\partial x\partial t^2}-\alpha\,(I_4-\alpha I_6)\,\frac{\partial^2\varphi}{\partial t^2}+\frac{C_0+3\alpha C_2}{2}+\right.$$

$$+(A_0-3\alpha A_2)\left(\varphi+\frac{\partial w}{\partial x}\right)+\frac{1}{4}\frac{\partial}{\partial x}\left[B_0\left(\frac{\partial\varphi}{\partial x}-\frac{\partial^2 w}{\partial x^2}\right)-3\alpha B_2\left(\frac{\partial\varphi}{\partial x}+\frac{\partial^2 w}{\partial x^2}\right)\right]+$$

$$+J_0\left(\frac{\partial u}{\partial x}+\frac{1}{2}\left(\frac{\partial w}{\partial x}\right)^2\right)\left(\frac{\partial w}{\partial x}\right)+\alpha\frac{\partial}{\partial x}\left[(J_4-\alpha J_6)\frac{\partial\varphi}{\partial x}-\alpha J_6\frac{\partial^2 w}{\partial x^2}\right]- \tag{6.115}$$

$$-3\alpha\,(A_2-3\alpha A_4)\left(\varphi+\frac{\partial w}{\partial x}\right)+9\alpha^2 B_2\left(\varphi+\frac{\partial w}{\partial x}\right)+$$

$$\left.+\frac{3}{4}\alpha\frac{\partial}{\partial x}\left[B_2\left(\frac{\partial\varphi}{\partial x}-\frac{\partial^2 w}{\partial x^2}\right)-3\alpha B_4\left(\frac{\partial\varphi}{\partial x}+\frac{\partial^2 w}{\partial x^2}\right)\right]\right\}\Bigg|_{x=0}^{x=1}=\bar{Q}\quad\text{or}\quad w|_{x=0}^{x=1}=\bar{w},$$

$$\left(\frac{\partial\varphi}{\partial x}-\frac{\partial^2 w}{\partial x^2}\right)\Bigg|_{x=0}^{x=1}=0\quad\text{or}\quad\frac{\partial w}{\partial x}\Bigg|_{x=0}^{x=1}=\overline{\frac{\partial w}{\partial x}}. \tag{6.116}$$

Here I_0, I_2, I_4, I_6, J_0, J_2, J_4, J_6 and A_0, A_2, A_4, B_0, B_2, B_4 are defined through (6.88) and in the case of panel with rectangular cross section, we have

$$\begin{aligned}
&J_0=I_0=1,\quad J_2=I_2=\frac{1}{12},\quad J_4=I_4=\frac{1}{80},\quad J_6=I_6=\frac{1}{448},\\
&A_0=\frac{1}{2\,(1+\nu)},\quad B_0=\frac{1}{2\,(1+\nu)}\left(\frac{l}{h}\right)^2,\\
&A_2=\frac{1}{24\,(1+\nu)},\quad B_2=\frac{1}{24\,(1+\nu)}\left(\frac{l}{h}\right)^2,\\
&A_4=\frac{1}{160\,(1+\nu)},\quad B_4=\frac{1}{160\,(1+\nu)}\left(\frac{l}{h}\right)^2,\quad \alpha=\frac{4}{3}.
\end{aligned} \tag{6.117}$$

In the reported below numerical case studies, we consider still clamping of the beam ends as the boundary conditions, i.e.

$$
\begin{aligned}
&w(0,t)=w(1,t)=0, \qquad \frac{\partial w(0,t)}{\partial x}=\frac{\partial w(1,t)}{\partial x}=0,\\
&u(0,t)=u(1,t)=0, \qquad \varphi(0,t)=\varphi(1,t)=0\,.
\end{aligned}
\tag{6.118}
$$

The initial conditions are as follows:

$$
\begin{aligned}
&w(x,0)=\left.\frac{\partial w(x,t)}{\partial t}\right|_{t=0}=0, \quad u(x,0)=\left.\frac{\partial u(x,t)}{\partial t}\right|_{t=0}=0,\\
&\varphi(x,0)=\left.\frac{\partial \varphi(x,t)}{\partial t}\right|_{t=0}=0.
\end{aligned}
\tag{6.119}
$$

In the case of a homogeneous Timoshenko beam, the governing equations follows:

$$
\lambda^2 J_0\frac{\partial}{\partial x}\left(\frac{\partial u}{\partial x}+\frac{1}{2}\left(\frac{\partial w}{\partial x}\right)^2\right)=I_0\frac{\partial^2 u}{\partial t^2},
\tag{6.120}
$$

$$
\lambda^2 J_2\frac{\partial^2\varphi}{\partial x^2}-\lambda^4 A_0\left(\varphi+\frac{\partial w}{\partial x}\right)+\frac{\lambda^2}{4}B_0\left(\frac{\partial^2\varphi}{\partial x^2}-\frac{\partial^3 w}{\partial x^3}\right)=I_2\frac{\partial^2\varphi}{\partial t^2},
\tag{6.121}
$$

$$
\begin{aligned}
&\lambda^2 A_0\left(\frac{\partial\varphi}{\partial x}+\frac{\partial^2 w}{\partial x^2}\right)+J_0\frac{\partial}{\partial x}\left\{\left[\left(\frac{\partial u}{\partial x}+\frac{1}{2}\left(\frac{\partial w}{\partial x}\right)^2\right)\right]\left(\frac{\partial w}{\partial x}\right)\right\}+\\
&+\frac{B_0}{4}\left(\frac{\partial^3\varphi}{\partial x^3}-\frac{\partial^4 w}{\partial x^4}\right)+q=I_0\frac{\partial^2 w}{\partial t^2}+\varepsilon\frac{\partial w}{\partial t}.
\end{aligned}
\tag{6.122}
$$

The corresponding boundary conditions take the following form:

$$
J_0\left.\left(\frac{\partial u}{\partial x}+\frac{1}{2}\left(\frac{\partial w}{\partial x}\right)^2\right)\right|_{x=0}^{x=1}=0 \quad \text{or} \quad u|_{x=0}^{x=1}=0,
\tag{6.123}
$$

$$
J_2\frac{\partial\varphi}{\partial x}+\frac{1}{4}B_0\left.\left(\frac{\partial\varphi}{\partial x}-\frac{\partial^2 w}{\partial x^2}\right)\right|_{x=0}^{x=1}=0 \quad \text{or} \quad \varphi|_{x=0}^{x=1}=0,
\tag{6.124}
$$

$$
\lambda^2 A_0\left(\varphi+\frac{\partial w}{\partial x}\right)+\frac{B_0}{4}\left(\frac{\partial^2\varphi}{\partial x^2}-\frac{\partial^3 w}{\partial x^3}\right)+J_0\left(\frac{\partial u}{\partial x}+\frac{1}{2}\left(\frac{\partial w}{\partial x}\right)^2\right)\left.\left(\frac{\partial w}{\partial x}\right)\right|_{x=0}^{x=1}=0
\tag{6.125}
$$

$$w|_{x=0}^{x=1} = 0,$$

$$\left(\frac{\partial \varphi}{\partial x} - \frac{\partial^2 w}{\partial x^2}\right)\Bigg|_{x=0}^{x=1} = 0 \quad \text{or} \quad \frac{\partial w}{\partial x}\Bigg|_{x=0}^{x=1} = 0. \tag{6.126}$$

Analogously, in the case of the Bernoulli–Euler beam, we consider the following governing PDEs:

$$\lambda^2 \frac{\partial}{\partial x}\left[J_0\left(\frac{\partial u}{\partial x} + \frac{1}{2}\left(\frac{\partial w}{\partial x}\right)^2\right)\right] = I_0 \frac{\partial^2 u}{\partial t^2}, \tag{6.127}$$

$$\frac{\partial^2}{\partial x^2}\left[-(J_2 + B_0)\frac{\partial^2 w}{\partial x^2}\right] + \frac{\partial}{\partial x}\left\{\left[J_0\left(\frac{\partial u}{\partial x} + \frac{1}{2}\left(\frac{\partial w}{\partial x}\right)^2\right)\right]\left(\frac{\partial w}{\partial x}\right)\right\} + q =$$
$$= I_0 \frac{\partial^2 w}{\partial t^2} - \frac{I_2}{\lambda^2}\frac{\partial^4 w}{\partial t^2 \partial x^2} + \varepsilon \frac{\partial w}{\partial t} \tag{6.128}$$

and the associated boundary conditions

$$J_0\left(\frac{\partial u}{\partial x} + \frac{1}{2}\left(\frac{\partial w}{\partial x}\right)^2\right)\Bigg|_{x=0}^{x=1} = 0 \quad \text{or} \quad u|_{x=0}^{x=1} = 0, \tag{6.129}$$

$$\left\{\frac{\partial}{\partial x}\left[-J_2\frac{\partial^2 w}{\partial x^2}\right] - \frac{\partial}{\partial x}\left[B_0\frac{\partial^2 w}{\partial x^2}\right] + \left[J_0\left(\frac{\partial u}{\partial x} + \frac{1}{2}\left(\frac{\partial w}{\partial x}\right)^2\right)\right]\left(\frac{\partial w}{\partial x}\right)\right\}\Bigg|_{x=0}^{x=1} = \bar{Q} \tag{6.130}$$

or $w|_{x=0}^{x=1} = \bar{w}$,

$$-(J_2 + B_0)\frac{\partial^2 w}{\partial x^2}\Bigg|_{x=0}^{x=1} = \bar{M} \quad \text{or} \quad \frac{\partial w}{\partial x}\Bigg|_{x=0}^{x=1} = \overline{\frac{\partial w}{\partial x}}. \tag{6.131}$$

As in the previous models, we take into account the boundary conditions corresponding to the stiff clamping of the beam ends

$$w(0,t) = w(1,t) = 0, \quad \frac{\partial w(0,t)}{\partial x} = \frac{\partial w(1,t)}{\partial x} = 0, \quad u(0,t) = u(1,t) = 0. \tag{6.132}$$

The initial conditions are analogous to the initial conditions for the Timoshenko and Sheremetev–Pelekh models

$$w(x, 0) = \left. \frac{\partial w\,(x, t)}{\partial t} \right|_{t=0} = 0, \qquad u(x, 0) = \left. \frac{\partial u\,(x, t)}{\partial t} \right|_{t=0} = 0. \tag{6.133}$$

The problem of solving PDEs is reduced to the Cauchy problems employing the FDM of the second-order accuracy. In the beginning, we have carried out a numerical experiment aimed at comparison of the numerical results obtained through the fourth and sixth Runge–Kutta methods, and based on the obtained results the fourth-order Runge–Kutta method has been chosen which stands also in agreement with the results reported in [54]. Based on the Runge principle, the following "optimal" parameters are chosen: number of beam length partition $n = 100$ (it corresponds to a step along the spatial coordinate $c = 0.01$, and the time step $\Delta t = 3.90625 \cdot 10^{-3}$ corresponding to the Runge–Kutta method. The investigations have been carried out for the following parameters: $L/h = 30$, $\nu = 0.3$, $l/h = 0$ and 0.3.

The computational results are compared with regard to the used three models: Bernoulli–Euler, Timoshenko and Sheremetev–Pelekh for the beam with an account of the size-dependent behaviour ($l/h = 0.3$) and without an account of the size-dependent behaviour ($l/h = 0$).

In order to get solutions to static problems, the relaxation (set-up) method has been employed which allows to get static solution after damped transitional dynamic processes.

In what follows we briefly describe the main idea of the set-up method. For this purpose, we present the system of nonlinear dynamic equations in the following matrix form:

$$\mathbf{T_2}\,[\bar{z}] + \varepsilon\,\mathbf{T_1}\,[\bar{z}] = \mathbf{B}\,[\bar{z}] + [\bar{q}] + L\,[\bar{z}], \tag{6.134}$$

where $\mathbf{T_2}$—matrix of the second time derivatives; $\mathbf{T_1}$—matrix of the first time derivatives; $\mathbf{B}$—matrix of linear terms; $L\,[\bar{z}]$—nonlinear operator either differential or algebraic; $[\bar{z}]$—vector of the searched variables; $[\bar{q}]$—given vector function; and ε—matrix of dissipation.

If both the stationary problem and the load $[\bar{q}]$ do not depend on time, then a possibility of getting solution to the static problem based on dynamic approach can be relatively easily realized. We introduce a certain non-zero initial condition, which plays a role of excitation to the static problem, and we take the dissipation $\varepsilon \neq 0$. This implies occurrence of damping of the perturbed solution. On the set-up method allows to get the static stationary solutions using arbitrary well-known and tested methods aimed at solving the Cauchy problem. As it has been already mentioned, the fourth-order Runge–Kutta method has been used.

An idea of getting stationary solutions a part of non-stationary solutions was pointed out in 30 years of twentieth century by Tikhonov. The set-up method possesses a wide palette of application. From one side, the set-up method can be viewed as the iterative method for solution of linear/nonlinear algebraic, transcendental equations, where on each step of approximation in time we obtain new approximation while looking for the roots of equations (see also [55–57]).

The set-up method has been used to get solutions to the differential and integral equations in the works [46, 58, 59]. As an iteration method, it exhibits a high accu-

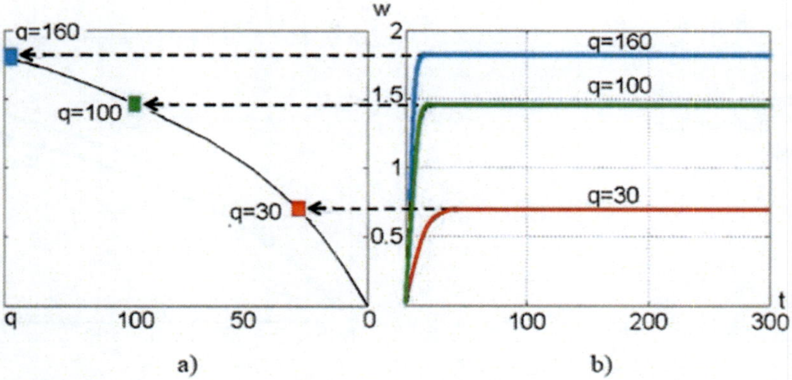

Fig. 6.5 Dependence of the beams load-deflection (**a**), time histories of the beam deflection (**b**) [reprinted with permission from International Journal of Non-Linear Mechanics publishers]

racy order. On the other hand, employment of the set-up method to solve nonlinear PDEs may be associated with linearization process as well as with order decrease of the studied systems. This can be achieved through the reduction of getting solution of evolutionary PDEs to that of the Cauchy problems of linear ODEs with respect to time. One more benefit of the set-up method is given by its simplicity due to numerical realization with the help of computers. Namely, nowadays, there are available numerous effective algorithms devoted to solve the Cauchy problem.

Finally, it seems that from the methodological point of view while solving a static problem use of the dynamic approach is more naturally established since all physical processes in our universe are associated with time.

Results of solution of the static problem for the Sheremetev–Pelekh model (6.110)–(6.118) for $l/h = 0.3$ and for the dissipations coefficient $\varepsilon = 1.5$ are shown in Fig. 6.5. Figure 6.5a reports the dependence load-deflection, based on the set-up method for the given values of $q(x) = q = const$. Figure 6.5b presents time series set-up of the dynamical process for the deflection function w for the load $q = 30, 100, 160$.

Time histories of the set-up method show that the set-up process is fastly convergent under fixed dissipation coefficient for all loads. For $q = 30$, the set-up process of the solution takes place at $t = 49$. In the cases of loads $q = 100, 160$ it takes place at $t = 28$ and $t = 19$, respectively. Therefore, the set-up method is stable and allows to get solution to nonlinear static problems without numerical difficulties.

Figures 6.6, 6.7, 6.8 report the dependencies load-deflection in the beam centre of the mathematical models of Bernoulli–Euler (Fig. 6.6), Timoshenko (Fig. 6.7) and Sheremetev–Pelekh (Fig. 6.8). The results comparison obtained via mentioned three mathematical models under the same size-dependent coefficient $l/h = 0.3$ are presented in Fig. 6.9. Dashed curve corresponds to $l/h = 0$, solid curve corresponds to $l/h = 0.3$, whereas dashed-dotted curve corresponds to $l/h = 0.5$. One may conclude based on Figs. 6.6, 6.7, 6.8 that the influence of the size-dependent

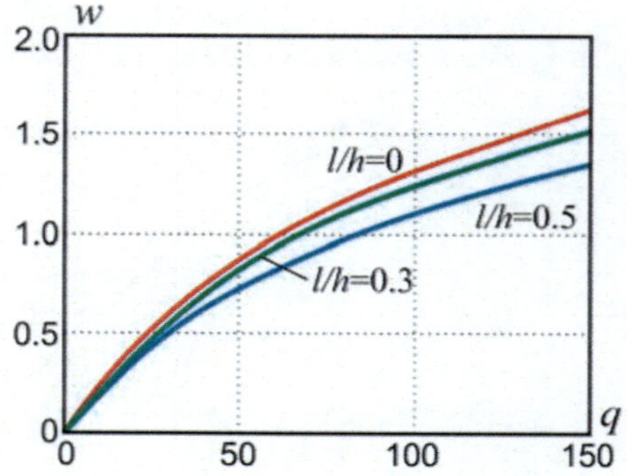

Fig. 6.6 Bernoulli–Euler model [reprinted with permission from International Journal of Non-Linear Mechanics publishers]

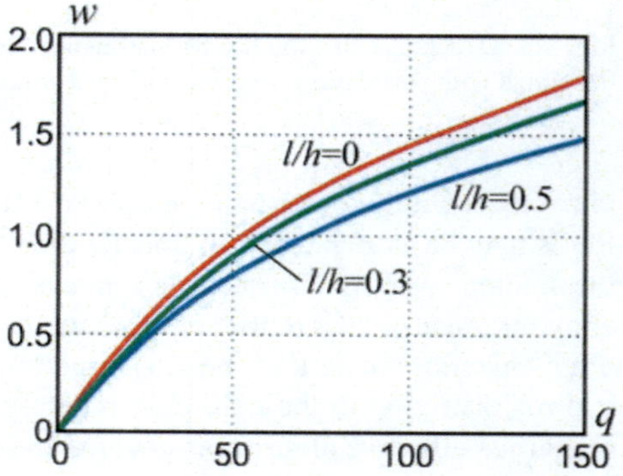

Fig. 6.7 Timoshenko model [reprinted with permission from International Journal of Non-Linear Mechanics publishers]

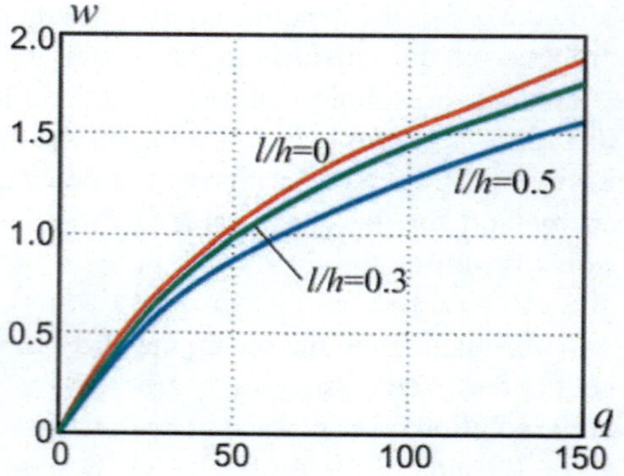

Fig. 6.8 Sheremetev–Pelekh model [reprinted with permission from International Journal of Non-Linear Mechanics publishers]

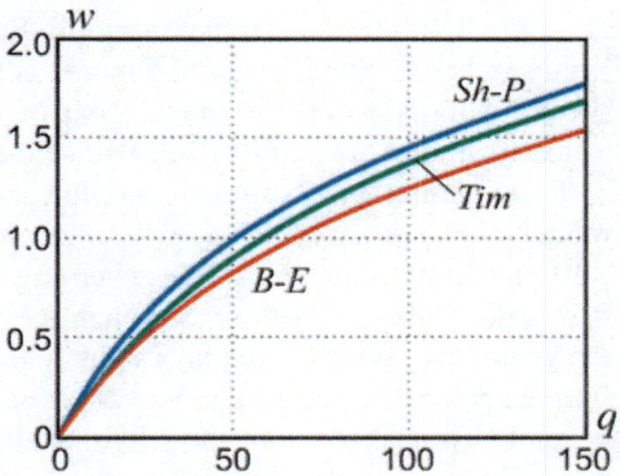

Fig. 6.9 Comparison of the Bernoulli–Euler, Timoshenko and Sheremetev–Pelekh models for $l/h = 0.3$ [reprinted with permission from International Journal of Non-Linear Mechanics publishers]

effects in three mathematical beam models implies that the decrease of deflection for all models under the same loads through the changes is different depending on the model. In the case of the Bernoulli–Euler model, the difference between deflection for $q = 150$ for $l/h = 0$ and for $l/h = 0.3$ achieves 5.8%. In the case of $l/h = 0$ and for $l/h = 0.5$, the difference is 16.4%. In the case of the Timoshenko model, the result differences are of 6.9 and 17.3%. In the case of the Sheremetev–Pelekh model, the result differences are of 8.2 and 18%.

It should be mentioned that the results of the Bernoulli–Euler model essentially differ from the results yielded by the Timoshenko and Sheremetev–Pelekh models (Fig. 6.9). Since the first model does not take into account both rotation of the normal to the middle beam curve (on contrary to the Timoshenko model) and curving (on contrary to the Sheremetev–Pelekh model).

6.6 Chaotic Dynamics of the Size-Dependent Flexible Bernoulli–Euler, Timoshenko and Sheremetev–Pelekh Beams Within the Modified Couple Stress Theory of Elasticity

Investigation of dynamics of nanomechanical constructions requires the employment of various mathematical models to account the size-dependent behaviour: couple stress theory of elasticity, nonlocal theory of elasticity, gradient theory of elasticity and surface elasticity.

Analysis of vibrations based on various mathematical models and nonlocal theory of elasticity carried out by Soltani et al. [60] and Xiao Dong and Lim [61]. In the first case transversal vibrations of layer single-walled carbon nanotubes (SWCNT) based on the Bernoulli–Euler and Timoshenko beam models interacting with the surrounding medium modelled on a basis of the Winkler and Pasternak hypotheses, the linear statement has been worked out. The influence of curvature magnitude, boundary conditions, beam length on the vibration characteristics and frequencies are studied. It is found that the Bernoulli–Euler model is more suited for more length beams, whereas the Timoshenko model is preferred for the more short beams. The second work is devoted to the analysis of the natural frequencies of nonlinear free transversal vibrations of the Bernoulli–Euler nanobeam.

Investigation on vibrations of nonlinear nanobeam based on the Bernoulli–Euler model in the framework of the gradient theory is carried out by Rajabi and Ramezani [29] and Kong et al. [62]. The governing PDEs are obtained with the help of the first-order Galerkin method, and then they are reduced to ODEs. It is shown that the geometric nonlinearity plays a key role in increase of the beam natural vibration frequency for the case of relatively thick beams. However, when the beam thickness becomes compatible with the length scale material parameter, then a key role is played by the size-dependent behaviour.

Ke et al. [54] and Ke and Wang [63] reported results for the mathematical models constructed on a basis of the modified couple stress theory of elasticity. They investigated dynamical stability of linear Timoshenko microbeam and nonlinear beam free vibrations. The obtained results show that the size-dependent effect for the characteristics of the dynamic stability plays a crucial influence only if the beam thickness is comparable with the scale parameter of the material length.

Surface theory of elasticity is employed to construct the mathematical models of nanobeam vibrations in [64]. There were considered transversal and longitudinal vibrations of two nanobeams within framework of the Euler–Bernoulli model in linear statement. The mentioned theory is adopted to the case of nanoporous Timoshenko beam [65]. The obtained results indicate that the critical value of stability loss of nanoporous materials depends on the characteristic dimensions of the microstructures.

Besides the already mentioned works, dynamical problems of nanobeams were considered in [24, 54, 66–73].

The so far carried out overview of the state of the art points out that in spite of the large number of publications devoted to study of beams and plates in the framework of various models and different theories aimed at inclusion of the size-dependent parameters, there is a gap in reason of focused on chaotic dynamics of nanobeams.

In this section, we consider numerous problems of nonlinear dynamics of homogeneous straight-linear nanobeams in the framework of the modified couple theory of elasticity for the mathematical beam models of Bernoulli–Euler, Timoshenko and Sheremetev–Pelekh with an account of geometric nonlinearity. The comparison of the dynamic characteristics (Fourier and wavelet spectra, phase portraits, Poincaré maps, LLEs) for the given physical-geometric parameters with and without the size-dependent behaviour is carried out [74]. The charts of the character of vibrations are constructed and studied. Moreover, scenarios of transition from regular to chaotic vibrations are illustrated and discussed.

Derivation of the governing equations and boundary conditions is reported in Sect. 6.6. We begin with investigation aimed at the determination of convergence of the numerical algorithms and reliability of the obtained results. For this purpose,

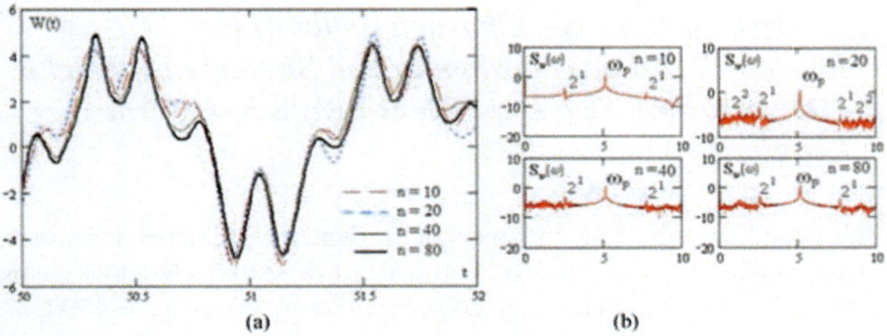

Fig. 6.10 Convergence of the results for different numbers of beam length partitions n for the beam deflection (**a**) and power spectrum (**b**) [reprinted with permission from International Journal of Non-Linear Mechanics publishers]

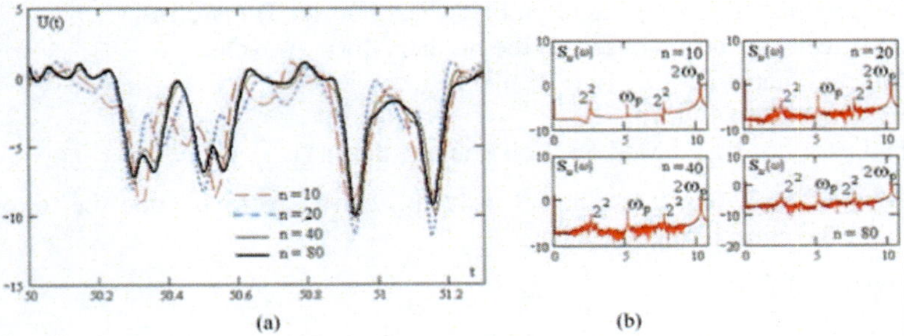

Fig. 6.11 Result convergence versus different numbers of beam length partition ($n = 10, 20, 40, 80$) for the longitudinal displacement (**a**) and Fourier power spectrum (**b**) [reprinted with permission from International Journal of Non-Linear Mechanics publishers]

time histories $w(0.5, t)$ (Fig. 6.10a) and $u(0.25, t)$ (Fig. 6.11a) and the Fourier spectra $S_w(\omega)$ (Fig. 6.10b) and $S_u(\omega)$ (Fig. 6.11b) for some values of the beam length partition n are compared. The investigations have been carried out for all models for the values of scale length material parameter $l/h = 0$ and 0.3. Figures 6.10, 6.11 report the results for the Bernoulli–Euler model (6.127) (6.128) with the boundary conditions (6.129)–(6.131), for $l/h = 0$. The sinusoidal transverse load has the form $q = 1050 \sin(5.14t)$.

Convergence of the results in the power spectrum is achieved already for $n = 20$ though the signals $w(0.5, t)$ and $u(0.25, t)$ exhibit uniform convergence for $n = 40$.

Therefore, the employed methods and stability algorithms allow to achieve reliable results.

6.6.1 *Analysis of Dynamic Characteristics of the Bernoulli–Euler, Timoshenko and Sheremetev–Pelekh Models for the Size-Dependent Beams Versus the Size Length*

In this subsection, we report a comparison of the dynamic problems of beams subject to the transversal load $q = q_0 \sin(\omega_p t)$ for the fixed value of the length scale material parameter $l/h = 0.3$. The following load parameters are taken: $q_0 = 4000$, $\omega_p = 9$. The frequency $\omega_p = 9$ is chosen to avoid possible resonances, i.e. if differs from the beam natural frequencies.

In Tables 6.1, 6.4, 6.7, 6.10, 6.13 and 6.16, the following results are presented:

(a) signal $w(0.5, t)$;
(b) Fourier spectrum $S(\omega)$ based on FFT (Fast Fourier Transform);
(c) 2D wavelet spectrum based on the mother Morlet wavelet;
(d) Poincaré map $w(t + T)$ $[w(t)]$;
(e) phase portrait $\dot{w}[w(t)]$;
(f) the history of LLE based on Wolf's algorithm [41].

Tables 6.2, 6.3, 6.5, 6.6, 6.8, 6.9, 6.11, 6.12, 6.14, 6.15 include the following results for the chosen time intervals:

(a) signal $w(0.5, t)$;

Table 6.1 Numerical results for the Bernoulli–Euler model ($l/h = 0.3$, $t \in [300; 2500]$) [reprinted with permission from International Journal of Non-Linear Mechanics publishers]

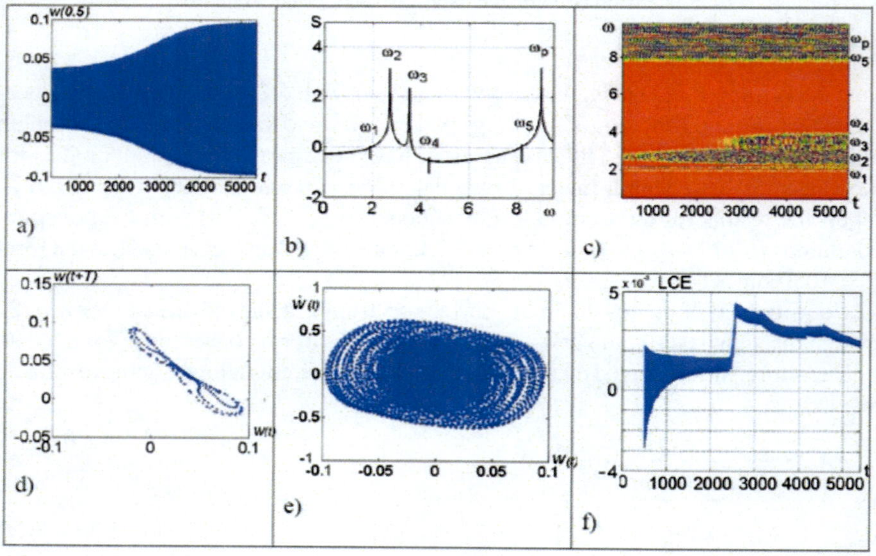

Table 6.2 Numerical results for the Bernoulli–Euler model ($l/h = 0.3, t \in [300; 5300]$) [reprinted with permission from International Journal of Non-Linear Mechanics publishers]

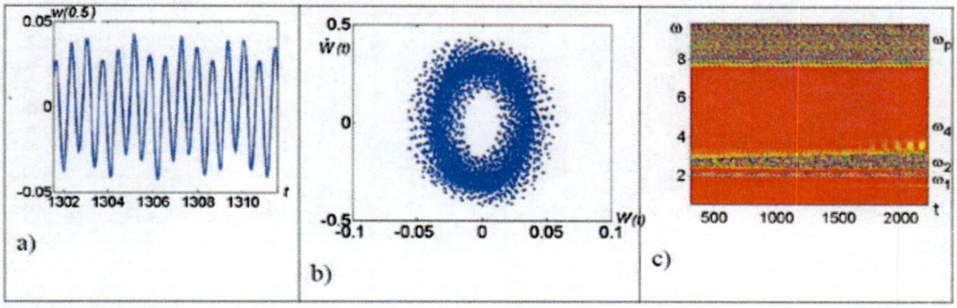

Table 6.3 Numerical results for the Bernoulli–Euler model ($l/h = 0.3, t \in [2500; 5300]$) [reprinted with permission from International Journal of Non-Linear Mechanics publishers]

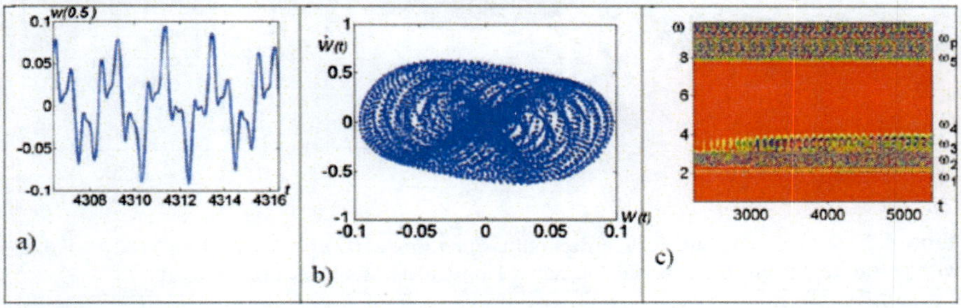

(b) phase portrait $\dot{w}[w(t)]$;
(c) 2D wavelet spectrum based on the mother Morlet wavelet.

6.6.1.1 The Bernoulli–Euler Model for l/h = 0.3

The Fourier spectrum exhibits six frequencies $\omega_p = 9$, $\omega_1 = 1.91$, $\omega_2 = 2.73$, $\omega_3 = 3.55$, $\omega_4 = 4.372$, $\omega_5 = 8.181$. They are coupled through the following resonance conditions: $2 \cdot \omega_2 + \omega_3 = \omega_p$, $\omega_2 - \omega_1 = \omega_3 - \omega_2 = \omega_4 - \omega_3 = \omega_p - \omega_5$.

Therefore, the frequency ω_2 plays a role of independent frequency, whereas the remaining frequency is linearly dependent, i.e. we have two-frequency regime. The

Table 6.4 Numerical results for the Bernoulli–Euler model ($l/h = 0$, $t \in [300; 5300]$) [reprinted with permission from International Journal of Non-Linear Mechanics publishers]

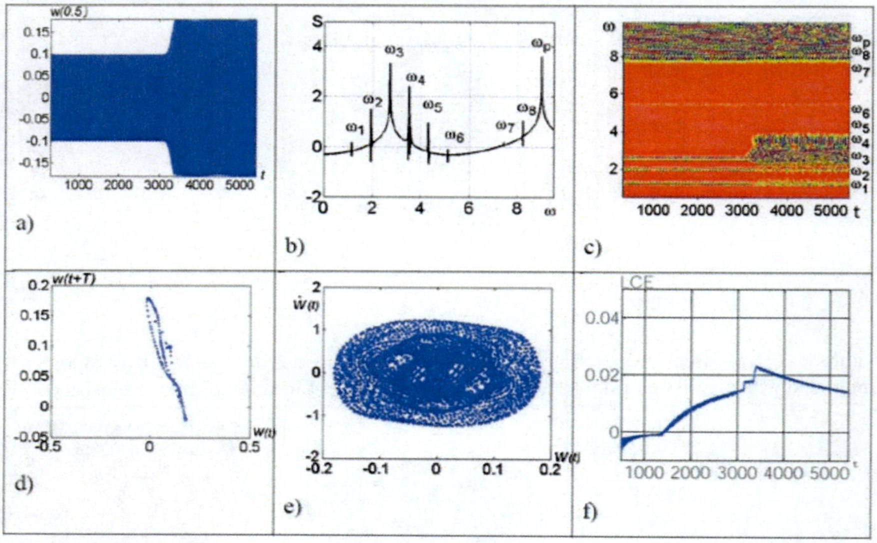

Table 6.5 Numerical results for the Bernoulli–Euler model ($l/h = 0$, $t \in [300; 3100]$) [reprinted with permission from International Journal of Non-Linear Mechanics publishers]

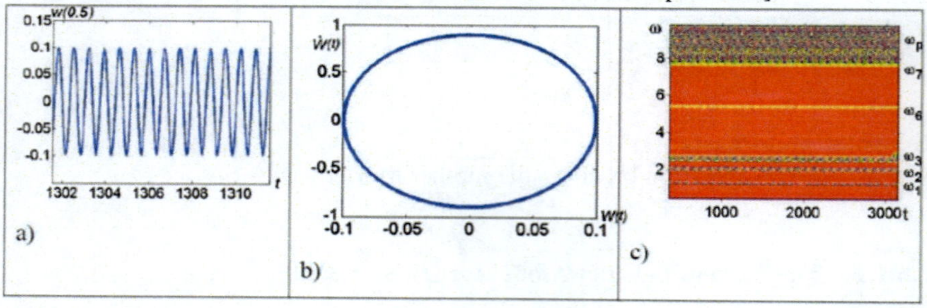

wavelet spectrum shows that the frequencies ω_2, ω_3 appear in different times. Occurrence of frequency ω_2 for $t = 2450$ implies a change of the system energy which is manifested by a jump of the LLE. Besides the wavelet spectrum exhibits a frequency in the region of $\omega \approx 5.5$, which is not visible in the Fourier spectrum. This is because it has almost zero power and it occurs in non-smooth way in time. The phase portrait of the function $w(t)$ presents a combination of node-limiting cycle interplay which is also validated by dengue points in vicinity of the centre of the Pincaré map.

Table 6.6 Numerical results for the Bernoulli–Euler model ($l/h = 0$, $t \in [3100; 5300]$) [reprinted with permission from International Journal of Non-Linear Mechanics publishers]

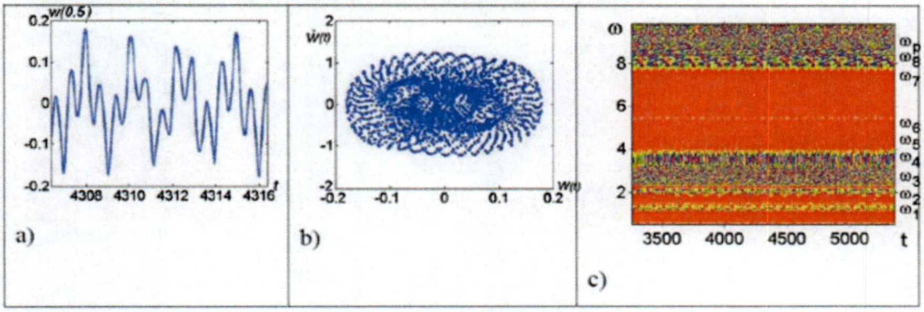

Table 6.7 Numerical results for the Timoshenko model ($l/h = 0.3$, $t \in [300; 5300]$) [reprinted with permission from International Journal of Non-Linear Mechanics publishers]

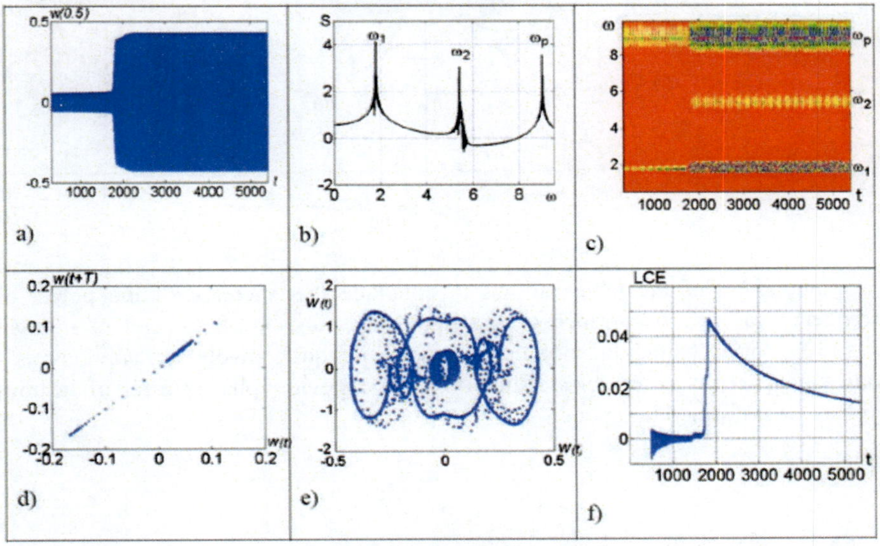

Table 6.8 Numerical results for the Timoshenko model ($l/h = 0.3$, $t \in [300; 1600]$) [reprinted with permission from International Journal of Non-Linear Mechanics publishers]

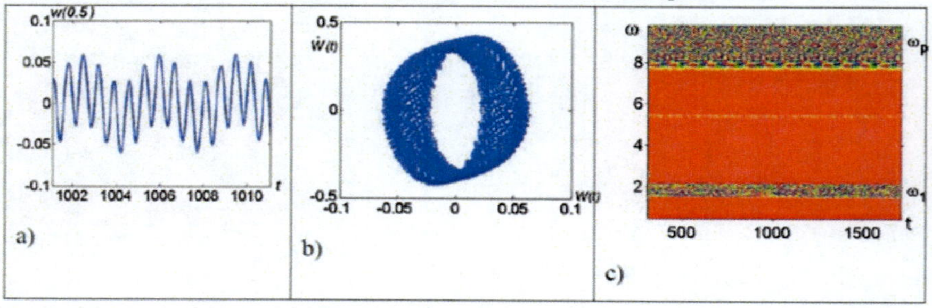

Table 6.9 Numerical results for the Timoshenko model ($l/h = 0.3$, $t \in [1600; 5300]$) [reprinted with permission from International Journal of Non-Linear Mechanics publishers]

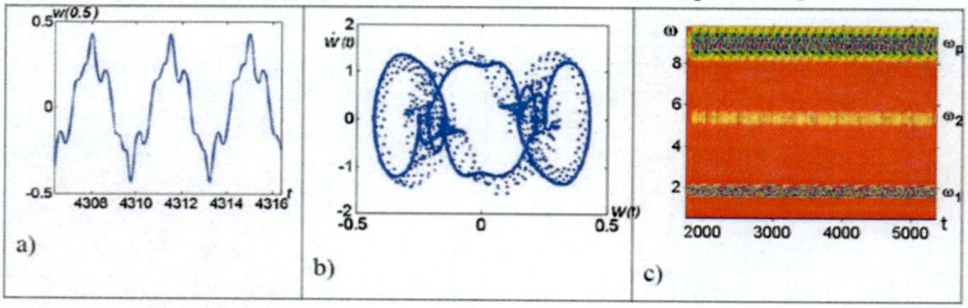

Comparison of the tables allows to conclude that increase of the power of frequency ω_2 and occurrence of the frequencies ω_2, ω_5 of the end of interval $t \in [300; 2500]$ changes the vibration characteristics qualitatively. The previous periodic signal becomes complex and unperiodic. The previous phase portrait of the limit cycle is transformed into node.

6.6.1.2 The Bernoulli–Euler Model for l/h = 0

Table 6.4 presents results associated with the study of time window $t \in [300; 5300]$, whereas Tables 6.5 and 6.6 report the results of time windows $t \in [300; 3100]$ and $t \in [3100; 5300]$, respectively.

The frequency spectrum contains the following frequencies: $\omega_1 = 1.15$, $\omega_2 = 1.941$, $\omega_3 = 2.732$, $\omega_4 = 3.523$, $\omega_5 = 4.315$, $\omega_6 = 5.11$, $\omega_7 = 7.42$, $\omega_8 = 8.21$

Table 6.10 Numerical results for the Timoshenko model ($l/h = 0, t \in [300; 5300]$) [reprinted with permission from International Journal of Non-Linear Mechanics publishers]

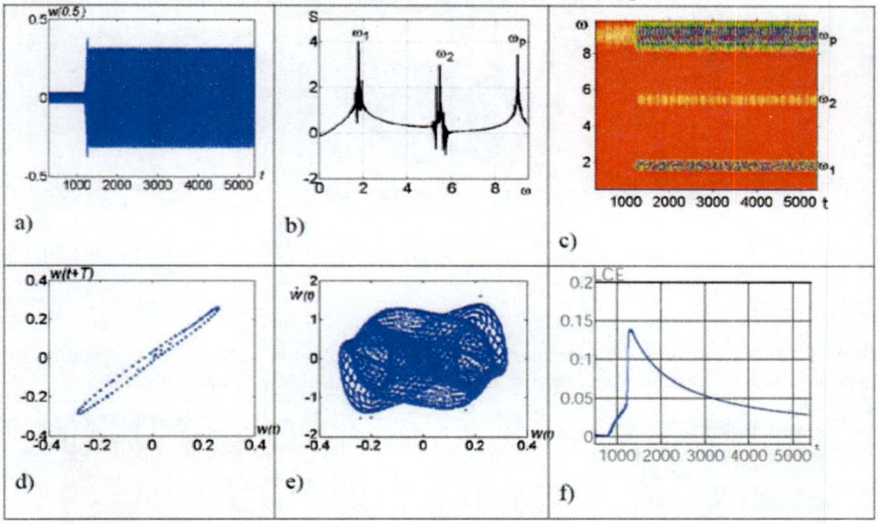

Table 6.11 Numerical results for the Timoshenko model ($l/h = 0, t \in [300; 1200]$) [reprinted with permission from International Journal of Non-Linear Mechanics publishers]

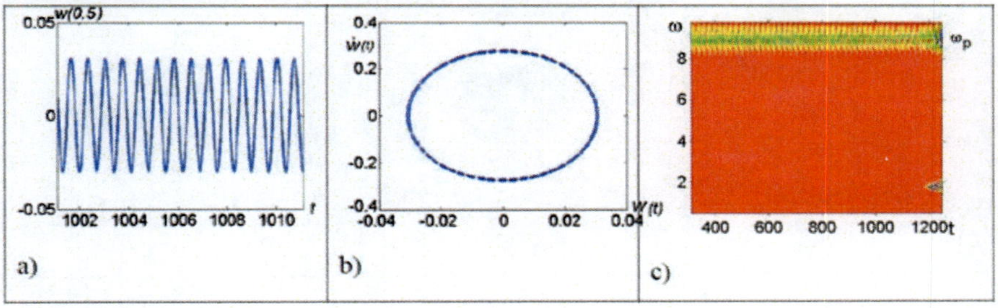

and $\omega_p = 9$. They are coupled through the following linear relations: $\omega_3 - \omega_2 = \omega_2 - \omega_1 = \omega_4 - \omega_3 = \omega_5 - \omega_4 = 0.791$, $\omega_p - \omega_8 = \omega_8 - \omega_7 = 0.791$. The spectrum contains two independent frequencies ω_3, ω_p, whereas the remaining frequencies are linearly dependent. Comparing the results shown in Table 6.4, one may conclude that the frequency of the given spectrum ($\omega_2 = 1.941$, $\omega_3 = 2.732$, $\omega_4 = 3.523$, $\omega_5 = 4.315$, $\omega_8 = 8.21$) also appear in the vibrational spectrum of the Bernoulli–Euler beam for $l/h = 0.3$.

Table 6.12 Numerical results for the Timoshenko model ($l/h = 0$, $t \in [1200; 5300]$) [reprinted with permission from International Journal of Non-Linear Mechanics publishers]

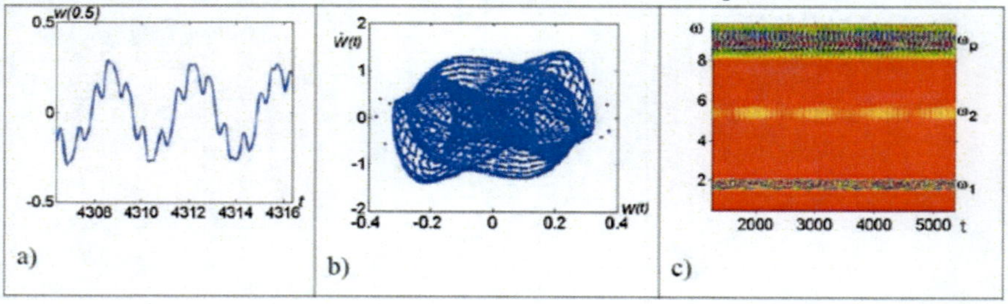

Table 6.13 Numerical results for the Sheremetev–Pelekh model ($l/h = 0.3$, $t \in [300; 5300]$) [reprinted with permission from International Journal of Non-Linear Mechanics publishers]

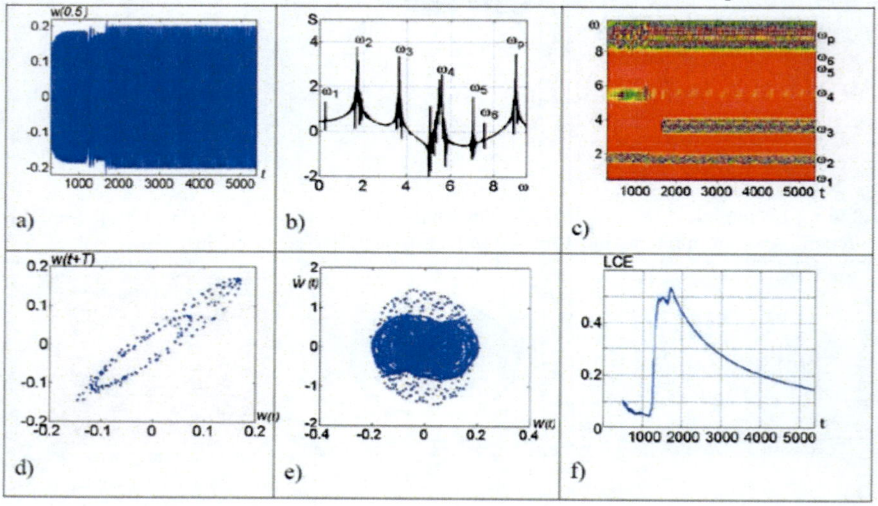

The time history (signal) exhibits sudden amplitude change corresponding to the birth of frequencies ω_3, ω_4 in time instant $t = 3200$. The power of the excitation frequencies $S(\omega_p)$ in that time instant decreases, and hence energy exchange of the beam takes place (c). The phase portrait (d) implies the existence of attractors symmetrically located with regard to the origin (0;0) which manifests the influence of frequencies ω_3, ω_p. LLE time history (e) in the interval $t \in [1300; 3100]$ takes positive values under the influence of the frequencies $\omega_1, \omega_2, \omega_6, \omega_7, \omega_8$. The birth of frequencies ω_3, ω_4 implies sudden increase of the LLE in time interval $t \in [3100; 3400]$.

Table 6.14 Numerical results for the Sheremetev–Pelekh model ($l/h = 0.3, t \in [300; 1100]$) [reprinted with permission from International Journal of Non-Linear Mechanics publishers]

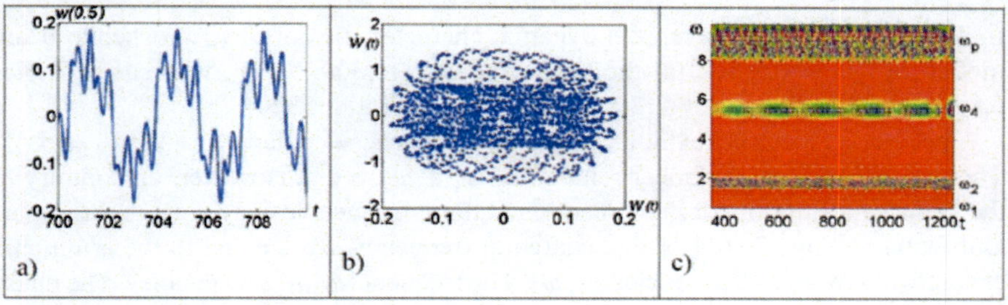

Table 6.15 Numerical results for the Sheremetev–Pelekh model ($l/h = 0.3, t \in [1100; 1600]$) [reprinted with permission from International Journal of Non-Linear Mechanics publishers]

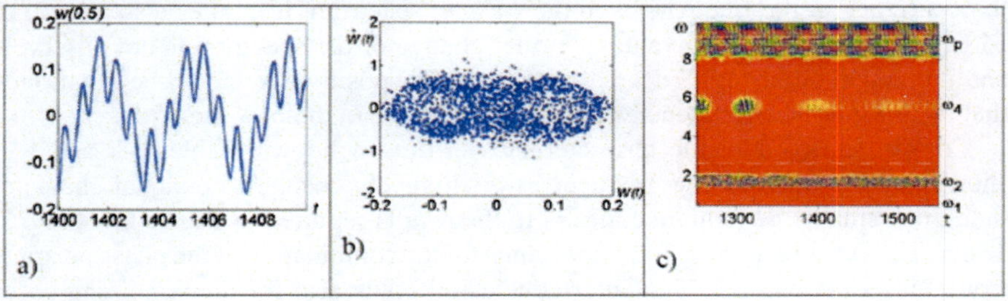

After transitional vibration process, the value of LLE is smoothly decreased though it remains positive, i.e. vibrations are chaotic.

Let us consider more deeply the vibration characteristics for two time windows. In spite of occurrence in the wavelet spectrum frequencies in the interval [300; 3100], the signal and phase portrait exhibit the periodic vibrations. Occurrence of new frequencies and increase of the corresponding frequencies in the time interval [3100; 5300] changes the picture qualitatively. The signals become unperiodic, and the phase portrait is analogous to that shown in Table 6.4. Analysis of the phase portraits presented in Tables 6.4, 6.5 and 6.6 indicates that the phase portrait of Table 6.4 includes phase portraits exhibited by those associated with time windows.

6.6.1.3 The Timoshenko Model for l/h = 0.3

In Table 6.7, there are given the dynamic characteristics of the Timoshenko beam model for $l/h = 0.3$ and for the time window $t \in [300; 5300]$, whereas in Tables 6.8, 6.9 for the windows $t \in [300; 1600]$ and $t \in [1600; 5300]$.

The Fourier spectrum exhibits three frequencies $\omega_p = 9$, $\omega_1 = 1.805$, $\omega_2 = 5.41$. They are coupled by the linear relation $2 \cdot \omega_1 + \omega_2 = \omega_p$. However, on contrary to the Bernoulli–Euler model, the mentioned three frequencies have one more important features. Power $S(\omega)$ of the excitation frequency ω_p is equal to the arithmetic average of powers of frequencies ω_1, ω_2. The frequencies ω_1, ω_2 are noisy. The phase portrait presents three-rotational stable cycle. Noisy frequencies imply divergence of the trajectories in the phase portrait. However, the Poincare sections do not exhibit divergence effects.

The signal (a) for $t \approx 1700$ exhibits a sudden amplitude increase. It corresponds to occurrence of the frequencies in the wavelet spectrum. It implies also a jump of LLE from zero to positive values. Further change of the spectrum is not observed, and the magnitude of LLE decreases but remains positive. It should be mentioned that the birth of new frequencies implies an increase in the power of frequency.

Comparison of dynamic characteristics in time windows (Tables 6.8 and 6.9) shows that the birth of new frequencies qualitatively changes the signal character though it remains periodic in contrary to the results reported in Tables 6.1 and 6.3. The phase portrait shown in Table 6.7 stands for the combination of the phase portraits from Tables 6.8 and 6.9. The latter conclusion holds also for the remaining cases considered in this section.

6.6.1.4 The Timoshenko Model for l/h = 0

Table 6.10 presents results of numerical analysis of nonlinear dynamics of the Timoshenko model for the time interval $t \in [300; 5300]$. In Tables 6.14 and 6.15, there are also given results for the time windows $t \in [300; 1200]$ and $t \in [1200; 5300]$.

The signal exhibits the amplitude jumps for $t \approx 1170$, which is a consequence of the birth of frequencies ω_3, ω_4. At the same time instants, the power of excitation frequency ω_p suddenly increases that is visible in 2D wavelet (c), the Fourier spectrum (b) shows occurrence of the frequencies $\omega_1 = 1.7696$, $\omega_2 = 5.4634$, which obey the following dependence $2 \cdot \omega_1 + \omega_2 = \omega_p$. The time history of LLE suffers for a sudden jump in time instant where both frequencies appear. The frequencies in the Fourier spectrum are noisy which is confirmed by positive Lyapunov exponent. The phase portrait possesses a complex structure in the form of the embedded in each other cylinder and forces. Observe that the frequency spectrum contains three frequencies, as it was observed for the Timoshenko beam with an account of the size-dependent behaviour. The qualitative similarity concerns also the LLE but the remaining characteristics (phase portrait and Poincaré map) are different.

The analysed results shown in Tables 6.11 and 6.12 allow to conclude that the change of the frequency spectrum, i.e. the birth of the frequencies ω_3, ω_4 and increase of the power of frequency ω_p implies strong change of all vibrational characteristics (see Figs. 6.11, 6.12a, b). As it happened in other cases, the whole phase portrait consists of information reported in figure associated with window intervals.

6.6.1.5 The Sheremetev–Pelekh model l/h = 0.3

Table 6.13 presents the characteristics for the time interval $t \in [300; 5300]$, whereas Tables 6.14, 6.15, 6.16 report the results for the chosen windows $t \in [300; 1100]$, $t \in [1100; 1600]$ and $t \in [1600; 5300]$. A sequence of the result presentation follows the sequence of the previous mathematical models.

The frequency spectrum exhibits the following frequencies: $\omega_1 = 0.26078$, $\omega_2 = 1.6966$, $\omega_3 = 3.6524$, $\omega_4 = 5.6082$, $\omega_5 = 7.044$, $\omega_6 = 7.5641$, $\omega_p = 9$, which satisfy the following relations: $\omega_4 - \omega_3 = \omega_3 - \omega_2 = 1.9558$, $\omega_p - \omega_6 = \omega_5 - \omega_4 = \omega_2 - \omega_1 = 1.436$. The beam vibrations are spanned by two independent frequencies and the excitation frequency. The Poincaré map is composed of three embedded nonsymmetric ellipses localized in the vicinity of centre and nonexhibiting action of three frequencies $\omega_p, \omega_2, \omega_3$ with the same power. The phase portrait presents with the surrounding rings which exhibit action of the frequencies ω_1, ω_4, ω_5 and ω_6. The time history and the wavelet spectrum show the intermittency effect being coupled with the time evolution of the frequency spectrum. Since the frequency characteristics change in time, we focus on consideration of the vibration characteristics for the time intervals shown in Tables 6.14, 6.15, 6.16 and we compare them with the results of Table 6.13.

The wavelet spectrum implies non-homogeneity of the frequency spectrum along time axis. In the interval [300; 1180], there are five frequencies $\omega_p, \omega_2, \omega_4, \omega_5, \omega_6$. The phase portrait has a complex symmetric shape with the visible one of attraction of the phase trajectories. The LLE decreases but remains in the positive area. In the interval [1180; 1300], the frequencies ω_5, ω_6 disappear, whereas the frequency ω_4 exhibits almost zero power. The LLE sharply increases.

Table 6.16 Numerical results for the Sheremetev–Pelekh model ($l/h = 0.3, t \in [1600; 5300]$) [reprinted with permission from International Journal of Non-Linear Mechanics publishers]

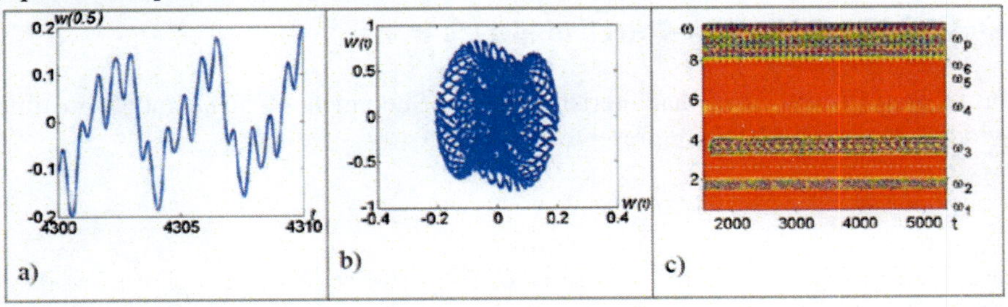

In the interval [1300; 1600], the frequency spectrum does not undergo changes in time and LLE is decreased. The signal possesses non-periodic structure on contrary to the previous interval, whereas the phase portrait has washed out shape since the frequency ω_4 undergoes periodic change.

Interval [1600; 1700] is associated with the occurrence of the frequency ω_3 in the wavelet spectrum which achieves in the beginning its maximum value (see the wavelet spectrum). It should be emphasized that the remaining frequencies of the spectrum perhaps without ω_2 loos their power $S(\omega)$. In this time interval, the second hum exhibited by LLE is observed. In the interval [1700; 5300], there are no changes in the frequency spectrum, and the LLE smoothly decreases though remain positive. The signal possesses complex character but regular. The phase portrait exhibits a tori with the surrounding rings.

Taking into account the same parameters, size coefficient $l/h = 0.3$, the sinusoidal load parameters $q_0 = 4000$, $\omega_p = 9$, the vibration characteristics (signals, Fourier and wavelet spectra, phase portrait, Poincaré sections, time histories of the Lyapunov exponents), they differ with each other depending on the chosen model. Increase of complexity of the mathematical model, i.e. an account of the normal rotation in the Timoshenko model and normal curving in the Sheremetev–Pelekh model, implies an increase of complexity of the vibration character.

Each of the models has its own, and different numbers of frequencies exhibited by the Fourier spectrum: Bernoulli–Euler—6, Timoshenko—3 and the Sheremetev–Pelekh—7 (including the excitation frequency). Each of the studied 2D wavelets of beam models (Tables 6.17c, 6.4c, 6.7c) shows that the frequency spectrum undergoes changes in time which indicates changes of energy in time.

Both phase portraits and Poincaré maps (Table 6.7d, e) of the Timoshenko model differ from the counterpart characteristics of the Bernoulli–Euler (Table 6.1d, e) and Sheremetev–Pelekh (7.13d, e) models. Pictures (Table 6.7d, e) exhibit essential role of three frequencies on the system dynamics.

In spite of the difference between all models with regard to frequency spectra, time histories of LLE are qualitatively similar; in a given time instant, there observed a sudden jump into positive, and then an LLE smoothly decreases though all the times keeping positive value.

6.6.1.6 The Sheremetev–Pelekh Model l/h = 0

In Table 6.17, dynamic characteristics of the Sheremetev–Pelekh model in time window $t \in [300; 5300]$ are presented.

Table 6.17 Numerical results for the Sheremetev–Pelekh model ($l/h = 0, t \in [300; 5300]$) [reprinted with permission from International Journal of Non-Linear Mechanics publishers]

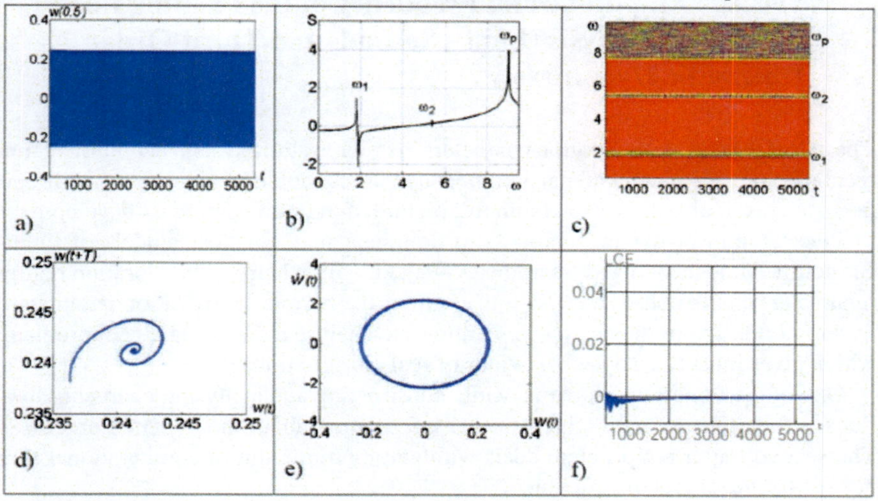

Time history/signal is homogeneous. The Fourier spectrum exhibits two frequencies: $\omega_1 = 1.8086$, $\omega_2 = 5.3812$ and the following frequency relation $2 \cdot \omega_5 + \omega_6 = \omega_p$. The wavelet spectrum also points out the spectrum homogeneity versus time. The phase portrait has the shape of an ellipse. The result difference occurs when the size-dependent behaviour is taken into account, in all characteristics of the dynamical process. The LLE is negative in the whole investigated time interval. Therefore, all dynamical characteristics point out the quasi-periodic vibrations. The wavelet spectrum shows that the frequency characteristics do not depend on time.

The general conclusions are as follows:

(i) dynamic regimes are different for different models, and the frequency spectra of the Bernoulli–Euler and Timoshenko models depend on time;
(ii) phase portraits for the whole time interval (Tables 6.10e and 6.13e) contain information given in time windows (Tables 6.11b, 6.12b, 6.14b, 6.15b);
(iii) time history of the LLE is in agreement with the time history and wavelet spectrum;
(iv) time histories of the LLE of the Euler–Bernoulli, Timoshenko and Sheremetev–Pelekh models are qualitatively different, on contrary to the case of $l/h = 0.3$.

6.7 Construction of the Charts of Vibration Characters Verusus Amplitude and Frequency of the Exciting Load and Their Analysis (First-, Second- and Third-Order Kinematic Hypotheses)

The use of the charts of vibration characters in order to study system evolution in time recalls Poincaré's statement that nonlinear dynamics should be viewed from different perspectives. It also allows to obtain full picture of the studied process development.

Investigation of dynamic process of nonlinear mathematical models of Euler–Bernoulli, Timoshenko and Sheremetev–Pelekh with a help of the vibration regime (character) charts constructed with the help of the harmonic excitation parameters, i.e. amplitude and frequency $\{q_0, \omega_p\}$, allows to analyse the wide palette of problems which gives hints to analyse behaviour of real constructions.

One of the important problems while constructing and analysing charts of vibrations is the accuracy with which we get information about the dynamic processes. The second important problem deals with getting minimum of computational time needed for the chart construction.

We have investigated suitability of the information mapping in vibration regimes for the charts of the vibration characters with the resolution of 100×100, 200×200, 300×300, 600×600 points. The charts with the resolutions 300×300 and 600×600 practically coincide. In order to obtain the charts with vibration regimes of 300×300, there is a need to solve $9 \cdot 10^4$ case studies, and to analyse the obtained results on the basis of frequency spectrum, autocorrelation functions and spectrum of Lyapunov exponents. In order to construct a chart with resolution of 600×600, we need to analyse $12 \cdot 10^4$ case studies, which requires long computational time. This is why, we have chosen the resolution 300×300. Table 6.18 presents the charts of the vibration regimes of the various resolutions obtained while solving problems of the Bernoulli–Euler beam model vibrations without an account of the size-dependent behaviour for the boundary clamping conditions.

$$w(0,t) = w(1,t) = 0, \qquad \frac{\partial w(0,t)}{\partial x} = \frac{\partial w(1,t)}{\partial x} = 0, \qquad u(0,t) = u(1,t) = 0.$$

The same notation is used in the remaining charts in this section.

6.7.1 *Analysis of the Charts of the Vibration Regimes for the Euler–Bernoulli, Timoshenko and Sheremetev–Pelekh Modes Versus λ*

We consider the charts of the vibration regimes of the mentioned models with and account of different values of the beam relative thickness λ for two different values of the size-dependent parameter $l/h = 0$ and $l/h = 0.3$.

Table 6.18 Graphs of types of vibrations for the Euler–Bernoulli beam $\gamma_2 = 0$ [reprinted with permission from International Journal of Non-Linear Mechanics publishers]

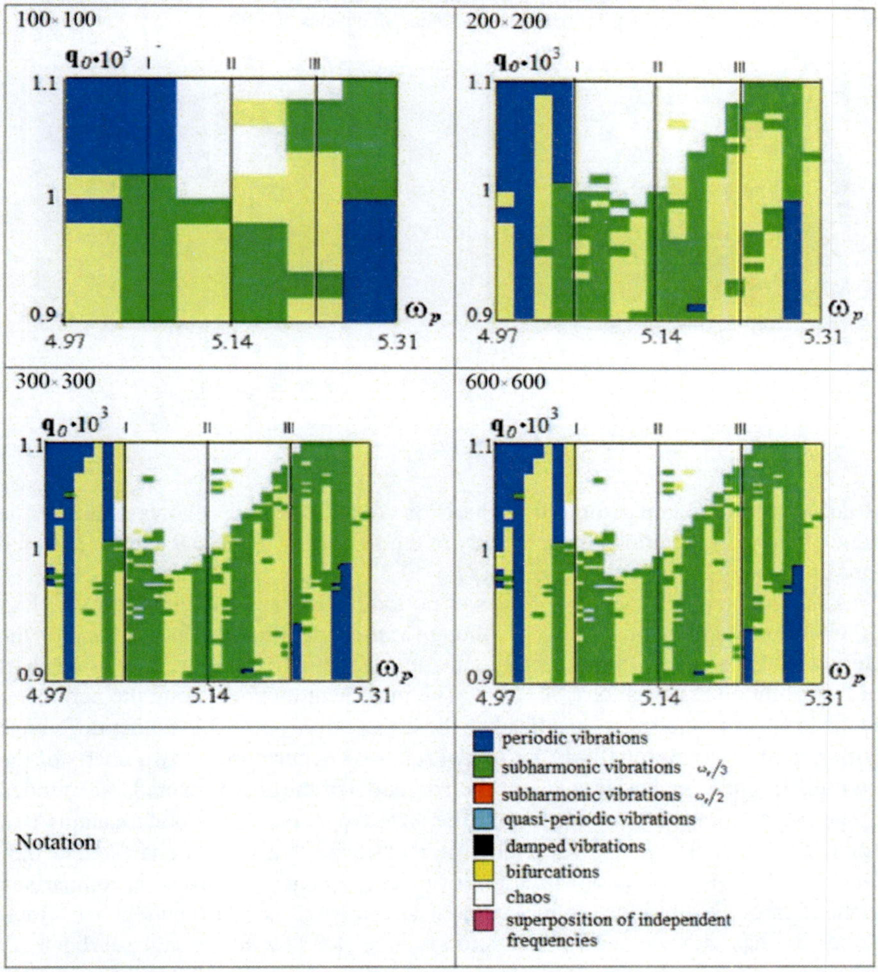

6.7.1.1 Charts of Vibration Regimes for λ = 30

Area of zones of periodic vibrations is larger in the case for account of the size-dependent behaviour $l/h = 0.3$, whereas zone of chaotic and other zone vibrations are larger on the chart of vibration regimes corresponding to $l/h = 0$. In the interval of regimes $\omega_p \in [3.45; 6.3]$ in chart Table 6.19a, there are islands of chaotic vibrations surrounded by zone of periodic vibrations. This states on contrary to the case of inclusion of the size-dependent behaviour, where the mentioned vibration distribution

Table 6.19 The Bernoulli–Euler model—charts of vibration regimes: (a) without the size-dependent behaviour $l/h = 0$; (b) with the size-dependent behaviour $l/h = 0.3$ [reprinted with permission from International Journal of Non-Linear Mechanics publishers]

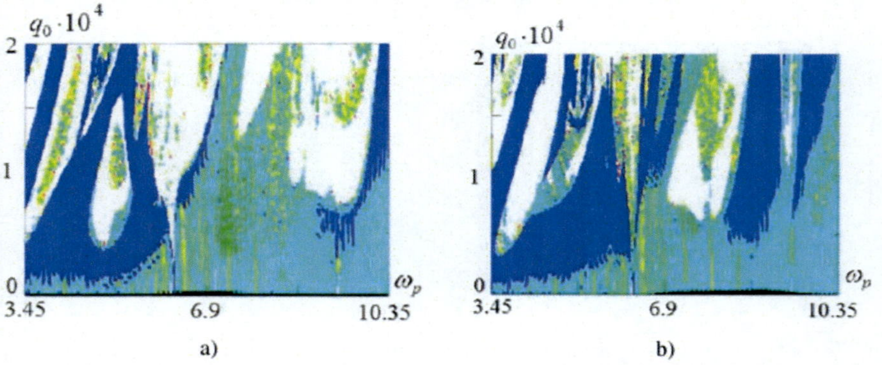

is not observed. Comparison of the charts of vibration regimes shows a shift of the zone of damped vibrations from vicinity of the frequency $\omega_p = 6.9$ and for $l/h = 0$ into the zone of high frequencies for $l/h = 0.3$.

As in the previous case, here zones of periodic vibrations are larger on the chart of vibration regime for $l/h = 0.3$, though rather marginally. On contrary to the Bernoulli–Euler model, the area of the chaotic zones of vibrations is larger in the chart of the vibration regimes for $l/h = 0.3$. The most lengthy zone along the parameter of the load, amplitude q_0, is localized on the frequency $\omega_p = 7.3$. Another difference with regard to the Bernoulli–Euler model relies on occurrence on two charts of the vibration regimes in the case of Timoshenko model of the chaotic islands surrounded by periodic vibrations. An account of the size-dependent behaviour transmits that zone into larger frequencies. It should be mentioned that for the chart $l/h = 0.3$, there is a shift of the periodic vibrations over the frequency $\omega_p = 6.9$, in comparison to the chart of the vibration regimes without account of the size-dependent behaviour. Zones of chaos for the chart of vibration regimes for $l/h = 0.3$ begin earlier with regard to q_0 then for the chart with $(l/h = 0)$. For the chart $l/h = 0.3$ in the frequency window $\omega_p \in [7.5; 9.3]$ for $q_0 \in [2000; 5000]$ in zone of vibrations spent on non-dependent frequency, there is a zone of subharmonic vibrations $\omega_p/3$. On contrary to the chart of vibration regimes $l/h = 0$ in the chart $l/h = 0.3$ in the subinterval of high frequencies $\omega > \omega_p = 9.5$, there is a zone of periodic vibration located in chaotic zone and being surrounded by a zone spanned on the non-dependent frequency.

Zones of periodic vibrations in the chart of vibration regimes for $l/h = 0.3$ possesses the enlarged, and they are located lower than in the case of the chart for $l/h = 0$ with regard to the parameter q_0 as it was happened earlier while compared with the vibration regimes of the Bernoulli–Euler model. The latter behaviour is implied by the occurrence of the biharmonic operator governing equation of deflection zones of periodic vibrations that have the shape of strips. In contrast to the chart of vibration regimes for $l/h = 0$, there is zone of periodic vibrations, with inclusion of chaotic zone (lower band of frequencies). In the analysed model, there is also coincidence of the obtained charts of vibration regimes with those associated with the Bernoulli–Euler model.

Generalizing the obtained results, one may conclude that zones of vibrations with $\omega_p \in [3.45; 6.9]$ differ marginally for all studied models. An account of the size-dependent behaviour for all mathematical models changes vibrations character, in particular, in zone of $\omega_p \in [6.9; 10.35]$.

6.7.1.2 Charts of Vibration Regimes for $\lambda = 50$

Areas of periodic and chaotic zones are increased in the chart of vibration regimes for $l/h = 0$. Increase of the periodic zones takes place in the interval $q_0 \in [30000; 60000]$.

Zones of periodic vibrations increase in the chart of the vibration regimes for $l/h = 0.3$. On contrary to other charts, zone corresponding to superposition of the independent frequencies occurs for $\omega_p = 7.1$. Chaotic zones are increased in the chart of vibration regimes without an account of the size-dependent behaviour.

Zones of periodic and chaotic vibrations are slightly larger on the chart of vibration regimes for $l/h = 0$ than for $l/h = 0.3$. In both cases, the beginning of chaos with regard to the loading parameter is the same. Inside the zone of periodic vibrations for $\omega_p \in [4.3; 5.2]$, there is zone of vibrations spanned on the non-dependent frequency by chaotic zone.

Comparison of the charts associated with various models yields the following conclusions. For $l/h = 0$ in the interval $\omega_p \in [3.45; 6.9]$, the characters of the picture is similar. In the chart of the Sheremetev–Pelekh model, there are zone characteristics only for the Bernoulli–Euler model (periodic vibrations) and only for the Timoshenko model (bifurcation zones). The vibration character for the interval $\omega_p \in [6.9; 10.35]$ for all models is different. For $l/h = 0.3$, in the window $\omega_p \in [3.45; 6.9]$, the charts of the vibration regimes of the Timoshenko and the Sheremetev–Pelekh models exhibit similar dynamic regimes (Tables 6.20, 6.21, 6.22, 6.23, 6.24).

6.7.1.3 Charts of Vibration Regimes for $\lambda = 100$

Zone of periodic vibrations shown in Table 6.25b is decreased due to increase of zones of vibration spanned on the independent frequency. Zones of damped vibrations are increased in the region corresponding to high frequencies $\omega_p \in [7.5; 10.1]$; the same holds for zone of periodic vibrations for $q_0 < 2000$. On contrary to the chart of vibration regimes for $l/h = 0$, zone of superposition of the independent frequencies has more sense.

The essential difference in comparison to the Bernoulli–Euler model is observed in the periodic larger amount of vibrations zones in the size-dependent case.

Table 6.20 The Timoshenko model—charts of vibration regimes: (a) without the size-dependent behaviour $l/h = 0$; (b) with the size-dependent behaviour $l/h = 0.3$ [reprinted with permission from International Journal of Non-Linear Mechanics publishers]

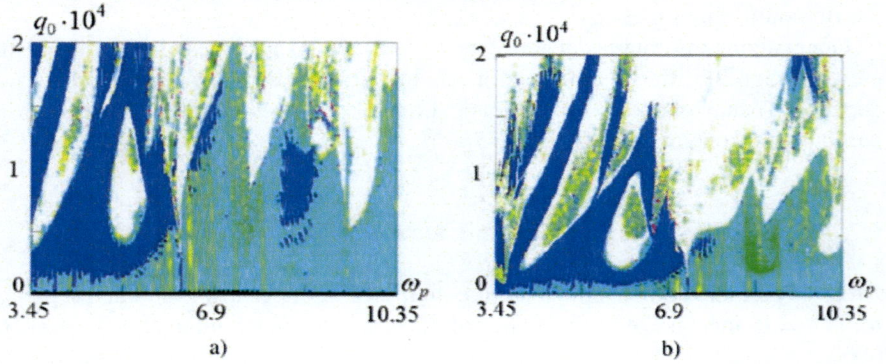

Table 6.21 The Sheremetev–Pelekh model—charts of vibration regimes: (a) without the size-dependent behaviour $l/h = 0$; (b) with the size-dependent behaviour $l/h = 0.3$ [reprinted with permission from International Journal of Non-Linear Mechanics publishers]

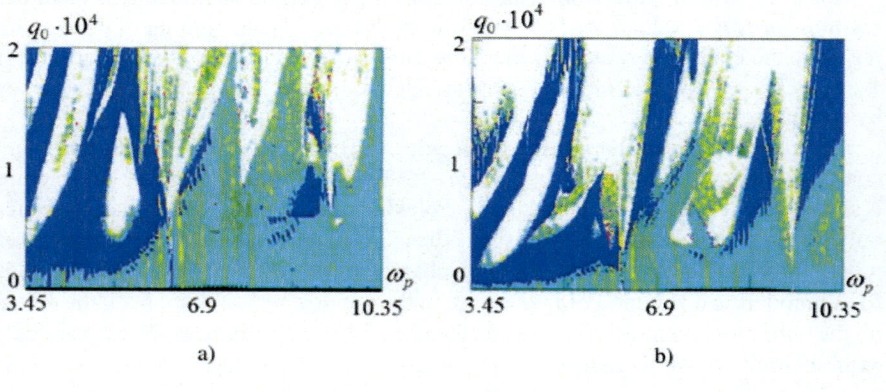

Area of zones of periodic vibrations is larger in the chart of the size-dependent vibration regimes. Localization of zones of damped vibrations differs essentially for two charts of the vibration regimes.

Table 6.22 Bernoulli–Euler model—charts of vibration regimes: (a) without the size-dependent behaviour $l/h = 0$; (b) with the size-dependent behaviour $l/h = 0.3$ [reprinted with permission from International Journal of Non-Linear Mechanics publishers]

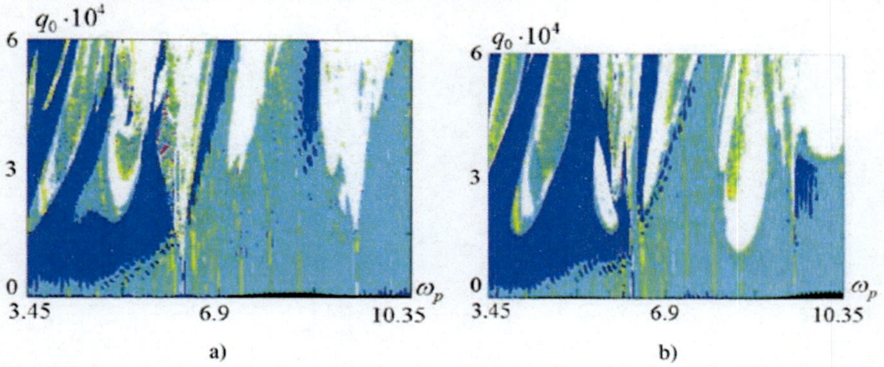

Table 6.23 Timoshenko model—charts of vibration regimes: (a) without the size-dependent behaviour $l/h = 0$; (b) with the size-dependent behaviour $l/h = 0.3$ [reprinted with permission from International Journal of Non-Linear Mechanics publishers]

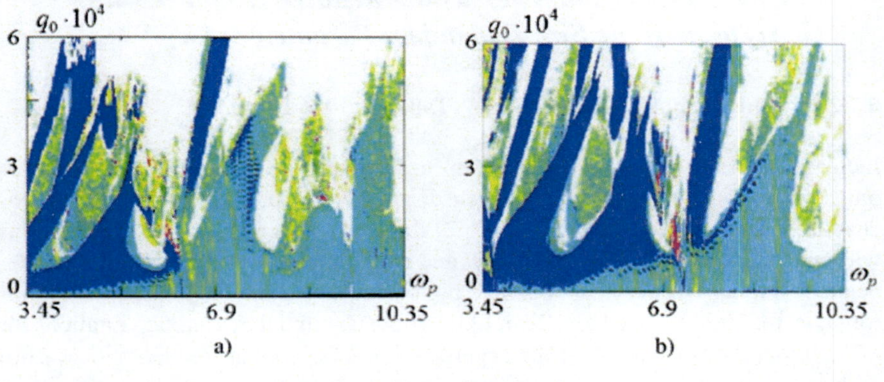

Comparison of the charts shows similarity in the character of vibrations in the charts for $l/h = 0$ and $l/h = 0.3$ for the Timoshenko and Sheremetev–Pelekh models in the whole considered frequency interval. This is due to the account of the shear deformation in the models.

Table 6.24 The Sheremetev–Pelekh model—charts of vibration regimes: (a) without the size-dependent behaviour $l/h = 0$; (b) with the size-dependent behaviour $l/h = 0.3$ [reprinted with permission from International Journal of Non-Linear Mechanics publishers]

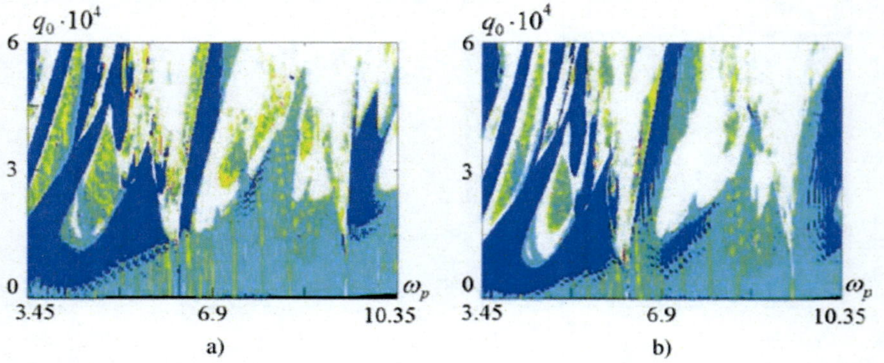

6.7.2 Comparison of the Charts of the Vibration Regimes for One Chosen Model Versus the Relative Length λ with Account of the Size-Dependent Parameter l/h

6.7.2.1 The Bernoulli–Euler Model (Tables 6.19, 6.22, 6.25)

Increase of the λ parameter from 30 up to 50 implies decrease of chaotic zones on the low frequencies [3.45; 6] which disappear and periodic vibrations appear chaotic zones $\omega_p \in [6.9; 8]$, $q_0 \in [0.7; 2]$ for $\lambda = 30$ are substituted by vibrations with independent frequency for $\lambda = 50$ and by regime of superposition of the independent frequencies for $\lambda = 100$. The high-frequency zone of periodic vibrations for $\omega_p \in [8; 10.35]$, $q_0 \in [0.3; 2]$ and $\lambda = 30$ is substituted by chaotic vibrations for $\lambda = 50$ vibrations spanned on the superposition-independent frequency ($\lambda = 100$) with crops of zone of the independent frequencies.

6.7.2.2 The Timoshenko Model (Tables 6.20, 6.23, 6.26)

Increase of the parameter λ implies sharp decrease of chaotic zones perhaps with the interval of high frequencies $\omega_p \in [9.3; 10.35]$. For $\lambda = 100$, the chaotic zone disappears and the periodic zone decreases. The latter zone has the largest area for $\lambda = 50$, and it is more compactly located in the chart of vibration regimes. There is also clearly exhibited zone of superposition of the independent frequencies for $\omega_p \in [7.1; 7.3]$.

Table 6.25 The Bernoulli–Euler model—charts of vibration regimes: (a) without the size-dependent behaviour $l/h = 0$; (b) with the size-dependent behaviour $l/h = 0.3$ [reprinted with permission from International Journal of Non-Linear Mechanics publishers]

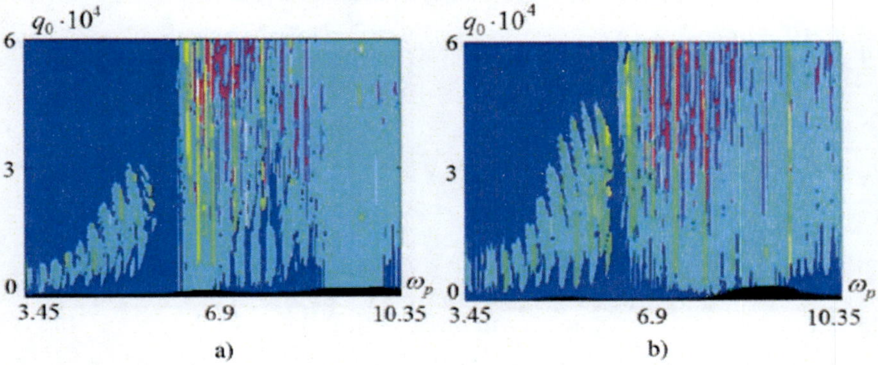

Table 6.26 Timoshenko model—charts of vibration regimes: (a) without the size-dependent behaviour $l/h = 0$; (b) with the size-dependent behaviour $l/h = 0.3$ [reprinted with permission from International Journal of Non-Linear Mechanics publishers]

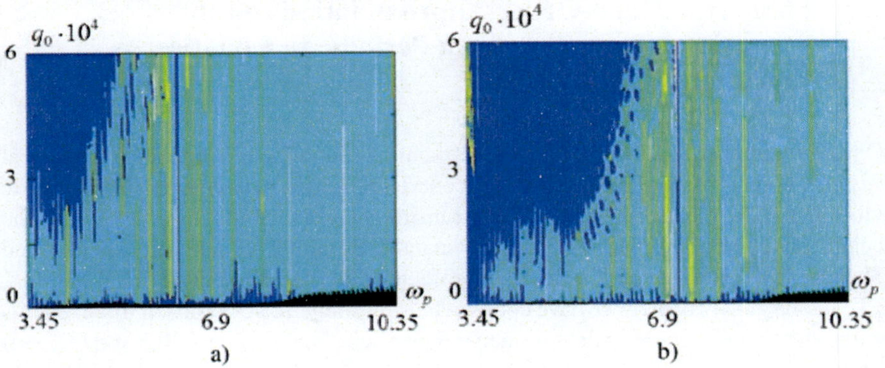

6.7.2.3 The Sheremetev–Pelekh Model (Tables 6.21, 6.24, 6.27)

As it was in the Timoshenko model, increase of λ implies sharp decrease of the chaotic zones. However, for $\lambda = 50$, in contrast to the Timoshenko model, they remain for $\omega_p \in [3.45; 6.9]$. The summed area of periodic vibrations is decreased with increase of the parameter λ, and those zones disappear for $\omega_p \in [6.9; 10.35]$ whereas for $\omega_p \in [3.45; 6.9]$ they are grouped and increased.

Table 6.27 The Sheremetev–Pelekh model—charts of vibration regimes: (a) without the size-dependent behaviour $l/h = 0$; (b) with the size-dependent behaviour $l/h = 0.3$ [reprinted with permission from International Journal of Non-Linear Mechanics publishers]

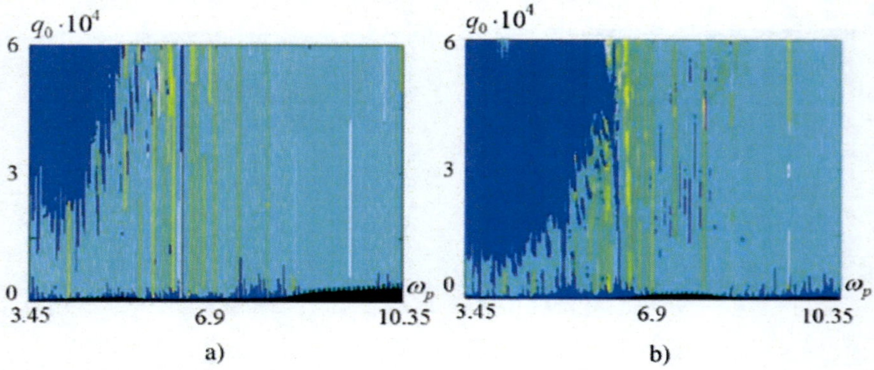

6.8 Influence of a Type of Kinematic Models of the Zero-, First- and Third-Order Approximations on the Scenario of Transition from Periodic to Chaotic Vibrations

One of the imputation aspects of the investigation of nonlinear dynamics of structural systems is construction and analysis of scenario of transition from periodic to chaotic vibrations. The mentioned scenarios of transition have been already analysed for the beam, plates and shell taking into account the Bernoulli–Euler, Timoshenko and Sheremetev–Pelekh models but without effects of the size-dependent behaviour [34, 75]. In this given chapter, we have constructed and analysed scenarios of three models with/without an account of the size-dependent effects, i.e. for $l/h = 0.3$ and $l/h = 0$ or the following fixed parameters $\omega_p = 9,\ \varepsilon = 1,\ \lambda = 30,\ \nu = 0.3$.

Scenarios exhibited by the Bernoulli–Euler, Timoshenko and Sheremetev–Pelekh beams followed the Ruelle-Taken-Newhouse scenario [76] for both cases, i.e. $l/h = 0.3$, $l/h = 0$.

References

1. Sheremetev, M.P., Pelekh, B.L.: To the construction of a refined theory of plates. Eng. J. **4**(3), 34–46 (1964)
2. Hanuska, A.: Contribution to the Reissnerian algorithm in the theory of bending of elastic plates. Aplik. Matemat. **12**(6), 449–467 (1967)
3. Krysko, A.V., Kirichenko, V.F.: Refined theory of orthotropic thermosensitive gently sloping shells in the framework of Pelekh-Sheremetev hypotheses. In: III International Symposium on 'Nonclassical problems of the theory of thin-walled structural elements of the physicochemical mechanics of composite materials', Ivano-Frankovsk, Ukraine, pp. 142–148 (1995)
4. Levinson, M.: A new rectangular beam theory. J. Sound Vibr. **74**, 81–87 (1981)
5. Reddy, J.N.: A simple higher-order theory for laminated composite plates. J. Appl. Mech. **51**, 745–752 (1984)
6. Fu, Y., Zhang, J.: Electromechanical dynamic buckling phenomenon in symmetric electric fields actuated microbeams considering material damping. Acta Mech. **212**, 29–42 (2010)
7. Moghimi, Z.M., Ahmadian, M.T.: Static pull-in analysis of electrostatically actuated microbeams using homotopy perturbation method. Appl. Math. Model. **34**, 1032–1046 (2010)
8. Jia, X.L., Yang, J., Kitipornchai, S.: Pull-in instability of geometrically nonlinear micro-switches under electrostatic and Casimir forces. Acta Mech. **218**, 161–174 (2011)
9. Fleck, N.A., Muller, G.M., Ashby, M.F., Hutchinson, J.W.: Strain gradient plasticity: theory and experiments. Acta Metall. Mater. **42**, 475–487 (1994)
10. Nix, W.D.: Mechanical properties of thin films. Metall. Trans. A **20**, 2217–2245 (1989)
11. Mazza, E., Abel, S., Dual, J.: Experimental determination of mechanical properties of Ni and Ni-Fe microbars. Micros. Technol. **2**(4), 197–202 (1996)
12. Lam, D.C.C., Yang, F., Chong, A.C.M., Wang, J., Tong, P.: Experiments and theory in strain gradient elasticity. J. Mech. Phys. Soli **51**, 1477–1508 (2003)
13. Ma, Q., Clarke, D.R.: Size dependent hardness of silver single crystals. J. Mater. Res. **10**(4), 853–863 (1995)
14. Scheible, D.V., Erbe, A., Blick, R.H.: Evidence of a nanomechanical resonator being driven into chaotic response via the Ruelle-Takens route. Appl. Phys. Lett. **81**, 1884–1886 (2002)
15. Mindlin, R.D., Tiersten, H.F.: Effects of couple-stresses in linear elasticity. Arch. Ration. Mech. Anal. **11**, 415–448 (1962)
16. Toupin, R.A.: Elastic materials with couple-stresses. Arch. Ration. Mech. Anal. **11**, 385–414 (1962)
17. Eringen, A.C.: Nonlocal polar elastic continua. Int. J. Eng. Sci. **10**, 1–16 (1972)
18. Aifantis, E.C.: Strain gradient interpretation of size effects. Int. J. Fract. **95**, 299–314 (1999)
19. Gurtin, M.E., Weissmuller, J., Larche, F.: The general theory of curved deformable inter-faces in solids at equilibrium. Philos. Mag. A **78**, 1093–1109 (1998)
20. Yang, F., Chong, M., Lam, D.C.C., Tong, P.: Couple stress based strain gradient theory for elasticity. Int. J. Solids Struct. **39**, 2731–2743 (2002)
21. Fleck, N.A., Hutchinson, J.W.: Strain gradient plasticity. In: Hutchinson, J.W., Wu, T.Y. (eds.) Advances in Applied Mechanics, vol. 33, pp. 295–366. Academic Press, New York (1997)
22. Asghari, M., Ahmadian, M.T., Kahrobaiyan, M.H., Rahaeifard, M.: On the size-dependent behavior of functionally graded micro-beams. Mater. Des. **31**, 2324–2329 (2010)
23. Asghari, M., Kahrobaiyan, M.H., Ahmadian, M.T.: A nonlinear Timoshenko beam formulation based on the modified couple stress theory. Int. J. Eng. Sci. **48**, 1749–1766 (2010)
24. Lazopoulos, K.A., Lazopoulos, A.K.: Bending and buckling of thin strain gradient elastic beams. Eur. J. Mech. A-Sol. **29**, 837–843 (2010)
25. Ma, H.M., Gao, X.L., Reddy, J.N.: A non-classical Mindlin plate model based on a modified couple stress theory. Acta Mech. **220**, 217–235 (2011)
26. Ma, H.M., Gao, X.L., Reddy, J.N.: A nonclassical Reddy-Levinson beam model based on a modified couple stress theory. Int. J. Multiscale Comput. Eng. **8**, 167–180 (2010)
27. Kirchhoff, G.: Vorlesungen Über Mathematische Physik. Mechanik, Leipzig (1876)

28. Park, S.K., Gao, X.L.: Bernoulli-Euler beam model based on a modified couple stress theory. Micromech. Microeng. **16**(11), 2355–2359 (2006)
29. Rajabi, F., Ramezani, S.: A nonlinear microbeam model based on strain gradient elasticity theory with surface energy. Arch. Appl. Mech. **82**, 363–376 (2012)
30. Ma, H.M., Gao, X.-L., Reddy, J.N.: A microstructure-dependent Timoshenko beam model based on a modified couple stress theory. J. Mech. Phys. Sol. **56**, 3379–3396 (2008)
31. Asghari, M., Kahrobaiyan, M.H., Nikfar, M., Ahmadian, M.T.: A size-dependent nonlinear Timoshenko microbeam model based on the strain gradient theory. Acta Mech. **223**, 1233–1249 (2012)
32. Ansari, R., Gholami, R., Darabi, M.A.: A nonlinear Timoshenko beam formulation based on strain gradient theory. J. Mech. Mater. Struct. **7**(2), 1749–1761 (2012)
33. Awrejcewicz, J., Krysko, V.A., Papkova, I.V., Krysko, A.V.: Deterministic Chaos in One-Dimentional Continuous Systems. World Scientific, Singapore (2016)
34. Awrejcewicz, J.A., Krysko, V.A., Zhigalov, M.V., Saltykova, O.A., Krysko, A.V.: Chaotic vibrations in flexible multilayered Bernoulli-Euler and Timoshenko type beams. Lat. Am. J. Sol. Struct. **5**(4), 319–363 (2008)
35. Krysko, A.V., Awrejcewicz, J., Saltykova, O.A., Zhigalov, M.V., Krysko, V.A.: Investigations of chaotic dynamics of multi-layer beams using taking into account rotational inertial effects. Comm. Nonlin. Sci. Num. Simul. **19**(8), 2568–2589 (2014)
36. Krysko, V.A., Awrejcewicz, J.: Nonclassical Thermoelastic Problems in Nonlinear Dynamics of Shells. Springer, Berlin (2003)
37. Krysko, V.A., Awrejcewicz, J.: Chaos in Structural Mechanics. Springer, Berlin (2008)
38. Krysko, V.A., Awrejcewicz, J., Krysko, A.V.: Thermo-Dynamics of Plates and Shells. Springer, Berlin (2007)
39. Krysko, V.A., Awrejcewicz, J., Vakakis, A.: Nonlinear Dynamics of Continuous Elastic Systems. Springer, Berlin (2004)
40. Simsek, M., Reddy, J.N.: Bending and vibration of functionally graded microbeams using a new higher order beam theory and the modified couple stress theory. Int. J. Eng. Sci. **64**, 37–53 (2013)
41. Wolf, A., Swift, J.B., Swinney, H.L., Vastano, J.A.: Determining Lyapunov exponents from a time series. Phys. D **16**, 285–317 (1985)
42. Nowacki, W.: Theory of Elasticity. Mir, Moscow (1975)
43. Krysko, A.V., Awrejcewicz, J., Zhigalov, M.V., Pavlov, S.P., Krysko, V.A.: Nonlinear behaviour of different flexible size-dependent beams models based on the modified couple stress theory. Part 1. Governing equations and static analysis of flexible beams. Int. J. Non-Lin. Mech. **93**, 96–105 (2017)
44. Ansari, R., Gholami, R.: Size-dependent vibration of functionally graded curved microbeams based on the modified strain gradient elasticity theory. Arch. Appl. Mech. **83**, 1439–1449 (2013)
45. Novozhilov, V.V.: Introduction to Nonlinear Theory of Elasticity. Gostechizdat, Moscow
46. Franklin, J.N.: On Tikhonov's method for ill-posed problems. Math. Comput. **28**(128), 889–907 (1974)
47. Muschtari, H.M.: Some generalizations of the theory of thin shells. Proc. Phys. Math. Soc. Kazan Univ. **XI**(8) (1938)
48. Muschtari, H.M.: On the stability of a round thin cylindrical shell under torsion. Proc. Kazan Aviat. Instit. **2** (1934)
49. Timoshenko, S.P.: On the correction for shear of differential equation for transverse vibration of prismatic bar. Philosoph. Mag. **41**, 744–746 (1921)
50. Bernoulli, J.: Essait theoretique sur les vibrations de plaques elastiques rectangulaires et libres. Nova Acta Acad. Petropolit. **5**, 197–219 (1789)
51. Euler, L.: Sur la force des colones. Memories de L'Academie de Berlin **13**, 252–282 (1757)
52. Love, A.: A Tretise on the Mathematical Theory of Eelasticity (1927)
53. Awrejcewicz, J., Krysko, A., Saltykova, O.A., Zhigalov, M.V., Soldatov, V.V.: Investigations of complex vibrations of beams within the framework of the Sheremet'ev-Pelekh kinematic model using the wavelet transform. J. Mach. Manufact. Reliab. **39**(4), 313–317 (2010)

54. Ke, L.L., Wang, Y.S., Yang, J., Kitipornchai, S.: Nonlinear free vibration of size-dependent functionally graded microbeams. Int. J. Eng. Sci. **50**(1), 256–267 (2012)
55. Krysko, V.A., Awrejcewicz, J., Komarov, S.A.: Nonlinear deformations of spherical panels ssubjected to transversal load action. Comput. Meth. Appl. Mech. Eng. **194**(27–29), 3108–3126 (2005)
56. Krysko, V.A., Zhigalov, M.V., Yakovleva, T.V., Papkova, I.V.: The method of establishing in nonlinear problems of beams and plates, taking into account the locality of loading. Bullet. STGU **2**, 7–17 (2012)
57. Krysko, V.A., Komarov, S.A., Egurnov, N.V.: Flexing of flexible plates under the action of longitudinal and transverse loads. Appl. Mech. **32**(9), 80–87 (1996)
58. Miller, K.: Least squares methods for ill-posed problems with a prescribed bound. SIAM J. Math. Anal. **1**(6), 52–74 (1970)
59. Phillips, D.L.: A technique for the numerical solution of certain integral equations of the first kind. J. Assoc. Comput. Mach. **9**(6), 84–97 (1962)
60. Soltani, P., Kassaei, A., Mehdi, M.T., Farshidianfar, A.: Vibration of wavy single-walled carbon nanotubes based on nonlocal Euler Bernoulli and Timoshenko models. Int. J. Adv. Struct. Eng. **4**(3) (2012)
61. Yang, X.D., Lim, C.W.: Nonlinear vibrations of nano-beams accounting for nonlocal ef-fect using a multiple scale method. Sci. China Ser. E-Tech. Sci. **52**(3), 617–626 (2009)
62. Kong, S.L., Zhou, S.J., Nie, Z.F., Wang, K.: Static and dynamic analysis of micro beams based on strain gradient elasticity theory. Int. J. Eng. Sci. **47**, 487–498 (2009)
63. Ke, L.L., Wang, Y.S.: Size effect on dynamic stability of functionally graded microbeams based on a modified couple stress theory. Compos. Struct. **93**, 342–350 (2011)
64. Wang, D.H., Wang, G.F.: Surface effects on the vibration and buckling of double-nanobeam-systems. J. Nanomater. **2011**, ID 518706 (2011)
65. Xia, R., Li, X., Qin, Q., Liu, J., Feng, X.Q.: Surface effects on the mechanical properties of nanoporous materials. Nanotechnology **22**, 265714 (2011)
66. Batra, R.C., Porfiri, M., Spinello, D.: Vibrations of narrow microbeams predeformed by an electric field. J. Sound Vib. **309**, 600–612 (2008)
67. Hassanpour, P.A., Esmailzadeh, E., Cleghorn, W.L., Mills, J.K.: Nonlinear vibration of micromachined asymmetric resonators. J. Sound Vib. **329**, 2547–2564 (2010)
68. Kitipornchai, S., Ke, L.L., Yang, J., Xiang, Y.: Nonlinear vibration of edge cracked functionally graded Timoshenko beams. J. Sound Vib. **324**, 962–982 (2009)
69. Kong, S., Zhou, S., Nie, Z., Wang, K.: The size-dependent natural frequency of Bernoulli-Euler micro-beams. Int. J. Eng. Sci. **46**, 427–437 (2008)
70. Krysko, A.V., Awrejcewicz, J., Kutepov, I.E., Zagniboroda, N.A., Dobriyan, V., Krysko, V.A.: Chaotic dynamics of flexible Euler-Bernoulli beams. Chaos **23**(4), 043130–1–043130–25 (2013)
71. Li, X.F.: A unified approach for analyzing static and dynamic behaviors of functionally graded Timoshenko and Euler-Bernoulli beams. J. Sound Vib. **318**, 1210–1229 (2008)
72. Shu, C.: Differential Quadrature and its Application in Engineering. Springer, London (2000)
73. Wang, B., Zhao, J., Zhou, S.: A micro scale Timoshenko beam model based on strain gradient elasticity theory. Europ. J. Mech. A/Sol. **29**, 591–599 (2010)
74. Krysko, A.V., Awrejcewicz, J., Zhigalov, M.V., Pavlov, S.P., Krysko, V.A.: Nonlinear behaviour of different flexible size-dependent beams models based on the modified couple stress theory. Part 2. Chaotic dynamics of flexible beams. Int. J. Non-Lin. Mech. **93**, 106–121 (2017)
75. Awrejcewicz, J., Krysko, V.A.: Feigenbaum scenario exhibited by thin plate dynamics. Nonlin. Dyn. **24**, 373–398 (2006)
76. Ruelle, D., Takehs, F.: On the nature of turbulence. Commun. Math. Phys. **20**, 167–192 (1976)

Chapter 7
Mathematical Models of Functionally Graded Beams in Temperature Field

7.1 Introduction

In this chapter, the analysis of nonlinear FGM (functionally graded material) straight and curved beams behaviour based on the modified couple stress theory is carried out. Defining the deflection curve in order to simplify the governing equations, we investigate the influence of scale length parameter and the non-homogeneity coefficient on the dynamic characteristics and the scenario of transition from periodic to chaotic beam vibrations. First, the properties of various FGM materials are revisited (Sect. 7.2) with an emphasis on the dependence of material properties on temperature. Then, regular and chaotic vibrations of size-dependent Timoshenko beams with functionally graded properties along their thickness are illustrated and analysed (Sect. 7.3). The following new properties regarding the research topic under consideration can be derived.

(i) We have considered the dynamics of nonlinear FG Timoshenko beams on the basis of the modified couple stress theory, using a novel concept of the bending line.
(ii) The influence of the size-dependent coefficient and the grading parameter on the load-deflection dependence is investigated for the static problem. In order to get solutions for nonlinear static problem, the relaxation method was employed. It has been shown that the minimum deflection is achieved by the beam when the functional grading process is taken into account and the stiffer layer is located on the upper side in both cases, i.e. with/without the size-dependent behaviour.
(iii) The functionally graded beam with the stiffer layer on the upper side is suitable for application to carry dynamic loads for a given frequency and amplitude of the harmonic excitation. This conclusion coincides with that formulated for static problems. In the case of the homogeneous beam and the beam with the stiffer layer located on the bottom side, the essential dependence of the obtained results on the size-dependent coefficient is observed.

J. Awrejcewicz et al., *Mathematical Modelling and Numerical Analysis of Size-Dependent Structural Members in Temperature Fields*, Advanced Structured Materials 142,
https://doi.org/10.1007/978-3-030-55993-9_7

(iv) In order to validate the reliability of the LLEs (largest Lyapunov exponents) computed with Wolf's algorithm, three qualitatively different methods have been applied, i.e. Rosenstein's, Kantz's and the neural network one. The analysis implies that all methods give qualitatively the same result, i.e. positive or negative values of the LLEs, over the whole time interval studied.

(v) For the considered values of the size-dependent and material grading parameters, the universal route to chaos, following the classical Ruelle-Takens-Newhouse scenario, has been detected.

In Sect. 7.4, the size-dependent model based on a modified coupled stress theory has been constructed for the geometrically nonlinear curvilinear functionally graded Timoshenko beams. To build the model, the concept of a reference line was used. Compared to previous models for functionally graded beams, this model is significantly simpler. In the case of the straight-line beams, the size-dependent effect decreases the value of the beam deflection for the same coefficient of nonhomogeneity. It has been reported that the localization of the most rigid layer on the upper side of the beam essentially decreases the deflections of the beam for the same value of the size-dependent coefficients. In the case of the curvature, the most loading ability is observed for the functionally graded beam with the most rigid layer located on the beam top. In the case of the LE computations, in all studied cases, we have observed a transition from the negative to positive LE value, which is associated with the transition from the pre-critical to post-critical beam state. In variants 2 and 8, there are two positive LEs. It means that in these two cases, stiffer stability loss is exhibited. In Sect. 7.5, investigation of the stability of flexible curvilinear Euler-Bernoulli beams in a temperature field has been carried out without any restrictions regarding the temperature field distribution. It has been shown that the occurrence of imperfections due to either beam curvature or external load implies a different form of beam stability loss while increasing the temperature intensity. The type of temperature field has an essential impact on the beam stability loss regarding the temperature intensity and external loading. Inclusion of the beam curvature in the heat transfer equation yields an increase of the critical load responsible for the stability loss as well as changes of the beam form regarding its pre-critical state. The last Sect. 7.6 is devoted to the study of the mathematical model of a three-layer micro- and nanobeams. Based on both Grigolyuk-Chulkov and modified couple stress theories, the new model validated by both static and dynamic analyses of the three-layer microbeams including only one scalar/length parameter has been constructed, which takes into account the size effect. The employed Hamilton principle yielded the governing equation of motion as well as general boundary and initial conditions regarding displacements formulated for the microbeams. The proposed model of the microbeam deformation is one of the simplest models and it includes the only one scalar length parameter. However, it allows us to take into account the microstructural effects in both external as well as internal beam layers for any boundary conditions. The finally formulated boundary value problem is of the sixth order, and in the case of the static problem, it is solved analytically. The numerical results show that the studied beam model can explain the scale effect exhibited by the microbeams. The obtained deflections and stresses based on the introduced modified couple stress model are smaller compared to the classical three-layer Grigolyuk-Chulkov beam model while increasing beam thickness.

7.2 Literature Review

The functionally graded materials (FGM) can be fabricated as non-homogenous composites, which are made from two or more materials with the required continuous change of properties along a given direction [1]. The continuous change of the FGM properties stands for their numerous engineering recognized advantages [2]. They allow to omit problems associated with the occurrence of large shear stresses which are likely to appear in multi-layer beams, stronger stability properties, the improved heat resistance and decreased intensity of stresses. Nowadays, fabricated structures made from FGM are widely applied in numerous branches of industry. The FGM structures can be used as thermoisolation of the space engineering objects, nuclear reactors, etc. This is why statics, bending and dynamic characteristics of structures made from FGM have been intensively investigated [3–5]. In the recent years, FGM found wide palette of applications in micro- and nanostructures such as thin films [6, 7] and micro- and nano-electromechanical systems [8, 9].

The micro- and nanosize beams, plates and shells are widely used in the electromechanical systems (MEMS, NEMS) as the vibration sensors [10], micro-tubes [11] and micro-switches [12]. The use of only one material does not allow to satisfy all material and economical requirements coming from the engineering world. Witvrouw and Mehta [9] used FGM in order to satisfy the required mechanical, thermal and electric properties. They analysed the case when the material properties of two-phase material made from poly-SiGe layers for MEMS/NEMS application of thickness of microns and submicrons, and hence influence of the size effects cannot be ignored. The size-dependent static and dynamic behaviour on the micro-level have been validated experimentally [13–16]. The carried out experimental investigations imply that the size-dependent behaviour is associated with internal materials property when the characteristic size (thickness or diameter) is close to the internal material length parameter [17]. Though numerous works devoted to that topic are based on the linear models, which is contradicted to the experimental results pointing out the necessity of inclusion of nonlinear features while investigating MEMS/NEMS [18].

Besides the existing experimental methods used for determination of mechanical properties in the nanoscale, there are also a few theoretical approaches including the molecular dynamics [19] and non-classical mechanics of continuum [20] for the purpose of modelling of micro- and nanostructures behaviour. Material properties in the nanoscale depend on the object size, and consequently, the effect of the size dependence of the mechanical behaviour of nanomaterials should be taken into account [21]. In order to overcome the mentioned problem, a few non-classical theories of continuum have been employed which include the size-dependent effects while modelling nanostructures. One of the widely employed theories with an account of the size-dependent phenomena was proposed by Eringen [22]. Based on the nonlocal Eringen's theory many investigations have been carried out aimed on work out of nonlocal continuum models and employ them to study general behaviour of nanostructures [23–28] as well as problems related to bending effects of nanostructures [29–32] and nanobeams [33, 34].

On the other hand, Toupin [35], Mindlin [36] and Koiter [37] developed the couple stress theory successfully used while modelling of nanobeams. More recently, the mentioned theory was employed by numerous researchers to study the size-dependent dynamical behaviour of microstructures [17, 38–44]. Owing to the huge increase of application of the micro-/nano-FGM structures [45–49] in recent years, the number of works devoted to analysis of the mechanical behaviour of such structures has essentially increased [50–62]. The mentioned references present a comparison of different values of the gradient material index between different beam models constructed on a basis of the classical, modified couple stress and gradient theories.

Asghari et al. [38] investigated the size-dependent static and dynamic behaviour of functionally graded microbeams based on the modified couple stress theory within the elastic regime. Asghari et al. [63] analysed dynamic characteristics of the size-dependent FGM microbeams based on the strain gradient Timoshenko beam theory. Asghari et al. [63] proposed the size-dependent governing equations of the FGM Timoshenko beam based on the modified couple stress theory. Janghorban and Zare [55] carried out free vibration study of functionally graded carbon nanotubes with variable thickness taking into account Timoshenko beam theory. Kahrobaiyan et al. [56] derived the size-dependent model of the FGM Euler-Bernoulli beam employing the theory of deformations gradient. The governing equations of motion were obtained and the classical and non-classical boundary conditions were considered. Both static and dynamic behaviour as well as free vibrations of the FGM beam simply supported were studied. Simsek and Yurtcu [61] considered the static bending of the FGM nanobeam based on the nonlocal Timoshenko and Euler-Bernoulli theories. They succeeded in getting Navier type analytical solution for the simply supported beam. Eltaher et al. [53] carried out an analysis of the free vibrations of FGM size-dependent nanobeams based on the nonlocal theory and Euler-Bernoulli beam theory. Kiani [58] proposed a novel mathematical model for study vibrations and transversal instabilities of moving nanoscale beam made from FGM. He derived equations of motion of Rayleigh nanobeam by using nonlocal theory of elasticity. The frequencies of both longitudinal and transverse nanobeam vibrations were found by utilizing Galerkin method and the associated modes of vibrations. Reddy [60] developed the microstructure-dependent couple stress theories of functionally graded beams. He used the principle of virtual displacements and he derived the size-dependent nonlinear PDEs for Euler-Bernoulli and Timoshenko beams with an account of von Kármán nonlinearity. The influence of the scalar length parameter, the rule of the properties of the beams change along with their thickness, shear deformation and geometric nonlinearity and two static beam deflections were investigated.

Arbind et al. [64] utilized Hamilton principle to derive nonlinear PDEs for the size-dependent stress-based third-order FGM beam theory. Ke et al. [65] investigated nonlinear free vibrations of FGM microbeams based on the modified couple stress theory and with an account of von Kármán nonlinearity. Non-classical beam model was developed in the frame of Timoshenko model with the inclusion of size-dependent length parameter.

Asgharifard and Yazdi [52] studied nonlinear free vibration of Euler-Bernoulli FGM beams with geometric nonlinearity and surface effects and their influence on the frequencies of vibrations. They derived nonlinear second-order ODE with square and cubic nonlinearity by employing free modes of vibrations of the corresponding linear system. Simsek [62] investigated nonlocal effects exhibited by the free longitudinal vibrations of functionally graded tapered nanorods. Two different boundary conditions were considered: both rod ends were clamped and one was clamped and one was free. Nateghi et al. [59] employed the modified couple stress theory to study size-dependent buckling of functionally graded microbeams. They considered different beam theories and different boundary conditions to determine the role of shear deformation on beams buckling. The derived differential equation was solved through the general method of differential quadratures. Ansari et al. [50] investigated size-dependent free vibration of FGM microbeam using the modified strain gradient elasticity theory. The authors worked out a new size-dependent model consisting of three internal length parameters in order to take into account the effects of small scale. Eltaher et al. [54] carried out the free vibration analysis of FGM size-dependent nanobeams based on the nonlocal theory and Bernoulli-Euler model by considering small deformation. Ansari et al. [51] employed the gradient elasticity theory to study small-scale effects influence of nonlinear vibrations of FGM Timoshenko microbeams with an account of von Kármán geometric nonlinearity.

Anjomshoa [66] derived a model of elastic medium based on the nonlocal elasticity theory to study stability loss of orthotropic circle and elliptic nanoplates under uniform compression in the plate plane. Ansari et al. [65] investigated axial buckling of single-walled carbon FGM nanotubes with different boundary conditions based on the Rayleigh-Ritz method and molecular dynamics.

Rahmani and Jandaghian [67] employed a nonlocal third-order shear deformation theory and Rayleigh-Ritz method to analyse the stability of FGM nanobeams. The proposed non-classical model includes the scale length parameter and takes into account the size effect. The beam was functionally graded in the thickness direction, and Poisson coefficient was assumed to be constant. The minimum of potential energy yielded the beam governing equations and boundary conditions. Influence of the size-dependent length parameter, gradient parameter and the ratio of length over thickness in the characteristics of stability of the FGM nanobeam with simple support ends and clamping ends was investigated.

Curved FGM beams in the transversal direction with an account of size-dependent behaviour are rather rarely investigated [68–70]. Xiang and Yang [71] considered free and forced vibration of three-layer laminate Timoshenko beam of variable thickness and under heat condition. However, the study did not include the size effect. Dehrouyeh-Semnani et al. [72] derived the size-dependent three-layer sandwich beam model based on symmetric-deviatoric couple stress theory and investigated influence of the length of layers on the damping characteristics of microbeams vibrations. The similar problem was also analysed in [73–75]. In all so far mentioned works, the same Bernoulli-Euler and Timoshenko hypotheses have been employed to the whole layers package, that did not allow to take into account a big difference of the layers' thickness (for instance, when upper and lower layers are made from thin films).

Three-layer constructions have been employed in engineering for a long time in aircraft/aviation engineering, ships design and civil engineering, machinery industries and space engineering. However, in recent years they are more often applied in the construction of various micro- and nano-devices, for instance, micro- and nano-biosensors and micro- and nano-electromechanical sensors. The general theory of material strength resistance of three-layer constructions of different types and subjected to external load have been worked out by Grigoluk and Chulkov [76]. Three-layer constructions require not only functional gradation along with their thickness but also require taking into account the filling layer work and transversal shear and transversal compression which implies to consider the coupling of layers. Grigolyuk and Chulkov introduced a hypothesis of linear distribution of tangent displacements along with the package height and lack of compression of the whole package what served for the obtaining three-layer thin-walled construction. On the contrary to Euler-Bernoulli hypothesis, a normal to the input surface undergoes rotation on the amount of a certain angle due to action of the transversal shear of the filler. In other words, external carrying load layers obey Bernoulli-Euler hypotheses, whereas the internal layer (filler) satisfies Timoshenko hypotheses. Grigoluk and Chulkov, by taking into account the straight cross section for the filler developed a general theory for beams. Namely, they derived the governing beam behaviour PDEs and considered stability and vibration of three-layer beams with carrying load by different layers made from materials of infinite stiffness with regard to shear and transversal compression, whereas the filler satisfied condition of the infinite stiffness against the transversal compression. Both static and dynamic behaviours of multi-layer beams on the basis of Euler-Bernoulli and Timoshenko hypotheses were analysed in [77, 78].

7.3 Laws of Properties Change of FGM

In this section, we consider the most known laws of changes of properties of FGM along either thickness or length of beam.

7.3.1 Properties of Material P-FGM

It is assumed that the volume amount of the FGM is graded along plate thickness and obeys the following function:

$$\rho\ (z) = \left(\frac{z + {}^{h}/_{2}}{h}\right)^{p}, \tag{7.1}$$

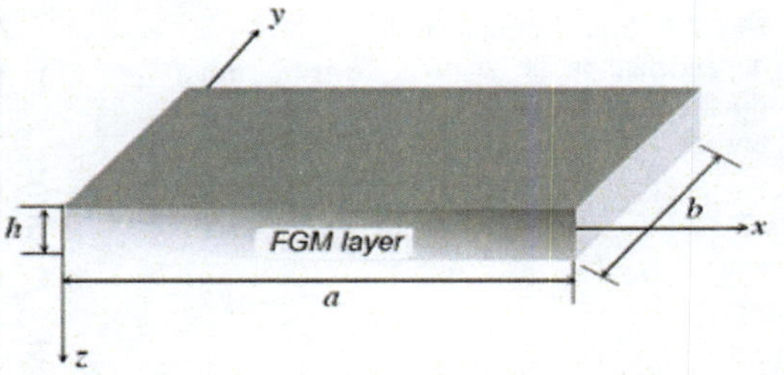

Fig. 7.1 Geometry of a plate made from FGM

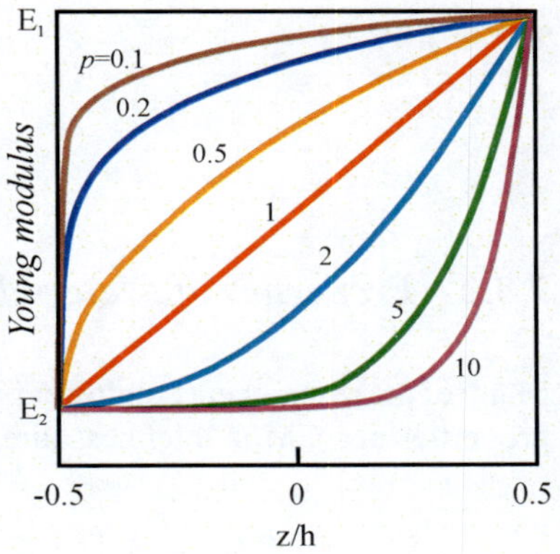

Fig. 7.2 Young modulus dependence on z based on formulas (7.1), (7.2)

where p stands for the order of material grading and h denotes thickness. Once the local volume amount $\rho\ (z)$ is defined, the properties of P-FGM such as temperature coefficient of linear expansion $\alpha\ (z)$, heat transfer coefficient $k\ (z)$ and Young modulus $E\ (z)$ can be obtained using the following formulas [34]:

$$\begin{aligned} &\alpha\ (z) = \rho\ (z)\,\alpha_1 + (1 - \rho\ (z))\,\alpha_2, \qquad k\ (z) = \rho\ (z)\,k_1 + (1 - \rho\ (z))\,k_2, \\ &E\ (z) = \rho\ (z)\,E_1 + (1 - \rho\ (z))\,E_2, \qquad \nu\ (z) = \nu_0, \end{aligned} \tag{7.2}$$

where index 1(2) defines material properties on the upper (lower) plate surface for $z = h/2\ \ (z = -h/2)$ (Fig. 7.1).

The change of Young modulus in the direction of the P-FGM plate thickness is shown in Fig. 7.2. It is seen that the Young modulus is quickly decreased in the neighbourhood of the lower surface for $p > 1$, and quickly increased in the neighbourhood of the upper surface for $p < 1$.

Equation (7.2) points out that the effective material properties are changed continuously along direction z, and can be approximated by the following relation [60, 61, 79–81]

$$P\ (z) = (P_1 - P_2)\left(\frac{z + {}^{h}/_{2}}{h}\right)^p + P_2.$$

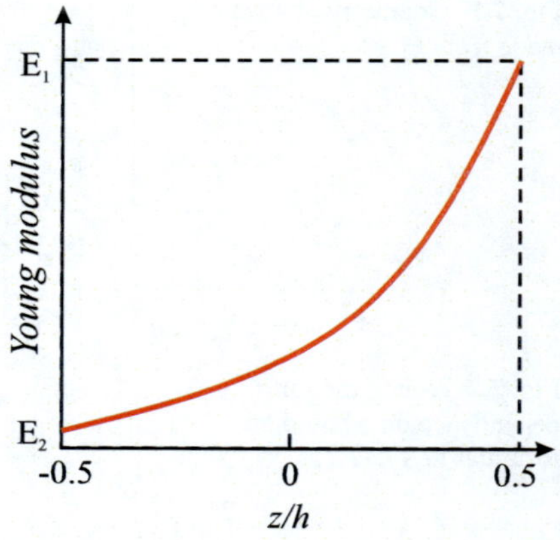

Fig. 7.3 Young modulus dependence on the beam thickness under exponential law

7.3.2 Properties of Material E-FGM

Many of the researchers employ exponential function to describe changes of material properties of FGM in thickness direction using the formula:

$$P(z) = Ae^{\beta z}, \ \ A = \sqrt{P_1 P_2}, \ \ \beta = \frac{1}{h}\ln\left(\frac{P_2}{P_1}\right), \tag{7.3}$$

where $P(z)$ denotes the material properties such as temperature coefficient of the linear expansion $\alpha(z)$, $k(z)$ stands for the heat transfer coefficient, and Young modulus is denoted by $E(z)$. We have $P(-h/2) = P_1$ and $P(h/2) = P_2$ [82, 83]. Then following relations hold for the being searched physical and mechanical properties of the gradual beam material

$$\alpha(z) = A_\alpha e^{\omega z}, \ \ A_\alpha = \sqrt{\alpha_1 \alpha_2}, \ \ \omega = \frac{1}{h}\ln\left(\frac{\alpha_2}{\alpha_1}\right),$$

$$k(z) = A_k e^{\beta z}, \ \ A_k = \sqrt{k_1 k_2}, \ \ \beta = \frac{1}{h}\ln\left(\frac{k_2}{k_1}\right),$$

$$E(z) = A_E e^{\lambda z}, \ \ A_E = \sqrt{E_1 E_2}, \ \ \lambda = \frac{1}{h}\ln\left(\frac{E_2}{E_1}\right).$$

Here, $(\bullet)_1$, $(\bullet)_2$ refer to material properties on the surfaces $z = -h/2$ and $z = h/2$, whereas the constant values A_α, A_k, A_E, ω, β and λ are yielded by the boundary conditions. Law of Young modulus $E(z)$ variation along beam thickness obeys the exponential function and it is shown in Fig. 7.3.

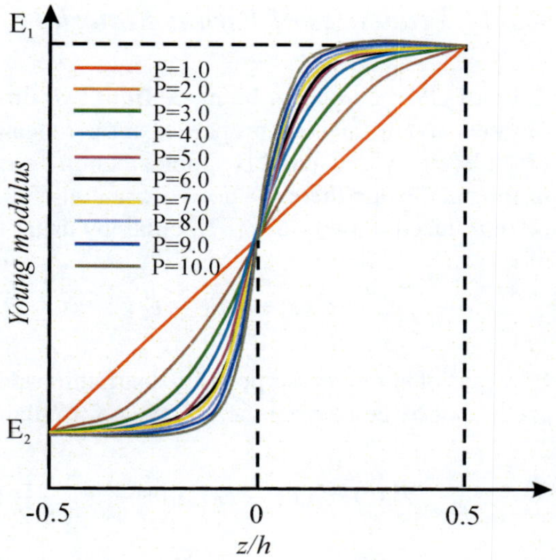

Fig. 7.4 Young modulus change along with thickness for sigma-shaped law

7.3.3 Properties of S-FGM Material

If one employs a stepwise function for FGM in a multi-layer composite then on a boundary where material continuity is violated its properties are fastly changed and the stress concentration appears [84, 85]. Therefore, Chung and Chi [86] used two step function to guarantee a smooth residual stresses distribution of functionally graded material in between all boundaries. The mentioned stepwise functions have the following form:

$$\rho_1(z) = 1 - \frac{1}{2}\left(\frac{h/2 - z}{h/2}\right)^p \quad \text{for} \quad 0 \leq z \leq h/2\,, \tag{7.4}$$

$$\rho_2(z) = \frac{1}{2}\left(\frac{h/2 + z}{h/2}\right)^p \quad \text{for} \quad -h/2 \leq z \leq 0. \tag{7.5}$$

Young's modulus of S-FGM can be found from the following formula:

$$E(z) = \rho_1(z)\,E_1 + (1 - \rho_1(z))\,E_1 \quad \text{for} \quad 0 \leq z \leq h/2\,, \tag{7.6}$$

$$E(z) = \rho_2(z)\,E_2 + (1 - \rho_2(z))\,E_2 \quad \text{for} \quad -h/2 \leq z \leq 0. \tag{7.7}$$

Figure 7.4 shows how Young modulus changes with regard to z computed based on Eqs. (7.6) and (7.7). Owing to the sigma shape of the curves, the material is named as S-FGM material.

7.3.4 Properties of Porous Materials

Let an FGM microbeam be made from two different materials. It is assumed that physical and mechanical properties of microbeam are graded along the stepwise law (7.1). Besides, if material is simple, then in order to achieve the required properties of the microbeam there is a need to account of a volume part of porosity ϑ. This can be introduced in two ways [87], either by using the following formula:

$$F(z) = (F_1 - F_2)\rho(z) + F_2 - \frac{\vartheta}{2}(F_1 + F_2) \tag{7.8}$$

for a case of homogenous porosity distribution along the microbeam transversal cross section, or by employing the following formula:

$$F(z) = (F_1 - F_2)\rho(z) + F_2 - \frac{\vartheta}{2}(F_1 + F_2)\left(1 - \frac{2|z|}{h}\right) \tag{7.9}$$

for a case of non-homogenous porosity distribution along the beam transversal cross section. In the Eqs. (7.8) and (7.9), $F(z)$ stands for the physical and mechanical properties in the given point z. Also F_1, (F_2) are the parameters characterizing the physical and mechanical properties of the pure material 1(2).

Owing to Eqs. (7.8), (7.9), the mass density $\gamma(z)$, Young modulus $E(z)$ and Poisson coefficient $\nu(z)$ can be described for homogenous/non-homogenous distribution of the porosity along the transversal microbeam cross section with a help of the following relations:

$$\begin{aligned}
\gamma(z) &= (\gamma_1 - \gamma_2)\rho(z) + \gamma_2 - \frac{\vartheta}{2}(\gamma_1 + \gamma_2), \\
E(z) &= (E_1 - E_2)\rho(z) + E_2 - \frac{\vartheta}{2}(E_1 + E_2), \\
\nu(z) &= (\nu_1 - \nu_2)\rho(z) + \nu_2 - \frac{\vartheta}{2}(\nu_1 + \nu_2),
\end{aligned} \tag{7.10}$$

and

$$\begin{aligned}
\gamma(z) &= (\gamma_1 - \gamma_2)\rho(z) + \gamma_2 - \frac{\vartheta}{2}(\gamma_1 + \gamma_2)\left(1 - \frac{2|z|}{h}\right), \\
E(z) &= (E_1 - E_2)\rho(z) + E_2 - \frac{\vartheta}{2}(E_1 + E_2)\left(1 - \frac{2|z|}{h}\right), \\
\nu(z) &= (\nu_1 - \nu_2)\rho(z) + \nu_2 - \frac{\vartheta}{2}(\nu_1 + \nu_2)\left(1 - \frac{2|z|}{h}\right).
\end{aligned} \tag{7.11}$$

7.3.5 Homogenization of Properties of Graded Material Based on Mori-Tanaka and Self-consistent Methods

Mori-Tanaka scheme belongs to one of the most commonly employed procedure of homogenization of the material properties and in particular for quantifying the effective properties of microbeams made from FGM. Owing to the scheme of Mori-Tanaka homogenization, formulas governing the effective elastic volume modulus K_e and the effective shear modulus G_e are reported in [88, 89], and they have the following form:

$$K_e = K_2 + \frac{(K_1 - K_2)\,\rho\,(z)}{1 + (1 - \rho\,(z))\,(K_1 - K_2)\,/\,(K_2 + 4G_2/3)}, \tag{7.12}$$

$$G_e = G_2 + \frac{(G_1 - G_2)\,\rho\,(z)}{1 + (1 - \rho\,(z))\,(G_1 - G_2)\,/\,[G_2 + G_2\,(9K_2 + 8G_2)\,/\,(6\,(K_2 + 2G_2))]}, \tag{7.13}$$

where K_1, G_1 and K_2, G_2 stand for the volume elasticity moduli on the lower $(z = h/2)$ and upper $(z = -h/2)$ FGM plate surface, respectively. Here, $\rho\,(z)$ is defined by Eq. (7.1) and defines the material concentration of the lower surface. The effective Young modulus and the effective Poisson coefficient for the FGM microbeam are given by the following equations:

$$E\,(z) = \frac{9K_eG_e}{3K_e + G_e}, \qquad \nu\,(z) = \frac{3K_e - 2G_e}{6K_e + 2G_e}. \tag{7.14}$$

The effective coefficient of a local heat transfer k_e is governed by the following formula [90]:

$$k_e = k_2 + \frac{(k_1 - k_2)\,\rho\,(z)}{1 + (1 - \rho\,(z))\,(k_1 - k_2)\,/3k_2}, \tag{7.15}$$

whereas the effective temperature coefficient of the linear expansion α_e is described by the following formula [91]:

$$\alpha_e = \alpha_2 + \frac{(\alpha_1 - \alpha_2)\,(1/K_e - 1/K_2)}{(1/K_1 - 1/K_2)}. \tag{7.16}$$

Therefore, we have $\beta_e = 3K_e\alpha_e$.

In the method of self-compliance [92], it is assumed that the reinforced inclusion is included in the material of a continuum, and its effective properties define the composite properties. The latter method does not allow to get differences between the matrix and reinforcing phase. The effective moduli are forecast in the composite in which the roles of phases are exchanged. The local effective moduli of self-consistent method are defined as follows:

$$\frac{\delta}{K_e} = \frac{\rho(z)}{K_e - K_1} + \frac{1-\rho(z)}{K_e - K_2}, \tag{7.17}$$

$$\frac{\eta}{G_e} = \frac{\rho(z)}{G_e - G_1} + \frac{1-\rho(z)}{G_e - G_2}, \tag{7.18}$$

where

$$\delta = 3 - 5\eta = K_e/(K_e + 4G_e/3).$$

Observe that formulas for the unknowns K_e and G_e are implicit.

Equation (7.17) can be solved with regard to K_e employing G_e defined by (7.18). In result, we obtain

$$K_e = 1/[\rho(z)/(K_1 + 4G_e/3) + (1-\rho(z))/(K_2 + 4G_e/3)] - 4G_e/3, \tag{7.19}$$

and G_e stands for a solution to the following quadratic equation:

$$\frac{\rho(z)K_1}{K_1 + 4G_e/3} + \frac{(1-\rho(z))K_2}{K_2 + 4G_e/3} + 5\left(\frac{\rho(z)G_2}{G_e - G_2} + \frac{(1-\rho(z))G_1}{G_e - G_1}\right) + 2 = 0. \tag{7.20}$$

Self-compliance estimation of the heat transfer coefficient is defined through the following equation [93]:

$$\frac{\rho(z)(k_1 - k_e)}{k_1 + 2k_e} + \frac{(1-\rho(z))(k_2 - k_e)}{k_2 + 2k_e} = 0. \tag{7.21}$$

Self-compliance estimation α_e can be obtained by a substitution of the self-compliance estimation of the volume elasticity modulus K_e from Eq. (7.19) into estimation (7.16). In other words, in order to get the shear modulus G_e and the heat transfer coefficient k_e, there is the need to solve the quadratic equations (7.20) and (7.21). This is why Tanaka-Mori scheme is recommended since it does not contain the mentioned drawback.

7.3.6 *Dependence of Material Properties on Temperature*

The functionally graded materials including ceramics and carbon nanotubes are structural members of constructions working in high-temperature regimes, and hence their properties can be changed depending on temperature. The material characteristics like Young modulus E, the density γ, the Poisson coefficient ν and the coefficient of thermal expansion α depend on temperature T (in K°) in the following nonlinear way [94–96]

$$P = P_0\left(P_{-1}T^{-1} + 1 + P_1T + P_2T^2 + P_3T^3\right), \tag{7.22}$$

where $P_{-1}, P_0, P_1, P_2, P_3$ are experimentally estimated.

Table 7.1 reports the values of the mentioned coefficients for some chosen materials. The density of zirconium dioxide and nickel is 5680 and 8908 kg/m^3, respectively.

The mentioned properties are obtained for each material component separately. The values of the material properties are used to establish the dependence of the beam material properties for different values z. In order to obtain the general material properties of the FGM beam, there is a need to take into account not only the temperature dependence of the properties but also the influence of the volume parts changes of separate materials with regard to thickness and length of the beam. The effective material properties can be estimated in a reliable way by taking into account the above-mentioned features.

Table 7.1 Constants of the cubic approximation (7.22) of properties of some materials

Property	Material	P_0	P_{-1}	P_1	P_2	P_3
Young modulus $E[Pa]$	Zirconium	244.27×10^9	0	-1.37×10^{-3}	1.21×10^{-6}	-3.68×10^{-10}
	Nickel	223.95×10^9	0	-2.79×10^{-4}	-4×10^{-9}	0
	Aluminium	349.55×10^9	0	-3.85×10^{-4}	4.03×10^{-7}	1.67×10^{-11}
	Steel SUS304	201.04×10^9	0	3.08×10^{-4}	-6.53×10^{-7}	0
	Ceramics Si3N4	348.43×10^9	0	-3.07×10^{-4}	2.16×10^{-7}	-8.94×10^{-11}
Thermal expansion coefficient $\alpha[K^{-1}]$	Zirconium	17.76×10^{-6}	0	-1.49×10^{-3}	1.00×10^{-5}	-6.78×10^{-11}
	Nickel	9.92×10^{-6}	0	8.71×10^{-4}	0	0
	Aluminium	6.827×10^{-6}	0	1.84×10^{-4}	0	0
	Steel SUS304	12.33×10^{-6}	0	8.08×10^{-4}	0	0
	Ceramics Si3N4	5.87×10^{-6}	0	9.09×10^{-4}	0	0
Poisson's coefficient ν	Zirconium	0.29	0	1.13×10^{-4}	0	0
	Nickel	0.31	0	0	0	0
	Aluminium	0.26	0	0	0	0
	Steel SUS304	0.32	0	-2.00×10^{-4}	-3.8×10^{-7}	0
	Ceramics Si3N4	0.24	0	0	0	0
Heat transfer [W/mK]	Aluminium	–14.08	–1123.6	-6.23×10^{-3}	0	0

7.4 Chaotic Dynamics of Size-Dependent Timoshenko Beams with Functionally Graded Properties Along Their Thickness

In this section, we consider the size-dependent model based on the modified couple stress theory of geometrically nonlinear Timoshenko beams made from FGM. We assume that the beam properties are changed along with the beam thickness. The influence of size-dependent and functionally graded coefficients on the vibration characteristics, scenarios of transition from regular to chaotic vibrations as well as a series of static problems with an emphasis put on the load-deflection behaviour are studied. Our theoretical/numerical analysis is supported by methods of nonlinear dynamics and the qualitative theory of differential equations supplemented by Fourier and wavelet spectra, phase portraits and Lyapunov exponents spectra estimated by different algorithms, including Wolf's, Rosenstein's, Kantz's and neural networks. The obtained results show that the beams modelled by the modified couple stress theory are stiffer in comparison to those modelled by the classical continuum theory (this is valid for any beam thickness distribution). We have also detected and numerically validated a general scenario governing transition into chaotic vibrations, which follows the classical Ruelle-Takens-Newhouse scenario for the considered values of the size-dependent and grading parameters.

7.4.1 Introduction

A functionally graded material can be made by mixing two or more materials with the required continuous properties along the desired direction [1]. Continuous changes in the material properties of the FGM yield a lot of engineering benefits since this approach allows one to avoid an occurrence of large shear stresses, which are typically observed while fabricating multi-layer beams. The mentioned materials can be directly applied in the cosmic industry, the nuclear reactors, etc. It should be emphasized that the static bending as well as the dynamic characteristics of the structural members made from FGM have been intensively studied recently [3–5, 71, 97–101].

Furthermore, in recent years, FGM have been widely applied in micro- and nanostructures, including thin films/layers [6, 7], and micro- and nano-electromechanical systems [8, 9]. The thickness of beams used in the above-mentioned structures is of the order of microns and submicrons, thus the influence of scale effects on the beam behaviour can be significant.

The experiments described in the works [13, 15, 102] show that size effects play an important role in materials with microscale structure, such as a thin copper wire, single crystal silver, nickel or steel epoxy polymeric beams. Namely, this phenomenon is exhibited when the characteristic dimension of the beam, i.e. either the beam thickness or size, is close to the internal material length scale parameter [17].

The classical theory of elasticity does not enable one to predict and account for the size effect that occurs in the micro- and submicro-scale structures. To explain these effects, various size-dependent continuum theories have been proposed, for example, the couple stress elasticity [35, 36], nonlocal elasticity theory [103], strain gradient elasticity [104] and surface elasticity [105].

In the couple stress theory, two additional material constants [35, 36] are distinguished. For isotropic elastic materials, they are related to the underlying microstructure of the material and are inherently difficult to determine. In Ref. [44], a modified stress couple theory with only one additional parameter of the length scale of the material has been developed. Such a model simplifies the use of a modified couple stress theory.

In Refs. [43, 99, 106–109], the modified couple stress theory has been employed to study static properties of Euler-Bernoulli beam. In Ref. [17], on the basis of the modified couple stress theory and Hamilton's principle, the equations of motion as well as the boundary/initial conditions have been derived. In Ref. [38], the size-dependent behaviour of microbeams made of FGM has been investigated using the modified couple stress theory for Euler-Bernoulli model.

Timoshenko beam model, contrarily Euler-Bernoulli model, allows one to take into account the shear deformation and is more accurate for relatively thick beams and beams of materials of low shear stiffness. The modified couple stress theory aimed at investigating the size-dependent behaviour of the homogeneous Timoshenko beam has been studied in Ref. [106]. In Ref. [63], these results have been extended for Timoshenko beams made from FGM. Furthermore, in the case of statics, a closed analytical formula has been obtained as well as the size-dependent frequencies have been detected and studied.

The beams which are non-homogeneous with respect to their thickness have been studied in Ref. [101] taking into account von Kármán nonlinearity and applying the modified couple stress theory. The nonlinear PDEs governing the dynamics of both Euler-Bernoulli and Timoshenko beams have been analysed. The investigations regarding the influence of the internal material length scale parameter, non-constant beam thickness properties, shear strain as well as the geometric nonlinearity on the static beam deflections have been analysed.

It should be mentioned that the nonlinear PDEs governing the microstructural dependences in beams made from FGM have been derived (up to the third-order terms) in Ref. [64].

However, the critical literature overview implies that there is a lack of investigations of the nonlinear/chaotic dynamics of the beams made from FGM exhibiting, in particular, the size-dependent behaviour. The above-mentioned stands for the motivation of the present section.

Problems devoted to solutions of nonlinear/chaotic continuous dynamical systems [110–117] have been studied. In addition, the analysis of the nonlinear dynamics of mechanical constructions has been conducted taking into account different types of the nonlinearity, including geometrical, physical and fabrication ones, and including actions of various physical fields like thermal, electric, magnetic and mechanical [118, 119] as well as white-noise interaction [120].

The monograph [121] focuses on the computational analysis of nonlinear vibrations of structural members (beams as well as axially symmetric plates and shells), where studying of dynamical problems can be reduced to considering one spatial variable and time. The simplification is introduced based on a formal mathematical approach aimed at reducing the problems with an infinite dimension to the problems of a finite one. The process includes also a transition from governing nonlinear partial differential equations to a set of a finite number of ordinary differential equations.

We report the influence of the length scale parameter and the grading parameter on the vibration characteristics and supplement it by a scenario of transition from regular to chaotic vibrations. As a result, it has been shown that beams modelled by the classical theory of continuum have less normalized deflection in comparison to the models yielded by the modified stress couple theory, independently of the material distribution along the beam thickness. It has been observed that the beam dynamics is most essentially influenced by the coefficient describing the heterogeneity of the material.

Finally, we have found that, for a set of fixed values of the parameters including the size and material grading parameters, regular vibrations of the beam tend to chaotic vibrations, following Ruelle-Takens-Newhouse scenario.

7.4.2 Mathematical Background

In the modified couple stress theory, the strain energy U in a linear elastic body occupying the volume V has the form [44]:

$$U = \frac{1}{2} \iiint\limits_V \left(\sigma_{ij}\varepsilon_{ij} + m_{ij}\chi_{ij}\right) dV. \tag{7.23}$$

Now, in the isotropic case, one can write

$$\sigma_{ij} = \lambda\varepsilon_{mm}\delta_{ij} + 2\mu\varepsilon_{ij}, \tag{7.24}$$

$$\varepsilon_{ij} = \frac{1}{2}\left(u_{i,j} + u_{j,i}\right), \tag{7.25}$$

$$m_{ij} = \beta\chi_{ij} = 2\mu l^2\chi_{ij}, \tag{7.26}$$

$$\chi_{ij} = \frac{1}{2}\left(\theta_{i,j} + \theta_{j,i}\right), \tag{7.27}$$

where σ_{ij}, ε_{ij}, m_{ij}, and χ_{ij} denote components the Cauchy stress tensor σ, the strain tensor ε, the deviator part of the couple stress tensor **m** and the symmetric curvature tensor χ, respectively. In addition, u_i are components of the displacement vector and **u**, θ stand for the infinitely small vector of rotation with components θ_i. Observe that $\theta_i = (rot\,(\mathbf{u}))_i/2$. Two Lamé constants are denoted by λ and μ, whereas l stands for the internal material length scale parameter.

It should be mentioned that the introduced theoretical background regarding the internal material length scale parameter l has been experimentally validated for many materials [15, 102, 107, 108].

The system of employed coordinates, kinematic parameters and loads acting on Timoshenko beam made from the FGM are taken to fit the modified stress couple theory, and they are presented in Fig. 7.5. It is assumed that the beam properties do not undergo changes along the axis OX. The kinematic relations of Timoshenko beam follow [122]:

$$u_x = u\,(x,t) + z\psi(x,t),\ \ u_y = 0,\ \ u_z = w\,(x,\ t)\,, \tag{7.28}$$

where $u\,(x,\ t)$, $w\,(x,\ t)$ and $\psi\,(x,\ t)$ denote the axial shift of the beam middle line, the transverse beam deviation and the angle of rotation of the transverse beam cross section with respect to the vertical direction, respectively.

It is assumed that all transverse cross sections remain flat after strain. However, they can undergo a rigid shift in the plane XOZ, as well as a rotation around the axis OY.

Coordinates z and $\tilde{z}$ (see Fig. 7.5) denote the distance between an arbitrarily chosen point and the reference line or the bottom surface, respectively. Also, the distance between the reference line and the bottom surface or the axial coordinate is represented by $\tilde{z}_c$ and x, respectively.

A special line, further called the bending line, is the line around which a pure rotation takes place in the cross section. $G(x,t)$ denotes a force per beam unit length,

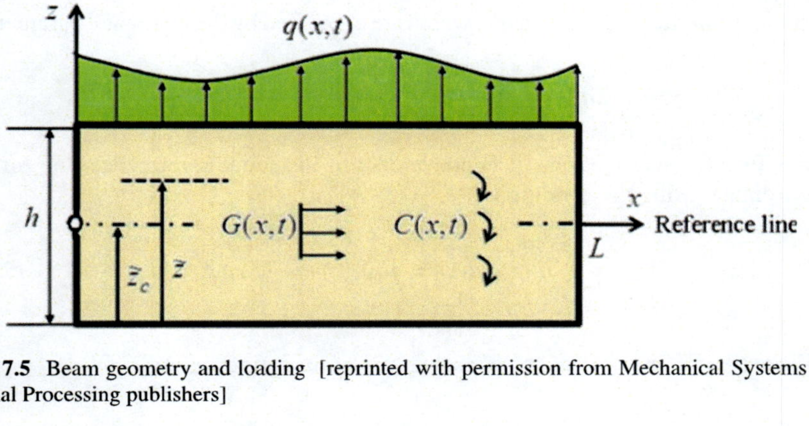

Fig. 7.5 Beam geometry and loading [reprinted with permission from Mechanical Systems and Signal Processing publishers]

acting on the axial cross section along the body axis OX. The resultant of the transverse stress acting on the upper side of the beam and the transverse volume forces per unit beam length is denoted by $q(x,t)$. The parameter $C(x,t)$ presents the y component of the resultant volume moment per unit beam length.

Using (7.28), von Kármán-type nonlinear strain-displacement relations are given by [57]

$$\varepsilon_{xx} = u_{,x} + \frac{1}{2}\left(w_{,x}\right)^2 + z\psi_{,x}, \quad \varepsilon_{xz} = \frac{1}{2}\left(\psi + w_{,x}\right). \tag{7.29}$$

Furthermore, $\theta_i = (rot\ \ (\mathbf{u}))_i/2$ yields

$$\theta_y = \frac{1}{2}\left(\psi - w_{,x}\right), \qquad \theta_x = 0. \tag{7.30}$$

Substituting (7.30) in (7.27), one obtains the following expression for a component of the symmetric part of the curvature tensor:

$$\chi_{xy} = \chi_{yx} = \frac{1}{4}\left(\psi_{,x} - w_{,xx}\right). \tag{7.31}$$

Assuming that the material properties are changed only with respect to the thickness, neglecting Poisson's effect as well as substituting (7.29) in (7.24), the following components of the stress tensor are obtained with respect to the kinematic parameters

$$\sigma_{xx} = E\left(\tilde{z}\right)\left(u_{,x} + \frac{1}{2}\left(w_{,x}\right)^2 + z\psi_{,x}\right), \qquad \sigma_{xz} = k_s\mu\left(\tilde{z}\right)\left(w_{,x} + \psi\right), \tag{7.32}$$

where $E(\tilde{z})$ and $\mu(\tilde{z})$ denote Young's modulus and shear modulus, respectively, and k_s stands for the correction coefficient, which is introduced due to the assumption of the graded shear strain regarding the beam transverse section, which depends on the form of the beam cross sections.

Substituting (7.31) in (7.26), one finds components of the deviator part of the tensor of the higher order moments, which are expressed by the kinematic parameters:

$$m_{xy} = m_{yx} = \frac{1}{4}\beta\left(\tilde{z}\right)\left(\psi_{,x} - w_{,xx}\right). \tag{7.33}$$

For further investigations, it is convenient to introduce a reference line for the z-coordinate, using the condition

$$B_{11} = \int_A E\left(\tilde{z}\right) z dA = \int_A E\left(\tilde{z}\right)\left(\tilde{z} - \tilde{z}\right) dA = 0. \tag{7.34}$$

From (7.34), one obtains the coordinate $\tilde{z}_c$ of the reference line

$$\tilde{z} = \frac{\int\limits_A E(\tilde{z})\,\tilde{z}\,dA}{\int\limits_A E(\tilde{z})\,dA}. \tag{7.35}$$

The expression (7.35) allows one to exclude the terms with coefficients B_{11} from the system of equations, which significantly simplifies the system of equations obtained in Refs. [57, 101].

7.4.3 Derivation of the Equations of Motion

In order to derive equations of motion of the FG Timoshenko beam on a basis of the modified couple stress theory, we substitute (7.23) in (7.29)–(7.33). As a result, the following strain energy U and kinetic energy K are obtained:

$$\begin{aligned} U &= \frac{1}{2}\int\limits_0^L \int\limits_A \left(\sigma_{ij}\varepsilon_{ij} + m_{ij}\chi_{ij}\right) dA\,dx = \\ &= \frac{1}{2}\int\limits_0^L \int\limits_A \left\{ E(\tilde{z})\left(u_{,x} + \frac{1}{2}(w_{,x})^2 + z\psi_{,x}\right)^2 + k_s\mu(\tilde{z})\left(\psi + w_{,x}\right)^2 + \right. \\ &\left. + \frac{1}{8}\beta(\tilde{z})\left(\psi_{,x} - w_{,xx}\right)^2 \right\} dA\,dx, \end{aligned} \tag{7.36}$$

$$K = \frac{1}{2}\int\limits_0^L \int\limits_A \rho(\tilde{z})\left\{\left(u_{,t} + z\psi_{,t}\right)^2 + \left(w_{,t}\right)^2\right\} dA\,dx, \tag{7.37}$$

where L stands for the beam length.

Variation of the work δW generated by the external forces, the distributed moments and stresses on a threshold surface takes the form:

$$\begin{aligned} \delta W &= \int\limits_0^L \left(G\delta u + q\,\delta w + c\delta\,\theta_y\right) dx + \\ &+ \left(\bar{N}\delta u + \bar{V}\delta w + \bar{M}_\sigma\delta\psi + \bar{M}_m\delta\,\theta_y\right)\Big|_{x=0}^{x=L}, \end{aligned} \tag{7.38}$$

where the external forces acting on the beam ends in the axial direction, the external transversal forces, the external bending moment generated by the normal stress σ_{xx} and the external bending moment caused by the moment m_{xy} are denoted by $\bar{N}$, $\bar{V}$, $\bar{M}_\sigma$ and $\bar{M}_m$, respectively.

Owing to (7.34), relations (7.36), (7.37) take the following form:

$$U = \frac{1}{2}\int\limits_0^L \left\{ k_1\left(u_{,x} + \frac{1}{2}\left(w_{,x}\right)^2\right)^2 + k_2\left(\psi_{,x}\right)^2 + \right. \\ \left. + k_3\left(\psi + w_{,x}\right)^2 + k_4\left(\psi_{,x} - w_{,xx}\right)^2 \right\} dx\,, \tag{7.39}$$

$$K = \frac{1}{2}\int\limits_0^L \left\{ m_0\left(u_{,t}\right)^2 + 2Q\,u_{,t}\,\psi_{,t} + \tilde{I}\left(\psi_{,t}\right)^2 + m_0\left(w_{,t}\right)^2 \right\} dx, \tag{7.40}$$

where

$$\begin{aligned} k_1 &= \int\limits_A E\left(\tilde{z}\right)\, dA,\ k_2 = \int\limits_A E\left(\tilde{z}\right) z^2\, dA = \int\limits_A E\left(\tilde{z}\right)\ (\tilde{z} - \tilde{z}_c)^2 dA\,, \\ k_3 &= k_s \int\limits_A \mu\left(\tilde{z}\right)\, dA,\ k_4 = \frac{1}{8}\int\limits_A \beta\left(\tilde{z}\right)\, dA = \frac{1}{4}\int\limits_A \mu\left(\tilde{z}\right) l^2\left(\tilde{z}\right)\, dA, \\ m_0 &= \int\limits_A \rho\left(\tilde{z}\right)\, dA,\ Q = \int\limits_A \rho\left(\tilde{z}\right) z\, dA = \int\limits_A \rho\left(\tilde{z}\right)\ (\tilde{z} - \tilde{z}_c)\, dA, \\ \tilde{I} &= \int\limits_A \rho\left(\tilde{z}\right) z^2\, dA = \int\limits_A \rho\left(\tilde{z}\right)\ (\tilde{z} - \tilde{z}_c)^2 dA\,. \end{aligned} \tag{7.41}$$

Hamilton's principle yields

$$\delta\int\limits_{t_1}^{t_2} (K - U + W)\ dt = 0. \tag{7.42}$$

The carried-out variation with respect to u, w and ψ, integration by parts, and setting the terms standing by δu, δw and $\delta\psi$ to be equal to zero yield the following equations of motion:

$$\begin{aligned} &\left[k_1\left(u_{,x} + \frac{1}{2}\left(w_{,x}\right)^2\right)\right]_{,x} + G = m_0 u_{,tt} + Q\,\psi_{,tt}, \\ &\left[k_2\psi_{,x} + k_4\left(\psi_{,x} - w_{,xx}\right)\right]_{,x} - k_3\left(\psi + w_{,x}\right) + \frac{C}{2} = Q\,u_{,tt} + \tilde{I}\psi_{,tt}\,, \\ &\left\{\left[k_1\left(u_{,x} + \frac{1}{2}\left(w_{,x}\right)^2\right)\right] w_{,x} + k_3\left(\psi + w_{,x}\right)\right\}_{,x} + \\ &+ \left[k_4\left(\psi_{,x} - w_{,xx}\right)\right]_{,xx} + q + \frac{C_{,x}}{2} = m_0 w_{tt}, \end{aligned} \tag{7.43}$$

along with the following boundary conditions

$$\left[k_1\left(u_{,x}+\frac{1}{2}(w_{,x})^2\right)-\bar{N}\right]\Big|_{x=0}^{x=L}=0 \quad \text{or} \quad \delta u|_{x=0}^{x=L}=0,$$

$$\left[k_2\psi_{,x}+k_4\left(\psi_{,x}-w_{,xx}\right)-\bar{M}_\sigma-\frac{\bar{M}_m}{2}\right]\Big|_{x=0}^{x=L}=0 \quad \text{or} \quad \delta\psi|_{x=0}^{x=L}=0, \tag{7.44}$$

$$\left\{\left[k_1\left(u_{,x}+\frac{1}{2}(w_{,x})^2\right)\right]w_{,x}+k_3\left(\psi+w_{,x}\right)+k_4\left(\psi_{,xx}-w_{,xxx}\right)+\frac{c}{2}-\bar{V}\right\}\Big|_{x=0}^{x=L}=0$$

or $\delta w|_{x=0}^{x=L}=0$

$$\left(k_4\left(\psi_{,x}-w_{,xx}\right)+\frac{\bar{M}_m}{2}\right)\Big|_{x=0}^{x=L}=0 \quad \text{or} \quad \delta w_{,x}\Big|_{x=0}^{x=L}=0$$

and the initial conditions

$$\begin{aligned} w(x,0)&=\varphi_1(x), \quad w(x,0)_{,t}=\varphi_2(x),\\ u(x,0)&=\varphi_3(x), \quad u(x,0)_{,t}=\varphi_4(x),\\ \psi(x,0)&=\varphi_5(x), \quad \psi(x,0)_{,t}=\varphi_6(x). \end{aligned} \tag{7.45}$$

The equations of motion (7.43) of the nonlinear Timoshenko beam made from FGM and the boundary conditions (7.44) are obtained using the relation (7.35). The resulting equations differ from those presented in [57, 101] and have a simpler form due to the vanishing of the terms/members with the coefficient B_{11}.

7.4.4 Statement of the Problem

For further numerical investigations and in order to simplify the problem, the following parameters are fixed: $G(x,t)=0$, $(x,t)=0$. To construct the rules of the change of the beam characteristics along the beam thickness, the following parameters are used: Young's modulus $E(\tilde{z})$, shear modulus $\mu(\tilde{z})$ and beam density $\rho(\tilde{z})$.

In order to choose an efficient numerical model of the beam material properties along with the beam thickness, the following linear rule is employed:

$$E(\tilde{z})=E_0+\frac{\tilde{z}+\frac{h}{2}}{h}\left(E_1-E_0\right), \qquad \mu(\tilde{z})=\mu_0+\frac{\tilde{z}+\frac{h}{2}}{h}\left(\mu_1-\mu_0\right),$$

$$\rho(\tilde{z})=\rho_0+\frac{\tilde{z}+\frac{h}{2}}{h}\left(\rho_1-\rho_0\right). \tag{7.46}$$

The required integration is carried out in formulas (7.35) and (7.41) by using the following coupling between the elastic and shear moduli:

$$E_1 = P_E E_0, \quad \mu_1 = \mu_E \mu_0, \quad \rho_1 = P_E \rho_0 \tag{7.47}$$

Substituting (7.46), (7.47) in (7.35), one obtains:

$$\tilde{z} = \frac{h}{12} \frac{P_E - 1}{1 + \frac{1}{2}(P_E - 1)}. \tag{7.48}$$

Formula (7.48) implies that for $P_E > 1$, ($\tilde{z} > 0$) and the neutral line is shifted above in comparison to the neutral line of the homogeneous beam ($P_E = 1$). In the case $P_E < 1$, $\tilde{z} < 0$, and the neutral line is shifted below the neutral line of the counterpart homogeneous beam.

Using (7.46)–(7.48), and taking into account (7.41), the following coefficients are defined:

$$k_1 = E_0 A \left[1 + \frac{1}{2}(P_E - 1)\right], \quad k_2 = \frac{1}{12} E_0\, A h^2 \left[\frac{3}{2} + \frac{(P_E - 1)^2}{12\,\left(1 + \frac{1}{2}(P_E - 1)\right)}\right],$$

$$k_3 = k_s \mu_0 A \left[1 + \frac{1}{2}(P_\mu - 1)\right], \quad k_4 = \frac{1}{4}\mu_0\, A\, l^2 \left[1 + \frac{1}{2}(P_\mu - 1)\right], \tag{7.49}$$

$$m_0 = \rho_0 A \left[1 + \frac{1}{2}(P_\mu - 1)\right], \quad Q = \rho_0\, A \frac{h}{12}\left[(P_\rho - 1) - \frac{1}{2}\frac{(P_E - 1)(P_\rho + 1)}{1 + \frac{1}{2}(P_E - 1)}\right],$$

$$\tilde{I} = \rho_0\, A \frac{h^2}{12}\left[\left(1 + \frac{(P_E - 1)^2}{12\,\left(1 + \frac{1}{2}(P_E - 1)\right)^2}\right)\left(1 + \frac{1}{2}(P_\rho - 1)\right) - \frac{1}{6}\frac{(P_E - 1)(P_\rho + 1)}{1 + \frac{1}{2}(P_E - 1)}\right].$$

The shear coefficient k_s is taken as $(5 + 5\nu)/(6 + 5\nu)$, which is most suitable for the description of the behaviour of beams with rectangular cross sections [76]. In what follows, we introduce the following dimensionless parameters:

$$\bar{w} = \frac{w}{h}, \quad \bar{u} = \frac{ua}{h^2}, \quad \bar{\psi} = \frac{\psi\, a}{h}, \quad \bar{x} = \frac{x}{a}, \quad \gamma_1 = \frac{a}{h}, \quad \gamma_2 = \frac{l}{h},$$

$$\bar{q} = q\frac{a^2}{h^2 E}, \quad \bar{t} = \frac{t}{\tau}, \quad \tau = \frac{a}{c}, \quad c = \sqrt{\frac{E}{\rho}}, \quad \bar{\varepsilon} = \varepsilon\frac{a}{c}, \quad \bar{k}_1 = \frac{k_1}{AE_0}, \tag{7.50}$$

$$\bar{k}_2 = \frac{k_2}{AE_0 h^2}, \quad \bar{k}_3 = \frac{k_3}{AE_0}, \quad \bar{k}_4 = \frac{k_4}{AE_0 l^2}.$$

Taking into account the introduced simplifications and notation, and omitting the bars over dimensionless parameters, the following dimensionless beam equations are eventually obtained:

$$k_1 \left[u_{,x} + \frac{1}{2}(w_{,x})^2 \right]_{,x} = u_{,tt},$$

$$k_2 \psi_{,xx} + 3k_4 \gamma_2^2 \left(\psi_{,xx} - w_{,xxx}\right) - 12\, k_s k_3 \gamma_1^2 \left(\psi + w_{,x}\right) = \psi_{,tt}, \tag{7.51}$$

$$\frac{1}{\gamma_1^2} \left\{ k_1 \left[\left(u_{,x} + \frac{1}{2}(w_{,x})^2 \right) \right] w_{,x} \right\}_{,x} + k_3 \left(\psi_{,x} + w_{,xx}\right) +$$

$$+ k_4 \frac{\gamma_2^2}{\gamma_1^2} \left(\psi_{,xxx} - w_{,xxxx}\right) + q = w_{,tt} + \varepsilon\, w_{,t}$$

We restrict further considerations to the following boundary conditions (rigid clamping of the beam ends):

$$\begin{aligned} w(0,t) &= w(1,t) = 0, \quad & w_{,x}(0,t) &= w_{,x}(1,t) = 0, \\ u(0,t) &= u(1,t) = 0, \quad & \psi(0,t) &= \psi(1,t) = 0, \end{aligned} \tag{7.52}$$

and the following initial conditions:

$$w(x,0) = w_{,t}(x,0), \quad u(x,0) = u_{,t}(x,0) = 0, \quad \psi(x,0) = \psi_{,t}(x,0) = 0. \tag{7.53}$$

Reduction of the PDEs (7.51)–(7.53) is carried out by means the finite difference method (FDM) of the second-order accuracy, and the finite element method (FEM). Both FDM and FEM are used to validate the results.

We have compared numerical results yielded by the fourth- and sixth-order Runge-Kutta methods. Owing to the coincidence of results and to the study developed in Ref. [123], we have further employed the fourth-order Runge-Kutta method. The optimal number of the spatial mesh elements regarding the beam length has been chosen on the basis of the Runge principle.

7.4.5 Results and Discussions

Numerical investigations of static and dynamic problems have been reported for the following fixed system parameters: relative beam length $\gamma_1 = \frac{a}{h} = 30$, size-dependent parameter $\gamma_2 = \frac{l}{h} = 0;\ 0.3$. Young's and shear coefficients regarding the thickness (7.47) are taken as $P_E = P_\mu = P_\rho = P = 1;\ 2;\ 0.5$. This choice fits with the formulas for the coefficients given by (7.49).

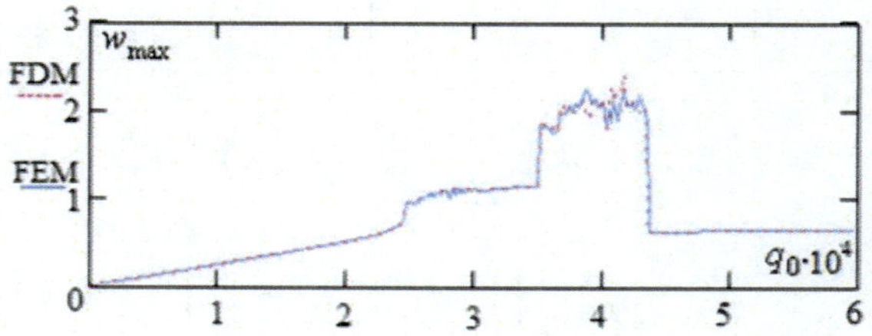

Fig. 7.6 Maximum deflection as a function of the excitation amplitude [reprinted with permission from Mechanical Systems and Signal Processing publishers]

7.4.5.1 Reliability and Validity of the Obtained Results

To check the reliability of the results obtained for Timoshenko beam model, solutions obtained with the use of FDM and FEM are compared. Then, the problem is studied for clamping-clamping boundary conditions. The curves of the dependence of the maximum deflection on the excitation amplitude, obtained with different computational methods (see Fig. 7.6), fully coincide in the case of regular beam vibrations, whereas slight differences appear only for chaotic dynamics (Fig. 7.6).

On the basis of results presented in Fig. 7.6, one can conclude that values of w_{max} fully coincide for both applied computational approaches if the excitation amplitude corresponds to regular vibrations. Small differences are noticeable in the case of chaotic dynamics. Vibration scales obtained by the two methods fully coincide. Transition to chaos appears later for FDM than for FEM computation and the chaotic regime is smaller if the FDM approach is applied.

7.4.5.2 Problems of Statics

In this section, we will solve the problems of statics by using equations governing the dynamics (7.51)–(7.53). This method will be further referred to as the relaxation method [121].

If the load $[\bar{q}]$ does not depend on time, a static solution to the problem can be yielded by the dynamic approach. Namely, the initial condition plays a role of an excitation, whereas the term with the first time derivative, including the dissipation/damping coefficient, presents a dissipative character of the excited solution. A solution to the so far defined dynamic problem can be found through the employment of an arbitrary method devoted to solving Cauchy problem. Once a stationary state is achieved, the counterpart static problem is solved.

The above-mentioned idea of finding solutions to stationary problems as parts of non-stationary problems for increasing time has been first illustrated by Tikhonov and Arsenin [124]. The applied relaxation technique can be derived as the iterational method to solve linear and nonlinear problems of the algebraic and transcendental equations, where a step in time defines a new approximation for a sought root of an equation. One more benefit of the relaxation method includes the simplicity of its

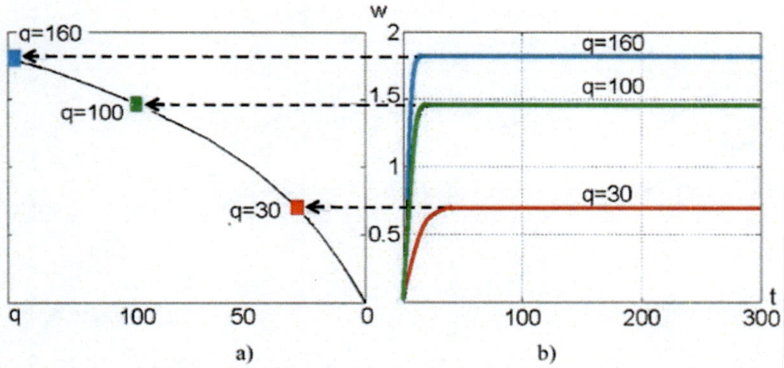

Fig. 7.7 The load-deflection (**a**) and the deflection-time dependencies (**b**) [reprinted with permission from Mechanical Systems and Signal Processing publishers]

Table 7.2 The values of the γ_2, P_E parameters versus the studied cases [reprinted with permission from Mechanical Systems and Signal Processing publishers]

Case number	1	2	3	4	5	6	7	8
Parameters	$\gamma_2 = 0.3$ $P_E = 1$	$\gamma_2 = 0.3$ $P_E = 0.5$	$\gamma_2 = 0.3$ $P_E = 2$	$\gamma_2 = 0$ $P_E = 1$	$\gamma_2 = 0$ $P_E = 0.5$	$\gamma_2 = 0$ $P_E = 2$	$\gamma_2 = 0$ $P_E = 1$ $E = 2E_0$	$\gamma_2 = 0.3$ $P_E = 1$ $E = 2E_0$

computer implementation since there exists a wide palette of effective algorithms devoted to solving Cauchy problem [125–128].

The results of solving the static problem for $\gamma_2 = l/h = 0.3$ and $\varepsilon = 3$ are reported in Fig. 7.7. In Fig. 7.7a the load-deflection dependence yielded by the relaxation method is shown for the given values of $q(x) = q = const$. On the other hand, in Fig. 7.7b, the stationary part of the dynamical process of the beam deflection $w(t)$ is reported for $q = 30,\ 100,\ 160$.

The $w(t)$ histories show that the steady/static state is achieved relatively fast for the fixed dissipation coefficient for all loads. For the load amplitude $q = 30$, the steady state is achieved at $t = 49$, whereas for $q = 100, 160$, at $t = 28$ and $t = 19$, respectively. Therefore, the employed method is highly stable and allows one to solve nonlinear static problems without any numerical difficulties.

The results of application of the relaxation method to study eight different combinations of the size-dependent parameter γ_2 and the material grading parameter P_E (the damping coefficient $\varepsilon = 3$) are shown in Table 7.2.

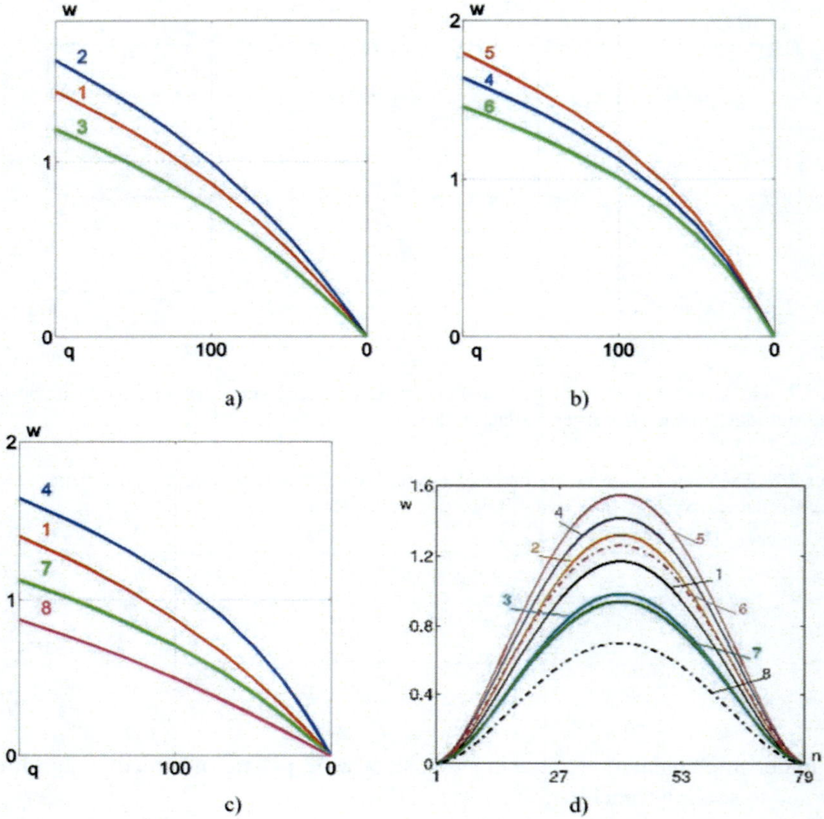

Fig. 7.8 Comparison of static solutions of Timoshenko beam: **a** $w(q)$ for the cases 1–3, $\gamma_2 = 0.3$; **b** $w(q)$ for the cases 4–6, $\gamma_2 = 0$; **c** $w(q)$ for the homogeneous beam; **d** $w(n)$ for $q = 150$ [reprinted with permission from Mechanical Systems and Signal Processing publishers]

The choice of parameters is as follows: $P = 1$—homogeneous material; $P = 2$—material with $E_1 = 2E$ on the upper side of the beam, and a material with $E_2 = E$ on the bottom side of the beam; $P = 0.5$—reverse material arrangement ($2E$ on the bottom side). The results of the carried-out investigations are given in Fig. 7.8.

Analysis of the results $w(q)$ for $\gamma_2 = 0.3$ (Fig. 7.8a) shows that, for the same load ($q > 100$), the minimum value of the deflection is observed for the non-homogeneous beam with the reinforcement (i.e. stiffer layer) on the upper side. The minimum value of the deflection is observed for the functionally graded beam in which the stiffer layer is located on the bottom side. If classical beams, i.e. without the size-dependent factor, are studied (Fig. 7.8b), the distribution is similar. The investigation of the results obtained for homogeneous beams with different values of stiffness and size-dependent coefficient (Fig. 7.8c) shows that the minimum deflection corresponds to the variant with the doubled stiffness and the size-dependent behaviour of the

Table 7.3 The frequencies associated with the studied cases

Case number	1	2	3	4	5	6	7	8
ω_0	4.9	4.5	5.4	3.9	3.7	4.05	5.9	6.7

material ($\gamma_2 = 0.3$, $P_E = 1$, $E = 2E_0$). A comparison of the results obtained for the same values of P (Fig. 7.8d) implies that the size-dependent behaviour results in a decrease of the deflection values.

Observe that for the case 2, the deflection curve changes its position, i.e. for $q = 50$ it is localized below the deflection of the case 6, for $q = 100$, it overlaps with the latter one, whereas, for $q = 150$, it is above the aforementioned deflection.

In general, the size-dependent behaviour reduces the deflection value for the same grading parameter. Placing the stiffer layer on the upper side of the beam essentially influences the deflection value for the same value of the size-dependent coefficient.

7.4.5.3 Dynamic Problems

The investigation of dynamic problems consists of the determination of the eigenfrequencies of the linear counterpart problem and the investigation of the chaotic dynamics with respect to different combinations of the parameters γ_2 and P.

(i) Determination of Timoshenko Beam Eigenfrequencies

In order to get the eigenfrequencies, we study the linear equations governing the size-dependent behaviour in the functionally graded beams, which are yielded by system (7.51) if nonlinear terms are neglected. The following linear PDEs are obtained regarding the functions ψ, w:

$$\begin{aligned} &k_2\,\psi_{,xx} + 3k_4\,\gamma_2^2\left(\psi_{,xx} - w_{,xxx}\right) - 12\,k_s k_3 \gamma_1^2\left(\psi + w_{,x}\right) = \psi_{,tt}\,, \\ &k_3\left(\psi_{,x} + w_{,xx}\right) + k_4 \frac{\gamma_2^2}{\gamma_1^2}\left(\psi_{,xxx} - w_{,xxxx}\right) - q = w_{,tt}\,. \end{aligned} \tag{7.54}$$

Equations (7.33) are solved using the algorithm presented earlier. Table 7.3 reports the results obtained numerically, and the studied cases correspond to those shown in Table 7.2.

The carried-out analysis yields the following conclusions. The smallest eigenfrequency is exhibited by the FG beam when the stiffer material is located on the beam bottom side (this is case 5 with a lack of the size-dependent behaviour). The largest eigenfrequency corresponds to case 8, i.e. the size-dependent behaviour is taken into account and the stiffness of the homogeneous beam is doubled. The size-dependent behaviour strongly reduces the value of the eigenfrequency. For the homogeneous beam (cases 1, 4 and 7, 8), the change in the frequency value is of 10 and 12%, respectively. For FG beam (cases 2, 5 and 3, 6)—18 and 25%, respectively. Therefore, the graded distribution of the beam thickness has an impact on the eigenfrequency value.

(ii) Investigation of Timoshenko Beam Chaotic Dynamics Versus the Combination of Parameters γ_2 and P

The investigation of the solutions to the dynamic problems, i.e. when the excitation $q = q_0 \sin(\omega_p t)$ is taken into account, has been carried out for the size-dependent coefficient $\gamma_2 = 0;\ 0.3$ and different values of P_E reported in Table 7.2.

The loading parameters are $q_0 = 17000,\ \omega_p = 8$. The frequency $\omega_p = 8$ is essentially higher comparing to the arbitrary eigenfrequency given in Table 7.3. All obtained results are shown in Tables 7.4, 7.5, 7.6, 7.7, 7.8, 7.9, 7.10 and 7.11:

Table 7.4 ($\gamma_2 = 0.3,\ P = 1,\ E = E_0$)—homogeneous beam with the size-dependent behaviour and an initial stiffness;

Table 7.5 ($\gamma_2 = 0.3,\ P = 2$)—functionally graded beam with the size-dependent behaviour for the case when the stiffer layer is located on the upper side of the beam;

Table 7.6 ($\gamma_2 = 0.3,\ P = 0.5$)—functionally graded beam with the size-dependent behaviour for the case when the stiffer layer is located on the bottom side of the beam;

Table 7.7 ($\gamma_2 = 0,\ P = 1,\ E = E_0$)—homogeneous beam without the size-dependent behaviour and with an initial stiffness;

Table 7.8 ($\gamma_2 = 0,\ P = 2$)—non-homogeneous beam without the size-dependent behaviour for the case when the stiffer layer is located on the upper side of the beam;

Table 7.9 ($\gamma_2 = 0,\ P = 0.5$)—non-homogeneous beam without the size-dependent behaviour for the case when the stiffer layer is located on the bottom side of the beam;

Table 7.10 ($\gamma_2 = 0,\ P = 1,\ E = 2E_0$)—homogeneous beam without the size-dependent behaviour and with a doubled stiffness;

Table 7.11 ($\gamma_2 = 0.3,\ P = 1,\ E = 2E_0$)—homogeneous beam with the size-dependent behaviour and a doubled stiffness.

The following results are reported in the above-listed tables: (a) time history $w(0.5; t)$; (b) Fourier spectrum based on the Fast Fourier Transform (FFT) $S(\omega)$; (c) 2D wavelet spectrum based on Morlet wavelet; (d) Poincaré section $w(t+T)\,[w(t)]$; (e) phase portrait $\dot{w}\,[w(t)]$; (f) time evolution of the largest Lyapunov exponent (LLE) based on Wolf's algorithm [129].

The results obtained for the functionally graded beam with the localization of the stiffer layer on the upper side (Table 7.5) essentially differ from the results shown in Tables 7.4, 7.6. This beam exhibits quasi-periodic vibrations at three linearly dependent frequencies and the LLE is negative. Results reported in Tables 7.4, 7.6 imply that the homogeneous beam and the functionally graded beam with the stiffer layer located on the bottom side vibrate chaotically. The difference between the results reported in Tables 7.4, 7.6 is as follows. In the case of the homogeneous beam (Table 7.4), the transition into chaos takes place at $t = 3800$, which has been also validated by

Table 7.4 Numerical results for $\gamma_2 = 0.3,\ P = 1$ [reprinted with permission from Mechanical Systems and Signal Processing publishers]

Time history	FFT	2D wavelet
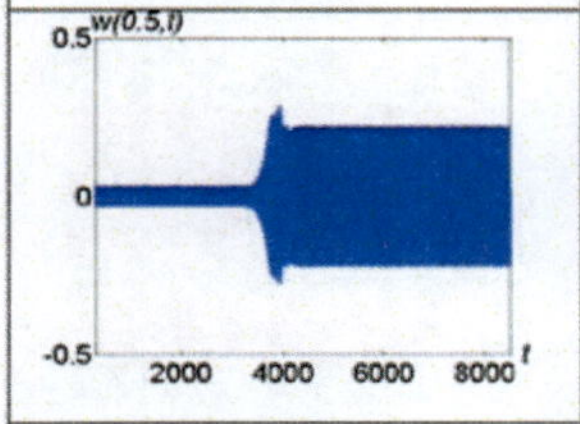	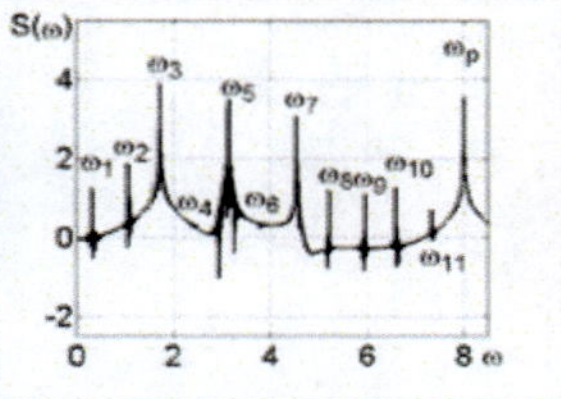	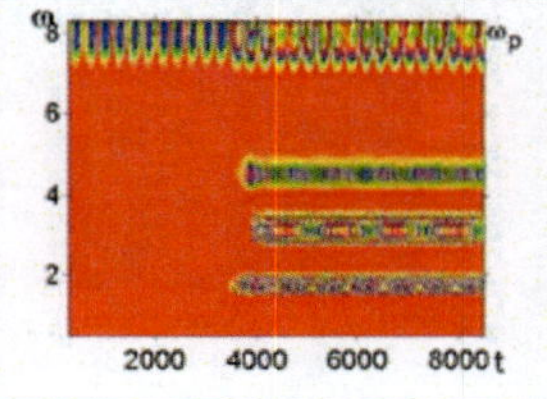
Phase portrait	**Poincaré section**	**LLE**
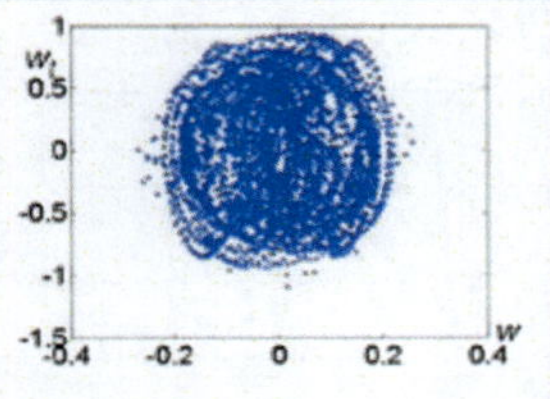	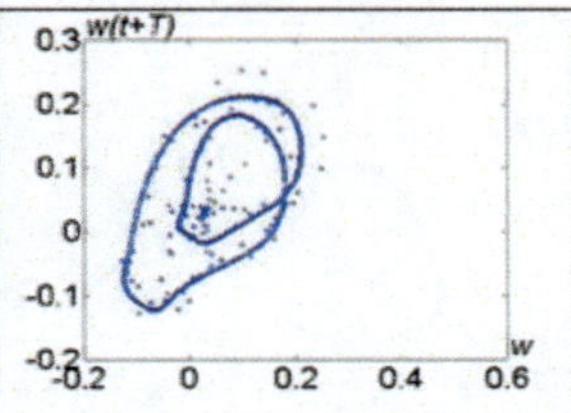	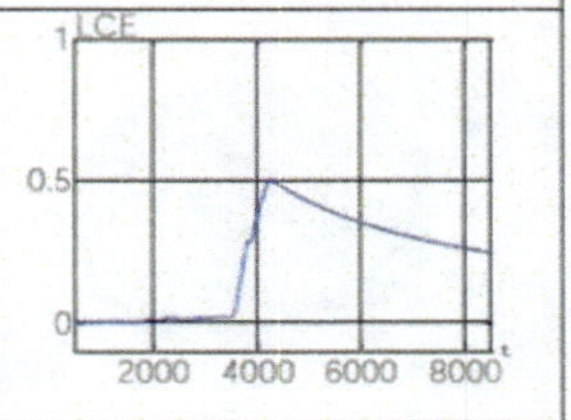

Table 7.5 Numerical results for $\gamma_2 = 0.3,\ P = 2$ [reprinted with permission from Mechanical Systems and Signal Processing publishers]

Time history	FFT	2D wavelet
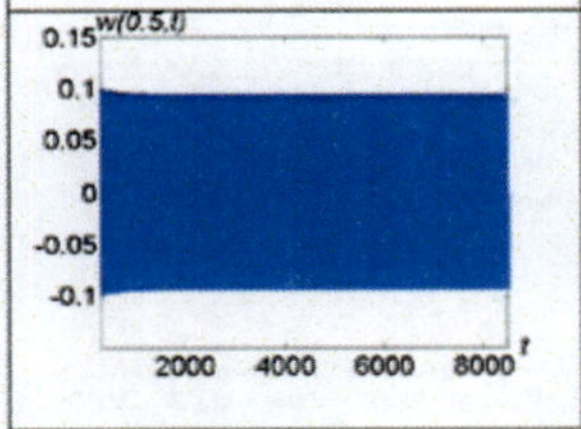	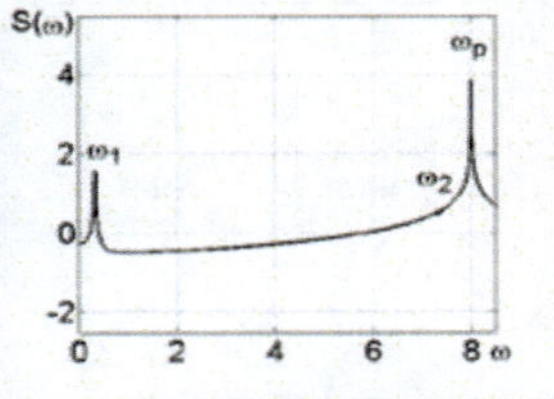	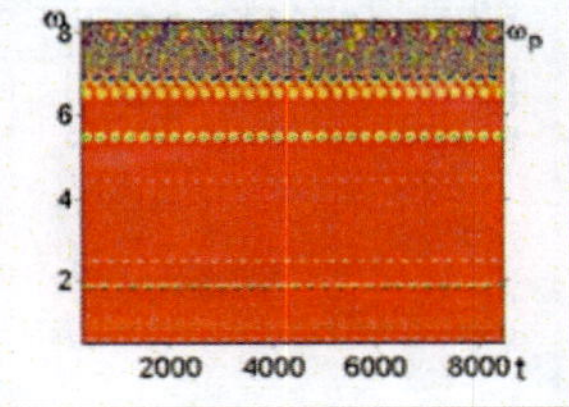
Phase portrait	**Poincaré section**	**LLE**
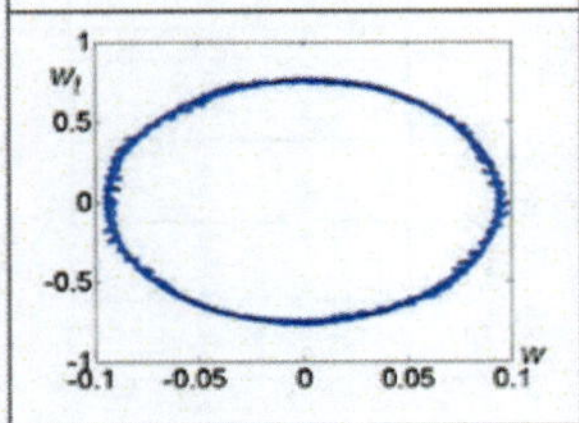	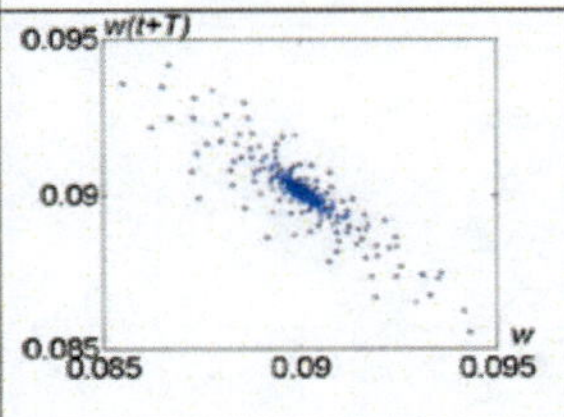	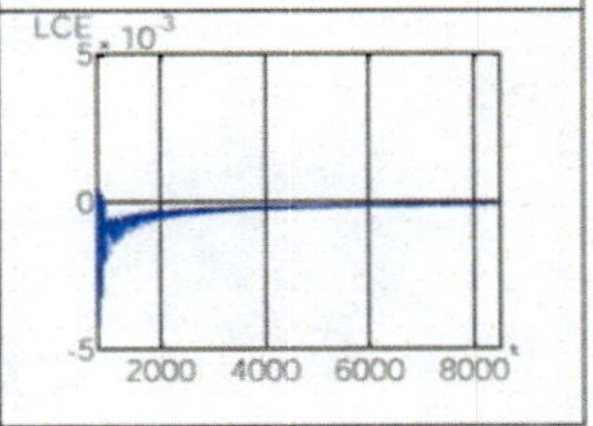

Table 7.6 Numerical results for $\gamma_2 = 0.3,\ P = 0.5$ [reprinted with permission from Mechanical Systems and Signal Processing publishers]

Time history	FFT	2D wavelet
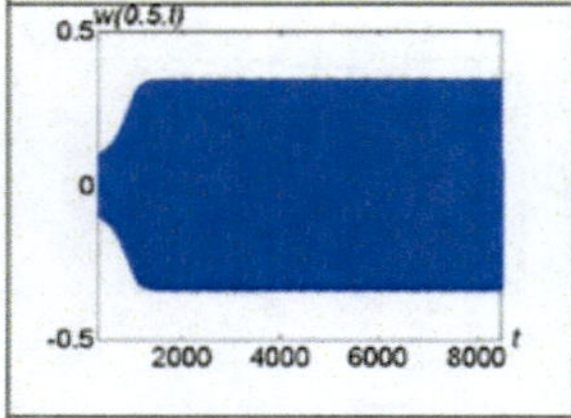	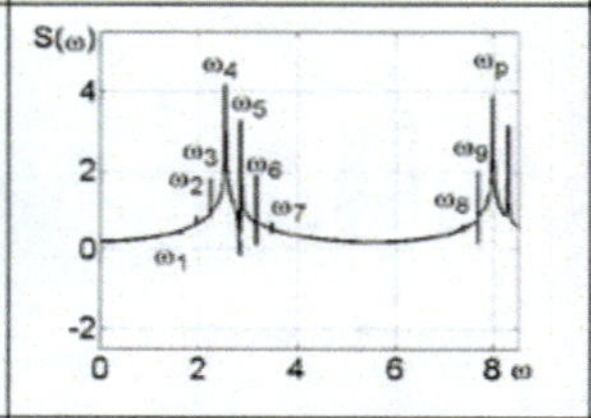	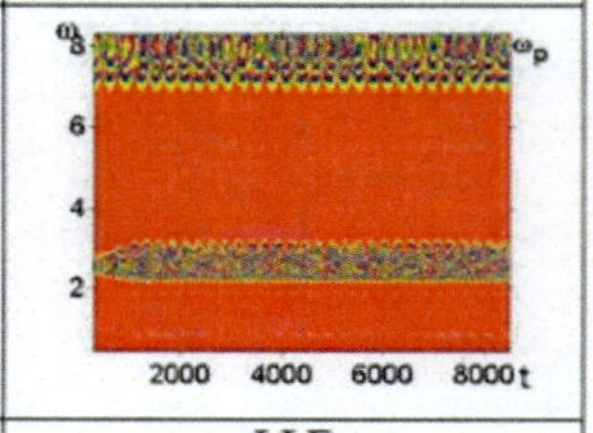
Phase portrait	Poincaré section	LLE
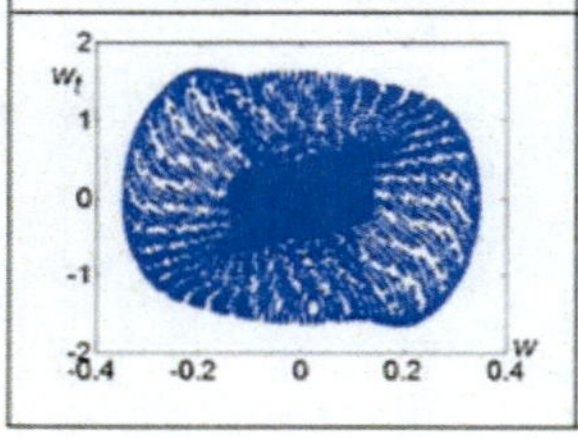	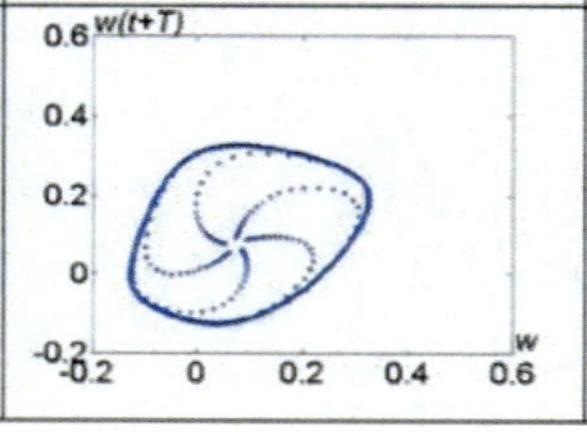	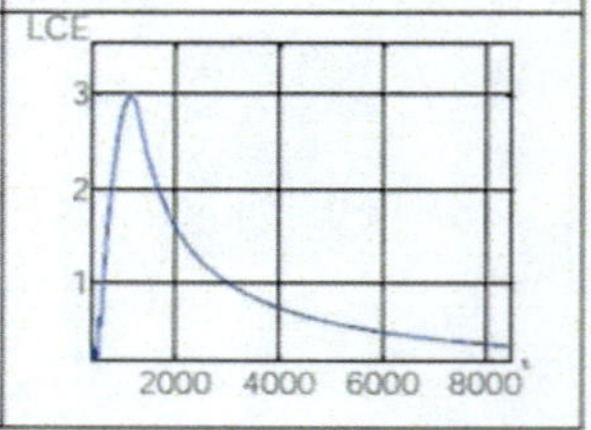

Table 7.7 Numerical results for $\gamma_2 = 0,\ P = 1$ [reprinted with permission from Mechanical Systems and Signal Processing publishers]

Time history	FFT	2D wavelet
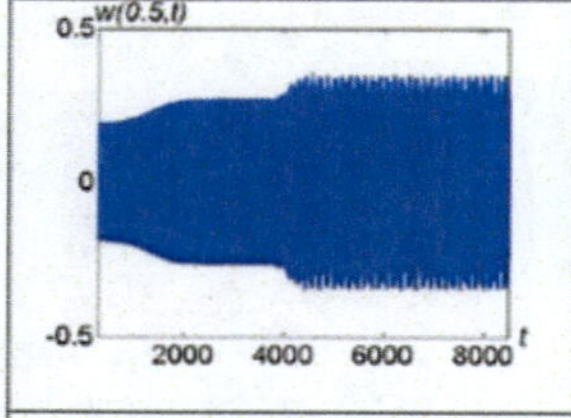	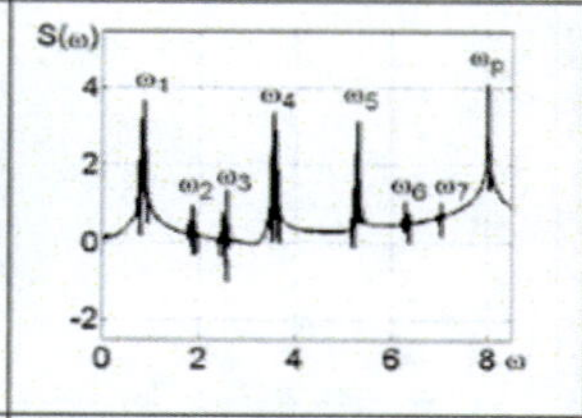	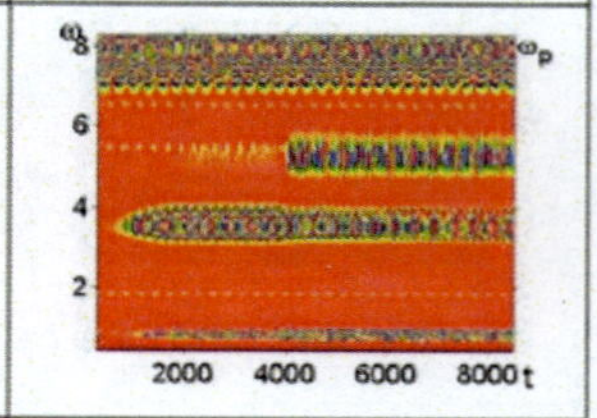
Phase portrait	Poincaré section	LLE
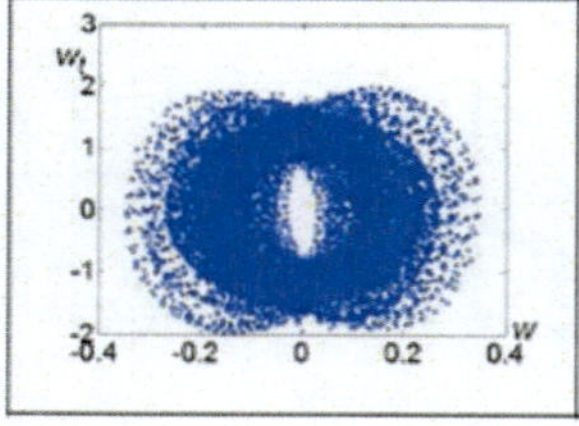	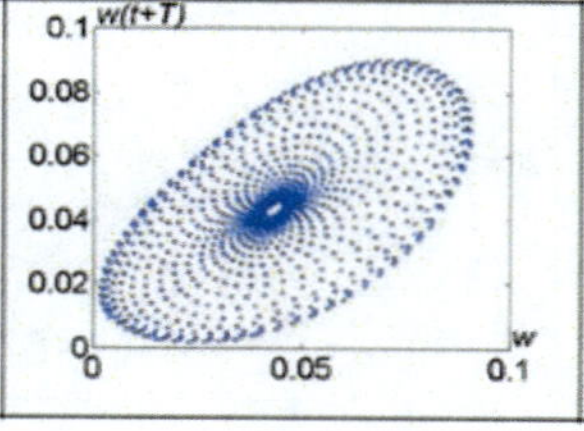	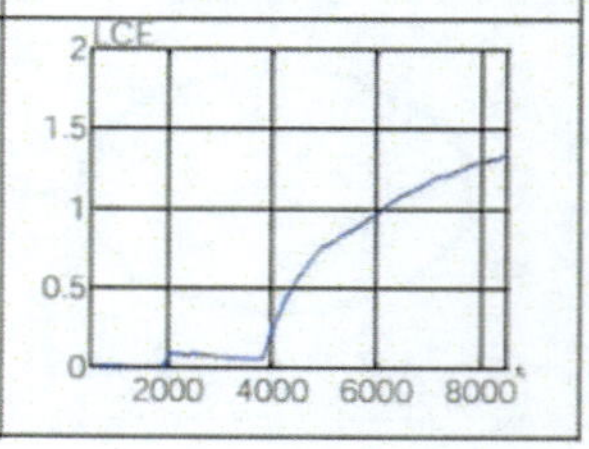

Table 7.8 Numerical results for $\gamma_2 = 0,\ P = 2$ [reprinted with permission from Mechanical Systems and Signal Processing publishers]

Time history	FFT	2D wavelet
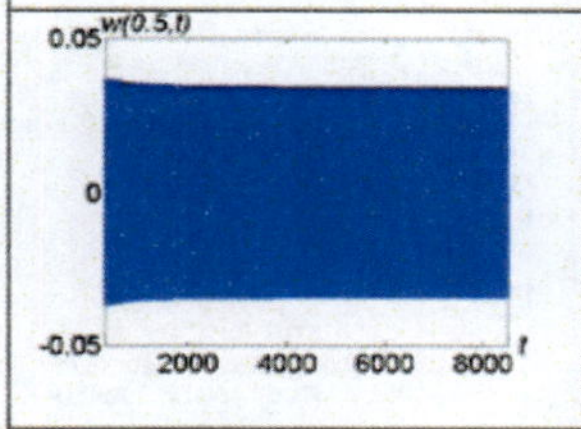	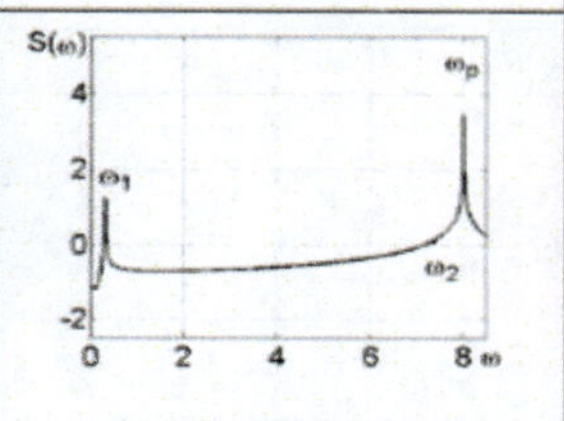	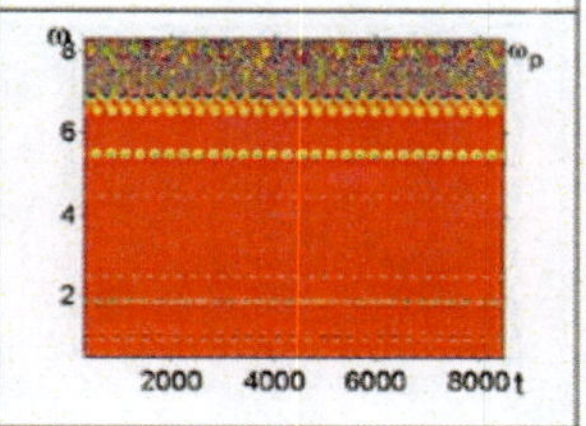
Phase portrait	Poincaré section	LLE
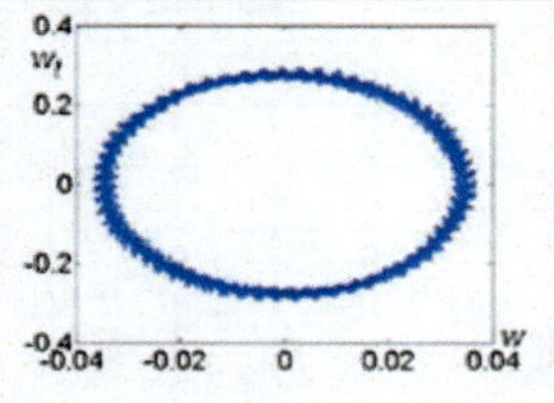	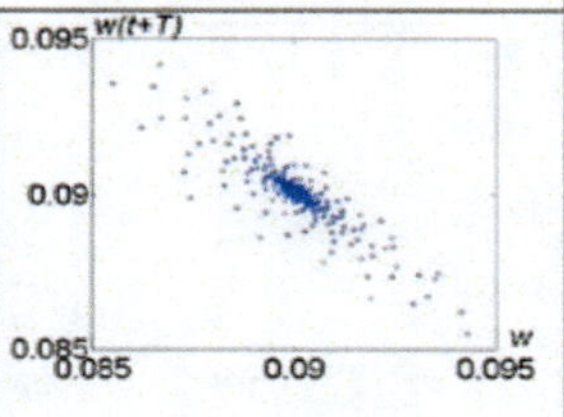	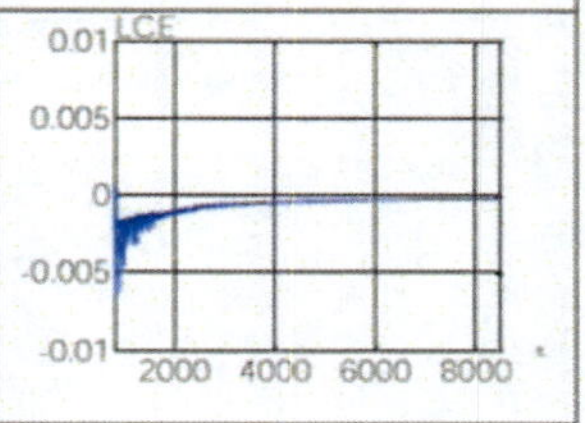

Table 7.9 Numerical results for $\gamma_2 = 0,\ P = 0.5$ [reprinted with permission from Mechanical Systems and Signal Processing publishers]

Time history	FFT	2D wavelet
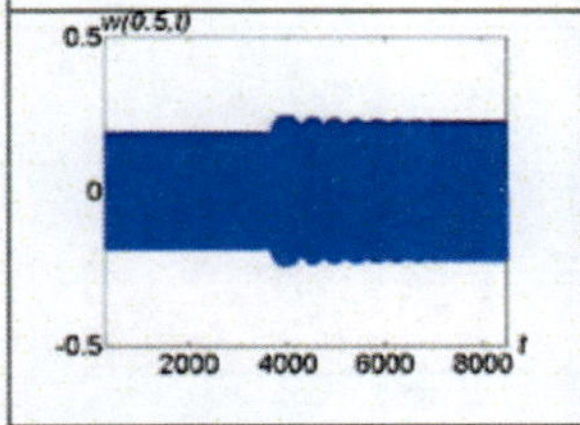	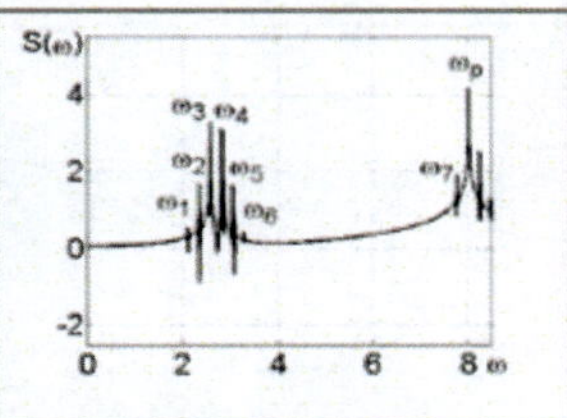	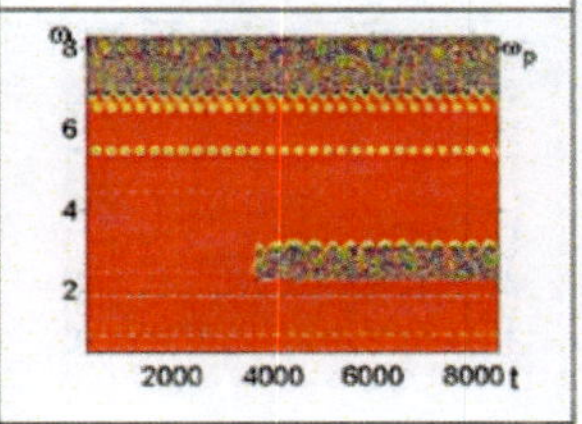
Phase portrait	Poincaré section	LLE
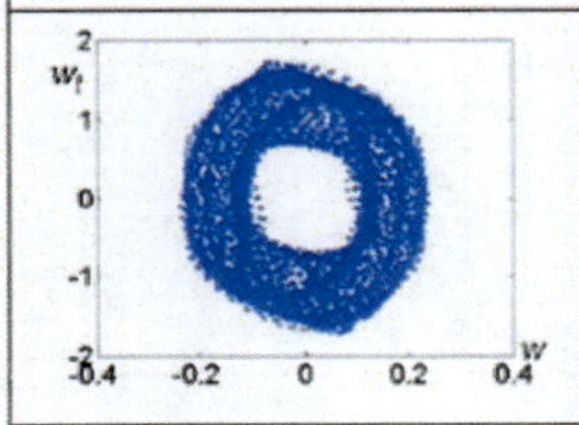	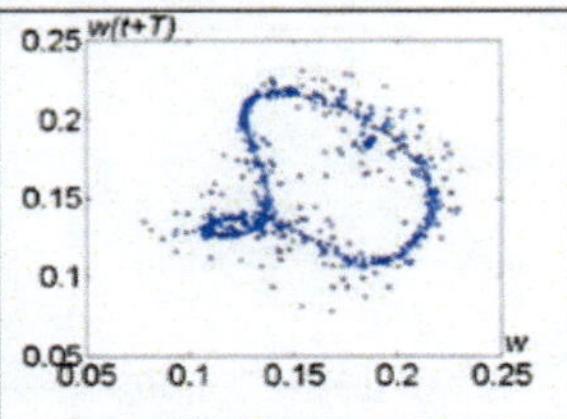	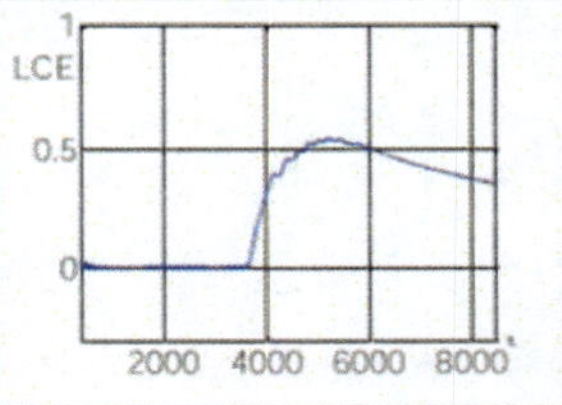

Table 7.10 Numerical results for $\gamma_2 = 0$, $P = 1$, $E = E_0$ [reprinted with permission from Mechanical Systems and Signal Processing publishers]

Time history	FFT	2D wavelet
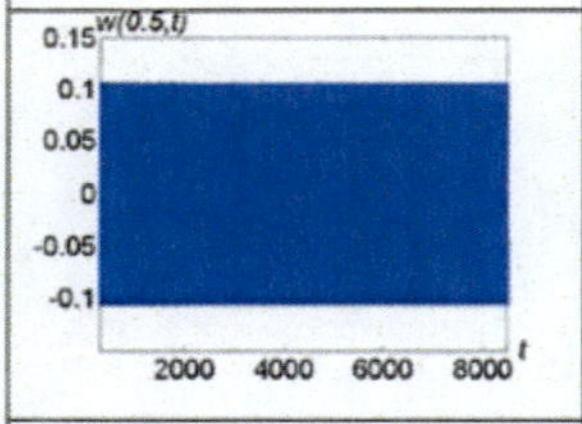	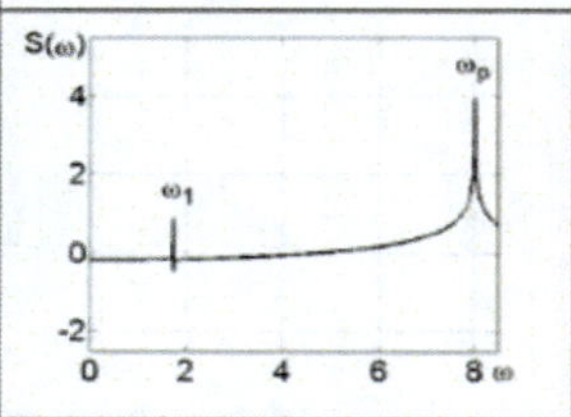	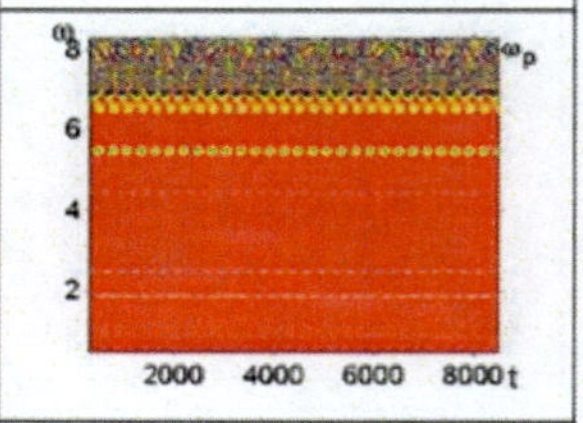
Phase portrait	**Poincaré section**	**LLE**
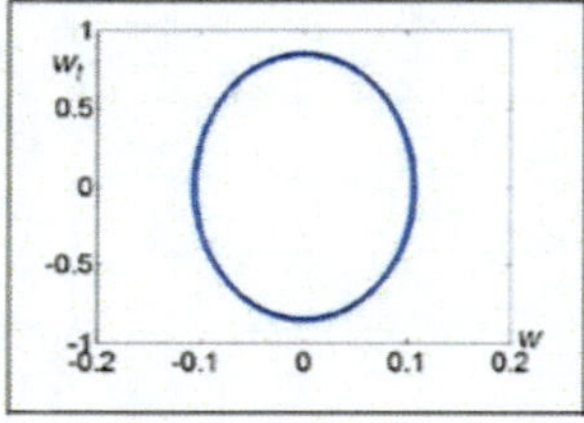	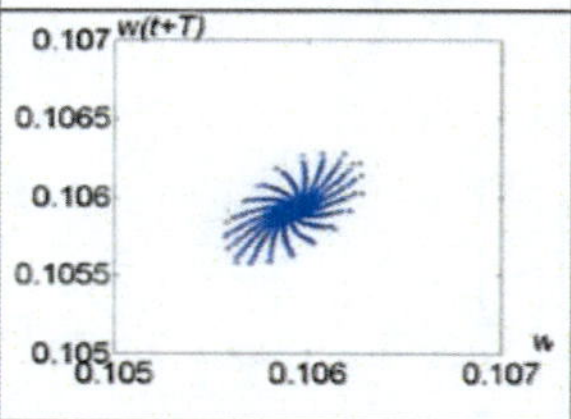	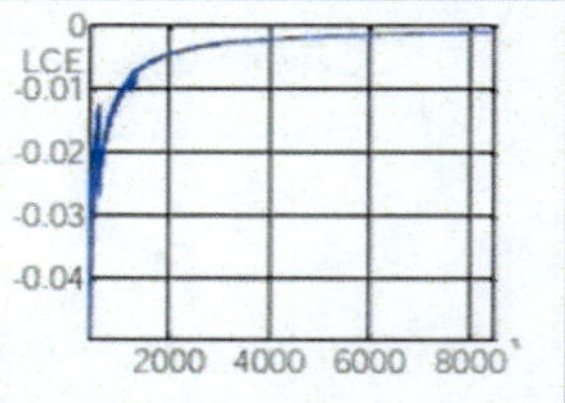

Table 7.11 Numerical results for $\gamma_2 = 0.3$, $P = 1$, $E = E_0$ [reprinted with permission from Mechanical Systems and Signal Processing publishers]

Time history	FFT	2D wavelet
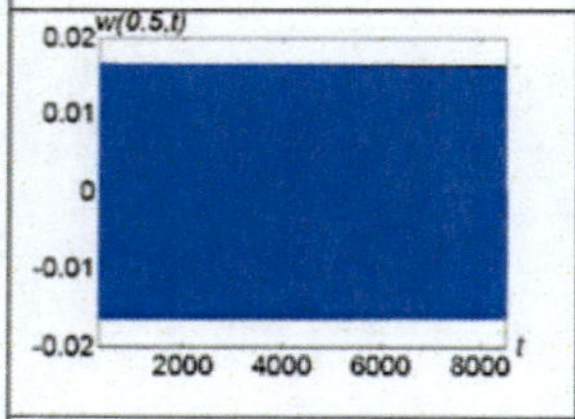	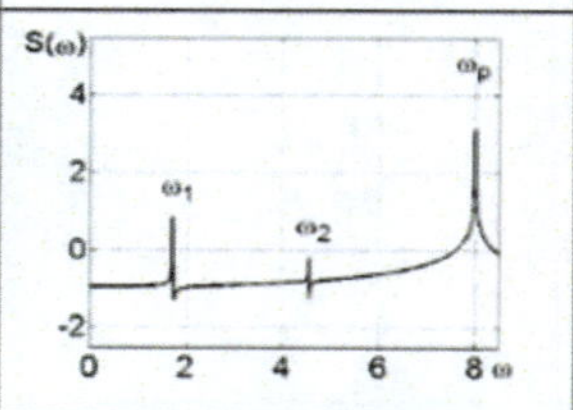	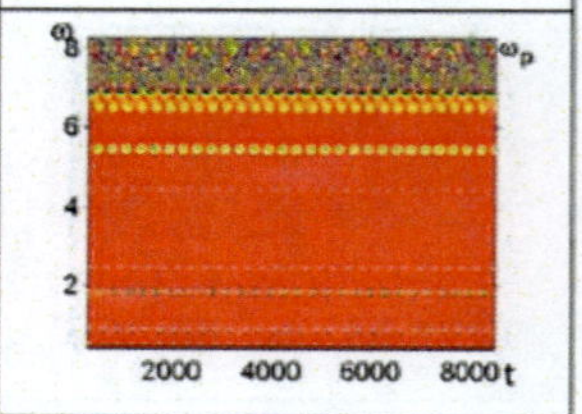
Phase portrait	**Poincaré section**	**LLE**
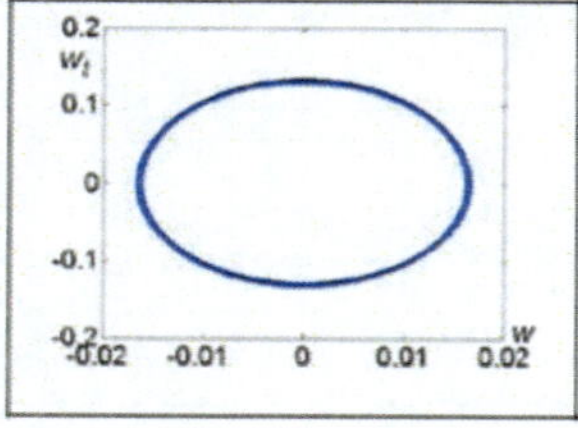	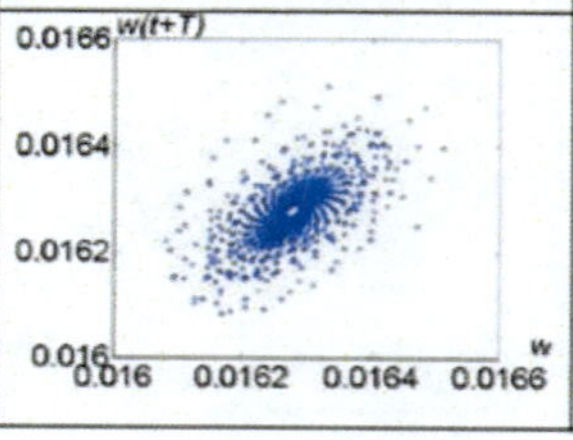	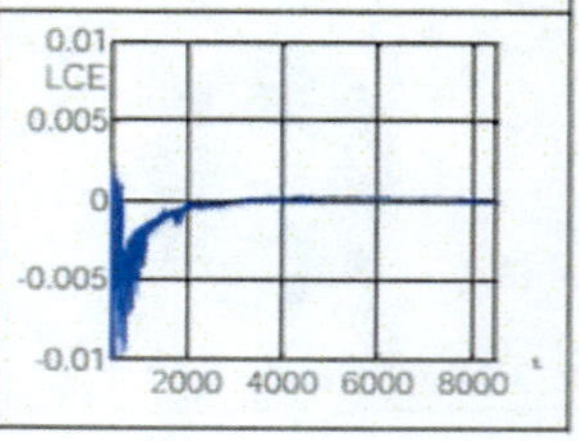

the wavelet spectrum and the LLE. On the contrary, the functionally graded beam (Table 7.6) starts to chaotically vibrate in the beginning of the time interval.

Therefore, the functionally graded beam with the stiffer layer located on the upper side can be employed for a given amplitude and frequency of the excitation load.

The analysis of the obtained results for the variants corresponding to a lack of the size-dependent behaviour ($\gamma_2 = 0$) validates a conclusion that the beam with the stiffer layer located on the upper side can be employed to carry the dynamic load. It should be mentioned that all characteristics of the vibration process, i.e. Fourier and wavelet spectra, Poincaré section, and the LLEs qualitatively coincide. The carried-out analysis and comparison with the previous variants ($\gamma_2 = 0.3$) for the homogeneous beam and the beam with the stiffer layer located on the bottom side implies the essential difference in all characteristics of the vibrational process. In other words, for the studied structures, the size-dependent behaviour plays an essential role.

The analysis of the results associated with the homogeneous beams (Tables 7.4, 7.7, 7.8, 7.9) allows one to extend the conclusions formulated with respect to static problems. Namely, application of the material of large stiffness has an essential influence on the character of vibrations.

In the case of the beams with a single (initial) stiffness, chaotic vibrations occur (Tables 7.4, 7.7), whereas in the case of beams with a doubled stiffness (Tables 7.8, 7.9), quasi-periodic vibrations are exhibited.

In order to validate the reliability of the LLEs computation using Wolf's algorithm [129], we have computed them using three other methods, i.e. Rosenstein's [130], Kantz's [131] and neural network (NW) approaches [121]. The investigations have been carried out for all mentioned variants. In what follows, we give exemplary results regarding the case 1 (see Table 7.4). The numbers of time intervals correspond to the following values: 1—$t \in [300; 2100]$, 2—$t \in [2105; 2160]$, 3—$t \in [2165; 3900]$, 4—$t \in [3901; 5000]$, 5—$t \in [5001; 8000]$.

All four methods yield positive values of the LLEs (λ_1) on all time intervals, which implies chaotic vibrations. However, there are some differences with respect to the computed values. The qualitative changes of λ_1 are similar for three methods (Wolf's, Rosenstein's, Kantz's) on the intervals 1–3. Beginning from the third time interval, all three characteristics (Rosenstein's, Kantz's, neural networks (NW)) shown in Fig. 7.9 approach each other, and hence essentially differ from the values obtained using Wolf's method.

Figure 7.10 reports time evolutions of Lyapunov spectrum obtained with the neural network approach. One may observe the qualitative similarity between the curves corresponding to the first four Lyapunov exponents.

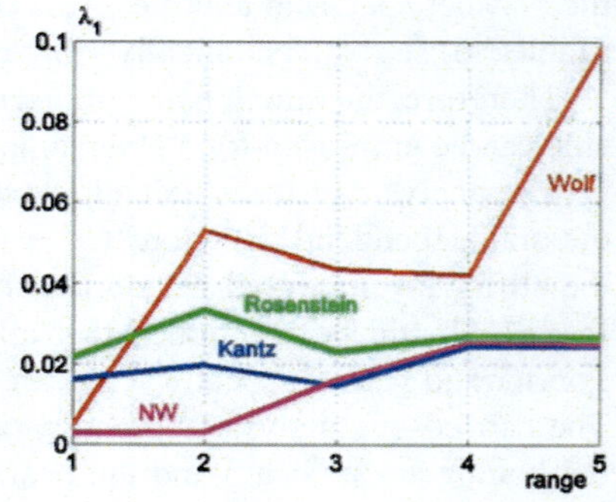

Fig. 7.9 Comparison of the LLE computed with four different algorithms [reprinted with permission from Mechanical Systems and Signal Processing publishers]

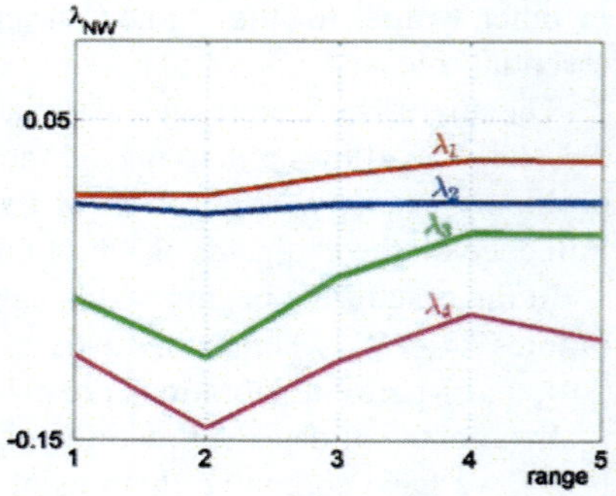

Fig. 7.10 Spectrum of the first four Lyapunov exponent obtained using the neural network method [reprinted with permission from Mechanical Systems and Signal Processing publishers]

(iii) Scenarios of Transition into Chaos

One of essential aspects while investigating the dynamics of continuous mechanical systems comprises the detection and analysis of the scenario of transition from regular to chaotic dynamics. Scenarios of such transitions for classical beams, plates and shells, including Bernoulli-Euler and Timoshenko theories, have been described in Refs. [132, 133]. In this chapter, we are aimed at analysing the scenario associated with Timoshenko model, taking into account two cases, i.e. the one concerning the size-dependent behaviour $\gamma_2 = 0.3$ and the through-thickness functionally graded beam for $P = 2$ or $P = 0.5$ as well as the one without the size-dependent effect ($\gamma_2 = 0$) and for the homogeneous beam $P = 1$ for $E = E_0,\ E = 2E_0$. The following parameters are fixed: $\omega_p = 8,\ \varepsilon = 1,\ \gamma_1 = 30,\ \nu = 0.3$.

On the basis of the obtained results, one may conclude that the scenarios of transition from regular to chaotic vibrations of Timoshenko beam follow the classical Ruelle-Takens-Newhouse scenario for the all considered cases.

7.5 Stability of the Size-Dependent and Functionally Graded Curvilinear Timoshenko Beams

The size-dependent model is studied, based on the modified couple stress theory for the geometrically nonlinear curvilinear Timoshenko beam made from a functionally graded material having its properties changed along the beam thickness. The influence of the size-dependent coefficient and the material grading on the stability of the curvilinear beams is investigated with the use of the set-up (relaxation) method.

The second-order accuracy FDM is used to solve the problem of nonlinear PDEs by reducing it to Cauchy problem. The obtained set of nonlinear ODEs is then solved by the fourth-order Runge-Kutta method. The relaxation method is employed to solve numerous static problems based on the dynamic approach. Eight different combinations of size-dependent coefficients and the functionally graded material coefficient are used to study the stress-strain responses of Timoshenko beams.

Stability loss of the curvilinear Timoshenko beams is investigated using Lyapunov criterion based on the estimation of Lyapunov exponents. Beams with/without the size-dependent behaviour, homogeneous beams and functionally graded beams having the same stiffness are investigated.

It is shown that in straight-line beams, the size-dependent effect decreases the beam deflection. The same is observed if the most rigid layer is located on the top of the beam. In the curvilinear Timoshenko beam, such a location of the most rigid layer essentially improves the beam strength against stability loss.

The observed transition of the largest Lyapunov exponent from a negative to positive value corresponds to the transition from a pre-critical to post-critical beam state.

7.5.1 Introduction

As already mentioned in Sect. 7.3.1, FGM can be composed by mixing two or more materials with the functionally changed properties along a desirable direction [1]. They have a wide range of applications in both theoretical and industrial areas. They can be used as materials for thermoisolation of cosmic structural elements in the constructions of the nuclear reactors, and for fabrication of numerous sensors and gyroscopes. In recent years, high interest has been observed in modelling and investigating the functionally graded structures, in particular including their statics, bending processes as well as dynamical characteristics [3–5].

It should be emphasized that in recent years the FGM have found direct applications in the micro- and nanostructures, like thin films/layers [7, 33], as well as in the electromechanical/mechatronics systems [8, 9]. The size-dependent static and dynamics behaviour of the microstructures has been approved experimentally [14, 16].

Unfortunately, no theory of the classical continuum can explain the size-dependent behaviour exhibited on the micro- and submicrostructural scales. However, the non-classical theories of continuum, like the strain gradient theory and the modified couple stress theory, can be employed to study the size-dependent properties.

Timoshenko beam model [122], on contrary to Euler-Bernoulli model, takes into account the shear deformation and can be modified to fit the real behaviour of the constructions.

In Ref. [101], the functionally graded beams have been investigated taking into account their thickness with and von Kármán-type nonlinearity. The size-dependent behaviour has been studied with the help of the modified couple stress theory. Nonlinear PDEs governing the dynamics of Euler-Bernoulli and Timoshenko beams have been derived and studied. An influence of the length parameter, the variation of the beam properties along its thickness, the shear deformations as well as the influence of the geometric nonlinearity on the beam static deflection have been investigated.

Free nonlinear vibrations of microbeams made of FGMs has been investigated in Ref. [57] based on the modified couple stress theory accompanied by Timoshenko beam theory and von Kármán geometric nonlinearity. It has been found that both linear and nonlinear frequencies significantly increase when the thickness of the FGM microbeam is comparable to the material length scale parameter. In Ref. [134] dynamic stability of microbeams made of FGMs has been investigated based on the modified couple stress theory and Timoshenko beam theory. The obtained results show that the influence of the size-dependent effect on the dynamic stability characteristics is significant only when the thickness of the beam has a similar magnitude as the material length scale parameter. The so far reported and overviewed studies considered only straight beams and a displacement of the deflection curve has not been taken into account that resulted in studying more complicated PDEs. On the contrary, this problem has been addressed in this section and the obtained governing PDEs have a simpler form comparing to those studied so far.

On the other hand, only a few investigations [50, 69, 70] present the analysis of the curvilinear beams made from FGM in their transverse direction, taking into account the size-dependent behaviour. In these works, the authors have studied only the linear behaviour of FGM curved microbeams.

It should be emphasized that, according to the authors' knowledge, there is a lack of investigations of models of nonlinear curved microbeams made of functionally graded materials in the existing literature.

The mentioned gaps in the research devoted to the size-dependent FGM structures stand for motivation to analyse nonlinear vibrations of the curvilinear Timoshenko-type beams on a basis of the modified couple stress theory. The PDEs governing the beam dynamics are derived and the influence of the size-dependent coefficient and the material grading coefficient on the character of the load-deflection curve and the plots of the beam forms are reported.

7.5.2 Theoretical Background

In the modified couple stress theory, for infinitely small deformations, the energy of deformation U accumulated in a linear elastic body of the volume V is as follows:

$$U = \frac{1}{2} \iiint\limits_V \left(\sigma_{ij}\varepsilon_{ij} + m_{ij}\chi_{ij}\right) dV. \tag{7.55}$$

In the isotropic case, we have

$$\sigma_{ij} = \lambda\varepsilon_{mm}\delta_{ij} + 2\mu\varepsilon_{ij}, \tag{7.56}$$

$$\varepsilon_{ij} = \frac{1}{2}\left(u_{i,j} + u_{j,i} + u_{m,i}u_{m,j}\right), \tag{7.57}$$

$$m_{ij} = \beta\chi_{ij} = 2\mu l^2\chi_{ij}, \tag{7.58}$$

$$\chi_{ij} = \frac{1}{2}\left(\theta_{i,j} + \theta_{j,i}\right). \tag{7.59}$$

The coordinate system, kinematic parameters and the load for the curvilinear Timoshenko FG beam are shown in Fig. 7.11. It is assumed that the properties of the curvilinear beam do not change along the axis OX. The kinematic relations regarding Timoshenko beam follow [122]

$$u_x = u(x,t) + z\psi(x,t), \quad u_y = 0, \quad u_z = w(x,t). \tag{7.60}$$

It is assumed that transverse beam cross sections remain flat after a deformation process. Nevertheless, they may undergo a rigid displacement in the plane XOZ as well as the rotation around the axis OY.

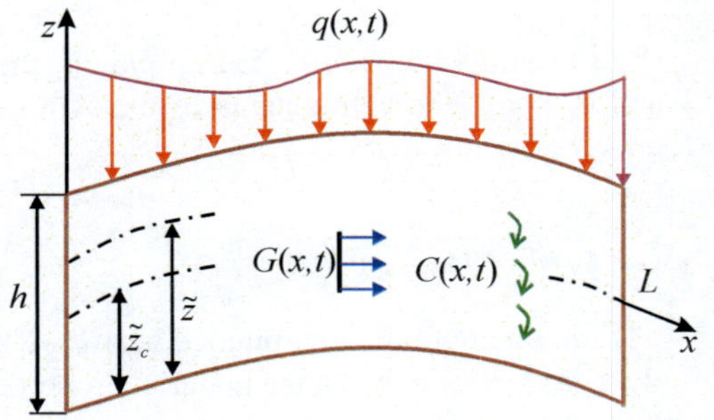

Fig. 7.11 Geometry of Timoshenko beam and the acting load [reprinted with permission from the Journal of Computational and Nonlinear Dynamics publishers]

The introduced line, further referred to as the bending line, is the line around which a pure rotation occurs in a cross section. The coordinate z in Fig. 7.11 denotes the distance between a cross section point and the bending line. In Fig. 7.11 $G(x,t)$ denotes the force per unit length acting on the axial cross section along the body axis OX. The $q(x,t)$ stands for the resultant of the transverse forces located on the upper beam surface per beam unit length. The parameter $C(x,t)$ stands for the y component of the resultant volume moment per the unit length of the curvilinear beam being adjusted to the beam section. A distance between an arbitrary point and the beam surface is denoted by $\tilde{z}$. The distance between the bending line and the lower surface is denoted by $\tilde{z}_c$.

Here, we consider the geometrically nonlinear shallow beam, i.e. the beam with small deformation and rotations but with possibly large deflection w, which imply nonlinear effects. Employment of (7.3), (7.6) yields. The non-zero physical components of the strain tensor ε_{ij} of the curved beam in the curvilinear coordinate system of the following form [50]

$$\varepsilon_{xx} = (1 + k_x z)^{-1} \left(u_{,x} + \frac{1}{2}(w_{,x})^2 + z\psi_{,x} + k_x w \right),$$

$$\varepsilon_{xz} = \frac{1}{2}(1 + k_x z)^{-1} \left(w_{,x} + \psi - k_x u \right),$$

where $k_x = 1/R_x$ denotes curvature and R_x is a radius of the beam curvature.

Assuming that the thickness of the curvilinear beam h is small in comparison to curvature radius R_x, i.e. $h/R_x << 1$ and $|z/R_x| << 1$, and assuming $(1 + k_x z) \approx 1$, as well considering the low-pitched beams $k_x u \approx 0$ only, the deformations depend on the displacements in the following way [135]

$$\varepsilon_{xx} = \left(u_{,x} + \frac{1}{2}(w_{,x})^2 + z\psi_{,x} + k_x w \right), \quad \varepsilon_{xz} = \frac{1}{2}\left(w_{,x} + \psi \right). \tag{7.61}$$

The relation $\theta_i = (rot\ (\mathbf{u}))_i/2$ yields one non-zero rotation vector of components of the form

$$\theta_y = \frac{1}{2}\psi + \frac{1}{2}(1 + k_x z)^{-1} \left(k_x u - w_{,x} + k_x z\psi \right). \tag{7.62}$$

Substituting (7.62) into (7.59), employing the above-given assumption $(1 + k_x z) \approx 1$ and $k_x u \approx 0$, the components of the symmetric part of the curvature tensor are found:

$$\chi_{xy} = \chi_{yx} = \frac{1}{4}\left(\psi_{,x} - w_{,xx} \right), \quad \chi_{yz} = \chi_{zy} = \frac{1}{4}k_x \left(\psi + w_{,x} \right). \tag{7.63}$$

Again, owing to the low-pitched beam, we have $\chi_{yz} \approx 0$.

In a general case, when the material properties are changed along the beam length and thickness, after substitution of (7.61) into (7.56), the following components of the symmetric part of the stress tensor are obtained (Poisson's effect is neglected)

$$\sigma_{xx} = E(x, \tilde{z}) \left(u_{,x} + \frac{1}{2}(w_{,x})^2 + z\psi_{,x} + k_x w \right), \quad \sigma_{xz} = k_s \mu(x, \tilde{z}) \left(w_{,x} + \psi \right),$$

where $E(x, \tilde{z})$ and $\mu(x, \tilde{z})$ denote Young's and shear moduli, respectively, k_s is the shear corrector factor, which has been introduced due to the grading shear deformation with respect to the transverse cross section of the curvilinear beam (it depends on the form of the cross section of the curvilinear beam) [57, 135].

In what follows, we assume that the beam properties are changed only along the beam thickness, and the components of the symmetric part of the stress tensor curve are expressed in terms of the kinematic parameters in the following way

$$\sigma_{xx} = E(\tilde{z}) \left(u_{,x} + \frac{1}{2}(w_{,x})^2 + z\psi_{,x} + k_x w \right), \quad \sigma_{xz} = k_s \mu(\tilde{z}) \left(w_{,x} + \psi \right). \tag{7.64}$$

Substituting (7.63) into (7.58), the components of the derivative part of the higher order moment can be expressed by the kinematic parameter χ_{xy}:

$$m_{xy} = m_{yx} = \frac{1}{4}\beta(\tilde{z}) \left(\psi_{,x} - w_{,xx} \right). \tag{7.65}$$

For further analysis, it is convenient to introduce a reference line for the z-coordinate, using the following condition

$$B_{11} = \int_A E(\tilde{z})\, z dA = \int_A E(\tilde{z})(\tilde{z} - \tilde{z})\, dA = 0. \tag{7.66}$$

From Eq. (7.66), we obtain the coordinate $\tilde{z}_c$ of the reference line

$$\tilde{z} = \frac{\int_A E(\tilde{z})\, \tilde{z} dA}{\int_A E(\tilde{z})\, dA}. \tag{7.67}$$

The obtained formula (7.67) allows one to exclude the terms with coefficients B_{11} from the system of equations, which significantly simplifies the system of equations obtained in Refs. [57, 101].

7.5.3 Derivation of Equations of Motion

In order to define the equation governing the dynamics of the FG curvilinear Timoshenko beam based on the modified couple stress theory, we substitute (7.61)–(7.65) into (7.1). As a result, the following expressions for the potential (U) and kinematic (K) energies are obtained:

$$U = \frac{1}{2}\int_0^L \int_A \left(\sigma_{ij}\varepsilon_{ij} + m_{ij}\chi_{ij}\right) dA\, dx =$$
$$= \frac{1}{2}\int_0^L \int_A \left\{ E(\bar{z}) \left(u_{,x} + \frac{1}{2}(w_{,x})^2 + z\psi_{,x} + k_x w\right)^2 + \right. \tag{7.68}$$
$$\left. + k_s \mu(\bar{z}) \left(\psi + w_{,x}\right)^2 + \frac{1}{8}\beta(\bar{z}) \left(\psi_{,x} - w_{,xx}\right)^2 \right\} dA\, dx,$$

$$K = \frac{1}{2}\int_0^L \int_A \rho(\bar{z}) \left\{ \left(u_{,t} + z\psi_{,t}\right)^2 + \left(w_{,t}\right)^2 \right\} dA dx, \tag{7.69}$$

where L denotes the length of the beam. The variation of the work δW is implied by the external body forces G and q per unit length along the x axis as well as by C, which states for the y component of the body couple per unit length. The applied axial force, the transverse shear force and bending moments at the ends of the beam are denoted by $\bar{N}$, $\bar{V}$, $\bar{M}$, respectively, whereas the force exerted due to the viscous damping of the surrounding medium is as follows [136, 137]

$$\delta W = \int_0^L \left(G\delta u - q\delta w + C\delta\theta_y\right) dx + \left(\bar{N}\delta u + \bar{V}\delta w + \bar{M}\delta\psi\right)\Big|_{x=0}^{x=L}, \tag{7.70}$$
$$\delta W_\varepsilon = -\varepsilon \int_0^L \left(w_{,t}\, \delta w\right) dx,$$

where ε represents the viscous damping coefficient.

Owing to (7.66), the relations (7.68) and (7.69) can be rewritten as follows:

$$U = \frac{1}{2}\int_0^L \left\{ k_1 \left(u_{,x} + \frac{1}{2}(w_{,x})^2 + k_x w\right)^2 + k_2 \left(\psi_{,x}\right)^2 + \right. \tag{7.71}$$
$$\left. + k_3 \left(\psi + w_{,x}\right)^2 + k_4 \left(\psi_{,x} - w_{,xx}\right)^2 \right\} dx,$$

$$K = \frac{1}{2}\int_0^L \left\{ m_0 \left(u_{,t}\right)^2 + 2Q\, u_{,t}\, \psi_{,t} + \tilde{I}\left(\psi_{,t}\right)^2 + m_0 \left(w_{,t}\right)^2 \right\} dx, \tag{7.72}$$

where

$$
\begin{aligned}
k_1 &= \int_A E(\tilde{z})\, dA, \quad k_2 = \int_A E(\tilde{z})\, z^2\, dA = \int_A E(\tilde{z})\, (\tilde{z}-\tilde{z}_c)^2 dA\,,\\
k_3 &= k_s \int_A \mu(\tilde{z})\, dA, \quad k_4 = \frac{1}{8}\int_A \beta(\tilde{z})\, dA = \frac{1}{4}\int_A \mu(\tilde{z})\, l^2(\tilde{z})\, dA,\\
m_0 &= \int_A \rho(\tilde{z})\, dA, \quad Q = \int_A \rho(\tilde{z})\, z\, dA = \int_A \rho(\tilde{z})\, (\tilde{z}-\tilde{z}_c)\, dA,\\
\tilde{I} &= \int_A \rho(\tilde{z})\, z^2\, dA = \int_A \rho(\tilde{z})\, (\tilde{z}-\tilde{z}_c)^2 dA\,.
\end{aligned}
\tag{7.73}
$$

Employing Hamilton's principle

$$
\delta \int_{t_1}^{t_2} (K - U + W)\, dt = 0, \tag{7.74}
$$

varying with respect to the variable u, w, and ψ, carrying out the integration by parts and setting the terms standing by δu, δw and $\delta,\ \psi$ to zero, the following solving equations:

$$
\begin{aligned}
&\left[k_1\left(u_{,x} + \frac{1}{2}(w_{,x})^2 + k_x w\right)\right]_{,x} + G = m_0 u_{,tt} + Q\,\psi_{,tt}\\
&\left[k_2\psi_{,x} + k_4\left(\psi_{,x} - w_{,xx}\right)\right]_{,x} - k_3\left(\psi + w_{,x}\right) + \frac{C}{2} = Q\,u_{,tt} + \tilde{I}\psi_{,tt}\,,\\
&\left\{\left[k_1\left(u_{,x} + \frac{1}{2}(w_{,x})^2 + k_x w\right)\right] w_{,x} + k_3\left(\psi + w_{,x}\right)\right\}_{,x} -\\
&-k_x k_1\left(u_{,x} + \frac{1}{2}(w_{,x})^2 + k_x w\right) +\\
&+\left[k_4\left(\psi_{,x} - w_{,xx}\right)\right]_{,xx} - q + \frac{C_{,x}}{2} = m_0 w_{,tt} + \varepsilon w_{,t},
\end{aligned}
\tag{7.75}
$$

along with the boundary

$$
\left[k_1\left(u_{,x} + \frac{1}{2}(w_{,x})^2 + k_x w\right) - \bar{N}\right]\Bigg|_{x=0}^{x=L} = 0 \quad \text{or} \quad \delta u|_{x=0}^{x=L} = 0,
$$

$$\left[k_2\psi_{,x}+k_4\left(\psi_{,x}-w_{,xx}\right)-\bar{M}\right]\Big|_{x=0}^{x=L}=0 \quad \text{or} \quad \delta\,\psi|_{x=0}^{x=L}=0,$$

$$\left\{\left[k_1\left(u_{,x}+\frac{1}{2}(w_{,x})^2+k_x w\right)\right]w_{,x}+\ k_3\left(\psi+w_{,x}\right)+\right.$$

$$\left.+k_4\left(\psi_{,xx}-w_{,xxx}\right)+\frac{C}{2}-\bar{V}\right\}\Bigg|_{x=0}^{x=L}=0 \quad \text{or} \quad \delta\,w|_{x=0}^{x=L}=0,$$

$$\left[k_2\psi_{,x}+k_4\left(\psi_{,x}-w_{,xx}\right)\right]_{,x}-k_3\left(\psi+w_{,x}\right)+\frac{C}{2}=Q\,u_{,tt}+\tilde{I}\psi_{,tt}\,, \tag{7.76}$$

and initial conditions

$$\begin{aligned} w(x,0)&=\varphi_1(x); \quad & w\,(x,0)\,_{,t}&=\varphi_2(x);\\ u(x,0)&=\varphi_3(x)\,; \quad & u\,(x,0)\,_{,t}&=\varphi_4(x);\\ \psi(x,0)&=\varphi_5(x); \quad & \psi\,(x,0)\,_{,t}&=\varphi_6(x), \end{aligned} \tag{7.77}$$

are obtained.

The governing nonlinear partial differential equations of the system (7.75) of the FGM curvilinear Timoshenko beam and the boundary conditions (7.76) are obtained by taking into account (7.67). This system differs from that reported in Refs. [50, 69, 70].

7.5.4 The Methods of Analysis

The results regarding statics are obtained based on investigation of dynamic equations (7.74)–(7.76). This method, known as a set-up method, has been first proposed by Tikhonov [124] and employed to the problems of theory of shells by Feodos'ev [138]. This method is described in detail in the paper [125].

In a case when the load $[\bar{q}]$ does not depend on time, we can solve the static problem on a basis of the dynamic approach. The initial conditions play the role of the excitation in the case of statics, whereas the term with the first time derivative plays a role of a damping coefficient of the excited solution.

A solution to the problem of dynamics can be found by applying an arbitrary method aimed at solving a Cauchy problem. After achieving a stationary state by a solution excited by initially conditions, the static problem is solved.

In this section, the reduction of PDEs (7.75)–(7.77) to ODEs has been made on a basis of FDM of the second-order accuracy. Then, the problem has been solved using the RK4 method.

7.5.5 Numerical Results

7.5.5.1 Static Problem of the FGM Size-Dependent Curvilinear Timoshenko Beam

In order to study the stated problem and to simplify considerations, we assume $G(x, t) = 0$, $(x, t) = 0$, whereas in order to trace the ways of changing the beam properties with respect to the beam thickness, Young's modulus $E(\tilde{z})$, shear modulus $\mu(\tilde{z})$ and beam thickness $\rho(\tilde{z})$ are employed.

The properties of the beam material along the beam thickness are defined using the following linear functions:

$$E(\tilde{z}) = E_0 + \frac{\tilde{z} + \frac{h}{2}}{h}(E_1 - E_0), \quad \mu(\tilde{z}) = \mu_0 + \frac{\tilde{z} + \frac{h}{2}}{h}(\mu_1 - \mu_0),$$
$$\rho(\tilde{z}) = \rho_0 + \frac{\tilde{z} + \frac{h}{2}}{h}(\rho_1 - \rho_0). \tag{7.78}$$

In order to carry out the integration in formulas (7.66) and (7.73), the following relationship between moduli is employed:

$$E_1 = P_E E_0, \quad \mu_1 = P_\mu \mu_0, \quad \rho_1 = P_\rho \rho_0, \tag{7.79}$$

Substituting (7.78), (7.79) into (7.67), we find $\tilde{z}$:

$$\tilde{z} = \frac{h}{12} \frac{P_E - 1}{1 + \frac{1}{2}(P_E - 1)}. \tag{7.80}$$

Formula (7.80) implies that for $P_E > 1$, we have $\tilde{z} > 0$, and the neutral line moves above the neutral line of the counterpart homogeneous beam $(P_E = 1)$. In the case $P_E < 1$, we have $\tilde{z} < 0$, and the neutral line moves below the neutral line of the counterpart homogeneous beam.

Using (7.78)–(7.80), the coefficients of formula (7.73) follow:

$$k_1 = E_0 A\left[1 + \frac{1}{2}(P_E - 1)\right], \quad k_2 = \frac{1}{12}E_0 A h^2\left[\frac{3}{2} + \frac{(P_E - 1)^2}{12\left(1 + \frac{1}{2}(P_E - 1)\right)}\right],$$

$$k_3 = k_s \mu_0 A\left[1 + \frac{1}{2}(P_\mu - 1)\right], \quad k_4 = \frac{1}{4}\mu_0 A l^2\left[1 + \frac{1}{2}(P_\mu - 1)\right], \tag{7.81}$$

$$m_0 = \rho_0 A\left[1 + \frac{1}{2}(P_\mu - 1)\right], \quad Q = \rho_0 A \frac{h}{12}\left[(P_\rho - 1) - \frac{1}{2}\frac{(P_E - 1)(P_\rho + 1)}{1 + \frac{1}{2}(P_E - 1)}\right],$$

$$\tilde{I} = \rho_0 A \frac{h^2}{12}\left[\left(1 + \frac{(P_E-1)^2}{12\left(1+\frac{1}{2}(P_E-1)\right)^2}\right)\left(1+\frac{1}{2}(P_\rho - 1)\right) - \frac{1}{6}\frac{(P_E-1)(P_\rho+1)}{1+\frac{1}{2}(P_E-1)}\right].$$

Observe that the shear coefficient k_s being equal to $(5+5\nu)/(6+5\nu)$, is considered to be most suitable for description of the beams with the rectangular cross section [57].

Let us introduce the following dimensionless parameters

$$\begin{aligned}
&\bar{w} = \frac{w}{h}, \quad \bar{u} = \frac{ua}{h^2}, \quad \bar{\psi} = \frac{\psi a}{h}, \quad \bar{x} = \frac{x}{a}, \quad \gamma_1 = \frac{a}{h},\\
&\gamma_2 = \frac{l}{h}, \quad \bar{q} = q\frac{a^2}{h^2 E}, \quad \bar{t} = \frac{t}{\tau}, \quad \tau = \frac{a}{c}, \quad c = \sqrt{\frac{E}{\rho}}, \\
&\bar{\varepsilon} = \varepsilon\frac{a}{c}, \quad \bar{k}_1 = \frac{k_1}{AE_0}, \quad \bar{k}_2 = \frac{k_2}{AE_0 h^2}, \quad \bar{k}_3 = \frac{k_3}{AE_0}, \quad \bar{k}_4 = \frac{k_4}{AE_0 l^2}.
\end{aligned} \tag{7.82}$$

Taking into account the introduced simplifications and neglecting the bars over the non-dimensional parameters, the following equation governing the dynamics of the curvilinear Timoshenko beam are obtained:

$$\begin{aligned}
&k_1\left[u_{,x} + \frac{1}{2}(w_{,x})^2 + k_x w\right]_{,x} = u_{,tt},\\
&k_2\,\psi_{,xx} + 3k_4\,\gamma_2^2\left(\psi_{,xx} - w_{,xxx}\right) - 12\,k_s k_3 \gamma_1^2\left(\psi + w_{,x}\right) = \psi_{,tt}\,,\\
&\frac{1}{\gamma_1^2}\left\{k_1\left[\left(u_{,x} + \frac{1}{2}(w_{,x})^2 + k_x w\right)\right] w_{,x}\right\}_{,x} - \\
&-\frac{k_x k_1}{\gamma_1^2}\left(u_{,x} + \frac{1}{2}(w_{,x})^2 + k_x w\right) + \\
&+k_3\left(\psi_{,x} + w_{,xx}\right) + k_4\frac{\gamma_2^2}{\gamma_1^2}\left(\psi_{,xxx} - w_{,xxxx}\right) - q = w_{,tt} + \varepsilon\, w_{,t}.
\end{aligned} \tag{7.83}$$

As an example, we take the rigid clamping to describe the boundary conditions

$$\begin{aligned}
&w(0,t) = w(1,t) = 0; \quad w_{,x}(0,t) = w_{,x}(1,t) = 0;\\
&u(0,t) = u(1,t) = 0; \quad \psi(0,t) = \psi(1,t) = 0,
\end{aligned} \tag{7.84}$$

and the following initial conditions

$$w(x,0) = w_{,t}(x,0)\,; \quad u(x,0) = u_{,t}(x,0) = 0; \quad \psi(x,0) = \psi_{,t}(x,0) = 0. \tag{7.85}$$

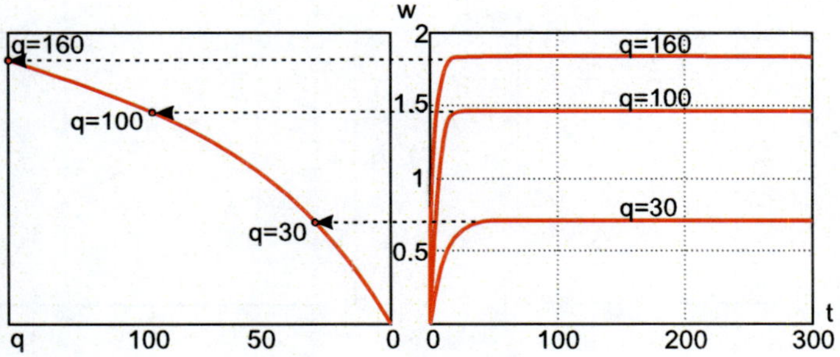

Fig. 7.12 Load-deflection curve $w(0.5; q)$ (**a**) and time histories of the deflection $w(0.5; t)$ (**b**) [reprinted with permission from the Journal of Computational and Nonlinear Dynamics publishers]

Table 7.12 The studied variants and corresponding parameters [reprinted with permission from the Journal of Computational and Nonlinear Dynamics publishers]

Variant	1	2	3	4	5	6	7	8
Parameter	$\gamma_2 = 0.3$ $P_E = 1$	$\gamma_2 = 0.3$ $P_E = 0.5$	$\gamma_2 = 0.3$ $P_E = 2$	$\gamma_2 = 0$ $P_E = 1$	$\gamma_2 = 0$ $P_E = 0.5$	$\gamma_2 = 0$ $P_E = 2$	$\gamma_2 = 0$ $P_E = 1$ $E = 2E_0$	$\gamma_2 = 0.3$ $P_E = 1$ $E = 2E_0$

The numerical investigations of static problems associated with Timoshenko model have been carried out for the following fixed parameters: the relative length $\gamma_1 = \frac{a}{h} = 30$, the size-dependent parameter $\gamma_2 = \frac{l}{h} = 0;\ \ 0.3$; the coefficients of Young's, shear moduli and beam density (7.79) are the same along the thickness $\left(P_E = P_\mu = P_\rho = P = 1;\ \ 2;\ \ 0.5\right)$. Therefore, the coefficients in formula (7.81) take much simpler form. Finally, the curvature coefficient is taken as $k_x = 0;\ \ 24$.

The employed relaxation method yielded the results shown in Fig. 7.12. Figure 7.12a reports the load-deflection dependence obtained by the relaxation method for the given $q(x) = q = constant$ values. Time histories of the dynamical processes $w(t)$ are shown in Fig. 7.12b for $q = 30;\ \ 100;\ \ 160$, and for $k_x = 0$, $P = 1$, $\gamma_2 = 0$.

The time histories $w(t)$ imply that the oscillation processes approach their counterpart stationary states relatively fast for the fixed dissipation coefficients for all studied loads. In the case of $q = 30$, the steady state begins at $t = 49$, whereas for $q = 100,\ \ 160$, it begins at $t = 28$ and $t = 19$, respectively.

Therefore, the employed relaxation method is not only stable, but it allows one to find solution to nonlinear static problems in a rather simple manner. Based on the relaxation method, we used eight different combinations of the size-dependent parameter γ_2 and the functionally graded material P (the damping coefficient has been fixed, i.e. $\varepsilon = 3$).

The chosen eight combinations of the parameters γ_2, P_E are reported in Table 7.12.

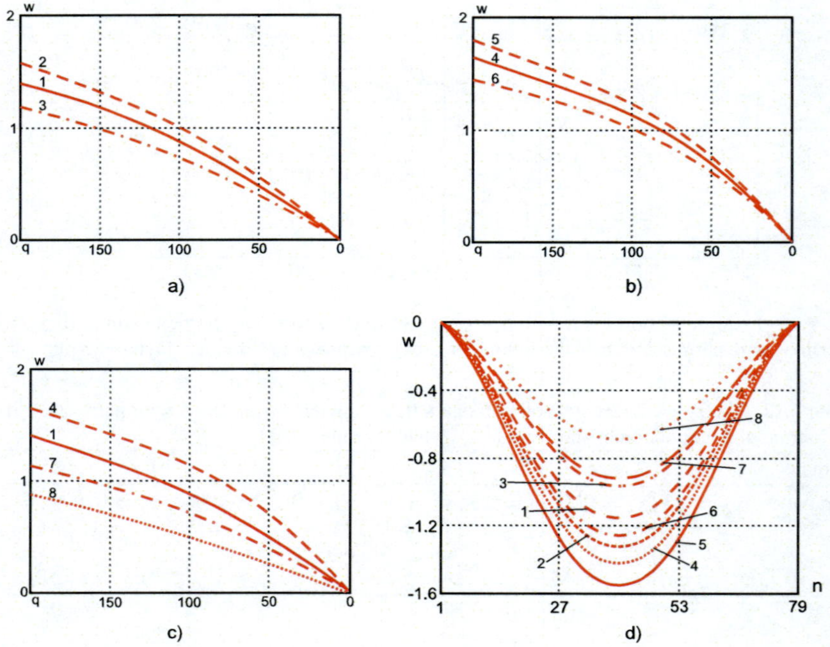

Fig. 7.13 Solutions to static problems of Timoshenko beam: **a** $w(0.5; q)$ for variants 1–3, $\gamma_2 = 0.3$; **b** $w(0.5; q)$ for variants 4–6, $\gamma_2 = 0$; **c** $w(0.5; q)$ for homogenous beams; **d** solution for $w(n)$ and $q = 150$ [reprinted with permission from the Journal of Computational and Nonlinear Dynamics publishers]

Let us discuss the meaning of the so far introduced parameters. $P = 1$ corresponds to a homogeneous material, $P = 2$ states for a functionally graded material, i.e. the material with $2E_0$ located on the upper part of the beam. For $P = 0.5$, the beam is also made of a functionally graded material, but the situation is reversed ($2E_0$ is now on the bottom beam part). Figure 7.13a reports the deflection-load dependence for variants 1–3; Fig. 7.13b shows the deflection-load relations for the variants 4–6; Fig. 7.13c presents the deflection-load curves for the variants 1, 4, 7, 8; Fig. 7.13d shows deflections for $q = 50$, 100, 150 and for all studied variants.

Figure 7.13a shows that for the same loads ($q > 100$), the largest deflection values are observed for the functionally graded beam with location of the stiffest layer on the bottom side of the beam. The difference between the homogenous beam and the functionally graded beam with stiffer layer located on the bottom (upper) side of the beam is on amount of 15% (20%) for $q = 200$. The same observation holds for the beam without the size-dependent effect (Fig. 7.13b). The difference between the values of deflection reaches 13 and 18%, respectively. On the other hand, the investigation of homogenous beams with various stiffness and size-dependent coefficients (Fig. 7.13c) shows that the largest deflection corresponds to the variant with the initial stiffness and no size-dependent coefficient ($\gamma_2 = 0,\ P_E = 1,\ E = E_0$). The difference with respect to the remaining values is as follows: ($\gamma_2 = 0.3,\ P_E = 1$)—15%, ($\gamma_2 = 0,\ P_E = 1,\ E = 2E_0$)—35%, ($\gamma_2 = 0.3,\ P_E = 1,\ E = 2E_0$)—55%.

Finally, plots of the beam forms are reported in Fig. 7.13d for the same P values. The smaller and larger values of the deflection are observed for the variant 8 ($\gamma_2 = 0.3,\ P_E = 1,\ E = 2E_0$) and 5 ($\gamma_2 = 0,\ P_E = 0.5$), respectively.

We have shown that the size-dependent behaviour decreases the deflection value for the same value of the functional grading parameter. Furthermore, if the most rigid layer is located in the upper part of the beam, the decrease in the deflection value is implied, taking into account the same size-dependent coefficient.

7.5.5.2 Stability Loss

In what follows, we consider the problems of a static stability loss of the curvilinear beam subjected to a constant load. In literature, there exists a large number of stability loss criterions. We will consider only the simple Lyapunov criterion, i.e. the beam is stable/unstable if the LLE estimated for the governing equation is negative/positive.

Due to the fact that there is no $G\ (x, t)$ and (x, t), the buckling occurs when the beam is exposed to transverse forces $q\ (x)$ just because of the initial curvature. This means that the transverse buckling is not taken into account (pitchfork bifurcation), and our study is restricted to follow the snap-through bifurcation.

In what follows, we study the dependence of the beam centre on the external load. We have employed all of the mentioned stability criterions, which coincide with each other well. The computational examples for $k_x = 24$ and for all eight variants of Table 7.12 are shown in Fig. 7.14.

The carried out numerical analysis of the curvilinear Timoshenko beams for $k_x = 24$ yields the following observations.

1. **Beams with the size-dependent behaviour** (Fig. 7.13a). The smallest value of the critical load, at which the stability loss occurs, takes place for the variant 2 of the functionally graded beam, where the most rigid layer ($2E_0$) is located on the bottom side of the beam. If the most rigid layer is located on the upper beam part, the stability loss occurs for a larger value of the load (500 instead of 190). It means that a change in the layers position essentially changes the beam stress-strain state.
2. **Beams without the size-dependent behaviour** (Fig. 7.13b). The qualitative picture of the change in the critical load follows the previously described case. However, a difference in the critical load values between the two locations of the rigid layer is decreased.
3. In the case of the **homogenous beams**, the largest load, at which stability loss occurs, is exhibited by the homogenous beam with a doubled stiffness ($2E_0$) and with the size-dependent effect (see variant 8). The difference between the largest and smallest values of the critical load estimations, at which stability loss occurs, for the variants 8 and 4, achieves the value of 200%.

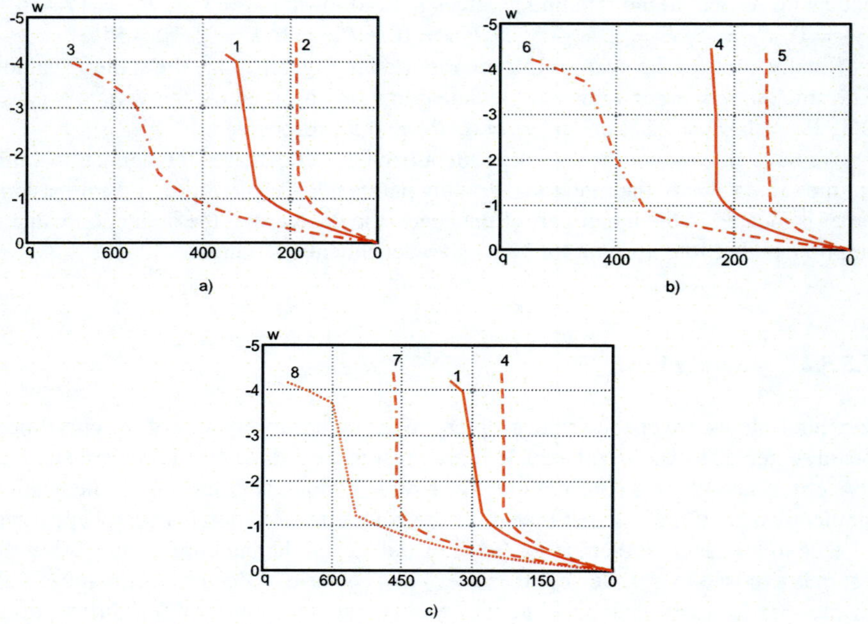

Fig. 7.14 Deflection-load functions for $k_x = 24$: **a** variants 1–3, $\gamma_2 = 0.3$; **b** variants 4–6, $\gamma_2 = 0$; **c** homogenous beam [reprinted with permission from the Journal of Computational and Nonlinear Dynamics publishers]

4. Comparison of the results of the functionally graded **beams with the same stiffness** shows that the size-dependent behaviour implies an increase in the load q_{cr} associated with the occurrence of the stability loss. The estimated difference for the variants 1–4 is of 21.7%, for the variants 2–5 is 24%, for the variants 3–6 is 31.6% and for the variants 7–8 is 19.6%. Therefore, the largest difference between the obtained results is observed for the functionally graded beam, where the stiffest layer is located on the upper side of the beam.
5. In the case of **all investigated variants**, all described criterions of stability loss give similar results.

In Table 7.13, the beam deflection form $w(n)$ as well as Lyapunov exponents obtained based on the employed neural networks algorithm [139] have been presented. Solid/dashed curves correspond to the q^- pre-critical /q^+ post-critical loads.

In the case of all eight studied variants, the occurred stability loss of the beam corresponds to the change of LE from its negative values (pre-critical load q^-) to positive ones (post-critical values q^+). Observe that, in the case of variants 2 and 8, we have detected two positive Lyapunov exponents. In the case of dynamical system, one may deal with the hyper-chaos [121].

For the studied static problem, the above-mentioned phenomenon implies stiffer stability loss.

Table 7.13 The beam forms $w(n)$ for the considered variants and the associated Lyapunov exponents [reprinted with permission from the Journal of Computational and Nonlinear Dynamics publishers]

<table>
<tr><td>1st variant $\gamma_2 = 0.3, P_E = 1$</td><td>2nd variant $\gamma_2 = 0.3, P_E = 0.5$</td></tr>
<tr><td>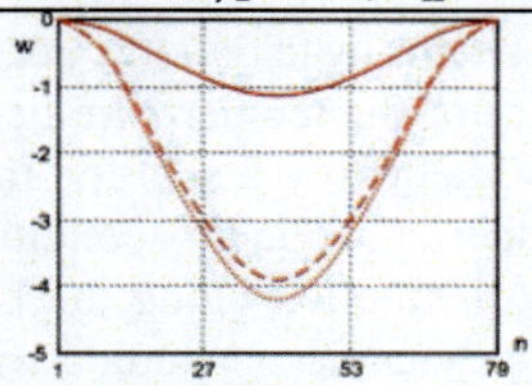

LE Spectrum:
$q^- = 270 \leftrightarrow -0.1860 - 0.0066 - 0.7486 - 0.7487$
$q^+ = 350 \leftrightarrow 0.2362 - 0.2390 - 0.9705 - 2.4735$</td><td>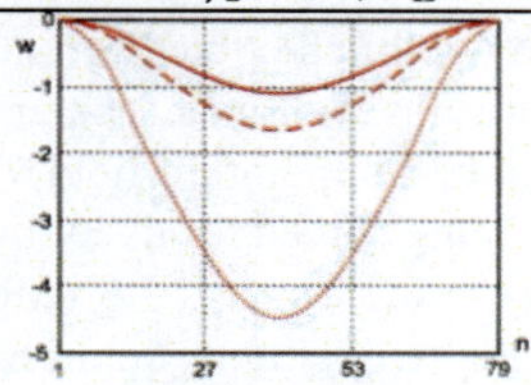

LE Spectrum:
$q^- = 170 \leftrightarrow -0.0622 - 0.7210 - 0.7706 - 1.0949$
$q^+ = 188 \leftrightarrow 0.37880.0166 - 1.0965 - 1.0966$</td></tr>
<tr><td>3rd variant $\gamma_2 = 0.3, P_E = 2$</td><td>4th variant $\gamma_2 = 0, P_E = 1$</td></tr>
<tr><td>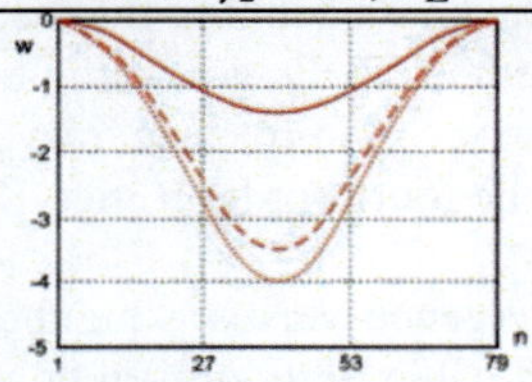

LE Spectrum:
$q^- = 490 \leftrightarrow -0.0114 - 0.6803 - 0.6803 - 0.9179$
$q^+ = 700 \leftrightarrow 0.0301 - 0.2700 - 0.27 - 2.1791$</td><td>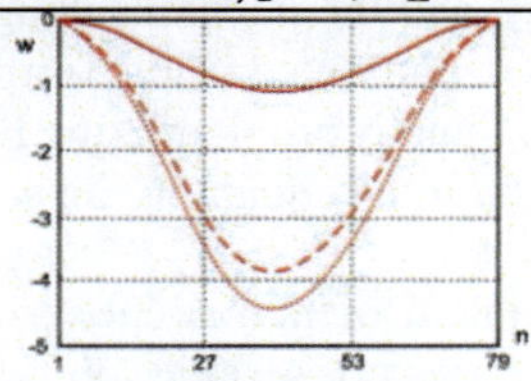

LE Spectrum:
$q^- = 220 \leftrightarrow -0.0155 - 0.2680 - 0.268 - 0.8603$
$q^+ = 240 \leftrightarrow 0.0490 - 0.0033 - 0.3499 - 0.3499$</td></tr>
<tr><td>5th variant $\gamma_2 = 0, P_E = 0.5$</td><td>6th variant $\gamma_2 = 0, P_E = 2$</td></tr>
<tr><td>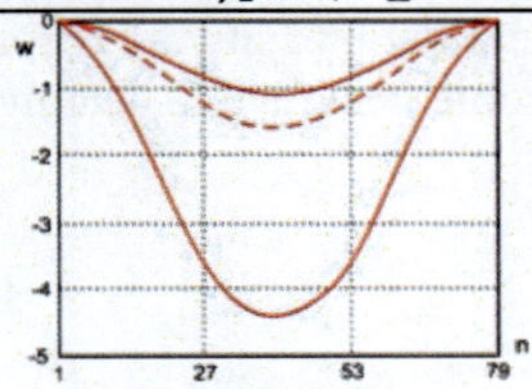

LE Spectrum:
$q^- = 140 \leftrightarrow 0.0007 - 0.5851 - 0.5851 - 3.0574$
$q^+ = 146 \leftrightarrow 0.0015 - 0.2092 - 0.2092 - 2.8491$</td><td>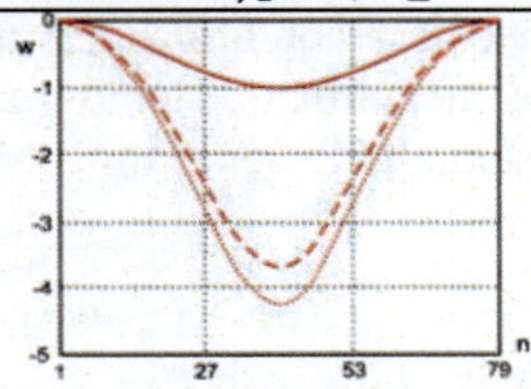

LE Spectrum:
$q^- = 370 \leftrightarrow 0.0695 - 0.0026 - 0.4382 - 0.4382$
$q^+ = 550 \leftrightarrow 0.0097 - 0.6236 - 0.6236 - 0.6613$</td></tr>
<tr><td>7th variant $\gamma_2 = 0, P_E = 1, E = 2E_0$</td><td>8th variant $\gamma_2 = 0.3, P_E = 1, E = 2E_0$</td></tr>
<tr><td>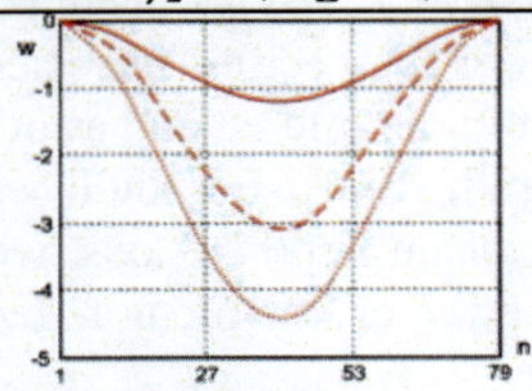

LE Spectrum:
$q^- = 450 \leftrightarrow -0.0181 - 0.0318 - 1.3612 - 1.3613$
$q^+ = 470 \leftrightarrow 0.1583 - 0.0426 - 0.1467 - 2.5499$</td><td>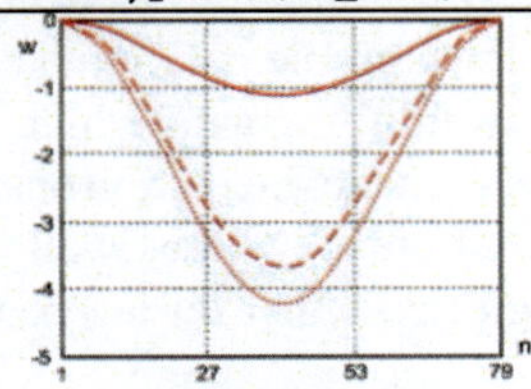

LE Spectrum:
$q^- = 540 \leftrightarrow -0.0608 - 0.8089 - 0.8134 - 1.5303$
$q^+ = 700 \leftrightarrow 0.12150.1217 - 0.00800 - 2.71332$</td></tr>
</table>

7.6 Stability of Curvilinear Euler-Bernoulli Beams in Temperature Fields

In this section, stability of thin flexible Bernoulli-Euler beams by taking into account a geometric nonlinearity as well as a type and intensity of the temperature field is investigated. The applied temperature field $T(x, z)$ is yielded by a solution to the 2D Laplace equation being solved for five kinds of the thermal boundary conditions, and there are not any restriction on the temperature distribution along the beam thickness. Action of the temperature field on the beam dynamics is studied with a help of Duhamel theory, whereas the motion of the beam subjected to the thermal load is yielded employing the variational principles [140].

The heat transfer Laplace equation is solved via the FDM of the third order with respect to its accuracy, and the integrals along beam thickness defining the thermal stress and moments are computed using Simpson's method. PDEs governing beam motion are reduced to Cauchy problem using FDM of the second-order accuracy. The obtained ordinary differential equations are solved by the fourth-order Runge-Kutta method.

Moreover, the problem of numerical results convergence versus a number of beam partitions is investigated. A static solutions for a flexible Bernoulli-Euler beam using the dynamic approach based on employment of the relaxation/set-up method is obtained.

Novel stability loss beam phenomena under the thermal field versus the beam geometric parameters, boundary conditions and the temperature intensity are reported. In particular, we have shown that a stability of the flexible beam while heating the face beam surface essentially depends on its thickness.

7.6.1 Introduction

Although the classical models of beams based on either Euler-Bernoulli or Timoshenko theories and their assumptions have been successively used for many years, there still are present open problems originated from both engineering and science. This section is aimed at fulfilling gaps which still exist in the offered and available results, putting emphasis on the interaction between vibrating Euler-Bernoulli beams and thermal fields of different types as well as on the beam imperfections, which usually cannot be neglected either during beam fabrication or in structural design requirements.

Euler-Bernoulli and Timoshenko beams employing the geometrically nonlinear theory and the thermally-induced post-buckling behaviour have been intensively studied over the recent years [141–145]. Li et al. [142] have employed the accurate geometrically nonlinear theory for Euler-Bernoulli beams, taking into account the longitudinal and transverse motions of uniformly heated beams with/without static thermal post-buckling deformations [24]. It has been shown that all frequencies of an unbuckled beam continuously decrease when the temperature rises.

Dinzart et al. [146] have applied cyclic bending to study the thermo-mechanical response of beams made of a thermoplastic polymer. The chosen beam material was dependent on temperature and frequency, while the inertia effects have been neglected. Stability of the steady-state solutions and the conditions for thermal run-away (the so-called thermo-mechanical instability) have been investigated.

Abbasi et al. [147] have found the analytical solution for a beam made of a functionally graded material, based on the first-order shear deformation theory with lateral thermal shock loads. It has been assumed that the material properties across the beam thickness are within the volume fraction of the constitutive materials and the solution obeys the coupled thermoelastic theory.

Beams made of FGM are widely employed in mechanical, aerospace, civil and nuclear engineering, and hence their adaptability to thermal loading and thermal shocks plays a crucial role in both engineering and science. Although the FGM beams were initially aimed at constructing thermal barriers in aerospace structures and fusion reactors, where one deals with the extremely high temperature and large thermal gradients, this research area includes many open problems still waiting to be solved.

Nonlinear mechanical properties of the FGM beams subjected to in-plane thermal loads have been analysed by Ma and Lee [148], whereas Gupta et al. [149] have studied the post-buckling behaviour of slender columns by means of employing the concept of coupled displacement field.

One of the research directions while studying vibrating beams in a temperature field is based on inclusion of the temperature dependence of the elastic moduli of materials (usually, the majority of metals exhibit a decrease in the elastic moduli when the temperature increases). Peng He et al. [150] have used an improved Timoshenko model to study beams of various tapered shapes under linear axial temperature distribution. The influence of the axial temperature difference on natural frequencies and modal shapes of the beams have been analysed. Static stability of a sandwich beam with both the viscoelastic core and supports subjected to an axial pulsating load and 1D temperature gradient has been investigated by Nayak et al. [151]. Hamilton's energy principle yielded the governing equations and boundary conditions, and then a set of Hill's equations have been derived. The authors have studied effects of shear parameter, geometric parameters and thermal gradient on the non-dimensional static buckling zones.

The classical Euler-Bernoulli and Timoshenko models have been recently employed while investigating the FGM beams, carbon nanotubes or micromechanical resonators taking into account thermal effects. Benzair et al. [152] have employed the nonlocal Timoshenko beam model and Euler beam model to study vibrations of single-walled carbon nanotubes (CNTs) with emphasis put on the thermal effect. The research has been aimed at the wave dispersion caused by the rotary inertia, the shear deformation and the nonlocal elasticity characterizing the microstructure of CNTs.

Free vibrations of statically thermal post-buckled FGM beams with surface-bounded piezoelectric layers subject to both temperature rise and voltage have been investigated by Li et al. [142]. Employing Euler-Bernoulli beam theory, geometrically nonlinear dynamic governing equations (PDEs) have been derived taking into account the thermo-electromechanical loadings, and then the problem has been reduced to two sets of coupled ODEs. Thermoelastic post-buckling equilibrium paths and characteristic curves of the first three natural frequencies versus the temperature, the electricity and the gradient parameters have been reported. In particular, it has been illustrated how the tensional force produced in the piezoelectric layers can increase the critical buckling temperature/natural frequency.

Kiani and Eslami [153] have investigated thermo-mechanical buckling of temperature-dependent FGM beams. They have used the Timoshenko model in which properties of the constituents depended on the temperature and thickness. Various types of boundary conditions have been investigated and closed forms of the solutions for the critical buckling temperature of the beams have been presented.

The model of gold nanobeam vibrations induced by laser pulse heating has been derived by Youssef et al. [154] in the context of two-temperature generalized thermoelasticity and non-Fourier heat conduction. It has been shown that an increase in the value of the two-temperature parameter yields a decrease in the values of the stress-strain energy, and the latter process has been damped.

Prabhakar and Vengallatore [155] have studied the effects of beam geometry, natural frequency, flexural mode shapes and structural boundary conditions on thermoelastic damping of single-crystal silicon microbeam resonators. A Green's function method has been applied to solve the 2D heat conductions equation, and a formula for thermoelastic damping has been derived in the form of infinite series.

Free vibration characteristic of an axially loaded FGM cantilever Euler-Bernoulli beam subjected to temperature rising have been studied by Akbas [156], where beam material properties have been assumed to be temperature-dependent and changed according to a low-power function.

Esfahani et al. [157] have investigated vibrations of an FGM beam under in-plane thermal loading in the pre-buckling and post-buckling regimes, taking into account material properties versus both position and temperature. It has been shown that, depending on the type of boundary conditions and loading, free vibrations of the beam under in-plane thermal loading may achieve zero at a certain temperature indicating existence of a bifurcation-type instability.

The research results reported in this chapter extend many earlier investigations on thermal field influence on the thermodynamics of structural members including Euler-Bernoulli beams [118, 119, 158–161].

7.6.2 Mathematical Model

The studied beam consists of a curvilinear body of the length l, height h and curvature $k_x = 1/R_x$ (see Fig. 7.15).

The mathematical model of the beam is introduced based on the following assumptions:

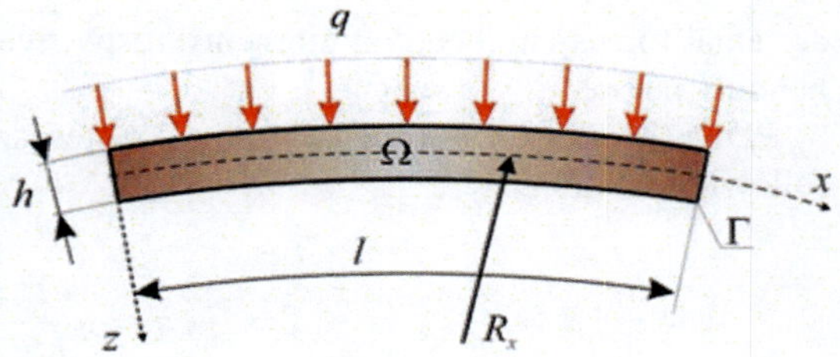

Fig. 7.15 The investigated beam [reprinted with permission from International the Journal of Nonlinear Mechanics publishers]

(i) Euler-Bernoulli hypothesis is taken [162];
(ii) nonlinear relation between deformations and displacements of von Kármán form is introduced (see [163]);
(iii) curvature condition is employed owing to Vlasov's theory [164];
(iv) elastic and isotropic beam satisfying Duhamel-Neuman principle is considered (see [165]);
(v) heat transfer coefficient does not depend on the temperature;
(vi) beam material properties do not depend on the temperature;
(vii) the temperature field distribution along the beam thickness can be taken in an arbitrary way.

Owing to the introduced hypotheses and assumptions, the mathematical model of the beam is governed by the following PDEs with respect to the beam displacements:

$$\frac{\partial N_x}{\partial x} - \frac{\gamma}{g} h \frac{\partial^2 u}{\partial t^2} = 0, \tag{7.86}$$

$$\frac{\partial^2 M_x}{\partial x^2} + k_x N_x + \frac{\partial}{\partial x}\left\{N_x \frac{\partial w}{\partial x}\right\} + q - \frac{\gamma}{g} h \frac{\partial^2 w}{\partial t^2} - \varepsilon \frac{\gamma}{g} h \frac{\partial w}{\partial t} = 0, \tag{7.87}$$

where

$$N_x = \int_{-\frac{h}{2}}^{\frac{h}{2}} \sigma_x dz = \int_{-\frac{h}{2}}^{\frac{h}{2}} E\varepsilon_x dz - \int_{-\frac{h}{2}}^{\frac{h}{2}} E\alpha T dz = Eh\varepsilon_x - N_x^T =$$

$$= Eh\left(\frac{\partial u}{\partial x} - k_x w + \frac{1}{2}\left(\frac{\partial w}{\partial x}\right)^2\right) - N_x^T,$$

and

$$M_x = \int_{-\frac{h}{2}}^{\frac{h}{2}} \sigma_x z dz = \int_{-\frac{h}{2}}^{\frac{h}{2}} E\left(-z\frac{\partial^2 w}{\partial x^2}\right) z dz - \int_{-\frac{h}{2}}^{\frac{h}{2}} E\alpha T z dz = -\frac{Eh^3}{12}\frac{\partial^2 w}{\partial x^2} - M_x^T$$

are axial forces and bending moments, respectively (T refers to thermal force and bending moment).

PDEs (7.86)–(7.87) are recast to the counterpart non-dimensional form using the following relations

$$\bar{w} = \frac{w}{h}, \bar{u} = \frac{ul}{h^2}, \bar{x} = \frac{x}{l}, \bar{z} = \frac{z}{h}, \bar{q} = q\frac{l^4}{h^4 E}, c = \sqrt{\frac{Eg}{\gamma}}, \bar{\varepsilon} = \frac{\varepsilon l}{c},$$
$$\bar{t} = \frac{tc}{l}, \overline{N_x^T} = \frac{N_x^T l^2}{Eh^3}, \overline{M_x^T} = \frac{M_x^T}{Eh^2}, \overline{k_x} = \frac{k_x l^2}{h}, \overline{T} = \alpha T, \tag{7.88}$$

where $w\,(x, t)$—beam normal deflection in the normal direction; $u\,(x, t)$—beam element displacement in the longitudinal direction; ε—damping coefficient; γ—beam weight per unit volume; g—Earth acceleration; E—Young's modulus; t—time; q—external conditions load; N_x^T and M_x^T—thermal force and torque, respectively.

In addition, one of the boundary conditions are as follows

Rigid clamping:

$$\begin{aligned} w\,(0, t) = u\,(0, t) = w'_x\,(0, t) = 0 \quad &\text{for } x = 0; \\ w\,(1, t) = u\,(1, t) = w'_x\,(1, t) = 0 \quad &\text{for } x = 0; \end{aligned} \tag{7.89}$$

Simple fixed support:

$$\begin{aligned} w\,(0, t) = u\,(0, t) = M_x\,(0, t) = 0 \quad &\text{for } x = 0; \\ w\,(1, t) = u\,(1, t) = M_x\,(1, t) = 0 \quad &\text{for } x = 0; \end{aligned} \tag{7.90}$$

Simple fixed support:

$$\begin{aligned} w\,(0, t) = u\,(0, t) = M_x\,(0, t) = 0 \quad &\text{for } x = 0; \\ w\,(1, t) = u\,(1, t) = w'_x\,(1, t) = 0 \quad &\text{for } x = 0; \end{aligned} \tag{7.91}$$

The employed initial conditions are

$$w\,(x, 0) = f_1\,(x)\,; \quad \bar{w}\,(x, 0) = f_2\,(x)\,; \quad u\,(x, 0) = f_3\,(x)\,; \quad \bar{u}\,(x, 0) = f_4\,(x)\,. \tag{7.92}$$

Observe that in the case of lack of the temperature, Eqs. (7.86), (7.87) overlap with Volmir's 1D model [166]. The thermal stresses M_x^T and N_x^T are governed by the following equations:

$$N_x^T = \int_{-1/2}^{1/2} T\,(x, z)\,dz; \quad M_x^T = \int_{-1/2}^{1/2} T\,(x, z)\,zdz. \tag{7.93}$$

Bolotin [167] has proposed employing the temperature field $T(x, z)$, assumed to be linearly distributed along the body thickness, using the Lagrange polynomial and taking into account a linear approximation. However, in the latter approach, limitations regarding the temperature distribution along the beam thickness should be additionally proposed, while the temperature field has been defined by the following PDE:

$$\nabla^2 T(x, z) = \frac{\partial^2 T(x, z)}{\partial x^2} + \lambda^2 \frac{\partial^2 T(x, z)}{\partial z^2} = 0. \tag{7.94}$$

In the case of curvilinear beams made of isotropic materials, the stationary transfer PDE (7.94) takes the following form:

$$\frac{\partial^2 T}{\partial x^2} + \lambda^2 \frac{\partial^2 T}{\partial z^2} + 2k_x \frac{\partial T}{\partial x} = \frac{-W_0}{\lambda}, \tag{7.95}$$

where λ stands for a heat transfer coefficient. Since, in our further investigation, we do not consider an internal heat source, we take $W_0 = 0$ yielding

$$\frac{\partial^2 T}{\partial x^2} + \lambda^2 \frac{\partial^2 T}{\partial z^2} + 2k_x \frac{\partial T(x, z)}{\partial x} = 0. \tag{7.96}$$

In what follows, we consider either the first-kind

$$T(x, z)|_\Gamma = g_1(x, z), \tag{7.97}$$

or the second-kind

$$\left.\frac{\partial T(x, z)}{\partial n}\right|_\Gamma = g_2(x, z) \tag{7.98}$$

heat boundary conditions, where $\frac{\partial}{\partial n}$ states for differentiation along an external normal to the beam boundary Γ.

7.6.3 Numerical Solution

Owing to the earlier experience and method advantages described in Ref. [168], we have employed the FDM to solve PDEs (7.86), (7.87). The beam space is meshed using i nodes. Partial derivatives are substituted by central finite-difference approximations, and the problem is reduced to study the following system of nonlinear ODEs:

$$\begin{cases} \ddot{u} = \Lambda_{x^2}(u)_i - \Lambda_x(w)_i\Lambda_{x^2}(w)_i - \Lambda_x(N_T)_i \\ \ddot{w} + \varepsilon w = \dfrac{1}{\lambda^2}\left\{-\dfrac{1}{12}\Lambda_{x^4}(w)_i + k_x\left[\Lambda_x(u)_i - k_x w_i - \dfrac{1}{2}(\Lambda_x(w)_i)^2 - \right.\right. \\ \left.\left. -w_i\Lambda_{x^2}(w)_i\right] + \Lambda_{x^2}(u)_i\Lambda_x(w)_i + \Lambda_x(u)_i\Lambda_{x^2}(w)_i + \dfrac{3}{2}(\Lambda_x(w)_i)^2\Lambda_{x^2}(w)_i\right\} - \\ -\Lambda_{x^2}(M_T)_i - \Lambda_x(N_T)_i\Lambda_{x^2}(w)_i + q, \end{cases} \tag{7.99}$$

where $\Lambda_x, \Lambda_{x^2}, \Lambda_{x^4}$ stand for difference operators of the first, second and third orders, respectively.

The so far described finite difference approximations are also applied to boundary conditions (7.89)–(7.91) and initial conditions (7.92). Then, as usually, the task is reduced to Cauchy problem regarding an evolutionary variable, and is finally solved by the fourth-order Runge-Kutta method. The results have been validated through application of the sixth-order Runge-Kutta method. Since the latter one requires essentially longer computational time, the fourth-order Runge-Kutta method has been employed in all our computations (see [168] for the motivation of our choice).

In order to study stability of flexible Euler-Bernoulli beams, we employ the method originally proposed by Tikhonov [124], and then successfully employed by Feodos'ev for the shell problems (see [138, 158]). The formulated dynamical problems regarding stability have been solved using the relaxation/set-up method.

Though numerous approximate methods can be employed, an important role play iterational methods, which allow to obtain the required solutions with a high accuracy. If the iterational process can be treated as the result of a steady state of a certain process, then the iterational methods can be treated as the continuation methods of solutions with respect to a parameter.

Now, giving a physical meaning to the parameter (here beam deflection), we get a meaning of the steady-state process in time, which is analysed by the relaxation/set-up method. The so far described idea is illustrated in the attached Fig. 7.16.

Its right-hand side presents a steady state of the beam centre deflection $w(0.5)$ versus time for the load $q = 135$, for the boundary conditions (7.90), and initial conditions (7) for $f_i(x) = 0$; we take also $k_x = 24$ and the temperature $T = 50$. Its left-hand side reports the beam deflection versus q, which has been obtained through the employed successive solution to the equations of beam motion via the relaxation method by increasing the load parameter q.

It should be emphasized that in our case we analyse a dynamic dissipative process. Now, increasing the dissipation factor (damping) ε, the studied dynamical process is more fast damped and it is set-up on the corresponding value of the static load. Observe that a transition from a stable state into unstable state for $\varepsilon = 0$ appears earlier than for $\varepsilon = 0.05$, and then the solution tends to its steady/static state.

The so far introduced description explains how we have studied the problem of stability though in static context but using the dynamical approach.

This method has been widely used earlier. In particular, let us refer to the monograph [158], where numerous theorems with their proofs have been reported, as well

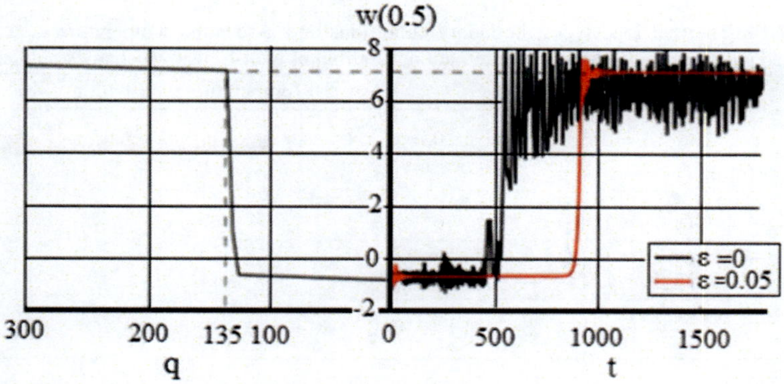

Fig. 7.16 Deflection of the beam centre versus time t and load q [reprinted with permission from International the Journal of Non-linear Mechanics publishers]

as to the paper [125], where problems of different kinds including those governed by linear/nonlinear sets of algebraic/differential equations have been studied.

Computations of M_x^T and N_x^T defined by formulas (7.93) is carried out with the help of Simpson method. The heat transfer equations are solved with the FDM using a standard approach.

7.6.3.1 Reliability and Convergence of the Results

The results reliability is verified based on Runge principle. Namely, the FDM of the second order is employed while solving PDEs (7.86), (7.87) and the space is meshed into 4×10, 6×20, 8×40, 10×80, $\times 12 - 120$ mesh cells regarding coordinates z and x, respectively. It occurred that the most optimal case is that of 10×80 mesh. In addition, numerical results associated with solving the heat transfer equations has been validated by means of the analytical solution reported by Carslou and Jeger [169] for the boundary condition given in Table 7.14 (Type 1).

7.6.3.2 Numerical Experiment

It should be emphasized that our developed algorithms and programs allow to study static stability of flexible Euler-Bernoulli beams in a temperature field for the following parameters: relative beam thickness, beam curvature, boundary conditions (7.89), (7.91) and thermal conditions types given in Table 7.14 (Type 1–5).

The 2D beam space given in the rectangular coordinates is defined as follows: $\Omega = \{(x, z) \in [0, l] \times [-h/2, h/2]\}$. We have employed the explicit numerical scheme of the fourth-order Runge-Kutta method, whereas the associated theorem regarding

Table 7.14 Thermal load types, boundary conditions and associated temperature distributions [reprinted with permission from the International Journal of Non-Linear Mechanics publishers]

Type no.	Boundary condition			Temperature distribution
Type 1	$T(x, z) = T$	$z = -1/2$	$0 < x < 1$	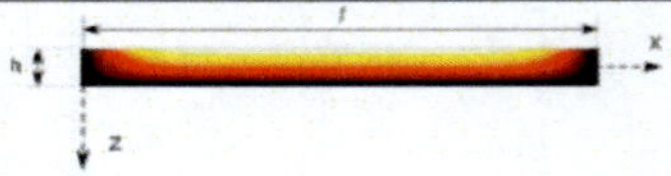
	$T(x, z) = 0$	$z = 1/2$	$0 < x < 1$	
	$T(x, z) = 0$	$x = 1$	$-1/2 < z < 1/2$	
	$T(x, z) = 0$	$x = 0$	$-1/2 < z < 1/2$	
Type 2	$T(x, z) = T$	$z = -1/2$	$0 < x < 1$	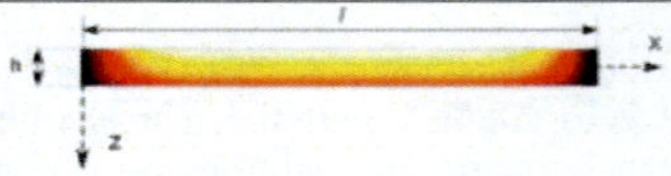
	$\frac{\partial T}{\partial z} = const$	$z = 1/2$	$0 < x < 1$	
	$T(x, z) = 0$	$x = 1$	$-1/2 < z < 1/2$	
	$T(x, z) = 0$	$x = 0$	$-1/2 < z < 1/2$	
Type 3	$T(x, z) = T$	$z = -1/2$	$0 < x < 0.5$	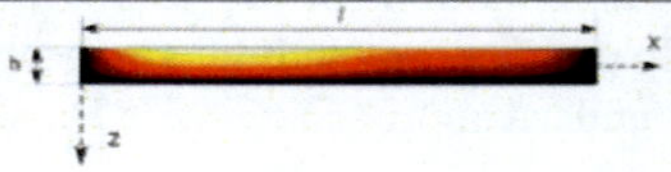
	$\frac{\partial T}{\partial z} = const$	$z = -1/2$	$0.5 < x < 1$	
	$T(x, z) = 0$	$x = 1$	$-1/2 < z < 1/2$	
	$T(x, z) = 0$	$x = 0$	$-1/2 < z < 1/2$	
Type 4	$T(x, z) = T$	$z = -1/2$	$0 < x < 1$	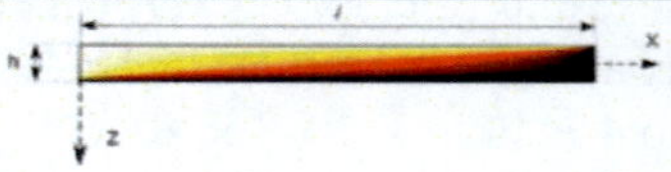
	$T(x, z) = 0$	$z = 1/2$	$0 < x < 1$	
	$T(x, z) = 0$	$x = 1$	$-1/2 < z < 1/2$	
	$T(x, z) = T$	$x = 0$	$-1/2 < z < 1/2$	
Type 5	$T(x, z) = 0$	$z = -1/2$	$0 < x < 1$	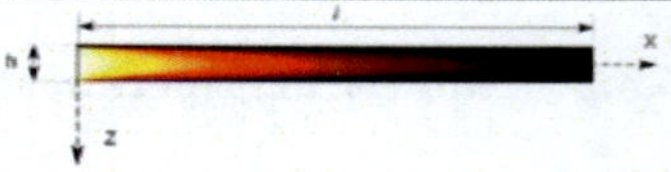
	$T(x, z) = 0$	$z = 1/2$	$0 < x < 1$	
	$T(x, z) = 0$	$x = 1$	$-1/2 < z < 1/2$	
	$T(x, z) = T$	$x = 0$	$-1/2 < z < 1/2$	

stability of that method have been proved in numerous works devoted to numerical analysis. The determination of a suitable step regarding the evolutionary coordinate as well as examples of numerical experiments with respects to the results validity are reported in Ref. [168]. Furthermore, estimation of the employed method accuracy has been carried out with a help of the Runge method. Namely, two different computations have been carried out with respect to time step $\Delta t = 1/256$, and with respect to twicely decreased step $\Delta t = 1/512$. As the accuracy criterion, we have used a small relative change of the obtained solution with respect to time step decrease.

Combinations of the boundary conditions (7.97)–(7.98) and the beam boundary Γ allow to study different cases of the temperature loading. In this work, the temperature field is employed through five methods reported in Table 7.14 together with the graphical temperature field interpretation. The temperature intensity has been changed in the interval $0 \leq T \leq 100\,^{\circ}\mathrm{C}$. Observe that here and further the temperature T is taken in the non-dimensional form for steel, and for the studied temperature interval, the physical beam material characteristics does not change. In order to achieve dimensional values of the temperature, one may use the following fixed data: $\alpha = 12.5 \times 10^{-6}$ and $T = \Delta T + T_0$, where $T_0 = 22\,^{\circ}\mathrm{C}$.

We have focused on the following investigations:

1. Influence of *small beam curvature* and *small loading* understood as *initial imperfections* influencing the beam stability loss;
2. Analysis of the temperature field type depending on $\Delta T\ (w; 0.5)$;
3. Influence of temperature *field type* on the stationary loss of stability of curvilinear beams.
4. Analysis of the temperature *field intensity* influence on the loss of stability of curvilinear beams.

7.6.4 *Influence of Imperfections*

It is assumed that the beam with boundary conditions (7.91) and initial conditions (7.92) is in a temperature field of type 1 (Table 7.14) and has small curvature parameter $k_x = 0.001$. Investigation of the dependence $\Delta T\ (w; 0.5)$ for the beam with $k_x = 0$ and $k_x = 0.001$ is shown in Fig. 7.17 meaning that an action of small imperfections in the form of the introduced curvature yields a change in the beam deflection sign. In the case of a straight beam $k_x = 0$, stability loss occurs for $\Delta T = 16\,^{\circ}\mathrm{C}$, while the deflection increases and moves in the positive side. In the case of $k_x = 0.001$, stability loss occurs for $\Delta T = 12\,^{\circ}\mathrm{C}$ and the beam deflection is located on the negative side. Analogous results have been obtained for the beam with external load imperfection.

Namely, for the load $q = 0.1$ and $q = -0.1$, loss of stability occurs for $\Delta T = 6\,^{\circ}\mathrm{C}$ with an associated change in the deflection into positive and negative side, respectively. These investigations are within the physical expectations, which validates the appropriate choice of both beam and the the temperature field models.

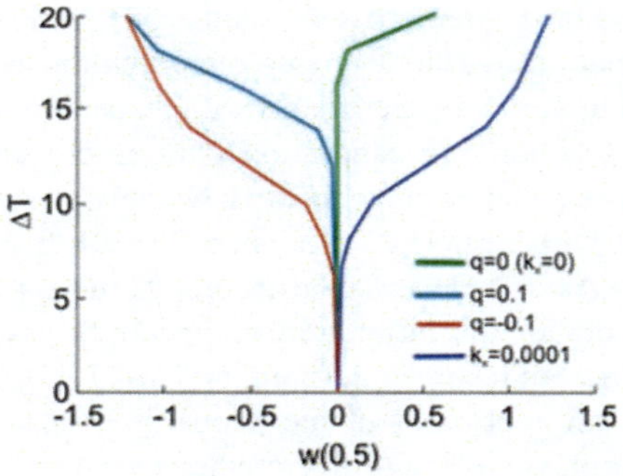

Fig. 7.17 Beam stability loss for different k_x and q [reprinted with permission from the International Journal of Non-linear Mechanics publishers]

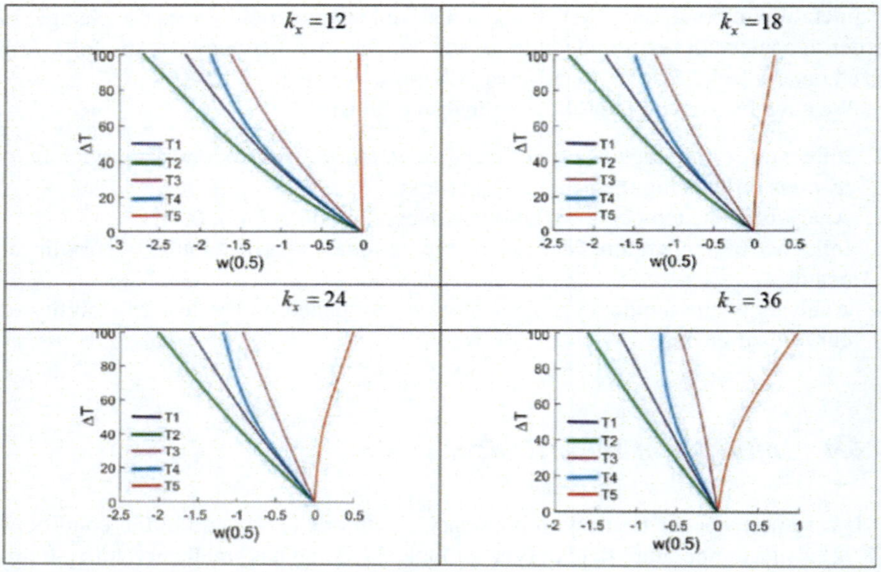

Fig. 7.18 Dependencies $\Delta T\ (w\ (0.5))$ for different k_x [reprinted with permission from the International Journal of Non-linear Mechanics publishers]

7.6.4.1 Temperature Field Influence on $\Delta T\ (w;\ 0.5)$

In this section, we consider the influence of the temperature field type on the deflection of the central beam point versus the intensity of the thermal load for various values of the parameter k_x (Fig. 7.18). The following parameters/functions are fixed: $\lambda = 50$, $\varepsilon = 1$, boundary conditions (7.91), initial condition $f_i\ (x) = 0$ for $i = 1, \ldots, 4$ and we take $q = 0$.

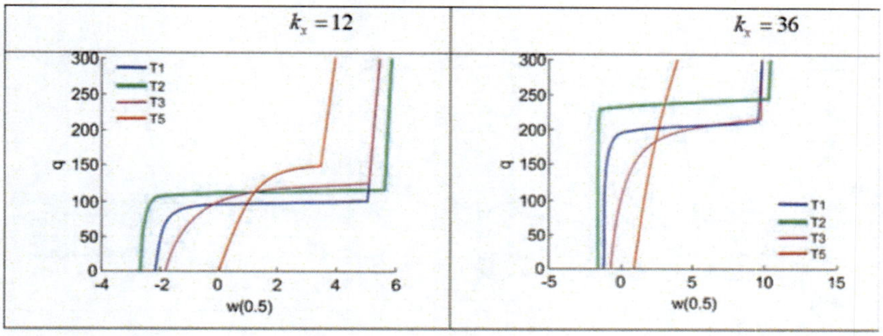

Fig. 7.19 Beam deflection versus external load for different temperature field types T_i and fixed k_x [reprinted with permission from the International Journal of Non-linear Mechanics publishers]

Results presented in Fig. 7.18 show that a change in the temperature field type has an essential influence on the beam deflection for all investigated curvature values. The influence of the temperature field type 5 (Table 7.14) yields an essential increase in the curvature and thermal load intensity.

7.6.4.2 Type of the Temperature Field Versus Stationary Stability Loss

As it has been already mentioned, the type of the temperature field for a fixed thermal load $\Delta T = 100\,°\mathrm{C}$ yields different beam deflections. In order to study the influence of mixed boundary conditions of the simple support-clamping type, we analyse the change in the deflections of a beam in a thermal field and subjected to external load q.

It has been shown through functions $q(w)$ in Fig. 7.19 that an increase in the curvature parameter k_x implies an increase in the critical load, and a rapid increase in the beam deflection occurs. On the contrary, in the case of employing the temperature field of type 5 (Table 7.14) for $k_x = 36$, the stability loss in the investigated external load interval $q \in [0; 300]$ has not been detected.

7.6.4.3 Influence of Temperature Field Intensity

In order to investigate the influence of the temperature field intensity on loss of stability of a curvilinear beam, we study a beam with mixed boundary conditions and subjected to external load. The beam is in the temperature field of type 1 (Table 7.14) with the intensity $T = 0$, $\Delta T = 40\,°\mathrm{C}$, $\Delta T = 80\,°\mathrm{C}$ and $\Delta T = 100\,°\mathrm{C}$.

The reported dependencies $q\ (w)$, shown in Fig. 7.20, imply that an increase in the thermal load intensity yields an increase in the critical load q_{cr} exhibited by a sudden

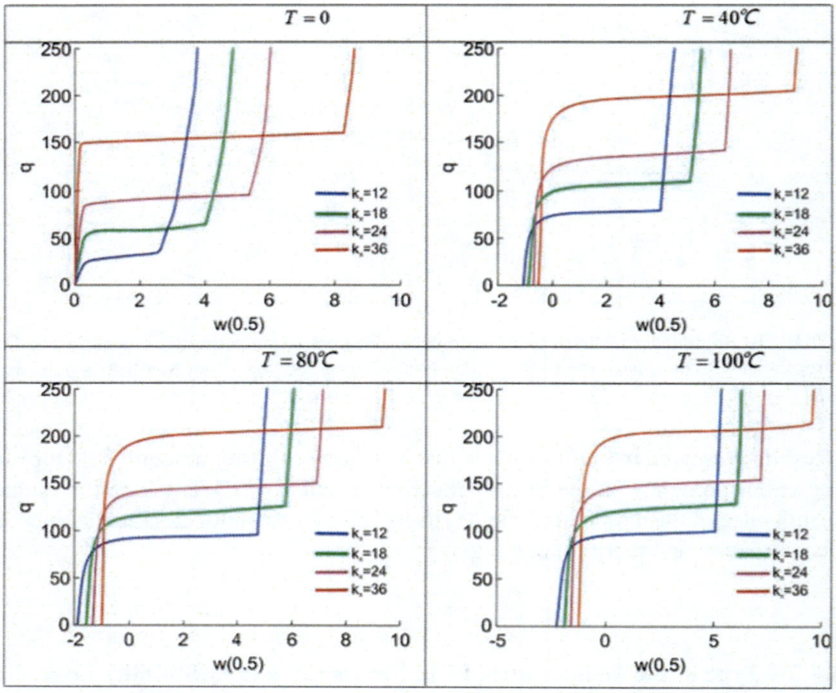

Fig. 7.20 Beam deflection versus external load for different k_x and fixed temperature T [reprinted with permission from the International Journal of Non-linear Mechanics publishers]

beam deflection. Observe that the type of the temperature load and its intensity have an influence on the initial beam deflection for $q = 0$, which yields an increase in the beam deflection while losing stability.

7.6.4.4 Influence of the Beam Curvature

In Fig. 7.21, dependencies $\Delta T\ (w)$ for the boundary conditions (7.89) and for different types of temperature fields (Table 7.14) are reported. The used notations "−" and "+" refer to the dependencies construction for the heat transfer equation with and without curvature parameter k_x, respectively. It has been found that occurrence of the beam curvature in the heat transfer equations implies a change in the beam deflection for the same values of the temperature load.

The most remarkable influence of the beam curvature is given by the type 5 temperature field (Table 7.14), where also an essential difference comparing to the case $k_x = 36$ has been detected. In the case of temperature fields of types 1, 2 and

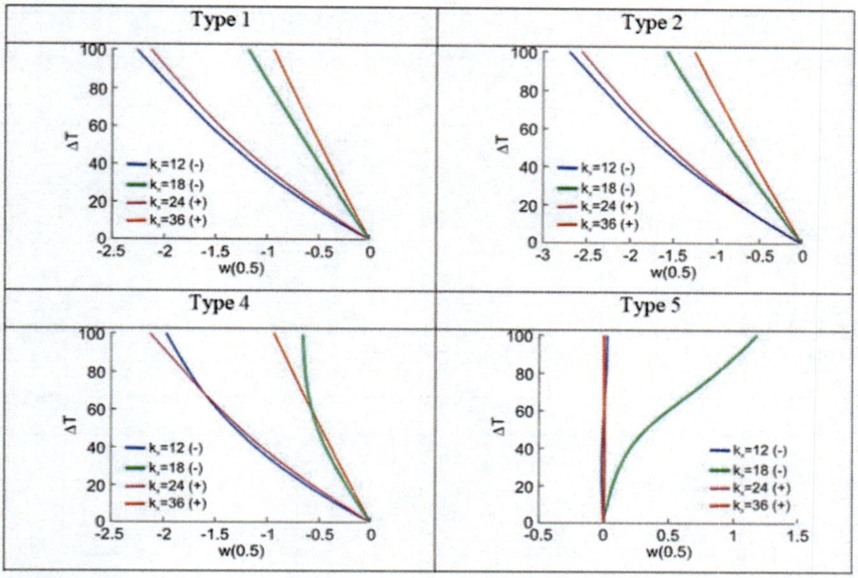

Fig. 7.21 Temperature load intensity versus $w(0.5)$ and heat boundary conditions type for different curvature parameter k_x [reprinted with permission from the International Journal of Non-linear Mechanics publishers]

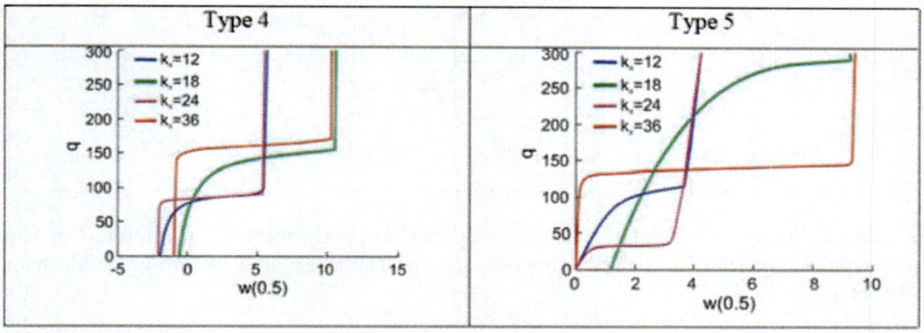

Fig. 7.22 Dependencies q versus $w(0.5)$ for different k_x [reprinted with permission from the International Journal of Non-linear Mechanics publishers]

4, we have found that with an increase in the curvature parameter k_x, the maximum beam deflection w decreases.

In what follows, we study the dependencies q (w) reported in Fig. 7.22 for different types of the temperature field (types 4, 5 in Table 7.14) and the corresponding boundary conditions of the simple-simple support for the intensity of the temperature load $\Delta T = 100\,^{\circ}\mathrm{C}$.

It has been found that for the type 4 temperature field, for $k_x = 12, \ldots, 36$, the curvature parameter implies an increase in the critical load q_{cr}. On the contrary, for

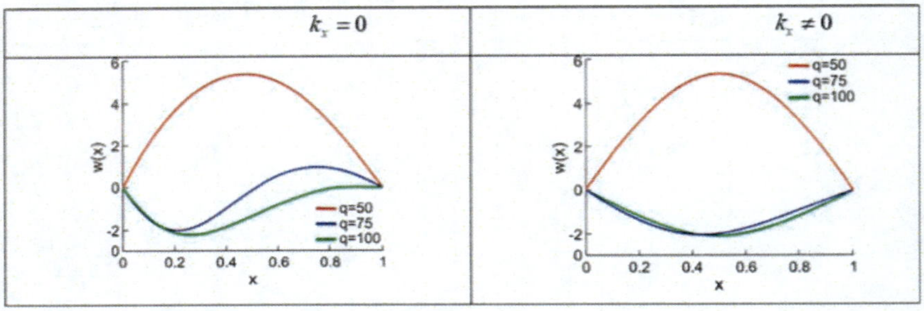

Fig. 7.23 Graphs of $w(x)$ for different q ($k_x = 0,\ k_x \neq 0$) [reprinted with permission from the International Journal of Non-linear Mechanics publishers]

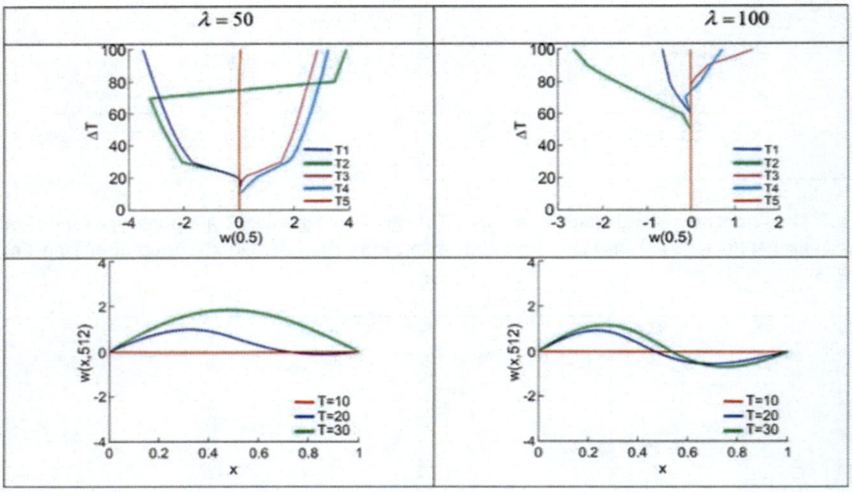

Fig. 7.24 ΔT versus $w(0.5)$ for different λ (**a**) and graphs of the beam form for different T and time instant $t = 512$ [reprinted with permission from the International Journal of Non-linear Mechanics publishers]

the beam associated with the type 5 temperature field (Table 6.14), an increase in k_x yields a decrease in q_{cr}.

For the type 4 temperature field (Table 7.14) and boundary conditions (7.90), the graphs of the beam deflection for $k_x = 12$ and $\Delta T = 100\,^{\circ}\mathrm{C}$ are reported in Fig. 7.23, and for the case of pre-stability $q = 50$ and post-stability ($q = 100$) loss.

It has been detected that an increase in the external load q implies a change in the beam form. If we consider a straight beam $k_x = 0$, the $w(x)$ graph exhibits a shape of a half-wave after stability loss ($q = 100$) and one wave before stability loss ($q = 50.75$). For a curvilinear beam, in both cases the graphs exhibit occurrence of half-wave shape.

Figure 7.24 reports the influence of the temperature field intensity $\Delta T\ (w)$ on the beam stability loss for different geometric parameters $\lambda = 50$ and $\lambda = 100$.

In the case of the types 1–5 of the temperature field (see Table 7.14), for $q = 0$ and $k_x = 0$, an increase in the temperature action implies a rapid change in the beam deflections. An increase in the parameter λ yields an increase in the critical intensity of the temperature input for which loss of stability occurs. For the type 4 of the temperature field, one may conclude based on the reported results that the intensity of the temperature action as well as the external load imply change in the beam form.

7.7 Mathematical Model of Three-Layer Micro- and Nano-Beams Based on the Hypotheses of the Grigolyuk-Chulkov and the Modified Couple Stress Theory

The mathematical model of three-layered beams developed based on the hypothesis of the Grigolyuk-Chulkov and the modified couple stress theory and the size-dependent equations governing the layers motions on the micro- and nanoscales is constructed. Hamilton's principle yields the novel equations of motion as well as the boundary/initial conditions regarding beams displacement. The latter ones clearly exhibit the size dependent dynamics of the studied micro- and nanobeams, and the introduced theory overlaps with the classical beam equations for large enough layer thickness. In particular, a three-layer beam with the microlayer thickness has been investigated with respect to the classical theory of Grigolyuk-Chulkov. The derived boundary problem is of sixth order and can be solved analytically in the case of statics. The carried out numerical experiments allowed to detect and explain size-dependent effects exhibited by the microbeams. The beam deflections and stress yielded by the employed couple stress model are less than those predicted by the classical Grigolyuk-Chulkov theory, while the estimated eigenfrequencies are higher, respectively. It has been shown that the proposed model can be reduced to the classical three-layer Grigolyuk-Chulkov beam through increase of the layers thickness, which validates our approach [170].

7.7.1 Introduction

It is well known that the three-layer structures have found applications in the mechanical and civil engineering, aviation, ship design, as well as in the airplanes and cosmic industries. In recent years, they are employed in the design of numerous micro- and nano-devices, including the micro- and nano-sensors as well as the electromagnetic sensors.

A general theory of static/dynamic behaviour of the three-layer structural construction with respect to their size dependent behaviour subjected to various external loading has been proposed by Grigolyuk and Chulkov [76].

Remarkably, three-layer constructions do not only imply the structural nonhomogeneity regarding the beam thickness, but they require inclusions of the middle layer due to occurrence of the transversal shear and compression, as well as coupling effects of the layers should be taken into account. Grigolyuk and Chulkov [76] introduced a hypothesis of the linear distribution of the tangent displacements with respect to both weight of the beam package and condition of the lack of compression in the structural package while constructing their theory of three-layered structures. On contrary to Bernoulli-Euler hypothesis, in Grigolyuk-Chulkov model a perpendicular line to the initial surface becomes not perpendicular to the deformed surface, since it rotates on amount of a certain angle due to transversal shear generated by the middle layer. On the other hand, the external layers fit with Bernoulli-Euler hypotheses, whereas the internal layer follows Timoshenko hypothesis. It should be emphasized that the mentioned theory generalizes the classical beam theory. Grigolyuk-Chulkov, using the hypothesis of straight cross sections for the internal layer, constructed the equilibrium equations and studied the stability and vibrations of the carrying load by the three-layer beam. Namely, the external layers have been made from materials of infinite stiffness against shear and transversal compression, whereas the internal layer had the infinite stiffness against the transversal compression. The static and dynamic behaviour of multi-layer beams based on Bernoulli-Euler and Timoshenko hypotheses have been employed for the whole structural package in Refs. [43, 44, 102, 117, 133, 171–174]. For multi-layer anisotropic plates and shells, Andreev and Nemirovsky developed a general theory based on a broken line [175]. A regular and chaotic contact/no-contact nonlinear dynamics of the multi-layer structure composed of one plate and three Euler-Bernoulli beams coupled only by boundary condition have been studied in the work of Awrejcewicz et al. [176].

The functional nanomaterials used as plies and layers put on the surfaces of the rigid bodies essentially improve the exploitation characteristics of the industrial products. If the multi-layer beams with the plane thin external layers do not have the required loading ability, the latter can be increased/improved via application of the reinforced external layering, i.e. employing the layers/plies/films having large Young's moduli in the form of nano-layers and the micro-layers made from carbon.

On the basis of the standard computations regarding the material resistance, the thickness of the layers of a microbeam should achieve a tenth of microns in order to satisfy the industrial requirements. However, as the experimental investigations show, the mechanical properties of the micro- and nano-size elements depend on their sizes. This is why a novel name has been introduced emphasizing the size-dependent effect and characterizing the change of the properties of the structures composed of elements of the size of microns and nanometres. Different features of the size-dependent effects exhibited by the micro- and nano-elements are widely described in the existing literature, and among them, the gradient effects play a significant role.

The size-dependent behaviour of elastic elements have been observed experimentally while bending and turning of the microbeams [14, 16, 102, 173]. In Ref. [14], it has been reported that the torsion stiffness of a copper made wires increases simultaneously decreasing their diameter from 170 down to 12 microns (observe that the size decrease should imply zero stiffness of a wire). In Ref. [16], the increase in the bending stiffness of a nickel foil corresponding to its height yielded the decrease in its thickness from 50 to 12.5 microns. The micro- and nano-oriented investigations [44, 172] show that the stiffness of the pure and poly-crystal metallic materials can be doubled while decreasing the thickness from 10 to 1 microns. In order to estimate the material resistance and keep the optimal design of the mentioned products, the key role is played by the reliable and highly accurate analysis of the stress-strain states of the investigated micro- and nanomechanical objects.

Importance of the size-dependent behaviour being an inherent property of materials has been presented experimentally by McFarland and Colton [15] and Kong et al. [17]. It has been demonstrated that in micron/submicron-scale regions, i.e. when the characteristic diameter of thickness is close to the internal material length scale parameter, the classical continuum mechanical theories cannot be employed to study the beam static/dynamic behaviour.

In Refs. [173, 174], investigations regarding the explanation of the so far mentioned effects on a basis of the strain gradient theory for the problems of bending and torsions have been carried out.

More recently, Srinivasa and Reddy [177] proposed a systematic treatment of higher gradient theories in a nonlinear context. The size-dependent models of Bernoulli-Euler beam have been studied in Ref. [43, 176], whereas Timoshenko beam has been analysed in [178], and Reddy-Levinson beam has been studied in Refs. [42, 60] using the modified couple stress theory proposed in Ref. [179] (the latter mainly employs the only size-dependent material length).

The origin and development of the couple stress elasticity theory can be found in the seminal works of Koiter [37], Toupin [35], Mindlin and Tiersten [36] and Mindlin [180].

As it has been already mentioned, a modified couple stress theory has been developed by Yang et al. [44], where only one scale parameter of material length as well as symmetric couple stress tensor have been employed. This concept has been expanded in Ref. [178], where the variational formalism associated with this theory has been introduced.

Park and Gao [43] utilized the concepts of the modified couple stress theory (MCST) to develop a new model for the bending of a Bernoulli-Euler beam. It contains an internal material length scale parameter and in contrary to the classical Bernoulli-Euler beam model, it captures the size scale effects. Considering a cantilever beam, it has been shown, that the bending rigidity estimated through their model is larger than that yielded by the classical model.

The seminal work of Eringen [181] serves as a source to develop microstructure-dependent nonlocal theories of beams being based on Hamilton's principle and the nonlocal constitutive relations including Bernoulli-Euler, Timoshenko, Reddy/Levinson models [26, 44, 182–184]. Ma et al. [106] constructed a microstructure-dependent Timoshenko beam model based on modified couple stress bending and axial principle, where both bending and axial deformations, as well as Poisson effect, have been taken into account.

In Ref. [101], solving equations for the models of functionally graded Bernoulli-Euler, Timoshenko and Reddy-Levinson beams with respect to their thickness on the basis of the modified couple stress theory and taking into account the size-dependent state equations have been derived and analysed.

Asghari et al. [39] utilized the MCST to study nonlinear behaviour of Timoshenko hinged-hinged beam including the mid-plane stretching. In the case of static bending, the nonlinear size-dependent phenomena has been studied numerically, whereas a solution to the free vibrations problem has been solved analytically. The latter work has been extended by Ke and Wang [134] and Ke et al. [57] to study nonlinear beam vibrations with an emphasis put to the stability estimation while including the axial displacement in their study.

Reddy [60] extended theories related to nonlinear Euler-Bernoulli and Timoshenko beams taking into account through-thickness power-law vibration of a two-constituent material and moderate rotation of transverse normal through the von Kármán nonlinear strain. The proposed model, based on a modified couple stress theory, power-law variation of the material and von Kármán geometric nonlinearity uses only one material length scale parameter and captures the size effect in a functionally graded material.

Santos and Reddy [185] presented comparison among classical elasticity, nonlocal elasticity and modified couple stress theories for free vibration analysis of the Timoshenko beams, where the rotary inertia and nonlocal parameter have been taken into account. The convergence of the theories has been demonstrated taking into account an increase of the beam global dimension.

Reddy and Arbind [186] reformulated the classical beam theories, i.e. Bernoulli-Euler and Timoshenko, using a modified couple stress theory and employing thickness power-law variation of a functionally graded material. The algebraic relationships have been derived for the beam deflections, slopes and stress resultants. They have been validated through examples of straight beams with simply supported and clamped boundary conditions.

The nonlinear resonant dynamics of a microscale beam based on the modified couple stress theory has been analysed numerically in Ref. [137]. First, Hamilton's principle has been employed to derive a PDE governing motion using the modified couple stress theory, and then Galerkin technique has been utilized to obtain a set of coupled nonlinear ODEs. The effect of different system parameters on the resonant dynamics system response has been studied.

Kahrobaiyan et al. [187] proposed a new comprehensive Timoshenko beam element based on the modified couple stress theory. Then the mass and stiffness matrices have been computed using the energy approach and Hamilton's principle. In particular, the static deflection of a short microbeam and pull-in voltage of an electrostatically actuated microcantilever made of silicon are estimated using this new beam element.

A microstructure-dependent nonlinear third-order beam theory which accounts for through-thickness power-law variation of a two-constituent material has been developed by Arbind et al. [64] based on modified couple stress theory the influence of the material length has been investigated.

Thai and Vo [188] studied static bending, buckling and free vibration behaviours of size-dependent functionally graded sandwich microbeams based on the modified couple stress theory and Timoshenko beam theory. Two kinds of the sandwich beams have been analysed: functionally graded skins and homogeneous core and functionally graded core and homogeneous skins. It has been shown that inclusion of the size effect resulted in an increase in the beam stiffness.

In Ref. [189], a size-dependent formulation for Bernoulli-Euler beam based on the couple stress theory has been given. It has been shown that the natural frequencies obtained using the utilized couple stress model are higher than these predicted by the classical theory.

A new modified couple stress theory containing three material length scale parameters has been developed by Chen and Li [190] for anisotropic elasticity and microscale laminated Kirchhoff plate model. The principle of minimum total potential energy has been employed, and the curvature tensor has been taken as asymmetric, whereas the couple stress moment tensor has been used as symmetric. The carried out numerical simulation have validated the proposed Kirchhoff plate model, which captures the scale effect of the microstructures.

Mohammad-Abadi and Daneshmehr [191] carried out the vibration analysis of the composite laminated beams in order of micron based on the modified couple stress theory. In particular, Euler-Bernoulli, Timoshenko and Reddy beam models have been studied with respect to the differences in the estimation of shear deformation. The governing equations have been solved using three-boundary conditions and four types of lamination.

The transverse vibration of rotary tapered microbeam has been analysed by Shafiei et al. [192] using a modified couple stress theory and Euler-Bernoulli beam model. In particular, the effect of the small-scale parameter, beam length, rate of cross-sectional change, hub radius and non-dimensional angular velocity on the microbeam vibration process have been illustrated and discussed.

In Ref. [72], the size-dependent model of a three-layer beam employing the modified version of the couple stress theory and length parameter influence associated with all layers and its impact on the damping characteristics of the microbeam vibrations have been studied.

Rajneesh [193] solved the problem of thermoelastic beam using the modified couple stress theory. Both governing equations for the modified couple stress theory and heat conduction equation for coupled thermoelasticity have been employed to study the vibrations in a homogenous isotropic thin beam by applying Euler-Bernoulli theory. The lateral deflection, thermal moment, axial stress average due to normal heat flux have been derived and studied numerically.

In Ref. [194], shear deformable functionally graded nanobeams in post-buckling based on modified couple stress theory have been studied. The governing equations and boundary conditions are yielded by the principle of minimum potential energy. Exact and generalized differential quadrature solutions for the static post-buckling response of the functionally graded nanobeams under different boundary conditions have been derived. Effects of length scale parameter, material gradient, length-to-thickness ratio and Poisson's ratios have been illustrated and analysed, among others.

In all mentioned works, on contrary to Grigolyuk-Chulkov hypothesis, only one hypothesis, i.e. either the Bernoulli-Euler or Timoshenko assumptions (for all layers) have been employed. It does not allow to take into account a large difference between the thickness of the layers (e.g. when the external layers are made from the thin emulsion membranes).

We have employed the modified couple stress theory in order to take into account the size-dependent effects in the three-layer beam with both stiffening and softening external layers. This choice is motivated by an observation that in all works using the mentioned theory, the stiffening effect is exhibited to explain the size-dependent effects (see, for instance, the Refs. [14, 16, 44, 172]). Owing to the experimental results published in the work [195], where, in Table 1, the size dependency of micro/nanostructures is reported for different materials, the majority of the analysed materials show stiffening effects.

In this section, we study the beam having the external layer made from copper (Cu) and the internal layer made from epoxide tar. Owing to the mentioned Table 1, those materials exhibit stiffening effects. On the other hand, a comparison of the couple stress theory and nonlocal Eringen's theory has been carried out by Tsiatas and Yiotis [196], where the effects of the beam softening based on nonlocal Eringen's theory and the beam stiffening due to the modified couple stress theory have been reported. In Ref. [185], it is pointed out while investigating Timoshenko beams dynamics that the use of nonlocal Eringen's theory implies a decrease in the frequencies of free beam vibrations, which stands in contrast to the results obtained using the modified couple stress theory.

Relevance of the gradient and nonlocal (integral) elastic models to include the size-dependent effects is widely illustrated and discussed, for instance, in Ref. [197]. The gradient models are considered as weak nonlocal models. Furthermore, in many cases, the nonlocal models may yield a paradox, since the obtained solutions based on the classical theory are the same as in the case of the nonlocal theory, i.e. there is a lack of the size-dependent effect. The mentioned paradox does not appear while using the modified couple stress theory.

Here, we propose the theory of three-layer beams based on the hypotheses of Grigolyuk and Chulkov [76] as well as the modified couple stress theory and the derived size-dependent equations of motion for the micro- and nano-order thickness of the layers. Hamilton principle yields new equations of motion as well as the boundary and initial conditions regarding displacements for the microbeams. The obtained equations allow to explain the size-dependent behaviour of a microbeam and they coincide with the classical equations if the layer thickness becomes large enough. A numerical example of computation of the three-layer beam with micro-level thickness of the layers has been given, and a comparison of its behaviour versus the classical Grigolyuk-Chulkov theory has been conducted.

7.7.2 Theory of Bending Including Shear Effects and the Modified Couple Stress Theory

In the process of the beam deformation the transversal cross sections of the middle layer being perpendicular to the beam axis, in contrary to Bernoulli-Euler hypothesis, rotate as a rigid body on amount of angle ψ. Here, on contrary to the hypothesis of the plane cross sections, we do not require cross sections to be perpendicular to the bended beam axis while deformation process, but in general, this behaviour is not excluded. This stands for a general hypothesis, which is not contradicting Bernoulli-Euler one. Namely, it tends to the latter one if the middle layer stiffness against shear is of infinite magnitude. The external layers material is assumed to be linearly elastic, and they obey Hooke's law. The homogenous internal layer also obeys Hooke's law.

The beam is analysed using the rectangular coordinates Oxz (see Fig. 7.25). The axis x overlaps with the middle beam layer line, whereas the axis z goes in the direction opposite to the Earth gravity. The carrying out load layer located in the positive z axis direction is called the first layer and the middle layer is called the third layer.

Let h_k $(k = 1, 2, 3)$ denote the layer thickness $(h_3 = 2c)$; $h = h_1 + h_2 + h_3$ stands for the thickness and b is the width of a beam wall, respectively; E_k is the elasticity material modulus, whereas ν_k denotes Poisson's coefficient of a layer with number k; G_3 is the modulus regarding the transversal shear of the middle beam, l_k stands for the internal material length scale parameter. In order to keep a compact formulation of the problem, we introduce the following averaged elasticity modulus:

$$E = (E_1 h_1 + E_2 h_2 + E_3 h_3) / h, \tag{7.100}$$

as well as the non-dimensional stiffness characteristics γ_k and the non-dimensional thickness of the layer t_k as follows:

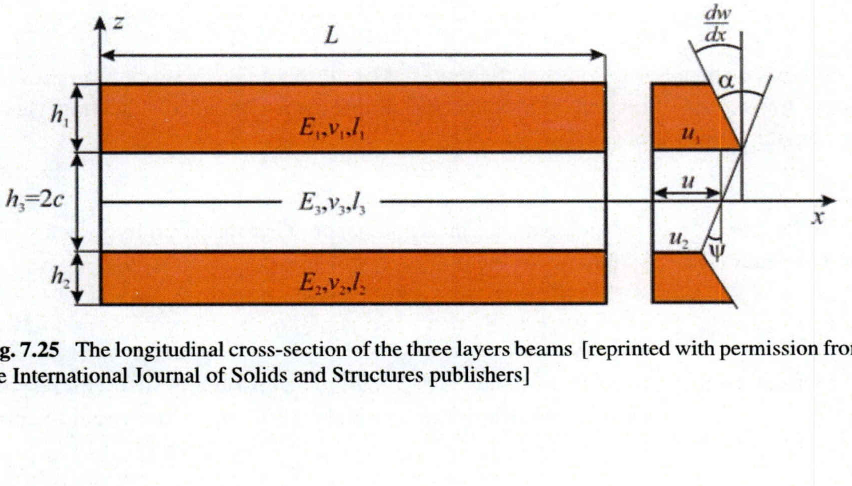

Fig. 7.25 The longitudinal cross-section of the three layers beams [reprinted with permission from the International Journal of Solids and Structures publishers]

$$\gamma_k = E_k h_k / Eh, \quad t_k = h_k / h. \tag{7.101}$$

Owing to the concept of continuum mechanics, the deformations ε_{ij}, displacements u_i and the curvatures χ_{ij} satisfy the following relations:

$$\varepsilon_{ij} = \frac{1}{2}\left(u_{i,j} + u_{j,i}\right), \qquad \chi_{ij} = \frac{1}{2}\left(\theta_{i,j} + \theta_{j,i}\right), \tag{7.102}$$

where θ is the infinite small vector of rotations with components θ_i. Observe that $\theta = (rot\,(\mathbf{u}))/2$.

In order to simplify our analysis and in order to get explicit results, we employ the couple stress theory [43, 44, 178], in which the deformation energy contains only one parameter of the material length and is governed by a symmetric function of deformation and is associated with the symmetric curvatures. Relations between deformations and symmetric components of the curvatures and coupled with them moments of higher orders follow

$$\sigma_{ij} = \frac{E\nu}{(1+\nu)(1-2\nu)}\delta_{ij}\varepsilon_{kk} + 2G\varepsilon_{ij}, \tag{7.103}$$

$$m_{ij} = 2l^2 G\chi_{ij}, \tag{7.104}$$

where E, G and ν is the Young modulus, the shear modulus and the Poisson's coefficient of a rigid body, respectively, and l stands for the internal material length scale parameter.

In the modified couple stress theory, the summed energy of deformation of an elastic body of space Ω is governed by the following formula (it differs from the classic formula only with respect to the second term)

$$U = (1/2)\int\limits_{\Omega}\left(\sigma_{ij}\varepsilon_{ij} + m_{ij}\chi_{ij}\right)d\Omega. \tag{7.105}$$

Owing to the assumption that the material of all three layers is non-compressed in the transversal direction, the deflection w does not depend on the transversal coordinate z, i.e. we have

$$w = w(x). \tag{7.106}$$

In the case of the middle layer, we take into account Timoshenko hypothesis and the longitudinal displacements of the points read

$$u_3 = u + z\psi, \qquad -c \le z \le c. \tag{7.107}$$

Material of the carrying load layers is assumed to be absolutely stiff with respect to shear, and hence the shear angles in the first and second layers are equal to zero, i.e.

$$
\begin{aligned}
\alpha_1 &= \frac{\partial u_1}{\partial z} + \frac{\partial w}{\partial x} = 0, \quad (c \le z \le c + h_1) \,, \\
\alpha_2 &= \frac{\partial u_2}{\partial z} + \frac{\partial w}{\partial x} = 0, \quad (-c - h_2 \le z \le -c) \,.
\end{aligned} \tag{7.108}
$$

The last formula, taking into account (7.106), (7.107) and assuming lack of the relative sliding of the layers, yields the following formulas governing the longitudinal displacements of the points of the transversal beam cross section:

$$
u\,(x,\ z) = \begin{cases} u + c\psi - (z - c)\,\frac{\partial w}{\partial x}, & (c \le z \le c + h_1) \\ u + z\psi, & (-c \le z \le c) \\ u - c\psi - (z + c)\,\frac{\partial w}{\partial x}, & (-c - h_2 \le z \le -c)\,. \end{cases} \tag{7.109}
$$

Introducing α instead of the shear angle ψ, due to relations $\alpha = \psi + \partial w/\partial x$ Eq. (7.109) yields

$$
u\,(x,\ z) = \begin{cases} u + c\alpha - z\frac{\partial w}{\partial x}, & (c \le z \le c + h_1) \\ u + z\alpha - z\frac{\partial w}{\partial x}, & (-c \le z \le c) \\ u - c\alpha - z\frac{\partial w}{\partial x}, & (-c - h_2 \le z \le -c)\,. \end{cases} \tag{7.110}
$$

For the given displacements, the deformations of each layer lying in a distance z from the averaged line of the middle layer, have the following form:

$$
\varepsilon_{xx}\,(x,\ z) = \begin{cases} \frac{\partial u}{\partial x} + c\frac{\partial \alpha}{\partial x} - z\frac{\partial^2 w}{\partial x^2}, & (c \le z \le c + h_1) \\ \frac{\partial u}{\partial x} + z\frac{\partial \alpha}{\partial x} - z\frac{\partial^2 w}{\partial x^2}, & (-c \le z \le c) \\ \frac{\partial u}{\partial x} - c\frac{\partial \alpha}{\partial x} - z\frac{\partial^2 w}{\partial x^2}, & (-c - h_2 \le z \le -c)\,. \end{cases} \tag{7.111}
$$

$$
\begin{aligned}
\varepsilon_{xz} &= \begin{cases} \alpha_1 = 0, & (c \le z \le c + h_1) \\ \alpha_3\,(x) = \alpha\,(x)\,, & (-c \le z \le c) \\ \alpha_2 = 0, & (-c - h_2 \le z \le -c)\,, \end{cases} \\
\varepsilon_{xy} &= \varepsilon_{yz} = 0.
\end{aligned} \tag{7.112}
$$

Symmetric components of the curvature possess one non-zero component

$$
\chi_{xy} = \chi_{yx} = \begin{cases} -\frac{1}{2}\frac{\partial^2 w}{\partial x^2}, & (c \le z \le c + h_1) \\ \frac{1}{4}\left(\frac{\partial \alpha}{\partial x} - 2\frac{\partial^2 w}{\partial x^2}\right), & (-c \le z \le c) \\ -\frac{1}{2}\frac{\partial^2 w}{\partial x^2}, & (-c - h_2 \le z \le -c)\,. \end{cases} \tag{7.113}
$$

Now, having in hand the deformations and employing Hook's law, we find the normal stresses in the layers:

$$\sigma_{xx}(x, z) = \begin{cases} E_1\left(\frac{\partial u}{\partial x} + c\frac{\partial \alpha}{\partial x} - z\frac{\partial^2 w}{\partial x^2}\right), & (c \le z \le c + h_1) \\ E_3\left(\frac{\partial u}{\partial x} + z\frac{\partial \alpha}{\partial x} - z\frac{\partial^2 w}{\partial x^2}\right), & (-c \le z \le c) \\ E_2\left(\frac{\partial u}{\partial x} - c\frac{\partial \alpha}{\partial x} - z\frac{\partial^2 w}{\partial x^2}\right), & (-c - h_2 \le z \le -c) \end{cases} \tag{7.114}$$

the tangential stresses in the middle layer

$$\sigma_{xz} = \tau = \begin{cases} G_1\alpha_1 = 0, & (c \le z \le c + h_1) \\ G_3\alpha_3(x) = G_3\alpha(x), & (-c \le z \le c) \\ G_2\alpha_2 = 0, & (-c - h_2 \le z \le -c) \end{cases} \tag{7.115}$$

$$\sigma_{xy} = \sigma_{yz} = 0,$$

as well as the higher order moments

$$m_{xy} = \begin{cases} -l_1^2 G_1 \frac{\partial^2 w}{\partial x^2}, & (c \le z \le c + h_1) \\ \frac{l_3^2 G_3}{2}\left(\frac{\partial \alpha}{\partial x} - 2\frac{\partial^2 w}{\partial x^2}\right), & (-c \le z \le c) \\ -l_2^2 G_2 \frac{\partial^2 w}{\partial x^2}, & (-c - h_2 \le z \le -c). \end{cases} \tag{7.116}$$

Applying (7.114)–(7.116), we find the forces and moments in each beam layer. In the case of the first carrying load layer $c \le z \le c + h_1$, we have

$$N_1 = b\int_c^{c+h_1} \sigma_{xx} dz = B\gamma_1 \frac{\partial u}{\partial x} + K\gamma_1\left[t_3\frac{\partial \alpha}{\partial x} - (t_1 + t_3)\frac{\partial^2 w}{\partial x^2}\right], \tag{7.117}$$

$$M_1 = M_1^0 + M_1^h = b\int_c^{c+h_1}\left[\sigma_{xx}(z - c) + m_{xy}^1\right] dz = -cN_1 + K\gamma_1 t_1 \frac{\partial u}{\partial x} +$$
$$+ D\gamma_1 t_1 \Theta^{-1}\left[3t_3\frac{\partial \alpha}{\partial x} - \left(4t_1\left(1 + \frac{3}{2(1+\nu_1)}\left(\frac{l_1}{h_1}\right)^2\right) + 3t_3\right)\frac{\partial^2 w}{\partial x^2}\right]. \tag{7.118}$$

In the case of the middle layer $-c \le z \le c$, we obtain:

$$N_3 = b\int_{-c}^{c} \sigma_{xx} dz = B\gamma_3 \frac{\partial u}{\partial x}, \tag{7.119}$$

$$M_3 = M_3^0 + M_3^h = b\int_{-c}^{c}\left[\sigma_{xx}z + m_{xy}^3\right]dz =$$
$$= D\gamma_3 t_3^2\Theta^{-1}\left[\left(1+\frac{3}{1+\nu_3}\left(\frac{l_3}{h_3}\right)^2\right)\frac{\partial\alpha}{\partial x} - \left(1+\frac{6}{1+\nu_3}\left(\frac{l_3}{h_3}\right)^2\right)\frac{\partial^2 w}{\partial x^2}\right], \tag{7.120}$$

$$Q_3 = b\int_{-c}^{c} G_3\alpha dz = G_3 hbt_3\alpha. \tag{7.121}$$

Finally, for the second layer ($-c - h_2 \le z \le c - c$), we get

$$N_2 = b\int_{-c-h_2}^{-c}\sigma_{xx}dz = B\gamma_2\frac{\partial u}{\partial x} - K\gamma_2\left[t_3\frac{\partial\alpha}{\partial x} - (t_2+t_3)\frac{\partial^2 w}{\partial x^2}\right], \tag{7.122}$$

$$M_2 = M_2^0 + M_2^h = b\int_{-c-h_2}^{-c}\left[\sigma_{xx}(z+c) + m_{xy}^2\right]dz = cN_2 - K\gamma_2 t_2\frac{\partial u}{\partial x} +$$
$$+ D\gamma_2 t_2\Theta^{-1}\left[3t_3\frac{\partial\alpha}{\partial x} - \left(4t_2\left(1+\frac{3}{2(1+\nu_2)}\left(\frac{l_2}{h_2}\right)^2\right) + 3t_3\right)\frac{\partial^2 w}{\partial x^2}\right], \tag{7.123}$$

where in (7.117)–(7.123) the following notation has been introduced:

$$B = Ehb, \quad K = \frac{1}{2}Eh^2 b, \quad D = \frac{Eh^3 b}{12}\Theta. \tag{7.124}$$

Furthermore, D in (7.124) denotes a usual classical minimum beam bending stiffness, where the introduced parameter Θ is not defined yet. The full longitudinal force follows: $N = N_1 + N_2 + N_3$, and taking into account (7.117), (7.122), (7.119), we finally get

$$N = B\frac{\partial u}{\partial x} + K\left[c_{12}\frac{\partial\alpha}{\partial x} - c_{13}\frac{\partial^2 w}{\partial x^2}\right]. \tag{7.125}$$

If instead of the displacement u, we introduce a new generalized displacement V due to the formula

$$V = u + \frac{1}{2}h\left(c_{12}\alpha - c_{13}\frac{\partial w}{\partial x}\right), \tag{7.126}$$

then the full longitudinal force (7.125) takes the form

$$N = B\frac{\partial V}{\partial x}. \tag{7.127}$$

The full moment computed in relations to the averaged line of the third layer follows:

$$\hat{M} = D\left(\gamma\frac{\partial\alpha}{\partial x} - \frac{\partial^2 w}{\partial x^2}\right) + \frac{h}{2}c_{13}N + M^h, \tag{7.128}$$

where M^h stands for an additional moment due to the size dependent additives yielded by the occurred size effects. The latter is governed by the formula

$$\begin{gathered} M^h = D\Theta^{-1}\left\{\frac{3\gamma_3 t_3^2}{1+\nu_3}\left(\frac{l_3}{h_3}\right)^2\frac{\partial\alpha}{\partial x} - \right. \\ \left. -\left[\frac{6\gamma_1 t_1^2}{1+\nu_1}\left(\frac{l_1}{h_1}\right)^2 + \frac{6\gamma_3 t_3^2}{1+\nu_3}\left(\frac{l_3}{h_3}\right)^2 + \frac{6\gamma_2 t_2^2}{1+\nu_2}\left(\frac{l_2}{h_2}\right)^2\right]\frac{\partial^2 w}{\partial x^2}\right\}. \end{gathered} \tag{7.129}$$

On this step, we introduce a new moment $\hat{H}$, which does not have analogy in theory of homogeneous beams, and which defines the transversal shear in the middle layer. It is called the shear moment and it is defined in the following way:

$$\hat{H} = M_3 + cN_1 - cN_2 \tag{7.130}$$

or equivalently

$$\hat{H} = \frac{h}{2}c_{12}N + D\gamma\left(\frac{\gamma}{1-\vartheta}\frac{\partial\alpha}{\partial x} - \frac{\partial^2 w}{\partial x^2}\right) + H^h \tag{7.131}$$

where

$$H^h = \frac{D}{2\Theta}\left\{\frac{3\gamma_3 t_3^2}{1+\nu_3}\left(\frac{l_3}{h_3}\right)^2\frac{\partial\alpha}{\partial x} - \frac{6\gamma_3 t_3^2}{1+\nu_3}\left(\frac{l_3}{h_3}\right)^2\frac{\partial^2 w}{\partial x^2}\right\} \tag{7.132}$$

stands for the size dependent additive term to the shear moment.

In the previous formulas, the following notation has been introduced:

$$\begin{gathered} \Theta = c_{33} - 3c_{13}^2; \;\; \gamma = (c_{23} - 3c_{12}c_{13})\,\Theta^{-1}; \\ \vartheta = 1 - \gamma\frac{c_{23} - 3c_{12}c_{13}}{c_{22} - 3c_{12}^2}, \end{gathered} \tag{7.133}$$

where the explicit form of the coefficients c_{ij} read

$$\begin{aligned} &c_{12} = t_3(\gamma_1 - \gamma_2); \;\; c_{13} = \gamma_1(t_1 + t_3) - \gamma_2(t_2 + t_3); \\ &c_{22} = t_3^2(3\gamma_1 + 3\gamma_2 + \gamma_3); \\ &c_{23} = 3\gamma_1 t_3(t_1 + t_3) + 3\gamma_2 t_3(t_2 + t_3) + \gamma_3 t_3^2; \\ &c_{33} = \gamma_1\left(4t_1^2 + 6t_1t_3 + 3t_3^2\right) + \gamma_2\left(4t_2^2 + 6t_2t_3 + 3t_3^2\right) + \gamma_3 t_3^2. \end{aligned} \tag{7.134}$$

7.7.3 Equations of Motion and Boundary Conditions for the Three-Layer Beam

The state/motion equations as well as the associated boundary conditions corresponding to the assumed kinematic hypotheses are yielded directly from Hamilton's principle

$$\delta \int_{t_1}^{t_2} (T - U + W)\, dt = 0, \tag{7.135}$$

where U is the energy of deformation of the bended isotropic three-layer linear elastic beam, T is its kinetic energy, whereas W presents the work of external forces. Therefore, we study the three-layer beam of length L subjected to both external transversal load $q(x)$ and external stresses: normal $b\sigma$ and tangential $b\tau$ with intensities on boundaries

$$\sigma_0,\ \tau_0|_{x=0}, \qquad \sigma_L,\ \tau_L|_{x=L}. \tag{7.136}$$

We introduce virtual displacements, i.e. normal

$$\delta w, \quad (-c - h_2 \le z \le c + h_1) \tag{7.137}$$

and tangential

$$\delta u(x, z) = \begin{cases} \delta u + c\delta\alpha - z\frac{\partial \delta w}{\partial x}, & (c \le z \le c + h_1) \\ \delta u + z\delta\alpha - z\frac{\partial \delta w}{\partial x}, & (-c \le z \le c) \\ \delta u - c\delta\alpha - z\frac{\partial \delta w}{\partial x}, & (-c - h_2 \le z \le -c)\,. \end{cases} \tag{7.138}$$

The virtual displacements generate the following virtual deformations in the layers:

$$\delta\varepsilon_{xx}(x, z) = \begin{cases} \frac{\partial \delta u}{\partial x} + c\frac{\partial \delta\alpha}{\partial x} - z\frac{\partial^2 \delta w}{\partial x^2}, & (c \le z \le c + h_1) \\ \frac{\partial \delta u}{\partial x} + z\frac{\partial \delta\alpha}{\partial x} - z\frac{\partial^2 \delta w}{\partial x^2}, & (-c \le z \le c) \\ \frac{\partial \delta u}{\partial x} - c\frac{\partial \delta\alpha}{\partial x} - z\frac{\partial^2 \delta w}{\partial x^2}, & (-c - h_2 \le z \le -c)\,, \end{cases} \tag{7.139}$$

$$\delta\alpha = \begin{cases} 0, & (c \le z \le c + h_1) \\ \delta\alpha, & (-c \le z \le c) \\ 0, & (-c - h_2 \le z \le -c)\,, \end{cases} \tag{7.140}$$

$$\delta\chi_{xy} = \begin{cases} -\frac{1}{2}\frac{\partial^2 \delta w}{\partial x^2}, & (c \le z \le c + h_1) \\ \frac{1}{4}\left(\frac{\partial \delta\alpha}{\partial x} - 2\frac{\partial^2 \delta w}{\partial x^2}\right), & (-c \le z \le c) \\ -\frac{1}{2}\frac{\partial^2 \delta w}{\partial x^2}, & (-c - h_2 \le z \le -c)\,. \end{cases} \tag{7.141}$$

Since $\sigma_{yy} = \sigma_{zz} = 0$, then using (7.102)–(7.104) and (7.111)–(7.116), the internal energy U of the deformed isotropic three-layer beams takes the following form:

$$U = (1/2)\, b \int_0^L \int_{-c-h_2}^{c+h_1} \left(\sigma_{ij}\varepsilon_{ij} + m_{ij}\chi_{ij}\right) dx dz =$$
$$= (1/2)\, b \int_0^L \left[\int_c^{c+h_1} \left(E_1 \left(\frac{\partial u}{\partial x} + c\frac{\partial \alpha}{\partial x} - z\frac{\partial^2 w}{\partial x^2} \right)^2 + l_1^2 G_1 \left(\frac{\partial^2 w}{\partial x^2} \right)^2 \right) dz + \right.$$
$$+ \int_{-c}^{c} \left(E_3 \left(\frac{\partial u}{\partial x} + z\frac{\partial \alpha}{\partial x} - z\frac{\partial^2 w}{\partial x^2} \right)^2 + G_3\alpha^2 + \frac{l_3^2 G_3}{4} \left(\frac{\partial \alpha}{\partial x} - 2\frac{\partial^2 w}{\partial x^2} \right)^2 \right) dz +$$
$$\left. + \int_{-c-h_2}^{-c} \left(E_2 \left(\frac{\partial u}{\partial x} - c\frac{\partial \alpha}{\partial x} - z\frac{\partial^2 w}{\partial x^2} \right)^2 + l_2^2 G_2 \left(\frac{\partial^2 w}{\partial x^2} \right)^2 \right) dz \right] dx. \tag{7.142}$$

Let us introduce to the layers the virtual displacements δu, δw, $\delta\alpha$ owing to (7.137)–(7.141). Therefore, the variation of the internal energy takes the form

$$\delta U = b \int_0^L \left[\int_c^{c+h_1} \left(\sigma_{xx} \left(\frac{\partial \delta u}{\partial x} + c\frac{\partial \delta\alpha}{\partial x} - z\frac{\partial^2 \delta w}{\partial x^2} \right) - m_{xy} \left(\frac{\partial^2 \delta w}{\partial x^2} \right) \right) dz + \right.$$
$$+ \int_{-c}^{c} \left(\sigma_{xx} \left(\frac{\partial \delta u}{\partial x} + z\frac{\partial \delta\alpha}{\partial x} - z\frac{\partial^2 \delta w}{\partial x^2} \right) + \tau\delta\alpha + m_{xy}\frac{1}{2} \left(\frac{\partial \delta\, \alpha}{\partial x} - 2\frac{\partial^2 \delta\, w}{\partial x^2} \right) \right) dz +$$
$$\left. + \int_{-c-h_2}^{-c} \left(\sigma_{xx} \left(\frac{\partial \delta u}{\partial x} - c\frac{\partial \delta\alpha}{\partial x} - z\frac{\partial^2 \delta w}{\partial x^2} \right) - m_{xy} \left(\frac{\partial^2 \delta w}{\partial x^2} \right) \right) dz \right] dx. \tag{7.143}$$

Now, introducing the general displacement v instead if u, and taking into account (7.126), we get

$$\delta u = \delta V - \frac{1}{2} h \left(c_{12}\delta\alpha - c_{13}\frac{\partial \delta w}{\partial x} \right)$$

and hence (7.143) yields

$$\delta U = \int_0^L \left[\frac{\partial N}{\partial x}\delta V + \left(\frac{\partial H}{\partial x} - Q_3 \right) \delta\alpha + \frac{\partial^2 M}{\partial x^2}\delta w \right] dx -$$
$$- \left[N\delta V + M \left(\gamma\delta\alpha - \frac{\partial \delta w}{\partial x} \right) + \left(\frac{H}{\gamma} - M \right) \gamma\delta\alpha + \frac{\partial M}{\partial x}\delta w \right]_{x=0}^{x=L}. \tag{7.144}$$

In the above, the following notations have been introduced

$$M = \hat{M} - \frac{h}{2}c_{13}N = D\left(\gamma\frac{\partial\alpha}{\partial x} - \frac{\partial^2 w}{\partial x^2}\right) + M^h, \tag{7.145}$$

$$H = \hat{H} - \frac{h}{2}c_{12}N = D\gamma\left(\frac{\gamma}{1-\vartheta}\frac{\partial\alpha}{\partial x} - \frac{\partial^2 w}{\partial x^2}\right) + H^h. \tag{7.146}$$

Notice that M^h, H^h are defined by (7.129), (7.132). Since the kinetic energy T has the following form

$$T = (1/2)\,b\int_0^L\int_{-c-h_2}^{c+h_1}\rho\left\{\left[V_{,t}^2 + \left(z - \frac{1}{2}hc_{12}\right)\alpha_{,t} - \left(z - \frac{1}{2}hc_{13}\right)w_{,xt}\right]^2 + w_{,t}^2\right\}dxdz, \tag{7.147}$$

then

$$\begin{aligned}\delta T = -\int_0^L &\{[B^* V_{,tt} + K^*\left(c_{12}^* - c_{12}\right)\alpha_{,tt} - K^*\left(c_{13}^* - c_{13}\right)w_{,xtt}]\,\delta v + \\ &+\left[K^*\left(c_{12}^* - c_{12}\right)V_{,tt} + D^*\gamma^*\left(\frac{\gamma^*}{1-\vartheta^*}\alpha_{,tt} - w_{,xtt}\right)\right]\delta\vartheta + \\ &+\left[K^*\left(c_{13}^* - c_{13}\right)V_{,xtt} + B^* w_{,tt} + D^*\left(\gamma^*\alpha_{,xtt} - w_{,xxtt}\right)\right]\}\,dx + \\ &+\left[K^*\left(c_{13}^* - c_{13}\right)V_{,tt} + D^*\left(\gamma^*\alpha_{,tt} - w_{,xtt}\right)\right]\Big|_{x=0}^{x=L},\end{aligned} \tag{7.148}$$

where $\gamma_k^* = \rho_k h_k/\rho h$. Parameters c_{ik}^* are computed through (7.134), as it was with the parameters c_{ik} (we should use γ_k^* instead of γ_k). The remaining parameters introduced while computing variation of the inertia forces are as follows:

$$\begin{aligned}&B^* = \rho hb, \quad K^* = \frac{1}{2}\rho h^2 b, \quad D^* = \frac{\rho h^3 b}{12}\Theta^*, \\ &\Theta^* = c_{33}^* - 6c_{13}^* c_{13} + 3c_{13}^2, \\ &\gamma^* = \left(c_{23}^* - 3c_{12}^* c_{13} - 3c_{13}^* c_{12} + 3c_{12}c_{13}\right)/\Theta, \\ &\vartheta^* = 1 - \gamma^* \frac{c_{23}^* - 3c_{12}^* c_{13} - 3c_{13}^* c_{12} + 3c_{12}c_{13}}{c_{22}^* - 6c_{12}c_{12}^* + 3c_{12}^2}.\end{aligned} \tag{7.149}$$

The work of the external forces on the virtual displacement follows:

$$\delta W = \int_0^L q\delta w\,dx + \left[N_p\delta V + M_p\left(\gamma\delta\alpha - \frac{\partial\delta w}{\partial x}\right) + Q_p\delta w\right]_{x=0}^{x=L}. \tag{7.150}$$

Here, N_p, Q_p, M_p denote external forces and moments acting on the beam

$$N_p = b \int_{-c-h_2}^{c+h_1} \sigma_p dz, \;\; Q_p = b \int_{-c-h_2}^{c+h_1} \tau_p dz, \;\; M_p = b \int_{-c-h_2}^{c+h_1} z\sigma_p dz - \frac{1}{2} h c_{13} N_p, \tag{7.151}$$

and for $x = 0$ we have $\sigma_p = \sigma_0$, whereas for $x = L$ we have $\sigma_p = \sigma_L$.

Employing Hamilton principle, and comparing to zero multipliers standing by variation of the independent displacements, one gets

$$\begin{aligned}
&\frac{\partial N}{\partial x} = B^* \frac{\partial^2 V}{\partial t^2} + K^* \frac{\partial^2}{\partial t^2} \left[\left(c_{12}^* - c_{12}\right) \alpha - \left(c_{13}^* - c_{13}\right) \frac{\partial w}{\partial x} \right], \\
&\frac{\partial H}{\partial x} - Q_3 = K^* \left(c_{12}^* - c_{12}\right) \frac{\partial^2 V}{\partial t^2} + D^* \gamma^* \frac{\partial^2}{\partial t^2} \left[\frac{\gamma^*}{1 - \vartheta^*} \alpha - \frac{\partial w}{\partial x} \right], \\
&\frac{\partial^2 M}{\partial x^2} + q = K^* \left(c_{13}^* - c_{13}\right) \frac{\partial^3 V}{\partial x \, \partial t^2} + B^* \frac{\partial^2 w}{\partial t^2} + D^* \frac{\partial^2}{\partial t^2} \left[\gamma^* \frac{\partial \alpha}{\partial x} - \frac{\partial^2 w}{\partial x^2} \right].
\end{aligned} \tag{7.152}$$

The boundary conditions are

$$\begin{aligned}
&\left[N - N_p\right] \Big|_{x=0}^{x=L} = 0 \quad \text{or} \quad \delta u \Big|_{x=0}^{x+L} = 0, \\
&\left[M - M_p\right] \Big|_{x=0}^{x=L} = 0 \quad \text{or} \quad \delta \left(\gamma \alpha - \frac{\partial w}{\partial x} \right) \Bigg|_{x=0}^{x=L} = 0, \\
&\left[\frac{H}{\gamma} - M \right] \Bigg|_{x=0}^{x=L} = 0 \quad \text{or} \quad \delta \alpha \Big|_{x=0}^{x=L} = 0, \\
&\left[\frac{\partial M}{\partial x} - Q_p - K^* \left(c_{13}^* - c_{13}\right) \frac{\partial^2 V}{\partial t^2} - D^* \Theta^* \frac{\partial^2}{\partial t^2} \left(\gamma^* \alpha - \frac{\partial w}{\partial x} \right) \right] \Bigg|_{x=0}^{x=L} = 0 \\
&\text{or} \quad \delta w \Big|_{x=0}^{x=L} = 0.
\end{aligned} \tag{7.153}$$

7.7.4 Static Transversal Bending of the Three-Layer Beam

In order to get directly the counterpart static problem, we omit dynamic terms in Eq. (7.152), and the following system of equilibrium equations is obtained

$$\frac{dN}{dx} = 0, \quad \frac{dH}{dx} - Q_3 = 0, \quad \frac{d^2 M}{dx^2} + q = 0. \tag{7.154}$$

Now, proceeding to displacements, we obtain

$$\frac{d^2V}{dx^2} = 0, \tag{7.155}$$

$$\begin{aligned} &D\gamma^2 \left[\left(1 + \frac{3}{2(1+\nu_3)} \frac{(1-\vartheta)\,\gamma_3 t_3^2}{\gamma^2 \Theta} \left(\frac{l_3}{h_3} \right)^2 \right) \gamma \frac{d^2\alpha}{dx^2} - \right. \\ &\left. - (1-\vartheta) \left(1 + \frac{3\gamma_3 t_3^2}{\gamma \Theta} \left(\frac{l_3}{h_3} \right)^2 \right) \frac{d^3 w}{dx^3} \right] - (1-\vartheta)\, G_3 b h t_3 \gamma \alpha = 0, \end{aligned} \tag{7.156}$$

$$\begin{aligned} &D \left\{ \left(1 + \frac{3}{2(1+\nu_3)} \frac{\gamma_3 t_3^2}{\gamma \Theta} \left(\frac{l_3}{h_3} \right)^2 \right) \gamma \frac{d^3\alpha}{dx^3} - \right. \\ &\left. - \left(1 + \frac{6}{\Theta} \left[\frac{\gamma_1 t_1^2}{1+\nu_1} \left(\frac{l_1}{h_1} \right)^2 + \frac{\gamma_3 t_3^2}{1+\nu_3} \left(\frac{l_3}{h_3} \right)^2 + \frac{\gamma_2 t_2^2}{1+\nu_2} \left(\frac{l_2}{h_2} \right)^2 \right] \right) \frac{d^4 w}{dx^4} \right\} + qb = 0. \end{aligned} \tag{7.157}$$

After employment of the following simplifications

$$\begin{aligned} &a_h^2 = \frac{3}{2(1+\nu_3)} \frac{(1-\vartheta)\,\gamma_3 l_3^2}{\gamma^2 \Theta}, \quad b_h^2 = \frac{3\gamma_3 l_3^2}{\gamma \Theta}, \\ &c_h^2 = \frac{3}{2(1+\nu_3)} \frac{\gamma_3 l_3^2}{\gamma \Theta}, \quad d_h^2 = \frac{6}{\Theta} \left[\frac{\gamma_1 l_1^2}{1+\nu_1} + \frac{\gamma_3 l_3^2}{1+\nu_3} + \frac{\gamma_2 l_2^2}{1+\nu_2} \right], \end{aligned} \tag{7.158}$$

Equations (7.156), (7.157) take the following form

$$\begin{aligned} &D\gamma^2 \left[\left(1 + \left(\frac{a_h}{h} \right)^2 \right) \gamma \frac{d^2\alpha}{dx^2} - (1-\vartheta) \left(1 + \left(\frac{b_h}{h} \right)^2 \right) \frac{d^3 w}{dx^3} \right] - \\ &- (1-\vartheta)\, G_3 b h t_3 \gamma \alpha = 0, \end{aligned} \tag{7.159}$$

$$D \left\{ \left(1 + \left(\frac{c_h}{h} \right)^2 \right) \gamma \frac{d^3\alpha}{dx^3} - \left(1 + \left(\frac{d_h}{h} \right)^2 \right) \frac{d^4 w}{dx^4} \right\} + qb = 0. \tag{7.160}$$

Therefore, the system of equilibrium equations (7.155)–(7.157) split into Eq. (7.155) and the system of Eqs. (7.159)–(7.160) regarding the functions $\gamma\alpha$ and w. It is suitable for a further analysis to reduce the whole problem to only one equation by introducing w and $\gamma\alpha$ as a function χ being differentiable the required times:

$$\begin{aligned} w &= \left\{ 1 - \frac{h^2}{\beta} \left[1 + \left(\frac{a_h}{h} \right)^2 \right] \frac{d^2}{dx^2} \right\} \chi, \\ \gamma\alpha &= -(1-\vartheta) \frac{h^2}{\beta} \left[1 + \left(\frac{b_h}{h} \right)^2 \right] \frac{d^3\chi}{dx^3}. \end{aligned} \tag{7.161}$$

Substituting the expression (7.161) into Eq. (7.159), we find

$$\beta = \frac{12G_3t_3\,(1-\vartheta)}{E\gamma^2\Theta}. \tag{7.162}$$

Introducing (7.161) into (7.160), the following equation regarding the displacement function χ is obtained

$$D^h\left[1-\frac{h^2}{\beta}\left(\left(1+\left(\frac{a_h}{h}\right)^2\right)-(1-\vartheta)\,p_h\right)\frac{d^2}{dx^2}\right]\frac{d^4\chi}{dx^4}=qb, \tag{7.163}$$

where

$$\begin{aligned} D^h &= D\left(1+\left(\frac{d_h}{h}\right)^2\right),\\ p_h &= \left(1+\left(\frac{c_h}{h}\right)^2\right)\left(1+\left(\frac{b_h}{h}\right)^2\right)\Big/\left(1+\left(\frac{d_h}{h}\right)^2\right). \end{aligned} \tag{7.164}$$

Equation (7.163) describes the transversal bending of the three-layer beam. Since the function χ keeps the displacements, and consequently the moments and transverse forces, we call Eq. (7.163) the solving equation.

In order to choose the formulation of the problem regarding deformation of the three-layer beam, it is necessary to attach boundary conditions to the equilibrium equations (7.155) and (7.163). The latter express the influences of both the boundaries action on the beam edges and the boundary loads action. In the case of Eq. (7.155), where the stiffness is equal to zero (the edge is free), we have

$$\left.\frac{dV}{dx}\right|_{x=0}^{x=L}=0,\quad or\quad N|_{x=0}^{x=L}=0. \tag{7.165}$$

On the contrary, if the coupling possesses an infinite stiffness, then the boundary condition takes the form $V|_{x=0}^{x=L}=0$.

In what follows, we proceed to formulation of the boundary conditions for the moments. If there is no coupling, then the edge is free, and boundary conditions for χ have the form:

$$\chi=\frac{d^2\chi}{dx^2}=\left.\frac{d^4\chi}{dx^4}\right|_{x=0}^{x=L}=0, \tag{7.166}$$

which follow from the boundary conditions

$$w|_{x=0}^{x=L}=\left.\frac{\partial^2 w}{\partial x^2}\right|_{x=0}^{x=L}=\left.\frac{\partial\gamma}{\partial x}\right|_{x=0}^{x=L}=0.$$

The boundary conditions for clamped ends have the form:

$$\left\{1-\frac{h^2}{\beta}\left[1+\left(\frac{a_h}{h}\right)^2\right]\frac{d^2}{dx^2}\right\}\chi=\frac{d\chi}{dx}=\frac{d^3\chi}{dx^3}\Bigg|_{x=0}^{x=L}=0. \tag{7.167}$$

Introducing the following non-dimensional quantity

$$n^2=\frac{\beta l^2}{h^2\left(\left(1+\left(\frac{a_h}{h}\right)^2\right)-(1-\vartheta)\,p_h\right)}, \tag{7.168}$$

we reduce the solving Eq. (7.163) to the form

$$\left(\frac{d^2}{dx^2}-\frac{n^2}{l^2}\right)\frac{d^4\chi}{dx^4}=-\frac{qn^2}{D^h l^2}. \tag{7.169}$$

A general solution to the counterpart homogenous equation of (7.169) follows:

$$\chi_0\left(x\right)=A_0+A_1\frac{x}{l}+A_2\left(\frac{x}{l}\right)^2+A_3\left(\frac{x}{l}\right)^3+A_4 sh\frac{nx}{l}+A_5 ch\frac{nx}{l}, \tag{7.170}$$

whereas the particular equation, when $q\left(x\right)=q_0=constant$, takes the following form

$$\bar{\chi}\left(x\right)=\frac{q_0}{24D^h}x^4. \tag{7.171}$$

Let us find a solution to Eq. (7.169). For this purpose, we define coefficients A_i in general solution (7.170) employing the boundary conditions (7.166):

$$\begin{aligned}A_0&=-\frac{24q_0l^4}{D^h}, & A_1&=\frac{q_0l^4\left(12-n^2\right)}{D^h}, & A_2&=-\frac{12q_0l^4}{D^h},\\ A_3&=\frac{2q_0l^4n^2}{D^h}, & A_4&=-\frac{24q_0l^4\left[1-ch\left(n\right)\right]}{D^hn^2sh\left(n\right)}, & A_5&=\frac{24q_0l^4}{D^hn^2}.\end{aligned} \tag{7.172}$$

Therefore, $\chi\left(x\right)=\chi_0\left(x\right)+\bar{\chi}\left(x\right)$ and both functions $\gamma\alpha$ and w are defined now through formulas (7.161).

7.7.5 *Vibrations of a Three-Layer Beam*

Let us turn back to dynamic problem. The governing equations are based on the introduced hypotheses and since they take into account the full inertial force. The whole problem is governed by a relatively complex system of *parabolic equations*. Observe that assuming the layers carrying the load (layers 1, 2) are membranes ($t_1=t_2=0$), then the system becomes *hyperbolic*. In the general case, however, the studied system of PDEs can be reduced to only one PDE of eighth order with even derivatives. Its coefficients allow to carry out a deep analysis regarding the system behaviour. However, in the current study, our investigations will be limited to only the main inertial term, assuming that the influence of the remaining inertial terms can be neglected. Namely, we take $K^* = 0,\ D^* = 0$ in Eq. (7.152), and we obtain

$$\frac{\partial N}{\partial x} = B^* \frac{\partial^2 V}{\partial t^2}, \tag{7.173}$$

$$\frac{\partial H}{\partial x} - Q_3 = 0, \tag{7.174}$$

$$\frac{\partial^2 M}{\partial x^2} + q = B^* \frac{\partial^2 w}{\partial t^2}, \tag{7.175}$$

or the equivalent form with respect to displacements as follows:

$$B\frac{d^2 V}{dx^2} = B^* \frac{\partial^2 V}{\partial t^2}, \tag{7.176}$$

$$D\gamma^2 \left[\left(1 + \left(\frac{a_h}{h}\right)^2\right) \gamma \frac{d^2\alpha}{dx^2} - (1-\vartheta)\left(1 + \left(\frac{b_h}{h}\right)^2\right) \frac{d^3 w}{dx^3} \right] - \\ - (1-\vartheta)\, G_3 b h t_3 \gamma \alpha = 0, \tag{7.177}$$

$$D\left\{ \left(1 + \left(\frac{c_h}{h}\right)^2\right) \gamma \frac{d^3\alpha}{dx^3} - \left(1 + \left(\frac{d_h}{h}\right)^2\right) \frac{d^4 w}{dx^4} \right\} + \\ + qb = B^* \frac{\partial^2 w}{\partial t^2}. \tag{7.178}$$

We take $a^2 = E/\rho$, and hence (7.176) yields

$$\frac{d^2 V}{dx^2} = \frac{1}{a^2}\frac{\partial^2 V}{\partial t^2}, \tag{7.179}$$

i.e. we have got the PDE governing longitudinal vibrations of the beam.

Introducing, owing to (7.161), the function $\chi\ (x)$, two independent dynamic equations are obtained (the second Eq. (7.177) is satisfied identically). The third PDE (7.178) takes the form:

$$D\left(1 + \left(\frac{d_h}{h}\right)^2\right) \left[1 - \frac{h^2}{\beta}\left(\left(1 + \left(\frac{a_h}{h}\right)^2\right) - (1-\vartheta)\, p_h\right) \frac{\partial^2}{\partial x^2}\right] \frac{\partial^4 \chi}{\partial x^4} + \\ + \frac{Ehb}{a^2}\frac{\partial^2}{\partial t^2}\left[1 - \frac{h^2}{\beta}\left(1 + \left(\frac{a_h}{h}\right)^2\right)\frac{\partial^2}{\partial x^2}\right]\chi = qb. \tag{7.180}$$

We begin with a study of free vibrations of the beam $q \equiv 0$. For this purpose, we modify Eq. (7.180) using the non-dimensional coordinate $\xi = \pi x/L$ and presenting the function $\chi(\xi, t)$ in its counterpart form $\chi(\xi, t) = \frac{L}{\pi} X(\xi) e^{i\omega t}$. Dividing the equation for $X(\xi)$ by $e^{i\omega t}$ and by $D^h \frac{h^2}{\beta} \frac{\pi^6}{l^6} \tilde{\vartheta}$, where

$$\tilde{\vartheta} = \left(1 + \left(\frac{a_h}{h}\right)^2\right) - (1 - \vartheta)\, p_h, \tag{7.181}$$

we get

$$X^{VI} - \frac{1}{k\tilde{\vartheta}} X^{IV} - \frac{\omega_*^2 \left(1 + \left(\frac{a_h}{h}\right)^2\right)}{\tilde{\vartheta}\left(1 + \left(\frac{d_h}{h}\right)^2\right)} X^{II} + \frac{\omega_*^2}{k\tilde{\vartheta}\left(1 + \left(\frac{d_h}{h}\right)^2\right)} X = 0, \tag{7.182}$$

where the following non-dimensional parameters have been introduced

$$k = \frac{h^2\pi^2}{l^2\beta}, \qquad \omega_*^2 = \frac{12\, l^4\omega^2}{a^2h^2\pi^4\Theta}. \tag{7.183}$$

It has been shown in [76] that the characteristic equation associated with Eq. (7.182) reads:

$$s^3 - \frac{1}{k\tilde{\vartheta}} s^2 - \frac{\omega_*^2 \left(1 + \left(\frac{a_h}{h}\right)^2\right)}{\tilde{\vartheta}\left(1 + \left(\frac{d_h}{h}\right)^2\right)} s + \frac{\omega_*^2}{k\tilde{\vartheta}\left(1 + \left(\frac{d_h}{h}\right)^2\right)} = 0, \tag{7.184}$$

and consequently, it may have one real and negative root. Introducing the following notations $s_1 = -\lambda_1^2$, $s_2 = \lambda_2^2$, $s_3 = \lambda_3^2$, the general solution to Eq. (7.182) can be presented in the following form:

$$\begin{aligned} X(\xi) = {} & C_1 \sin(\lambda_1\xi) + C_2 \cos(\lambda_1\xi) + C_3 sh(\lambda_2\xi) + \\ & + C_4 ch(\lambda_1\xi) + C_5 sh(\lambda_3\xi) + C_6 ch(\lambda_3\xi). \end{aligned} \tag{7.185}$$

On this step, we need only to construct the characteristic equation to find the roots λ_1^2, λ_2^2, λ_3^2 by satisfying the homogeneous boundary conditions. Note that the boundary conditions are the same as in the case if the static bending, and hence we have to find the roots of rather a complex transcendental equation. For the case of $a_h = 0$, $d_h = 0$ (classical case) and for a $a_h \neq 0$, $d_h \neq 0$ (couple stress theory), the values of λ_1, λ_2, λ_3 are the same and they depend only on the boundary conditions. Since $-\lambda_1^2$ is a root, then (7.184) implies

$$\omega_*^2 = \lambda_1^4 \frac{\left(\lambda_1^2 k\tilde{\vartheta} + 1\right)\left(1 + \left(\frac{d_h}{h}\right)^2\right)}{k\lambda_1^2\left(1 + \left(\frac{a_h}{h}\right)^2\right) + 1}. \tag{7.186}$$

In the classical case, we have [76]

$$\omega_{clas}^2 = \lambda_1^4 \frac{\lambda_1^2 k\vartheta + 1}{k\lambda_1^2 + 1}. \tag{7.187}$$

Let us investigate the ratio $r(\lambda_1^2) = \omega_*^2/\omega_{clas}^2$ expressed explicitly in the form

$$r(\lambda_1^2) = \frac{\left(\lambda_1^2 k\tilde{\vartheta} + 1\right)\left(1 + \left(\frac{d_h}{h}\right)^2\right)}{k\lambda_1^2\left(1 + \left(\frac{a_h}{h}\right)^2\right) + 1} \frac{k\lambda_1^2 + 1}{\lambda_1^2 k\vartheta + 1}, \tag{7.188}$$

where λ_1^2 plays a role of the control parameter.

7.7.6 Numerical Results and Their Validation

Owing to the literature reports [117, 133, 174], the values of the size-dependent parameter l are quite different and still awaiting the estimation for many materials. Since the three-layer beams can be metallic, polymer and made from the biological tissues, therefore the scalar length l parameter can change within a rather large interval. For instance, in the case of Al and Si materials, the scalar parameter regarding length l equals to 10^{-10} m. The results of the molecular modelling reported in the Ref. [14] show that the gradient effects occur the thickness of amount $l \approx 10^{-9}$ m. In Ref. [44], in the case of investigation of the copper wire, the value of the length parameter achieves $l = 3 \cdot 10^{-3}$ m. On the other hand, in the case of the rubber epoxides materials, in Ref. [172] has been shown that the scalar parameter has been found experimentally as $l = 12 \cdot 10^{-3}$ m for Young's moduli $E = 1.44\cdot$ GPa and Poisson's coefficient $\nu = 0.38$.

As an example, we consider a three-layer beam with the microstructural effect by taking into account the following length parameters $l_1 = l_2$, l_3 for the given values $t_1 = t_2 = 0.125$, $t_3 = 0.75$, the thickness $h = 32 \cdot 10^{-6}$ m and length $L = 240 \cdot 10^{-6}$ m, being under action of the constant and uniformly distributed load $q_0 = 1$ N/m.

The numerical examples have been carried out for the three-layer microbeam composed of two copper made external layers and the middle layer made from the rubber epoxide material:

$$E_1 = E_2 = 120\,\text{GPa}, \quad \nu_1 = \nu_2 = 0.38, \quad l_1 = l_2 = 3 \cdot 10^{-3}\,\text{m},$$
$$E_3 = 1.44\,\text{GPa}, \quad \nu_3 = 0.38, \quad l_3 = 12 \cdot 10^{-3}\,\text{m}.$$

7.7.6.1 Static Bending

In the beginning, the deflections w_0^c in the three layer beam centre are computed where the simple boundary conditions are taken and the classical theory of Grigolyuk-Chulkov is employed ($l_1 = l_2 = l_3 = 0,\ D^h = D$). In the next step, we find the values of the deflection w_0^h in the beam centre, which have been obtained based on our introduced theoretical background for the different fixed values of $l_1/h_1 = l_2/h_2$ for $l_3 = 0$ and $l_3 = 12 \cdot 10^{-3}$ m.

In Fig. 7.26 the dependencies characterizing the relative values of the deflection of the beam centre w_0^h/w_0^c versus the ratio $l/h = l_1/h_1$ are obtained (solid/dashed curve corresponds to $l_3 = 0$ / $l_3 = 12 \cdot 10^{-3}$ m).

The reported results show that for $l_3 = 0$ and $l_3 = 12 \cdot 10^{-3}$ m the deflection decreases while l_1/h_1 increases. The similar results has been obtained in Ref. [14] for the plates modelled by Kirchhoff hypotheses and taking into account the gradient effects. Also in Ref. [15], where the microstructural effects for a simply supported Timoshenko beam with the employment of the modified couple stress theory of elasticity [198], the similar effect has been reported. Observe that for $l_3 = 12 \cdot 10^{-3}$ m the beam centre deflections become less, i.e. the beam becomes more stiff while taking into account the microstructural effect in the middle layer. The figure presents also the dependence (dotted-dashed curve for $l_1 = 0$ and dotted curve for $l_1 = 3 \cdot 10^{-3}$ m) w_0^h/w_0^c versus the parameter $l/h = l_3/h_3$. In the latter case, the deflection decreases versus increase of l_3/h_3.

Furthermore, in Fig. 7.27, the profiles regarding the deflection value $w(x)/h$ of the simply supported three-layer beam for $l_3/h_3 = 0$ and for the different values of the non-dimensional length parameters l_1/h_1 are reported ($l_1/h_1 = 0$ corresponds to the classical Grigolyuk-Chulkov solution).

Results reported in Fig. 7.27 imply that the obtained deflections of our model are less than those yielded by the classical theory of Grigolyuk-Chulkov on amount of six times for $l_1/h_1 = 0.8$. A decrease of the difference between two models while

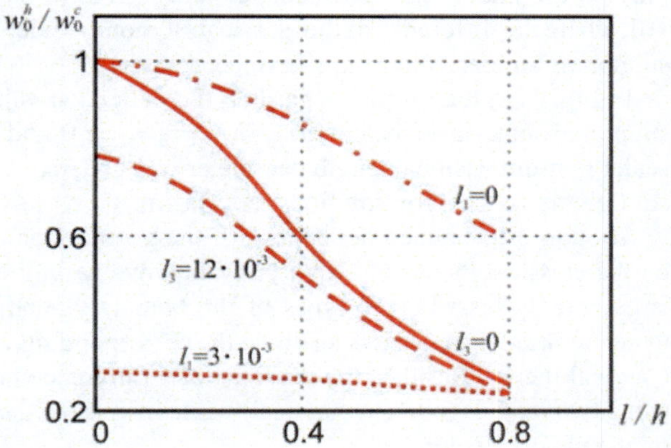

Fig. 7.26 Relative values of the beam centre deformation versus l/h [reprinted with permission from the International Journal of Solids and Structures publishers]

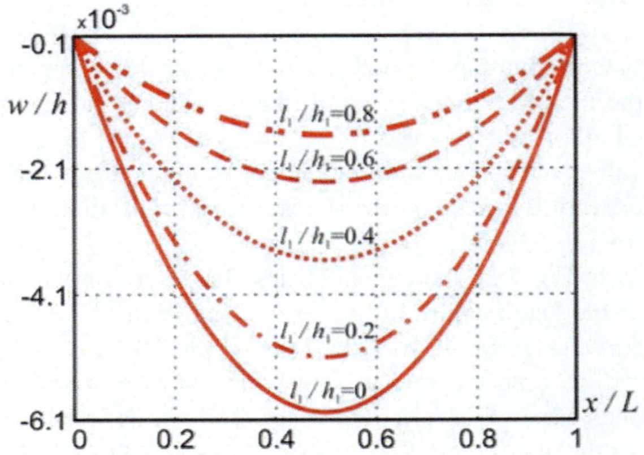

Fig. 7.27 Profiles of the relative beam deflection value for different values l_1/h_1 for $l_3/h_3 = 0$ [reprinted with permission from the International Journal of Solids and Structures publishers]

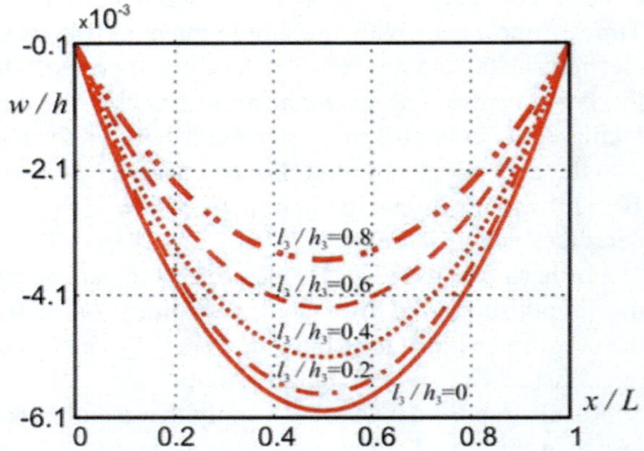

Fig. 7.28 Profiles regarding the beam deflection for different values of l_3/h_3 for $l_1/h_1 = 0$ [reprinted with permission from the International Journal of Solids and Structures publishers]

increasing the thickness (when l_1/h_1 is decreased) means that the size effect is visible only on the microscale. The same conclusions have been carried out in Refs. [15, 16], where the difference of the classical Bernoulli-Euler beams as well as the simply supported Timoshenko beams have been studied.

In Fig. 7.28, the profiles regarding the deflection values $w(x)/h$ of the simply supported three-layer beam for $l_1/h_1 = l_2/h_2 = 0$ and the different values of the scalar non-dimensional length parameter l_3/h_3 ($l_3/h_3 = 0$ corresponds to the classical Grigolyuk-Chulkov solutions) are shown.

Though the obtained deflections of the given model are less than those yielded by the classical theory of Grigolyuk-Chulkov, the influence of the parameter l_3/h_3 is essentially less. For all types of the boundary conditions, deflections obtained using our model are always less than the corresponding values obtained based on the classical theory. Besides, the decrease of a difference between the results obtained using two models (our and classical) while increasing the size effect is essential only on the nanoscale domain.

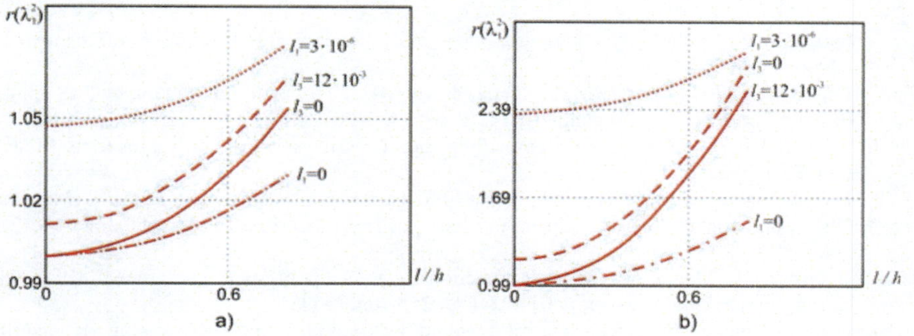

Fig. 7.29 Relative values $r(\lambda_1^2) = \omega_*^2/\omega_{clas}^2$ versus l/h : **a** $-\lambda_1^2 = 1$, **b** $\lambda_1^2 = 100$ [reprinted with permission from the International Journal of Solids and Structures publishers]

7.7.6.2 Free Vibrations

In what follows, we investigate qualitatively the influence of the size effects in the beam on the frequencies of its free vibrations taking into account the value of the negative root $\left(-\lambda_1^2\right)$ yielded by the characteristic equation (7.184). For this purpose, based on the formula (7.188), we find the ratio $r(\lambda_1^2)$.

In Fig. 7.29, graphs of the values $r(\lambda_1^2)$ of the three-layer beam versus the ratio $l/h = l_1/h_1$ are reported (solid curve corresponds to $l_3 = 0$, whereas the dashed curve to $l_3 = 12 \cdot 10^{-3}$ m; note that $l_1/h_1 = 0$ corresponds to the classical Grigolyuk-Chulkov solution).

The mentioned figure includes also dependencies $r(\lambda_1^2)$ against $l/h = l_3/h_3$ (dashed-dotted curve for $l_1 = 0$ and dotted curve for $l_1 = 3 \cdot 10^{-3}$ m for the parameter $\lambda_1^2 = 1,\ 100(l_3/h_3 = 0)$ corresponds to the classical Grigolyuk-Chulkov solution).

For all values of the boundary conditions, the ratio of the non-dimensional frequencies $r(\lambda_1^2)$ obtained through the classical Grigolyuk-Chulkov model always increases while increasing the parameter l/h in an arbitrary beam layer. However, the increase of the ratio l/h in the middle layer has a less essential consequence in comparison to the increase of the l/h in the remaining layers.

References

1. Koizumi, M.: The concept of FGM. Ceram. Trans. Funct. Graded Mater. **34**, 3–10 (1993)
2. Birman, V., Byrd, L.W.: Modeling and analysis of functionally graded materials and structures. Appl. Mech. Rev. **60**(5), 195–216 (2007)
3. Aydogdu, M., Taskin, V.: Free vibration analysis of functionally graded beams with simply supported edges. Mater. Des. **28**, 1651–1656 (2007)
4. Chakraborty, A., Gopalakrishnan, S., Reddy, J.N.: A new beam finite element for the analysis of functionally graded materials. Int. J. Mech. Sci. **45**, 519–539 (2003)

5. Sankar, B.V.: An elasticity solution for functionally graded beams. Compos. Sci. Technol. **61**, 689–696 (2001)
6. Fu, Y.Q., Du, H.J., Huang, W.M., Zhang, S., Hu, M.: TiNi-based thin films in MEMS applications: a review. Sens. Actuators A **112**(2–3), 395–408 (2004)
7. Fu, Y.Q., Du, H.J., Zhang, S.: Functionally graded TiN/TiNi shape memory alloy films. Mater. Lett. **57**(20), 2995–2999 (2003)
8. Lee, Z., Ophus, C., Fischer, L.M., Nelson-Fitzpatrick, N., Westra, K.L., Evoy, S.: Metallic NEMS components fabricated from nanocomposite Al-Mo films. Nanotechnology **17**(12), 3063–3070 (2006)
9. Witvrouw, A., Mehta, A.: The use of functionally graded poly-SiGe layers for MEMS applications. Funct. Graded Mater. VIII **492–493**, 255–260 (2005)
10. Fu, Y., Zhang, J.: Electromechanical dynamic buckling phenomenon in symmetric electric fields actuated microbeams considering material damping. Acta Mech. **212**, 29–42 (2010)
11. Moghimi, Z.M., Ahmadian, M.T.: Static pull-in analysis of electrostatically actuated microbeams using homotopy perturbation method. Appl. Math. Model. **34**, 1032–1046 (2010)
12. Jia, X.L., Yang, J., Kitipornchai, S.: Pull-in instability of geometrically nonlinear microswitches under electrostatic and Casimir forces. Acta Mech. **218**, 161–174 (2011)
13. Chong, A.C.M., Lam, D.C.C.: Strain gradient plasticity effect in indentation hardness of polymers. Mater. Res. **14**(10), 4103–4110 (1999)
14. Fleck, N.A., Muller, G.M., Ashby, M.F., Hutchinson, J.W.: Strain gradient plasticity: theory and experiments. Acta Metall. Mater. **42**, 475–487 (1994)
15. McFarland, A.W., Colton, J.S.: Role of material microstructure in plate stiffness with relevance to microcantilever sensors. Micromech. Microeng. **15**(5), 1060–1067 (2005)
16. Stolken, J.S., Evans, A.G.: Microbend test method for measuring the plasticity length scale. Acta Mater. **46**(14), 5109–5115 (1998)
17. Kong, S., Zhou, S., Nie, Z., Wang, K.: The size-dependent natural frequency of Bernoulli–Euler micro-beams. Int. J. Eng. Sci. **46**, 427–437 (2008)
18. Scheible, D.V., Erbe, A., Blick, R.H.: Evidence of a nanomechanical resonator being driven into chaotic response via the Ruelle-Takens route. Appl. Phys. Lett. **81**, 1884–1886 (2002)
19. Sanchez-Portal, D., Artacho, E., Soler, J.M., Rubio, A., Ordejo'n, P.: Ab initio structural, elastic, and vibrational properties of carbon nanotubes. Phys. Rev. B **59**, 12678 (1999)
20. Krishnan, A., Dujardin, E., Ebbesen, T., Yianilos, P., Treacy, M.: Young's modulus of single-walled nanotubes. Phys. Rev. B **58**, 14013 (1998)
21. Arash, B., Wang, Q.: A review on the application of nonlocal elastic models in modeling of carbon nanotubes and graphenes. Comput. Mater. Sci. **51**, 303–313 (2012)
22. Eringen, A.C.: Nonlocal Continuum Field Theories. Springer, New York (2002)
23. Ansari, R., Shahabodini, A., Rouhi, H.: Prediction of the biaxial buckling and vibration behavior of graphene via a nonlocal atomistic-based plate theory. Compos. Struct. **95**, 88–94 (2013)
24. Aydogdu, M.: A general nonlocal beam theory: its application to nanobeam bending, buckling and vibration. Phys. E **41**, 1651–1655 (2009)
25. Lee, H.L., Chang, W.J.: Free transverse vibration of the fluid-conveying single-walled carbon nanotube using nonlocal elastic theory. J. Appl. Phys. **103**, 024302 (2008)
26. Peddieson, J., Buchanan, G.R., McNitt, R.P.: Application of nonlocal continuum models to nanotechnology. Int. J. Eng. Sci. **41**, 305–312 (2003)
27. Reddy, J.N.: Nonlocal theories for bending, buckling and vibration of beams. Int. J. Eng. Sci. **45**, 288–307 (2007)
28. Thai, H.T., Vo, T.P.: A nonlocal sinusoidal shear deformation beam theory with application to bending, buckling, and vibration of nanobeams. Int. J. Eng. Sci. **54**, 58–66 (2012)
29. Adali, S.: Variational principles for multi-walled carbon nanotubes undergoing buckling based on nonlocal elasticity theory. Phys. Lett. A **372**, 5701–5705 (2008)
30. Setoodeh, A.R., Khosrownejad, M., Malekzadeh, P.: Exact nonlocal solution for postbuckling of single-walled carbon nanotubes. Phys. E **43**, 1730–1737 (2011)

31. Shen, H.S.: Buckling and postbuckling of radially loaded microtubules by nonlocal shear deformable shell model. Theor. Biol. **264**, 386–394 (2010)
32. Yang, Y., Lim, C.W.: A variational principle approach for buckling of carbon nanotubes based on nonlocal Timoshenko beam models. Nano **6**, 363–377 (2011)
33. Fu, Y., Du, H., Huang, W.M., Zhang, S.: A theoretical model for functionally graded shape memory alloy cylinders subjected to internal pressure. Mater. Lett. **57**, 2995–2999 (2003)
34. Praveen, G.N., Reddy, J.N.: Nonlinear transient thermoelastic analysis of functionally graded ceramic-metal plates. Int. J. Solids Struct. **35**(33), 4457–4476 (1998)
35. Toupin, R.A.: Elastic materials with couple-stresses. Arch. Ration. Mech. Anal. **11**, 385–414 (1962)
36. Mindlin, R.D., Tiersten, H.F.: Effects of couple-stresses in linear elasticity. Arch. Ration. Mech. Anal. **11**, 415–448 (1962)
37. Koiter, W.T.: Couple-stresses in the theory of elasticity, I and II. Proc. K. Ned. Akad. Wet. B **67**, 17–44 (1964)
38. Asghari, M., Ahmadian, M.T., Kahrobaiyan, M.H., Rahaeifard, M.: On the size-dependent behavior of functionally graded micro-beams. Mater. Des. **31**, 2324–2329 (2010)
39. Asghari, M., Kahrobaiyan, M.H., Ahmadian, M.T.: A nonlinear Timoshenko beam formulation based on the modified couple stress theory. Int. J. Eng. Sci. **48**, 1749–1766 (2010)
40. Lazopoulos, K.A., Lazopoulos, A.K.: Bending and buckling of thin strain gradient elastic beams. Eur. J. Mech. A-Solids **29**, 837–843 (2010)
41. Ma, H.M., Gao, X.L., Reddy, J.N.: A non-classical Mindlin plate model based on a modified couple stress theory. Acta Mech. **220**, 217–235 (2011)
42. Ma, H.M., Gao, X.L., Reddy, J.N.: A nonclassical Reddy-Levinson beam model based on a modified couple stress theory. Int. J. Multiscale Comput. Eng. **8**, 167–180 (2010)
43. Park, S.K., Gao, X.L.: Bernoulli-Euler beam model based on a modified couple stress theory. Micromech. Microeng. **16**(11), 2355–2359 (2006)
44. Yang, F., Chong, M., Lam, D.C.C., Tong, P.: Couple stress based strain gradient theory for elasticity. Int. J. Solids Struct. **39**, 2731–2743 (2002)
45. Batra, R.C., Porfiri, M., Spinello, D.: Vibrations of narrow microbeams predeformed by an electric field. J. Sound Vib. **309**, 600–612 (2008)
46. Jia, X.L., Yang, J., Kitipornchai, S., Lim, C.W.: Pull-in instability and free vibration of electrically actuated poly-SiGe graded micro-beams with a curved ground electrode. Appl. Math. Model. **36**, 1875–1884 (2012)
47. Lu, C., Lim, C., Chen, W.: Size-dependent elastic behavior of FGM ultra-thin films based on generalized refined theory. Int. J. Solids Struct. **46**, 1176–1185 (2009)
48. Mohammadi-Alasti, B., Rezazadeh, G., Borgheei, A.M., Minaei, S., Habibifar, R.: On the mechanical behavior of a functionally graded micro-beam subjected to a thermal moment and nonlinear electrostatic pressure. Compos. Struct. **93**, 1516–1525 (2011)
49. Zhang, J., Fu, Y.: Vibration of size-dependent functionally graded sandwich microbeams with different boundary conditions based on the modified couple stress theory. Meccanica **47**, 1649–1658 (2012)
50. Ansari, R., Gholami, R.: Size-dependent vibration of functionally graded curved microbeams based on the modified strain gradient elasticity theory. Arch. Appl. Mech. **83**, 1439–1449 (2013)
51. Ansari, R., Gholami, R., Sahmani, S.: Study of small scale effects on the nonlinear vibration response of functionally graded Timoshenko microbeams based on the strain gradient theory. J. Comput. Nonlinear Dyn. **7**, 031009 (2012)
52. Asgharifard, S.P., Haeri Yazdi, M.R.: Nonlinear free vibrations of functionally graded nanobeams with surface effects. Compos. B Eng. **45**, 581–586 (2013)
53. Eltaher, M.A., Emam, S.A., Mahmoud, F.F.: Free vibration analysis of functionally graded size-dependent nanobeams. Appl. Math. Comput. **218**, 7406–7420 (2012)
54. Eltaher, M.A., Emam, S.A., Mahmoud, F.F.: Static and stability analysis of nonlocal functionally graded nanobeams. Compos. Struct. **96**, 82–88 (2013)

55. Janghorban, M., Zare, A.: Free vibration analysis of functionally graded carbon nanotubes with variable thickness by differential quadrature method. Phys. E **43**, 1602–1604 (2011)
56. Kahrobaiyan, M.H., Rahaeifard, M., Tajalli, S.A., Ahmadian, M.T.: A strain gradient functionally graded Euler-Bernoulli beam formulation. Int. J. Eng. Sci. **52**, 65–76 (2012)
57. Ke, L.L., Wang, Y.S., Yang, J., Kitipornchai, S.: Nonlinear free vibration of size-dependent functionally graded microbeams. Int. J. Eng. Sci. **50**(1), 256–267 (2012)
58. Kiani, K.: Longitudinal and transverse instabilities of moving nanoscale beam-like structures made of functionally graded materials. Compos. Struct. **107**, 610–619 (2014)
59. Nateghi, A., Salamat-Talab, M., Rezapour, J., Daneshian, B.: Size dependent buckling analysis of functionally graded micro beams based on modified couple stress theory. Appl. Math. Model. **36**, 4971–4987 (2012)
60. Reddy, J.N.: Microstructure-dependent couple stress theories of functionally graded beams. Mech. Phys. Solids **59**, 2382–2399 (2011)
61. Simsek, M., Yurtcu, H.H.: Analytical solutions for bending and buckling of functionally graded nanobeams based on the nonlocal Timoshenko beam theory. Compos. Struct. **97**, 378–386 (2013)
62. Simsek, M.: Nonlocal effects in the free longitudinal vibration of axially functionally graded tapered nanorods. Comput. Mater. Sci. **61**, 257–265 (2012)
63. Asghari, M., Rahaeifard, M., Kahrobaiyan, M.H., Ahmadian, M.T.: The modified couple stress functionally graded Timoshenko beam formulation. Mater. Des. **32**, 1435–1443 (2011)
64. Arbind, A., Reddy, J.N., Srinivasa, A.R.: Modified couple stress-based third-order theory for nonlinear analysis of functionally graded beams. Lat. Am. J. Solids Struct. **11**, 459–487 (2014)
65. Ansari, R., Sahmani, S., Rouhi, H.: Rayleigh-Ritz axial buckling analysis of single-walled carbon nanotubes with different boundary conditions. Phys. Lett. A **375**, 1255–1263 (2011)
66. Anjomshoa, A.: Application of Ritz functions in buckling analysis of embedded orthotropic circular and elliptical micro/nano-plates based on nonlocal elasticity theory. Meccanica **48**, 1337–1353 (2013)
67. Rahmani, O., Jandaghian, A.A.: Buckling analysis of functionally graded nanobeams based on a nonlocal third-order shear deformation theory. Appl. Phys. A **119**, 1019–1032 (2015)
68. Ansari, R., Gholami, R., Sahmani, S.: Size-dependent vibration of functionally graded curved microbeams based on the modified strain gradient elasticity theory. Arch. Appl. Mech. **83**, 1439–1449 (2013)
69. Liu, Y.P., Reddy, J.N.: A nonlocal curved beam model based on a modified coupled stress theory. Int. J. Struct. Stab. Dyn. **11**, 495–512 (2011)
70. Zhang, B., He, Y., Liu, D., Gan, Z., Shen, L.: A novel size-dependent functionally graded curved microbeam model based on the strain gradient elasticity theory. Compos. Struct. **106**, 374–392 (2013)
71. Xiang, H.J., Yang, J.: Free and forced vibration of a laminated FGM Timoshenko beam of variable thickness under heat conduction. Compos. Part B: Eng. **39**(2), 292–303 (2008)
72. Dehrouyeh-Semnani, A.M., Dehrouyeh, M., Torabi-Kafshgari, M., Nikkhah-Bahrami, M.: A damped sandwich beam model based on symmetric-deviatoric couple stress theory. Int. J. Eng. Sci. **92**, 83–94 (2015)
73. Chen, W.J., Li, X.P.: Size-dependent free vibration analysis of composite laminated Timoshenko beam based on new modified couple stress theory. Arch. Appl. Mech. **83**(3), 431–444 (2013)
74. Chen, W.J., Li, X.P., Ma, X.: A modified couple stress model for bending analysis of composite laminated beams with first order shear deformation. Compos. Struct. **93**(11), 2723–2737 (2011)
75. Mohammadimehr, M., Shahedi, S., Navi, B.R.: Nonlinear vibration analysis of FG-CNTRC sandwich Timoshenko beam based on modified couple stress theory subjected to longitudinal magnetic field using generalized differential quadrature method. Arch. Proc. Inst. Mech. Eng. C J. Mech. Eng. Sci. **203–210**, 1989–1996 (2016)

76. Grigolyuk, E.I., Chulkov, P.P.: Stability and Oscillation of Sandwich Shells. Mashinostroenie, Moscow (1973) (in Russian)
77. Krysko, V.A., Koch, M.I., Krysko, A.V., Zhigalov, M.V.: Chaotic phase synchronization of vibration multilayer beam structures. J. Appl. Mech. Tech. Phys. **53**(3), 451–459 (2012)
78. Awrejcewicz, J., Krysko, A.V., Zhigalov, M.V., Saltykova, O.A., Krysko, V.A.: Chaotic vibrations in flexible multilayered Bernoulli-Euler and Timoshenko type beams. Lat. Am. J. Solids Struct. **5**(4), 319–363 (2008)
79. Alshorbagy, A.E., Eltaher, M.A., Mahmoud, F.F.: Free vibration characteristics of a functionally graded beam by finite element method. Appl. Math. Model. **35**(1), 412–425 (2011)
80. Li, L., Xiaobai, L., Yujin, H.: Free vibration analysis of nonlocal strain gradient beams made of functionally graded material. Int. J. Eng. Sci. **102**, 77–92 (2016)
81. Nazemnezhad, R., Hosseini-Hashemi, S.: Nonlocal nonlinear free vibration of functionally graded nanobeams. Compos. Struct. **110**, 192–199 (2014)
82. Rahimi, G.H., Davoodinik, A.R.: Thermal behavior analysis of the functionally graded Timoshenko's beam. IUST Int. J. Eng. Sci. **19**(5–1), 105–113 (2008)
83. Li, Y., Meguid, S.A., Fu, Y., Xu, D.: Nonlinear analysis of thermally and electrically actuated functionally graded material microbeam. Proc. R. Soc. A **470**, 20130473 (2014)
84. Bao, G., Wang, L.: Multiple cracking in functionally graded ceramic/metal coatings. Int. J. Solids Struct. **32**, 2853–2871 (1995)
85. Lee, Y.D., Erdogan, F.: Residual/thermal stress in FGM and laminated thermal barrier coatings. Int. J. Fract. **69**, 145–165 (1995)
86. Chung, Y.L., Chi, S.H.: The residual stress of functionally graded materials. J. Chin. Inst. Civ. Hydraul. Eng. **13**, 1–9 (2001)
87. Shafiei, N., Mousavi, A., Ghadiri, M.: On size-dependent nonlinear vibration of porous and imperfect functionally graded tapered microbeams. Int. J. Eng. Sci. **106**, 42–56 (2016)
88. Ansari, R., Gholami, R., Sahmani, S.: Free vibration analysis of size-dependent functionally graded microbeams based on the strain gradient Timoshenko beam theory. Compos. Struct. **94**, 221–228 (2011)
89. Vel, S.S., Batra, R.C.: Exact solution for thermoelastic deformations of functionally graded thick rectangular plates. AIAA J. **40**, 1421–1433 (2002)
90. Hatta, H., Taya, M.: Effective thermal conductivity of a misoriented short fiber composite. J. Appl. Phys. **58**(7), 2478–2486 (1985)
91. Rosen, B.W., Hashin, Z.: Effective thermal expansion coefficients and specific heats of composite materials. Int. J. Eng. Sci. **8**, 157–173 (1970)
92. Hill, R.: A self-consistent mechanics of composite materials. J. Mech. Phys. Solids **13**(2), 213–222 (1965)
93. Hashin, Z.: Assessment of the self consistent scheme approximation: conductivity of particulate composites. J. Compos. Mater. **2**, 284–300 (1968)
94. Azadi, M.: A finite element method (FEM) free and forced lateral vibration analysis of beams made of functionally graded materials (FGMs). J. Mech. Sci. Technol. **25**(1), 69–80 (2011)
95. Shen, H.: Functionally Graded Materials, Non-linear Analysis of Plates and Shells. CRC Press, New York (2009)
96. Young-Wann, K.: Temperature dependent vibration analysis of functionally graded rectangular plates. J. Sound Vib. **284**, 531–549 (2005)
97. Zhong, Z., Yu, T.: Analytical solution of a cantilever functionally graded beam. Compos. Sci. Technol. **67**, 481–488 (2007)
98. Ying, J., Li, C.F., Chen, W.Q.: Two-dimensional elasticity solutions for functionally graded beams resting on elastic foundations. Compos. Struct. **84**, 209–219 (2008)
99. Li, X.F.: A unified approach for analyzing static and dynamic behaviors of functionally graded Timoshenko and Euler-Bernoulli beams. J. Sound Vib. **318**, 1210–1229 (2008)
100. Kadoli, R., Akhtar, K., Ganesan, N.: Static analysis of functionally graded beams using higher order shear deformation theory. Appl. Math. Model. **32**, 2509–2525 (2008)
101. Arbind, A., Reddy, J.N.: Nonlinear analysis of functionally graded microstructure-dependent beams. Compos. Struct. **98**, 272–281 (2013)

102. Lam, D.C.C., Yang, F., Chong, A.C.M., Wang, J., Tong, P.: Experiments and theory in strain gradient elasticity. J. Mech. Phys. Solids **51**, 1477–1508 (2003)
103. Eringen, A.C.: Nonlocal polar elastic continua. Int. J. Eng. Sci. **10**, 1–16 (1972)
104. Aifantis, E.C.: Strain gradient interpretation of size effects. Int. J. Fract. **95**, 299–314 (1999)
105. Gurtin, M.E., Weissmuller, J., Larche, F.: The general theory of curved deformable inter-faces in solids at equilibrium. Philos. Mag. A **78**, 1093–1109 (1998)
106. Ma, H.M., Gao, X.-L., Reddy, J.N.: A microstructure-dependent Timoshenko beam model based on a modified couple stress theory. J. Mech. Phys. Solids **56**, 3379–3396 (2008)
107. Begley, M.R., Hutchinson, J.W.: The mechanics of size-dependent indentation. J. Mech. Phys. Solids **46**, 2049–2068 (1998)
108. Nix, W.D., Gao, H.: Indentation size effects in crystalline materials: a law for strain gradient plasticity. J. Mech. Phys. Solids **46**, 411–425 (1998)
109. Reddy, J.N.: Theory and Analysis of Elastic Plates and Shells. Taylor & Francis, Philadelphia (2007)
110. Awrejcewicz, J., Krysko, V.A., Papkova, I.V., Krysko, A.V.: Routes to chaos in continuous mechanical systems. Part 1: mathematical models and solution methods. Chaos Solitons Fractals **45**, 687–708 (2012)
111. Krysko, A.V., Awrejcewicz, J., Pavlov, S.P., Zhigalov, M.V., Krysko, V.A.: On the iterative methods of linearization, decrease of order and dimension of the Kármán-type PDEs. Sci. World J. **2014** (2014), 15 pp
112. Kirichenko, V.F., Awrejcewicz, J., Kirichenko, A.F., Krysko, A.V., Krysko, V.A.: On the non-classical mathematical models of coupled problems of thermo-elasticity for multi-layer shallow shells with initial imperfections. Int. J. Non-linear Mech. **74**, 51–72 (2015)
113. Awrejcewicz, J., Krysko, A.V., Soldatov, V.: On the wavelet transform application to a study of chaotic vibrations of the infinite length flexible panels driven longitudinally. Int. J. Bifurc. Chaos **19**(10), 3347–3371 (2009)
114. Awrejcewicz, J., Krysko, A.V., Kutepov, I., Zagniboroda, N., Zhigalov, M., Krysko, V.A.: Analysis of chaotic vibrations of flexible plates using fast Fourier transforms and wavelets. Int. J. Struct. Stab. Dyn. **13**(7), 1340005-1–1340004-12 (2013)
115. Awrejcewicz, J., Saltykova, O.A., Zhigalov, M.V., Hagedorn, P., Krysko, V.A.: Analysis of non-linear vibrations of single-layered Euler-Bernoulli beams using wavelets. Int. J. Aerosp. Lightweight Struct. **1**(2), 203–219 (2011)
116. Krysko, A.V., Awrejcewicz, J., Zhigalov, M.V., Krysko, V.A.: On the contact interaction between two rectangular plates. Nonlinear Dyn. **85**(4), 2729–2748 (2016)
117. Krysko, A.V., Awrejcewicz, J., Saltykova, O.A., Zhigalov, M.V., Krysko, V.A.: Investigations of chaotic dynamics of multi-layer beams using taking into account rotational inertial effects. Commun. Nonlinear Sci. Numer. Simul. **19**(8), 2568–2589 (2014)
118. Krysko, V.A., Awrejcewicz, J., Kutepov, I.E., Zagniboroda, N.A., Serebryakov, A.V., Krysko, A.V., Papkova, I.V.: Chaotic dynamics of flexible beams with piezoelectric and temperature phenomena. Phys. Lett. A **377**, 2058–2061 (2013)
119. Awrejcewicz, J., Krysko, V.A., Kutepov, I.E., Zagniboroda, N.A., Dobriyan, V., Papkova, I.V., Krysko, A.V.: Chaotic vibrations of flexible curvilinear beams in temperature and electric fields. Int. J. Non-linear Mech. **76**, 29–41 (2015)
120. Awrejcewicz, J., Krysko, A.V., Papkova, I.V., Zakharov, V.M., Erofeev, N.P., Krylova, E.Yu., Mrozowski, J., Krysko, V.A.: Chaotic dynamics of flexible beams driven by external white noise. Mech. Syst. Signal Process. **79**, 225–253 (2016)
121. Awrejcewicz, J., Krysko, V.A., Papkova, I.V., Krysko, A.V.: Deterministic Chaos in One-Dimensional Continuous Systems. World Scientific, Singapore (2016)
122. Timoshenko, S.P.: On the correction for shear of differential equation for transverse vibration of prismatic bar. Philos. Mag. **41**, 744–746 (1921)
123. Krysko, A.V., Awrejcewicz, J., Kutepov, I.E., Zagniboroda, N.A., Dobriyan, V., Krysko, V.A.: Chaotic dynamics of flexible Euler-Bernoulli beams. Chaos **23**(4), 043130-1–043130-25 (2013)

124. Tikhonov, A.N., Arsenin, V.Ya.: Methods of Solution of the Non-corrected Problems. Nauka, Moscow (1979) (in Russian)
125. Krysko, V.A., Awrejcewicz, J., Komarov, S.A.: Nonlinear deformations of spherical panels subjected to transversal load action. Comput. Methods Appl. Mech. Eng. **194**(27–29), 3108–3126 (2005)
126. Franklin, J.N.: On Tikhonov's method for ill-posed problems. Math. Comput. **28**(128), 889–907 (1974)
127. Miller, K.: Least squares methods for ill-posed problems with a prescribed bound. SIAM J. Math. Anal. **1**(6), 52–74 (1970)
128. Phillips, D.L.: A technique for the numerical solution of certain integral equations of the first kind. J. Assoc. Comput. Mach. **9**(6), 84–97 (1962)
129. Wolf, A., Swift, J.B., Swinney, H.L., Vastano, J.A.: Determining Lyapunov exponents from a time series. Phys. D **16**, 285–317 (1985)
130. Rosenstein, M.T., Collins, J.J., De Luca, C.J.: A practical method for calculating largest Lyapunov exponents from small data sets. Phys. D **65**, 117–134 (1993)
131. Kantz, H.: A robust method to estimate the maximal Lyapunov exponent of a time series. Phys. Lett. A **185**, 77–87 (1994)
132. Awrejcewicz, J., Krysko, V.A.: Feigenbaum scenario exhibited by thin plate dynamics. Nonlinear Dyn. **24**, 373–398 (2006)
133. Awrejcewicz, J.A., Krysko, V.A., Zhigalov, M.V., Saltykova, O.A., Krysko, A.V.: Chaotic vibrations in flexible multilayered Bernoulli-Euler and Timoshenko type beams. Lat. Am. J. Solids Struct. **5**(4), 319–363 (2008)
134. Ke, L.L., Wang, Y.S.: Size effect on dynamic stability of functionally graded microbeams based on a modified couple stress theory. Compos. Struct. **93**, 342–350 (2011)
135. Volmir, A.S.: Nonlinear Dynamics Plates and Shells. Science, Moscow (1972) (in Russian)
136. Gao, X.L., Zhang, G.Y.: A microstructure- and surface energy-dependent third-order shear deformation beam model. Z. Angew. Math. Phys. **66**, 1871–1894 (2015)
137. Ghayesh, M.H., Farokhi, H., Amabili, M.: Coupled nonlinear size-dependent behaviour of microbeams. Appl. Phys. A **112**, 329–338 (2013)
138. Fedoseyev, V.I.: On the method of finding solution to the non-linear stability problems of deformable systems. Appl. Math. Mech. **27**(2), 265–274 (1963) (in Russian)
139. Awrejcewicz, J., Krysko, V.A., Dobriyan, V., Papkova, I.V., Krysko, A.V.: On the Lyapunov exponents computation of coupled non-linear Euler-Bernoulli beams. In: Proceedings of the Fourteenth International Conference on Civil, Structural and Environmental Engineering Computing. Civil-Comp Press, Stirlingshire, UK, Paper 53 (2013)
140. Krysko, A.V., Awrejcewicz, J., Kutepov, I.E., Krysko, V.A.: Stability of curvilinear Euler-Bernoulli beams in temperature fields. Int. J. Non-linear Mech. **94**, 207–215 (2017)
141. Vaz, M.A., Solano, R.F.: Post-buckling analysis of slender elastic rods subjected to uniform thermal loads. J. Therm. Stress. **26**, 847–860 (2003)
142. Li, S.R., Cheng, C.J., Zhou, Y.H.: Thermal post-buckling of elastic beams subjected to a transversely non-uniform temperature rising. Appl. Math. Mech. **24**(5), 514–520 (2003)
143. Li, S.R., Cheng, C.J., Zhou, Y.H.: Thermal post-buckling analysis of heated elastic rods. Appl. Math. Mech. **21**(2), 133–140 (2000)
144. Li, S.R., Zhou, Y.H., Zheng, X.J.: Thermal post-buckling of heated elastic rods with pinned-fixed ends. J. Therm. Stress. **25**, 45–56 (2002)
145. Li, S.R., Zhou, Y.H.: Geometrically nonlinear analysis of Timoshenko beams under thermomechanical loadings. J. Therm. Stress. **26**, 861–872 (2003)
146. Dinzart, F., Molinari, A., Herbach, R.: Thermomechanical response of a viscoelastic beam under cyclic bending; self-heating and thermal failure. Arch. Mech. **60**(1), 59–85 (2008)
147. Abbasi, M., Sabbaghian, M., Eslami, M.R.: Exact closed-form solution of the dynamic coupled thermoelastic response of a functionally graded Timoshenko beam. J. Mech. Mater. Struct. **5**(1), 79–94 (2010)
148. Ma, L.S., Lee, D.W.: A further discussion of nonlinear mechanical behavior for FGM beams under in-plane thermal loading. Compos. Struct. **93**(2), 831–842 (2011)

149. Gupta, R.K., Jagadish, B.G., Ranga, J.G., Venkateswara, R.G.: Thermal post-buckling analysis of slender columns using the concept of coupled displacement field. Int. J. Mech. Sci. **52**(4), 590–594 (2010)
150. He, P., Liu, Z., Li, C.: An improved beam element for beams with variable axial parameters. Shock Vib. **20**, 601–617 (2013)
151. Nayak, S., Bisoi, A., Dash, P.R., Pradhan, P.K.: Static stability of a viscoelastically supported asymmetric sandwich beam with thermal gradient. Int. J. Adv. Struct. Eng. **65** (2014), 7 pp
152. Benzair, A., Tounsi, A., Besseghier, A., Heireche, H., Moulay, N., Boumia, L.: The thermal effect on vibration of single-walled carbon nanotubes using nonlocal Timoshenko beam theory. J. Phys. D: Appl. Phys. **41** (2008) 225404 (10 pp)
153. Kiani, Y., Eslami, M.R.: Thermomechanical buckling of temperature-dependent FGM beams. Lat. Am. J. Solids Struct. **10**, 223–246 (2013)
154. Youssef, H.M., El-Bary, A.A., Elsibai, K.A.: Vibration of gold nano beam in context of two-temperature generalized thermoelasticity subjected to laser pulse. Lat. Am. J. Solids Struct. **11**(13), 2460–2482 (2014)
155. Prabhakar, S., Vengallatore, S.: Theory of thermoelastic damping in micromechanical resonators with two-dimensional heat conduction. J. Microelectromech. Syst. **17**(2), 494–502 (2008)
156. Akbas, S.D.: Free vibration of axially functionally graded beams in thermal environment. Int. J. Eng. Appl. Sci. **6**(3), 37–51 (2014)
157. Esfahani, S.E., Kiani, Y., Komijani, M., Eslami, M.R.: Vibration of a temperature-dependent thermally pre/postbuckled FGM beam over a nonlinear hardening elastic foundation. J. Appl. Mech. **81**, 011004 (2014)
158. Awrejcewicz, J., Krysko, V.A.: Nonclassical Thermoelastic Problems in Nonlinear Dynamics of Shells. Springer, Berlin (2003)
159. Awrejcewicz, J., Krysko, V.A., Krysko, A.V.: Thermo-Dynamics of Plates and Shells. Springer, Berlin (2007)
160. Krysko, A.V., Awrejcewicz, J., Kuznetsova, E.S., Krysko, V.A.: Chaotic vibrations of closed cylindrical shells in a temperature field. Shock Vib. **15**(3–4), 335–343 (2008)
161. Krysko, A.V., Awrejcewicz, J., Kutepov, I.E., Krysko, V.A.: On a contact problem of two-layer beams coupled by boundary conditions in a temperature field. J. Therm. Stress. **38**, 468–484 (2015)
162. Love, A.E.H.: A Treatise on the Mathematical Theory of Elasticity. Dover Publications, New York (1944)
163. Kármán, T.L.: The Collected Works. Butterworths, London (1956)
164. Vlasov, V.Z.: General Theory for Shells and Its Application in Engineering. Gostekhizdat Publication, Moscow (1949)
165. Duhamel, J.M.C.: Second memoire sur les phenomenes thermoomecaniques. de l'Ecole Polytechnique **15** (1837)
166. Volmir, A.S.: Stability of Deformable Bodies. Nauka, Moscow (1968)
167. Bolotin, V.V., Novichkov, Yu.N.: Mechanics of Multilayer Constructions. Mashinostroenie, Moscow (1980)
168. Krysko, A.V., Awrejcewicz, J., Zagniboroda, N.A., Dobriyan, V., Krysko, V.A., Kutepov, I.E.: Chaotic dynamics of flexible Euler-Bernoulli beams. Chaos **34**(4), 043130 (2014)
169. Carslou, H., Jaeger, J.: Conduction of Heat in Solids. Oxford University Press, Oxford (1959)
170. Krysko, A.V., Awrejcewicz, J., Pavlov, S.P., Zhigalov, M.V., Krysko, V.A.: Mathematical model of a three-layer micro- and nano-beams based on the hypotheses of the Grigolyuk-Chulkov and the modified couple stress theory. Int. J. Solids Struct. **117**, 39–50 (2017)
171. Zenkour, A.M.: Transverse shear and normal deformation theory for bending analysis of laminated and sandwich elastic beams. Mech. Compos. Mater. Struct. **6**(3), 267–283 (1999)
172. Miller, R.E., Shenoy, V.B.: Size-dependent elastic properties of nanosized structural elements. Nanotechnology **11**, 139–147 (2000)
173. Chong, A.C.M., Yang, F., Lam, D.C.C., Tong, P.: Torsion and bending of micron-scaled structures. J. Mater. Res. **14**, 1052–1058 (2001)

174. Sun, Z.H., Wang, X.X., Soh, A.K., Wu, H.A., Wang, Y.: Bending of nanoscale structures: inconsistency between atomistic simulations and strain gradient elasticity solution. Comput. Mater. Sci. **40**, 108–113 (2007)
175. Andreev, A.N., Nemirovskii, Y.V.: Multilayered Anisotropic Shells and Plates: Bend, Stability, Vibration. Nauka, Novosibirsk (2001)
176. Awrejcewicz, J., Krysko-Jr., V.A., Yakovleva, T.V., Krysko, V.A.: Noisy contact interactions of multi-layer mechanical structures coupled by boundary conditions. J. Sound Vib. **369**, 77–86 (2016)
177. Srinivasa, A.R., Reddy, J.N.: A model for a constrained, finitely deforming, elastic solid with rotation gradient dependent strain energy, and its specialization to von Kármán plates and beams. J. Mech. Phys. Solids **61**(3), 873–885 (2013)
178. Park, S.K., Gao, X.L.: Variational formulation of a modified couple stress theory and its application to a simple shear problem. Z. Angew Math. Phys. **59**, 904–917 (2008)
179. Yang, J., Ono, T., Esashi, M.: Energy dissipation in submicrometer thick single-crystal silicon cantilevers. J. Microelectromech. Syst. **11**(6), 775–783 (2002)
180. Mindlin, R.D.: Influence of couple-stresses on stress concentrations. Exp. Mech. **3**, 1–7 (1963)
181. Eringen, A.C.: On differential equations of nonlocal elasticity and solutions of screw dislocation and surface waves. J. Appl. Phys. **54**, 4703–4710 (1983)
182. Polizzotto, C.: Gradient elasticity and nonstandard boundary conditions. Int. J. Solids Struct. **40**, 7399–7423 (2003)
183. Wang, C.M., Zhang, Y.Y., Ramesh, S.S., Kitipornchai, S.: Buckling analysis of micro- and nano-rods/tubes based on nonlocal Timoshenko beam theory. J. Phys. D: Appl. Phys. **39**, 3904–3909 (2006)
184. Wang, Q.: Wave propagation in carbon nanotubes via nonlocal continuum mechanics. J. Appl. Phys. **98**, 124301 (2005)
185. Santos, A., Reddy, J.N.: Vibration of Timoshenko beams using non-classical elasticity theories. Shock Vib. **19**(3), 251–256 (2012)
186. Reddy, J.N., Arbind, A.: Bending relationships between the modified couple stress-based functionally graded Timoshenko beams and homogeneous Bernoulli–Euler beams. A. Ann. Solid Struct. Mech. **3**(1), 15–26 (2012)
187. Kahrobaiyan, M.H., Asghari, M., Ahmadian, M.T.: A Timoshenko beam element based on the modified couple stress theory. Int. J. Mech. Sci. **79**, 75–83 (2014)
188. Thai, H.T., Vo, T.: A size-dependent functionally graded sinusoidal plate model based on a modified couple stress theory. Compos. Struct. **96**, 376–383 (2013)
189. Alashti, R.A., Abolghasemi, A.H.: A size-dependent Bernoulli-Euler beam formulation based on a new model of couple stress theory. IJE Trans. C: Asp. **27**(6), 951–960 (2014)
190. Chen, W., Li, X.: A new modified couple stress theory for anisotropic elasticity and microscale laminated Kirchhoff plate model. Arch. Appl. Mech. **84**, 323–341 (2014)
191. Mohammad-Abadi, M., Daneshmehr, A.R.: Modified couple stress theory applied to dynamic analysis of composite laminated beams by considering different beam theories. Int. J. Eng. Sci. **87**, 83–102 (2015)
192. Shafiei, N., Kazemi, M., Fatahi, L.: Transverse vibration of rotary tapered microbeam based on modified couple stress theory and generalized differential quadrature. Mech. Adv. Mater. Struct. **24**(3), 240–252 (2015)
193. Rajneesh, K.: Response of thermoelastic beam due to thermal source in modified couple stress theory. Comput. Methods Sci. Technol. **22**(2), 95–101 (2016)
194. Khorshidi, M.A., Shariati, M., Emam, S.A.: Postbuckling of functionally graded nanobeams based on modified couple stress theory under general beam theory. Int. J. Mech. Sci. **110**, 160–169 (2016)
195. Abazari, A.M., Safavi, S.M., Rezazadeh, G., Villanueva, L.G.: Modelling the size effects on the mechanical properties of micro/nano structures. Sensors **15**, 28543–28562 (2015)
196. Tsiatas, G.C., Yiotis, A.J.: Size effect on the static, dynamic and buckling analysis of orthotropic Kirchhoff-type skew micro-plates based on a modified couple stress theory: comparison with the nonlocal elasticity theory. Acta Mech. **226**, 1267–1281 (2015)

197. Challamel, N., Wang, C.M.: The small length scale effect for a non-local cantilever beam: a paradox solved. Nanotechnology **19**, 345703 (2008)
198. Kong, S., Zhou, S., Nie, Z., Wang, K.: Static and dynamic analysis of micro beams based on strain gradient elasticity theory. Int. J. Eng. Sci. **47**, 487–498 (2009)

Chapter 8
Thermoelastic Vibrations of Timoshenko Microbeams (Modified Couple Stress Theory)

8.1 Introduction

Thermoelastic vibrations of the Timoshenko microbeams based on the modified couple stress theory are studied. In particular, the dependence of the quality factor of nonlinear microbeam resonators under thermoelastic damping for Timoshenko beams with regard to geometric nonlinearity is analyzed. The constructed mathematical model is based on the modified couple stress theory which implies prediction of size-dependent effects in microbeam resonators. The Hamilton principle yields coupled nonlinear thermoelastic PDEs governing dynamics of the Timoshenko microbeams for both plane stresses and plane deformations. Nonlinear thermoelastic vibrations are investigated analytically and numerically and quality factors of the resonators versus geometric and material microbeam properties are estimated. Results are presented for gold microbeams for different ambient temperatures and different beam thicknesses, and they are compared with results yielded by the classical theory of elasticity in linear/nonlinear cases [1]. The most important conclusions of our study are summarized in the following three points.

1. Results reported in Figs. 8.4, 8.6 and 8.5, 8.7 with size-dependent behaviour and in the case of linear vibrations exhibit increase of the eigenfrequencies and increase of the quality factor of the beam resonator.
2. The occurrence of nonlinearity in both tested boundary (simple-simple) and (clamped-clamped) conditions implies a significant increase in the frequency of the fundamental vibration mode as well as the quality factor of the resonator (Figs. 8.8, 8.9, 8.10 and 8.11).
3. The analysis of the obtained results shows that in the case of the thermoelastic damping, it is necessary to take into account the nonlinear behaviour as well as size-dependent effects of the microbeams. Both mentioned features are crucial for the quality factor of the beam resonators.

J. Awrejcewicz et al., *Mathematical Modelling and Numerical Analysis of Size-Dependent Structural Members in Temperature Fields*, Advanced Structured Materials 142,
https://doi.org/10.1007/978-3-030-55993-9_8

8.2 Literature Review

Thermoelastic damping/dissipation (TED) occurs in all elastic materials subjected to cycling deformations, and particularly in the case when the period of an exciting cycle is close to the material relaxation [2–5]. For instance, in the case of vibrations of an elastic beam, a major part of mechanical work is transformed into elastic energy, but part of the work is converted into thermal energy. During bending, one side of the structure is extended and cooled, while the other side is compressed and heated. Temperature non-homogeneity described so far yields a transfer of heat from the more heated part to the more cooled part of the structure. According to the second law of thermodynamics, the latter heat transfer causes the increase of entropy and finally the loss of mechanical energy in the form of heat. There are various mechanisms of energy loss, such as thermoelastic damping [2, 6–8], energy loss in supports [9], internal energy loss [10] and air damping [11].

In almost all macroscopic systems with basic damping sources, thermoelastic coupling plays a secondary role because it is an order of magnitude lower than the damping caused by external and internal viscous effects. However, in the case of MEMS/NEMS devices and resonators, it plays a crucial role. It should be emphasized that micromechanical resonators with high quality factors are employed in a wide range of MEMS applications, such as sensors [12–16], electric filters [17, 18], accelerometers [19], giros [8], atomic force microscopes [20], MEMS type switches, and various other mechanisms [21–23]. From a technical point of view, the evaluation of thermoelastic damping (TED) of those structural elements at the design stage plays an important role in maintaining the required engineering characteristics. TED belongs to fundamental sources of quantification damping in the microelectromechanical systems (MEMS) [24] and nanoelectromechanical systems (NEMS) [4, 25] working in the vacuum regime. Evoy et al. [26] and Duwel et al. [8] have shown experimentally that TED is the dominant source of damping in MEMS and NEMS devices.

Zener [2, 27, 28] belongs to the first who defined the occurrence of TED and showed its crucial role in dissipation exhibited by bending resonators. Landau and Lifshitz [29] estimated analytically the TED coefficients of elastic vibrations. The second widely employed analytical model has been developed by Lifshitz and Roukes [3]. In that model, a change in the resonance frequency associated with TED was observed, and thus the classical Zener model was improved. The mentioned authors have also derived an analytical expression for quality factor Q in microbeam resonators and studied the influence of numerous geometric parameters on their quality factors. Nayfeh and Younis [30] constructed the analytical expression for Q of a microplate of arbitrary forms for different boundary conditions and studied the occurrence of TED. Sun et al. [31–33], Sharma et al. [34] and Rezazadeh et al. [35] investigated thermoelastic damping in capacity microbeam resonators using the hyperbolic heat transfer model. Besides, Vahdat and Rezazadeh [24] investigated the effect of axial and residual stresses on thermoelastic damping in a capacity microbeam resonator.

Berry [36] obtained experimentally Q factor as a function of frequency. Duwel et al. [8] presented experimental data illustrating the importance of TED in the resonance MEMS sensors. They also investigated the influence of various materials on TED behaviour. Zhang and Turner [37] developed a theory for TED for micro- and submicro-longitudinally vibrating beams. Vengallatore [38] studied TED in laminated composite micromechanical beam resonators. Prabhakar and Vengallatore [39] constructed an exact theory useful for TED computation in asymmetric two-layer micromechanical beam resonators. Yi [40] investigated the effect of geometric parameters on TED in the MEMS resonators. Analytical methods to study TED in nanobeams are presented in reference [41].

On the other hand, most of the models described in the literature use Euler-Bernoulli's beam theory which does not apply to a small ratio of beam length to its thickness. Massalas and Kalpakidis [42] investigated coupled thermoelastic vibrations of simply supported beams. They presented analytical solutions for the coupled problem of thermoelasticity of beams made from homogeneous and isotropic materials applicable for the Euler-Bernoulli and Timoshenko beam models. Shieh [43] studied TED of a simply supported Timoshenko beam with a circular cross section. He used 2D heat transfer equations and obtained damping coefficients. In reference [44], the authors proposed an analytical solution for the resonator quality factor under thermoelastic damping in beams based on the Timoshenko theory using the Lifshitz and Roukes method [3] for the Euler-Bernoulli beams.

There are known attempts to solve the problems of thermoelastic damping numerically. Prevost and Tao [45] employed finite element method (FEM) to solve a coupled thermoelastic problem. The authors used the second-order heat transfer PDE instead of the Fourier equation. Prabhakar and Vengallatore [46] matched 1D Euler-Bernoulli beam theory and 2D heat transfer PDE and obtained the quality factor using the introduced theory of series. Silver and Peterson [47] investigated elastic thermodynamic damping for the beam using FEM and also employed the method of perturbation. Sun et al. [48] obtained TED of microbeam resonators using the method of finite Fourier transform and modal analysis. Lepage [49] applied FEM to estimate the quality factor of TED of the Euler-Bernoulli beams. He employed the cubic approximation of temperature variation along with beam thickness. Guo et al. [50] used 2D FEM to study MEMS resonators. De and Aluru [51] employed a combined method of finite and boundary elements to validate the results of their TED-related modified theory.

In MEMS structures made of metals and polymers in which the smallest length of elements was in the micro-range, and the size-dependent phenomena were observed experimentally in many systems [52–56]. Effects of the dimensional dependence can be divided into two categories. The first one is implied by the boundary effect coupled with a molecular layer located on the specimen surfaces. Brezny and Green [57] and Onck et al. [58] observed that the effective Young modulus and resistance against compression of some materials essentially decreased while decreasing the size of the specimen. The second one known as the Kosser effect, or micropolar effect [59], occurs due to violation of the classical mechanical statements. Owing to the lack of eigen size of length, the classical mechanical theories cannot give proper interpretation and prediction of the observed size-dependent phenomena.

More recently, theories of the continuum of a higher order for prediction of the mentioned size-dependent relations were developed [60]. In the 1960s, some of researchers proposed the couple stress theory of elasticity which can be classified as the non-classical theory [61]. The theory made it possible to interpret the size-dependent effects using two higher order material constants in state equations. A simplification introduced by Yang et al. [62] to the scale complex relations of the couple stress theory of elasticity resulted in a modified couple stress theory which is suitable because of effects using only one scale parameter of the material length. Recently, many researchers investigated non-classical theorems of continuum in order to formulate and study the size-dependent mechanical behaviour of beams and plates.

Tsiatas [55] proposed a novel model of Kirchhoff plates based on a modified couple stress theory (MCST) which can take into account the size-dependent effects. Park et al. [63] proposed the Euler-Bernoulli beam model based on MCST, and they investigated the influence of the length material parameter on static mechanical microbeam properties.

Kong et al. [64, 65] solved the dynamical problem of Euler-Bernoulli beams based on the MCST and the gradient theory of elasticity and they demonstrated that eigenfrequencies of the microbeams obtained with the help of MCST are higher than those predicted by the classical theory of Euler-Bernoulli beams. Using MCST, Kahrobaiyan et al. [66] and Asghari et al. [67] studied analytically the static and dynamic behaviour of functionally graded microbeams and material of atomic force microscope microcantilevers.

Ma et al. [68] used Hamilton principle and relations of the nonlocal Eringer's theory to develop nonlocal theories of the Euler-Bernoulli, Timoshenko, Reddy and Levinson beams.

Besides, Wang et al. [60] proposed a microscale Timoshenko beam model based on the theory of gradient deformation. Using the Hamilton principle they derived the governing equations, initial and boundary conditions. The model took into account Poisson's effects, including three length material parameters, and consequently, it exhibited size effects.

In recent years, a series of works have been devoted to the study of MCST of beam and plates that also functionally graded taking into account shear deformations [35, 69–74]. Reddy [70] and Reddy and Kim [71] derived equations of beams and plates based on MCST.

However, in general, there are not many works reported in the existing literature devoted to the study of size-dependent effect in the problems of coupled thermoelasticity of microbeams. Guo and Rogerson [75] investigated the influence of the size on thermoelastic coupling in clamped elastic prismatic beams. Sun et al. [48] developed a theory for the coupled thermoelastic microbeam problem using the hyperbolic model of heat transfer. They demonstrated how the size effect generated by the thermoelastic coupling disappears when thickness of the microbeam exceeds its critical value depending on the material properties and boundary conditions. Rezazadeh et al. [76] derived analytical solutions to estimate the quality factor of TED obtained on the basis of MCST for the plane stress and plane deformation states of a linear Euler-Bernoulli beam. They found that when the beam thickness is close to the material length parameter, the results obtained by MCST are different compared to the results obtained using the classical beam theory.

In order to explain size dependence for microbeams, Taati et al. [77] combined the MCST and hyperbolic model of heat transfer which differs from the classical Fourier model. They showed that if one uses the hyperbolic heat transfer model then MCST shows more deviations from results compared to the classical theory.

In reference [78], the size-dependent model was proposed using the time of thermal relaxation for temporal analysis of thermoelasticity of Timoshenko microbeams. Besides, the size-dependent effects of a simply supported beam were studied under constant impulse force where two beam ends were exposed to constant ambient temperature.

Since eigenfrequencies of microbeam vibrations change when accounting for MCST, it may be expected that TED forecast by MCST will also differ from the results yielded by the classical theory of beams. Therefore, the influence of the size length material parameter on quality factor Q of microbeam resonators still requires a deeper analysis (in particular, when the characteristic size is compared with the internal size parameter of material length).

The microbeam resonators often work in nonlinear regimes with large amplitude while harvesting energy [79]. A key role in the studies of beam resonators plays energy dissipation in micro- and nanoscales [2, 4, 5, 80, 81]. In particular, TED belongs to one of the principal mechanism of energy dissipation for a large amount of the micro/nanomaterial resonators [3–5, 82, 83]. However, as it has already been mentioned, almost all published results deal with linearized regimes of vibrations of the microbeam resonators taking into account only vibrations with small amplitude.

Nonlinearity in micro-resonators may occur due to various reasons including large deflections (geometric nonlinearity or material nonlinearity). Geometric nonlinearity plays a key role in the microbeam resonator [84–86]. Though it was observed in high-frequency MEMS/NEMS resonators [79, 87–89], to our knowledge there are no reference studies on large deflection impact on thermoelastic dissipation. There are a few papers devoted to the analysis and quantitative estimation of the influence of thermoelastic damping in MEMS. Majority of the mentioned works deal with small amplitude linear vibrations [30, 32, 90]. For instance, in the work [83], though large static deflection generated by electrostatic forces under the action of constant stress is studied, thermoelastic investigation is based on the linearized vibration of small amplitude in the neighbourhood of static equilibrium with large deflection. An analogous approach can be found in reference [51], where large static deflection generated by electrostatic input affects the thermoelastic damping of linear small amplitude vibrations around static equilibrium. A series of analogous examples are reported in paper [41], where TEDs of resonators made from carbon tubes and vibrating around equilibrium were studied. Despite the thermoelastic dissipation analysis within linear theory, there are several studies devoted to nonlinear dynamics under large deflections of beam MEMS resonators [20, 91, 92]. Méndez et al. [92] studied large deflection nonlinearity at the resonant frequency localization and investigated the velocity of damping of cantilever microbeams but omitted the effect of large amplitude vibration on TED. In reference [93], the investigated thermoelastic damping with an account of nonlinear effects of the circle plate modelled with the use of the von Kármán hypothesis was analyzed. Employing the Kantorovich method and the methods of perturbation, a formula describing the resonator quality factor was obtained.

The authors of reference [94] studied nonlinear effects under large deflections and under thermoelastic dissipation of the microbeam resonators for adiabatic/isothermic beam surfaces. The models of thermoelasticity and thermoelastic dissipation were formulated for the case of large amplitude vibrations for the Euler-Bernoulli beam without the size-dependent effects. Mendez et al. [92] carried out numerical modelling for MEMS cantilever, and a difference between linear and nonlinear approximations was illustrated and discussed. It was shown that for large displacements, the beam becomes stiffer (frequency increase) and its vibrations are damped faster. Those observations made it possible to improve the fabrication of high-frequency MEMS resonators. In the work [95], the importance of nonlinear MEMS behaviour with a strong coupling between mechanical and electric fields was demonstrated.

The overview of the state of the art of the MEMS/NEMS beam vibrations carried out so far yields the following conclusions:

(i) thermoelastic damping of microbeam vibrations with large amplitudes plays a crucial role in a proper fabrication of the resonators;
(ii) the influence of geometric nonlinearity on the values of frequency, damping and on quality factors of the resonators is often omitted in the available literature;
(iii) both thermoelastic damping and large displacements may have an important effect not only on the high-frequency resonators but also on the MEMS/NEMS devices [39, 79, 88, 89];
(iv) in general, the thermoelastic PDEs governing the dynamics of microbeams were obtained based on the classical theory of continuum. There are no sufficient results following from the nonlinear model of coupled thermoelasticity of the Timoshenko microbeams based on the MCST.

This chapter is aimed at filling the gaps that exist in the research of MEMS/NEMS resonators. In this work, we employ the MCST to study quality factor of microbeam resonators with an account of thermoelastic damping and geometric nonlinearity. The associated mathematical model is constructed which, in contrary to the classical theory, consists of the internal scale length material parameter allowing for a reliable prediction of the size-dependent effect in microbeam resonators. The Hamilton principle yields coupled nonlinear thermoelastic equations of motion of the Timoshenko microbeams in the case of plane stress/deflection state. Based on the obtained equations both analytical and numerical investigations of the nonlinear thermoelastic vibrations of the microbeams are carried out and then the values of quality factors of the resonators versus geometric and material microbeam properties are estimated. The results are presented for gold microbeams at different ambient temperatures and

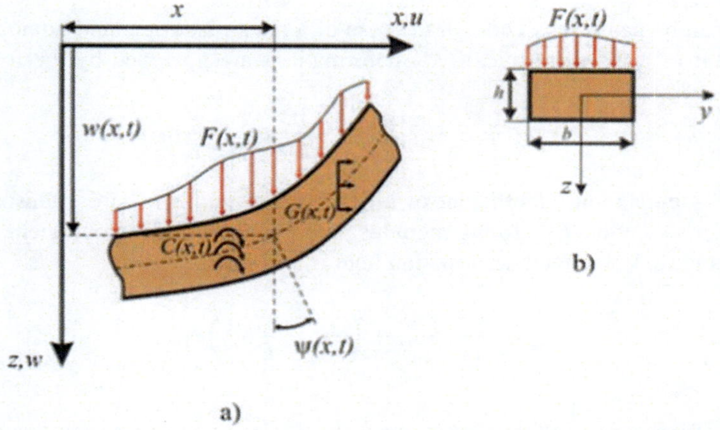

Fig. 8.1 Geometry of loading and the introduced system of coordinates of the beam resonator (**a**) and its cross section (**b**)

beam thicknesses which are compared with the results obtained based on the classical theory of elasticity in both linear and nonlinear cases.

The influence of temperature field on the statics and dynamics of nonlinear Euler-Bernoulli nanobeams [96] fabricated based on the optimization method [97] has been also investigated.

8.3 Formulation of the Problem

We introduce a rectangular system of coordinates to study a microbeam as shown in Fig. 8.1. The origin of coordinates is fixed on the left microbeam end. The microbeam has length L, thickness h, width b, and its transversal cross section parallel to the YOZ plane is denoted by A (see Fig. 8.1b).

8.3.1 Stress and Deformation Fields

Heat deformations occur in an elastic body due to a linear heat extension coefficient. Therefore, the total deformation composed of mechanical and heat components is as follows [76]:

$$\varepsilon_{ij} = \varepsilon_{ij}^{(M)} + \varepsilon_{ij}^{(T)} \ \ (i, j = 1,\ 2,\ 3)\,. \tag{8.1}$$

By $\theta = T - T_0$, we denote the temperature change with respect to the initial temperature T_0, and we assume that the shear deformations generated by the temperature change can be neglected. Then, in the case of a linear, isotropic and homogeneous body, heat (T) and mechanical (M) deformations are governed by the following formula [98]:

$$\varepsilon_{ij}^{(T)} = \alpha\,\theta\,\delta_{ij}, \qquad \varepsilon_{ij}^{(M)} = \frac{1+\nu}{E}\sigma_{ij} - \frac{\nu}{E}\sigma_{kk}\delta_{ij}, \tag{8.2}$$

where α—temperature coefficient of a linear extension; σ_{ij}—stress tensor; δ_{ij}—Kronecker's symbol; E—Young modulus, and ν is the Poisson coefficient. Owing to (8.1) and (8.2), the total deformation field is as follows:

$$\varepsilon_{ij} = \frac{1+\nu}{E}\sigma_{ij} + \left(\alpha\,\theta - \frac{\nu}{E}\sigma_{kk}\right)\delta_{ij}. \tag{8.3}$$

8.3.1.1 Plane Stress State Condition

In the case when beam thickness (in direction of the OZ axis) and its width (in direction of the OY axis) is sufficiently small in comparison to the beam length (in direction of the OX axis), then assumption of the plane stress state implies that all components of the stress tensor in directions of the axes OY and OZ are equal to zero ($\sigma_{12} = \sigma_{22} = \sigma_{32} = 0,\ \sigma_{23} = \sigma_{33} = 0$). However, in the case of the Timoshenko beams we have $\sigma_{13} \neq 0$ [98]. In the latter case, components of the deformations tensor can be recast to the following simple form:

$$\varepsilon_{11} = \frac{\sigma_{11}}{E} + \alpha\,\theta, \quad \varepsilon_{22} = \varepsilon_{33} = -\nu\,\varepsilon_{11} + (1+\nu)\,\alpha\,\theta, \quad \varepsilon_{12} = \varepsilon_{23} = 0. \tag{8.4}$$

A stress tensor under the plane stress state assumption has only two components different from zero, which is expressed by the components of the deformation tensor in the following way:

$$\sigma_{11} = E\varepsilon_{11} - E\alpha\,\theta,\ \sigma_{13} = 2k_s G\varepsilon_{13}, \tag{8.5}$$

where k_s—correction coefficient is introduced due to the assumption of non-homogeneous shear deformation with respect to the transverse beam cross section and it depends on the forms of the beam cross sections; G—shear modulus.

Owing to the known Cauchy-Green deformations

$$\varepsilon_{ij} = \frac{1}{2}\left(u_{i,j} + u_{j,i} + u_{m,i}u_{m,j}\right), \tag{8.6}$$

the kinematic relations of the theory of Timoshenko beams can be presented in the following form:

$$u_1 = u\,(x,t) + z\psi(x,t), \quad u_2 = 0,\ u_3 = w\,(x,\ t)\,, \tag{8.7}$$

where $u(x, t)$, $w(x, t)$ and $\psi(x, t)$ are the axial displacement of the beam middle line, transversal beam deflection and the angle of rotation of the transversal load with respect to the vertical direction, respectively.

However, in this work, we consider the geometrically nonlinear Kármán model, i.e. beam with small deformations and rotations, but with relatively large displacements w that cause the occurrence of geometric nonlinearity. The quantities ψ, $\partial u/\partial x$, $\partial \psi/\partial x$ under the mentioned conditions are very small. Employing the Timoshenko kinematic hypothesis and taking into account (8.4)–(8.7), the non-zero deformation and stress components can be defined via the displacements field in the following way:

$$\begin{aligned}
&\varepsilon_{11} = u_{1,x} = u_{,x} + \frac{1}{2}(w_{,x})^2 + z\psi_{,x}, \quad \varepsilon_{13} = \frac{1}{2}\left(\psi + w_{,x}\right), \\
&\varepsilon_{22} = \varepsilon_{33} = -\nu\left(u_{,x} + \frac{1}{2}(w_{,x})^2 + z\psi_{,x}\right) + (1+\nu)\,\alpha\,\theta, \\
&\sigma_{11} = E\left(u_{,x} + \frac{1}{2}(w_{,x})^2 + z\psi_{,x}\right) - E\alpha\,\theta, \; \sigma_{13} = k_s G\left(\psi + w_{,x}\right).
\end{aligned} \tag{8.8}$$

8.3.1.2 Plane Deformation Conditions

When beam thickness (in direction of the OZ axis) is sufficiently small in comparison to its length (in direction of the OX axis), but beam width (in direction of the OY axis) is sufficiently large, then based on the assumption of the plane deformation state one may conclude that components of the stress tensor in direction of the axis OZ and deformation components in direction of the axis OY are equal to zero ($\sigma_{23} = \sigma_{33} = 0$, $\varepsilon_{12} = \varepsilon_{22} = \varepsilon_{32} = 0$). The remaining non-zero components of the deformation and stress tensors can be expressed by the field of displacements (8.7) and temperature in the following way:

$$\begin{aligned}
&\varepsilon_{11} = u_{1,x} = u_{,x} + \frac{1}{2}(w_{,x})^2 + z\psi_{,x}, \\
&\varepsilon_{33} = -\frac{\nu}{1-\nu}\left(u_{,x} + \frac{1}{2}(w_{,x})^2 + z\psi_{,x}\right) + \frac{(1+\nu)}{1-\nu}\alpha\,\theta, \\
&\varepsilon_{13} = \frac{1}{2}\left(\psi + w_{,x}\right), \\
&\sigma_{11} = \frac{E}{1-\nu^2}\left(u_{,x} + \frac{1}{2}(w_{,x})^2 + z\psi_{,x}\right) - \frac{E}{1-\nu}\alpha\,\theta, \\
&\sigma_{22} = \frac{E\nu}{1-\nu^2}\left(u_{,x} + \frac{1}{2}(w_{,x})^2 + z\psi_{,x}\right) - \frac{E}{1-\nu}\alpha\,\theta, \\
&\sigma_{13} = k_s G\left(\psi + w_{,x}\right).
\end{aligned} \tag{8.9}$$

8.3.2 Beam Equation of Motion Based on the Modified Couple Stress Theory

In the modified couple stress theory [62], the accumulated energy of deformation U in a linear elastic body occupying space Ω is defined as follows:

$$U = \frac{1}{2} \iiint\limits_{V} \left(\sigma_{ij}\varepsilon_{ij} + m_{ij}\chi_{ij}\right) dV, \tag{8.10}$$

where V is the volume of elastic matter, m_{ij} are the components of a deviator part of the symmetric tensor momentum of higher order and χ_{ij} are the components of symmetric curvature tensor [53], which are defined in the following way:

$$m_{ij} = 2Gl^2\chi_{ij}, \quad \chi_{ij} = \frac{1}{2}\left(\varphi_{i,j} + \varphi_{j,i}\right), \quad \varphi_i = \frac{1}{2}(rot\,(u))_i, \tag{8.11}$$

where l is the material length parameter characterizing the influence of a higher order moment. Substituting (8.7) into (8.11) yields the following non-zero components of φ_i, χ_{ij}, m_{ij}:

$$\begin{gathered}\varphi_2 = -\frac{1}{2}\left(w_{,x} - \psi\right), \quad \chi_{12} = \chi_{21} = \frac{1}{4}\left(\psi_{,x} - w_{,xx}\right), \\ m_{12} = m_{21} = \frac{1}{2}Gl^2\left(\psi_{,x} - w_{,xx}\right). \end{gathered} \tag{8.12}$$

In the case of plane stress state after substitution of (8.8) and (8.12) into (8.10), the full deformation energy of the bended beam takes the following form:

$$\begin{aligned} U = \frac{1}{2}\int\limits_0^L \int\limits_{-h/2}^{h/2} \Bigg\{ & E\left(u_{,x} + \frac{1}{2}(w_{,x})^2 + z\psi_{,x}\right)^2 - E\left(u_{,x} + \frac{1}{2}(w_{,x})^2 + z\psi_{,x}\right)\alpha\,\theta + \\ & + k_s G\left(\psi + w_{,x}\right)^2 + \frac{1}{4}Gl^2\left(\psi_{,x} - w_{,xx}\right)^2\Bigg\}\, b\,dz\,dx\,. \end{aligned} \tag{8.13}$$

After computation of integrals with regard to z and a few simple transformations one gets

$$\begin{aligned} U = \frac{1}{2}\int\limits_0^L \Bigg\{ & k_1\left(u_{,x} + \frac{1}{2}(w_{,x})^2\right)^2 + k_2(\psi_{,x})^2 - 2N^T\left(u_{,x} + \frac{1}{2}(w_{,x})^2\right) - \\ & - 2M^T\psi_{,x} + k_3\left(\psi + w_{,x}\right)^2 + k_4\left(\psi_{,x} - w_{,xx}\right)^2\Bigg\}\, dx\,, \end{aligned} \tag{8.14}$$

where

$$k_1 = EA, \quad k_2 = EI, \quad k_3 = k_s GA, \quad k_4 = \frac{1}{4} l^2 GA,$$

$$I = \int\limits_{-h/2}^{h/2} bz^2\, dz = \frac{bh^3}{12}, \quad N^T = \int\limits_{-h/2}^{h/2} Eb\alpha\, \theta\, dz, \quad M^T = \int\limits_{-h/2}^{h/2} Eb\alpha\, \theta\, z\, dz\,. \tag{8.15}$$

It should be emphasized that in order to get a plane deformable state, it is sufficient to substitute E by $\tilde{E} = E/\left(1 - \nu^2\right)$ and α by $\tilde{\alpha} = \alpha\,(1 + \nu)]$ in formula (8.15).

Kinetic beam energy vibration K takes the following form:

$$\begin{aligned} K &= \frac{1}{2}\int\limits_{0}^{L}\int\limits_{-h/2}^{h/2} \rho\left\{\left(u_{,t} + z\psi_{,t}\right)^2 + \left(w_{,t}\right)^2\right\} bdz\, dx = \\ &= \frac{1}{2}\int\limits_{0}^{L} \left\{\rho A\left(u_{,t}\right)^2 + \rho I\left(\psi_{,t}\right)^2 + \rho A\left(w_{,t}\right)^2\right\} dx. \end{aligned} \tag{8.16}$$

External work associated with the forces, moments and stress on the boundary plane is governed by the formula

$$\begin{aligned} W &= \int\limits_{0}^{L} \left(G\,(x,\, t)\; u + F\,(x,\, t)\; w + C\,(x,\, t)\; \varphi_2\right)\, dx + \\ &+ \left(\bar{N}u + \bar{V}w + \bar{M}_\sigma \psi + \bar{M}_m w_{,x}\right)\Big|_{x=0}^{x=L}\,, \end{aligned} \tag{8.17}$$

where $\bar{N},\ \bar{V},\ \bar{M}_\sigma$ and $\bar{M}_m$ denote the axial resulting forms of normal stress σ_{11}, transversal resulting force of shear stress, resulting moment of normal stresses σ_{11} and the resulting moment about the OY axis due to action m_{12} in the transversal cross sections, respectively.

The Hamilton principle yields the following formula:

$$\delta \int\limits_{t_1}^{t_2} (K - U + W)\; dt = 0, \tag{8.18}$$

which after changing variables u, w and ψ, integration by parts and comparison of the terms standing by $\delta\, u$, $\delta\, w$ and $\delta\, \psi$ to zero, yields the following resulting equations of motion

$$\frac{\partial}{\partial x}\left\{k_1\left(\frac{\partial u}{\partial x}+\frac{1}{2}\left(\frac{\partial w}{\partial x}\right)^2\right)-N^T\right\}+G(x,t)=\rho A\frac{\partial^2 u}{\partial t^2},$$
$$\frac{\partial}{\partial x}\left\{\left[k_1\left(\frac{\partial u}{\partial x}+\frac{1}{2}\left(\frac{\partial w}{\partial x}\right)^2\right)-N^T\right]\frac{\partial w}{\partial x}\right\}+k_3\left(\frac{\partial^2 w}{\partial x^2}+\frac{\partial \psi}{\partial x}\right)-$$
$$-k_4\left(\frac{\partial^4 w}{\partial x^4}-\frac{\partial^3 \psi}{\partial x^3}\right)+F(x,t)+\frac{1}{2}\frac{\partial C}{\partial x}=\rho A\frac{\partial^2 w}{\partial t^2},$$
$$k_2\frac{\partial^2 \psi}{\partial x^2}-k_3\left(\frac{\partial w}{\partial x}+\psi\right)-k_4\left(\frac{\partial^3 w}{\partial x^3}-\frac{\partial^2 \psi}{\partial x^2}\right)-\frac{\partial M^T}{\partial x}+\frac{C}{2}=\rho I\frac{\partial^2 \psi}{\partial t^2}, \tag{8.19}$$

and the following boundary conditions

$$\left.\left(k_1\left(\frac{\partial u}{\partial x}+\frac{1}{2}\left(\frac{\partial w}{\partial x}\right)^2\right)-N^T-\bar{N}\right)\right|_{x=0,\,L}=0 \quad \text{or} \quad \delta u|_{x=0,\,L}=0,$$

$$\left(k_1\left(\frac{\partial u}{\partial x}+\frac{1}{2}\left(\frac{\partial w}{\partial x}\right)^2\right)-N^T\right)\frac{\partial w}{\partial x}+k_3\left(\frac{\partial w}{\partial x}+\psi\right)-$$
$$\left.-k_4\left(\frac{\partial^3 w}{\partial x^3}-\frac{\partial^2 \psi}{\partial x^2}\right)+\frac{1}{2}C-\bar{V}\right|_{x=0,\,L}=0$$
$$\text{or} \quad \delta w|_{x=0,\,L}=0,$$
$$\left.k_4\left(\frac{\partial^2 w}{\partial x^2}-\frac{\partial \psi}{\partial x}\right)-\bar{M}_m\right|_{x=0,\,L}=0 \quad \text{or} \quad \left.\delta\left(\frac{\partial w}{\partial x}-\psi\right)\right|_{x=0,\,L}=0, \tag{8.20}$$
$$\left.k_2\frac{\partial \psi}{\partial x}-M^T-\bar{M}_\sigma\right|_{x=0,\,L}=0 \quad \text{or} \quad \delta\psi|_{x=0,\,L}=0.$$

8.3.3 *Coupled Problem of Thermoelasticity for Timoshenko Beams*

Assuming small temperature variation, a coupled heat transfer equation for a thermoelastic body can be recast to the following form [98]:

$$\lambda\theta_{,ii}=\rho c_v\frac{\partial\theta}{\partial t}+\frac{E\alpha T_0}{(1-2\nu)}\frac{\partial\varepsilon_{ii}}{\partial t}, \tag{8.21}$$

where λ—heat transfer coefficient; c_v—heat capacity of the unit material volume.

Recall that the heat transfer equation is coupled with beam deformation by heat production through compression/extension of the beam. It is assumed that the volume deformation (trace of the deformation tensor) is the only source of heat production and hence there is a lack of curvature tensor in Eq. (8.21). In the case of plane stress state (8.8), the trace of deformation tensor ε_{ii} can be presented in the following form:

$$\varepsilon_{ii} = (1-2\nu)\left(u_{,x} + \frac{1}{2}\left(w_{,x}\right)^2 + z\psi_{,x}\right) + 2\,(1+\nu)\,\alpha\,\theta. \tag{8.22}$$

Substituting (8.22) into (8.21) and assuming that heat gradients along the OZ-axis are essentially larger than the gradient along the remaining OX- and OY-axes, Eq. (8.21) takes the following form

$$\begin{aligned} \frac{\partial^2\theta}{\partial z^2} = &\left(\frac{\rho\,c_v}{\lambda} + \frac{2\,(1+\nu)\,E\alpha^2\,T_0}{\lambda\,(1-2\nu)}\right)\frac{\partial\,\theta}{\partial\,t} + \\ &+ \frac{E\alpha\,T_0}{\lambda}\left(\frac{\partial^2 u}{\partial\,t\,\partial\,x} + \frac{\partial\,w}{\partial\,x}\frac{\partial^2 w}{\partial\,x\,\partial\,t\,\partial\,x} + z\frac{\partial^2\psi}{\partial\,t\,\partial\,x}\right). \end{aligned} \tag{8.23}$$

Let us introduce the coefficients of heat diffusion $D = \lambda/\rho\,c_v$ and relaxation $R = E\alpha^2 T_0/\rho\,c_v$. Then, Eq. (8.23) can be presented as follows:

$$(1+\Gamma)\frac{\partial\,\theta}{\partial\,t} = D\frac{\partial^2\theta}{\partial\,z^2} - \frac{R}{\alpha}\left(\frac{\partial^2 u}{\partial\,t\,\partial\,x} + \frac{\partial\,w}{\partial\,x}\frac{\partial^2 w}{\partial\,x\,\partial\,t\,\partial\,x} + z\frac{\partial^2\psi}{\partial\,t\,\partial\,x}\right), \tag{8.24}$$

where $\Gamma = 2R\,(1+\nu)/(1-2\nu)$.

In the case of plane deformation, the trace of the deformation tensor is yielded by relations (8.9) and takes the following form:

$$\varepsilon_{ii} = \frac{1-2\nu}{1-\nu}\left(u_{,x} + \frac{1}{2}\left(w_{,x}\right)^2 + z\psi_{,x}\right) + \frac{1+\nu}{1-\nu}\alpha\,\theta. \tag{8.25}$$

Substituting (8.25) into (8.21) and taking into account heat gradients only along the axis, the following equation is derived:

$$\begin{aligned} \frac{\partial^2\theta}{\partial z^2} = &\left(\frac{\rho\,c_v}{\lambda} + \frac{(1+\nu)}{(1-\nu)\,(1-2\nu)}\frac{E\alpha^2\,T_0}{\lambda}\right)\frac{\partial\,\theta}{\partial\,t} + \\ &+ \frac{E\alpha\,T_0}{\lambda\,(1-\nu)}\left(\frac{\partial^2 u}{\partial\,t\,\partial\,x} + \frac{\partial\,w}{\partial\,x}\frac{\partial^2 w}{\partial\,x\,\partial\,t\,\partial\,x} + z\frac{\partial^2\psi}{\partial\,t\,\partial\,x}\right), \end{aligned} \tag{8.26}$$

which can be rewritten in terms of heat diffusion and relaxation:

$$\left(1+\tilde{\Gamma}\right)\frac{\partial\,\theta}{\partial\,t} = D\frac{\partial^2\theta}{\partial\,z^2} - \frac{\tilde{R}}{\alpha}\left(\frac{\partial^2 u}{\partial\,t\,\partial\,x} + \frac{\partial\,w}{\partial\,x}\frac{\partial^2 w}{\partial\,x\,\partial\,t\,\partial\,x} + z\frac{\partial^2\psi}{\partial\,t\,\partial\,x}\right), \tag{8.27}$$

where $\tilde{R} = R/(1-\nu)$, $\quad \tilde{\Gamma} = \tilde{R}\,(1+\nu)/(1-2\nu)$.

As boundary conditions for PDEs (8.24) and (8.27), we take the thermal isolation condition with regard to the upper and lower microbeam surfaces, i.e. $\partial \theta / \partial z = 0$ for $z = \pm h/2$.

8.4 Quality Factor of the Resonator with Thermoelastic Damping

We consider the plane stress state. In what follows, we neglect the quality $\Gamma = 2E\alpha^2 T_0 (1+\nu)/(1-2\nu)\, \rho\, c_v$ with comparison to one (8.1) [3], and we recast equation (8.24) to the following form:

$$\frac{\partial \theta}{\partial t} + \frac{R}{\alpha}\frac{\partial}{\partial t}\left(\frac{\partial u}{\partial x} + \frac{1}{2}\left(\frac{\partial w}{\partial x}\right)^2\right) + \frac{R}{\alpha}\frac{\partial}{\partial t}\left(z\frac{\partial \psi}{\partial x}\right) = D\frac{\partial^2 \theta}{\partial z^2}. \quad (8.28)$$

It follows from (8.28) that temperature θ induced in the deformed beam and expressed in terms of displacements $u,\ \psi,\ w$ can be divided into two parts: bending deformation and extension deformations, i.e. we have

$$\theta(x, z, t) = \theta_1(x, t)\, f_1(z) + \theta_2(x, t)\, f_2(z), \quad (8.29)$$

where $\theta_1(x, t)$ and $\theta_2(x, t)$ are defined by the following two equations:

$$f_1(z)\frac{\partial \theta_1(x, t)}{\partial t} + \frac{R}{\alpha}\frac{\partial}{\partial t}\left(\frac{\partial u}{\partial x} + \frac{1}{2}\left(\frac{\partial w}{\partial x}\right)^2\right) = D\theta_1(x, t)\frac{\partial^2 f_1(z)}{\partial z^2}, \quad (8.30)$$

$$f_2(z)\frac{\partial \theta_1(x, t)}{\partial t} + \frac{R}{\alpha} z \frac{\partial}{\partial t}\left(\frac{\partial \psi}{\partial x}\right) = D\theta_2(x, t)\frac{\partial^2 f_2(z)}{\partial z^2}. \quad (8.31)$$

Let the transversal beam cross section be symmetric with respect to its neutral axis $z = 0$. Therefore, since non-homogeneous terms in Eqs. (8.30), (8.31) (secondary terms in these equations) are symmetric and anti-symmetric with regard to $z = 0$, respectively, functions $\theta_1(x, z, t) = \theta_1(x, t)\, f_1(z)$ should be even and function $\theta_2(x, z, t) = \theta_2(x, t)\, f_2(z)$ should be odd with respect to z.

Before integration along the area of the cross section A Eqs. (8.30), (8.31) we multiply equation (8.31) by z, and we get

$$G_1\frac{\partial \theta_1(x, t)}{\partial t} + \frac{RA}{\alpha}\frac{\partial}{\partial t}\left(\frac{\partial u}{\partial x} + \frac{1}{2}\left(\frac{\partial w}{\partial x}\right)^2\right) + D\theta_1(x, t)\, G_2 = 0, \quad (8.32)$$

$$G_3\frac{\partial \theta_2(x, t)}{\partial t} + \frac{RI}{\alpha}\frac{\partial}{\partial t}\left(\frac{\partial \psi}{\partial x}\right) + D\theta_2(x, t)\, G_4 = 0, \quad (8.33)$$

where four constants G_i are as follows:

$$G_1 \equiv \int_A f_1(z)\,dA, \quad G_2 \equiv -\int_A \frac{\partial^2 f_1(z)}{\partial z^2}\,dA,$$
$$G_3 \equiv \int_A z f_2(z)\,dA, \quad G_4 \equiv -\int_A z\frac{\partial^2 f_2(z)}{\partial z^2}\,dA. \tag{8.34}$$

If in the first equation of (8.19), we neglect longitudinal force $G(x, t)$ and longitudinal inertial force $\rho A\partial^2 u/\partial t^2$, then it can be double integrated with respect to x, and we get

$$k_1\left(\frac{\partial u}{\partial x}+\frac{1}{2}\left(\frac{\partial w}{\partial x}\right)^2\right)-N^T=$$
$$= C_1(t) \Rightarrow k_1\left(u+\int_0^x \frac{1}{2}\left(\frac{\partial w}{\partial x}\right)^2 dx\right)-\int_0^x N^T\,dx = C_1(t)x + C_2(t). \tag{8.35}$$

In the case of the beam with fixed ends ($u(0, t) = u(L, t) = 0$) we get $C_2(t) = 0$, and hence

$$C_1(t) = k_1\left(\frac{1}{L}\int_0^L \frac{1}{2}\left(\frac{\partial w}{\partial x}\right)^2 dx\right)-\frac{1}{L}\int_0^L N^T\,dx. \tag{8.36}$$

Therefore, longitudinal force N does not depend on x and remains constant along the beam

$$N = k_1\left(\frac{\partial u}{\partial x}+\frac{1}{2}\left(\frac{\partial w}{\partial x}\right)^2\right)-N^T = k_1\left(\frac{1}{L}\int_0^L \frac{1}{2}\left(\frac{\partial w}{\partial x}\right)^2 dx\right)-\frac{1}{L}\int_0^L N^T\,dx. \tag{8.37}$$

In what follows, we consider only coupled thermoelastic vibrations, and hence we take $F(x, t) = C(x, t) = 0$ in (8.19) and $\bar{N} = \bar{V} = \bar{M}_\sigma = \bar{M}_m = 0$ in (8.20). In these conditions and taking into account (8.37), the following two equations of motion are obtained:

$$k_3\left(\frac{\partial^2 w}{\partial x^2}+\frac{\partial \psi}{\partial x}\right)-k_4\left(\frac{\partial^4 w}{\partial x^4}-\frac{\partial^3 \psi}{\partial x^3}\right)+N\frac{\partial^2 w}{\partial x^2}=\rho A\frac{\partial^2 w}{\partial t^2},$$
$$k_2\frac{\partial^2 \psi}{\partial x^2}-k_3\left(\frac{\partial w}{\partial x}+\psi\right)-k_4\left(\frac{\partial^3 w}{\partial x^3}-\frac{\partial^2 \psi}{\partial x^2}\right)-\frac{\partial M^T}{\partial x}=\rho I\frac{\partial^2 \psi}{\partial t^2} \tag{8.38}$$

and the boundary conditions follow

$$N\frac{\partial w}{\partial x}+k_3\left(\frac{\partial w}{\partial x}+\psi\right)-k_4\left(\frac{\partial^3 w}{\partial x^3}-\frac{\partial^2 \psi}{\partial x^2}\right)\bigg|_{x=0,\,L}=0 \quad \text{or} \quad \delta w|_{x=0,\,L}=0,$$

$$k_4\left(\frac{\partial^2 w}{\partial x^2}-\frac{\partial \psi}{\partial x}\right)\bigg|_{x=0,\,L}=0 \quad \text{or} \quad \delta\left(\frac{\partial w}{\partial x}-\psi\right)\bigg|_{x=0,\,L}=0, \tag{8.39}$$

$$k_2\frac{\partial \psi}{\partial x}-M^T\bigg|_{x=0,\,L}=0 \quad \text{or} \quad \delta\psi|_{x=0,\,L}=0.$$

Let us express N and M^T by temperature $\theta_1(x,t)$ and $\theta_2(x,t)$ as follows:

$$N=\frac{k_1}{L}\int_0^L\frac{1}{2}\left(\frac{\partial w}{\partial x}\right)^2dx-\frac{E\alpha}{L}G_1\int_0^L\theta_1(x,t)\,dx,\quad M^T=E\alpha G_3\theta_2(x,t)\,. \tag{8.40}$$

As a result, we obtain the closed system of equations (8.32), (8.33) and (8.38)–(8.40) for the nonlinear analysis of TED of the Timoshenko microbeams.

Our goal is to derive a non-dimensional counterpart form of the governing PDEs. For this purpose, we introduce the following non-dimensional parameters

$$\begin{aligned}
&\bar{x}=\frac{x}{L},\quad \bar{z}=\frac{z}{h},\quad \bar{w}=\frac{w}{h},\quad \bar{\psi}=\frac{L}{h}\psi,\quad \omega_0=\sqrt{\frac{EI}{\rho AL^4}},\\
&\bar{t}=\omega_0 t,\quad \tau_0=\frac{h^2}{D},\quad \tau=\omega_0\tau_0,\quad \bar{\theta}=\frac{\theta}{T_0},\quad \bar{R}=R/\alpha T_0\,,\\
&P=\frac{AL^2}{I},\quad H=\frac{k_sGAL^2}{EI},\quad F=\frac{1}{4}\frac{l^2GA}{EI},\\
&\bar{G}_1=\frac{G_1}{A},\quad \bar{G}_2=\frac{G_2h^2}{A},\quad \bar{G}_3=\frac{G_3}{Ah},\quad \bar{G}_4=\frac{G_4h}{A}\,.
\end{aligned} \tag{8.41}$$

Formulas (8.40) yield

$$\begin{aligned}
&N=EA\bar{N},\quad \bar{N}=\left(\frac{h}{L}\right)^2\int_0^1\frac{1}{2}\left(\frac{\partial\bar{w}}{\partial\bar{x}}\right)^2d\bar{x}-\alpha T_0\bar{G}_1\int_0^1\bar{\theta}_1(x,t)\,dx,\\
&M=\frac{EIh}{L^2}\bar{M},\quad \bar{M}=\frac{\partial\bar{\psi}}{\partial\bar{x}}-\alpha T_0P\bar{G}_3\bar{\theta}_2(x,t)\,,
\end{aligned} \tag{8.42}$$

and hence Eqs. (8.32) and (8.33) take the following non-dimensional form (bars over non-dimensional quantities are omitted):

$$\frac{\partial\theta_1(x,t)}{\partial t}+\frac{R}{G_1}\left(\frac{h}{L}\right)^2\frac{\partial}{\partial t}\left[\frac{1}{2}\left(\frac{\partial w}{\partial x}\right)^2\right]+\frac{1}{\tau}\left(\frac{G_2}{G_1}\right)\theta_1(x,t)=0, \tag{8.43}$$

$$\frac{\partial \theta_2(x, t)}{\partial t} + \frac{R}{PG_3} \frac{\partial}{\partial t} \left(\frac{\partial \psi}{\partial x} \right) + \frac{G_4}{G_3} \theta_2(x, t) = 0, \tag{8.44}$$

whereas Eqs. (8.38) and (8.39) take the following counterpart non-dimensional form:

$$\begin{aligned} & H\left(\frac{\partial^2 w}{\partial x^2} + \frac{\partial \psi}{\partial x}\right) - F\left(\frac{\partial^4 w}{\partial x^4} - \frac{\partial^3 \psi}{\partial x^3}\right) + P \cdot N \frac{\partial^2 w}{\partial x^2} = \frac{\partial^2 w}{\partial t^2}, \\ & \frac{\partial^2 \psi}{\partial x^2} - H\left(\frac{\partial w}{\partial x} + \psi\right) - F\left(\frac{\partial^3 w}{\partial x^3} - \frac{\partial^2 \psi}{\partial x^2}\right) - \alpha T_0 P G_3 \frac{\partial \theta_2}{\partial x} = \frac{1}{P} \frac{\partial^2 \psi}{\partial t^2}. \end{aligned} \tag{8.45}$$

$$P \cdot N \frac{\partial w}{\partial x} + H\left(\frac{\partial w}{\partial x} + \psi\right) - F\left(\frac{\partial^3 w}{\partial x^3} - \frac{\partial^2 \psi}{\partial x^2}\right)\Bigg|_{x=0,1} = 0 \quad \text{or} \quad \delta w|_{x=0,1} = 0,$$

$$\left(\frac{\partial^2 w}{\partial x^2} - \frac{\partial \psi}{\partial x}\right)\Bigg|_{x=0,1} = 0 \quad \text{or} \quad \delta\left(\frac{\partial w}{\partial x} - \psi\right)\Bigg|_{x=0,1} = 0, \tag{8.46}$$

$$\frac{\partial \psi}{\partial x} - \frac{\alpha T_0 A L^2}{I} G_3 \theta_2 \Bigg|_{x=0,1} = 0 \quad \text{or} \quad \delta \psi|_{x=0,1} = 0.$$

We consider simple harmonic vibrations in the form $w(x, t) = w_0(x) e^{-i\omega t}$, $\psi(x, t) = \psi_0(x) e^{-i\omega t}$, where ω stands for circular frequency, and $w_0(x)$, $\psi_0(x)$ are approximated by a linear vibration mode. Observe that a large beam deflection may cause the frequency to depend on the amplitude of vibrations. Therefore, owing to relations (8.43), (8.44) and (8.42), other remaining functions have the following forms:

$$\begin{aligned} & \theta_1(x, t) = \theta_1(x) e^{-2i\omega t}, \ \theta_2(x, t) = \theta_2(x) e^{-i\omega t}, \\ & N(x, t) = N(x) e^{-2i\omega t}, \ M(x, t) = M(x) e^{-i\omega t}. \end{aligned} \tag{8.47}$$

Substituting (8.47) into (8.43) yields

$$-i\omega\theta_1(x) - 2i\omega \frac{R}{G_1}\left(\frac{h}{L}\right)^2 \left[\frac{1}{2}\left(\frac{\partial w}{\partial x}\right)^2\right] + \frac{1}{\tau}\left(\frac{G_2}{G_1}\right)\theta_1(x) = 0, \tag{8.48}$$

whereas integration along the beam length gives

$$\left(\frac{1}{\tau}\left(\frac{G_2}{G_1}\right) - i\omega\right) \int_0^1 \theta_1(x)\, dx = i\omega \frac{R}{G_1}\left(\frac{h}{L}\right)^2 \int_0^1 \left(\frac{\partial w}{\partial x}\right)^2 dx. \tag{8.49}$$

We have

$$\int_0^1 \theta_1(x)\, dx = \frac{i\omega R}{G_1\left(\frac{1}{\tau}\left(\frac{G_2}{G_1}\right) - i\omega\right)} \left(\frac{h}{L}\right)^2 \int_0^1 \left(\frac{\partial w}{\partial x}\right)^2 dx, \tag{8.50}$$

which is substituted into (8.42). Therefore

$$N=\left(1-\frac{2i\omega\alpha T_0 R}{\left(\frac{1}{\tau}\left(\frac{G_2}{G_1}\right)-i\omega\right)}\left(\frac{h}{L}\right)^2\right)\int_0^1\frac{1}{2}\left(\frac{\partial w}{\partial x}\right)^2dx=S_\omega\int_0^1\frac{1}{2}\left(\frac{\partial w}{\partial x}\right)^2dx, \tag{8.51}$$

where

$$S_\omega=\left(1-\frac{2i\omega\alpha T_0 R}{\left(\frac{1}{\tau}\left(\frac{G_2}{G_1}\right)-i\omega\right)}\left(\frac{h}{L}\right)^2\right). \tag{8.52}$$

Function S_ω can be presented in the following form:

$$S_\omega=\left(1+\frac{2\omega^2\alpha T_0 R}{\left(\frac{1}{\tau^2}\left(\frac{G_2}{G_1}\right)^2+\omega^2\right)}\left(\frac{h}{L}\right)^2-i\frac{2\omega\alpha T_0 R\left(\frac{1}{\tau}\left(\frac{G_2}{G_1}\right)\right)}{\left(\frac{1}{\tau^2}\left(\frac{G_2}{G_1}\right)^2+\omega^2\right)}\left(\frac{h}{L}\right)^2\right),$$

$$\mathrm{Re}\left[S_\omega\right]=1+\frac{2\omega^2\alpha T_0 R}{\left(\frac{1}{\tau^2}\left(\frac{G_2}{G_1}\right)^2+\omega^2\right)}\left(\frac{h}{L}\right)^2, \tag{8.53}$$

$$\mathrm{Im}\left[S_\omega\right]=-\frac{2\omega\alpha T_0 R\left(\frac{1}{\tau}\left(\frac{G_2}{G_1}\right)\right)}{\left(\frac{1}{\tau^2}\left(\frac{G_2}{G_1}\right)^2+\omega^2\right)}\left(\frac{h}{L}\right)^2.$$

We proceed in the same way with Eqs. (8.42) and (8.44) to get

$$\left(\frac{G_4}{G_3}-i\omega\right)\theta_2(x)=\frac{i\omega R}{PG_3}\left(\frac{\partial\psi}{\partial x}\right). \tag{8.54}$$

Hence

$$\theta_2(x)=\frac{i\omega R}{PG_3}\frac{1}{\left(\frac{G_4}{G_3}-i\omega\right)}\left(\frac{\partial\psi}{\partial x}\right) \tag{8.55}$$

is substituted into (8.42) to yield

$$M=\left[1-\frac{i\omega\alpha T_0 R}{\left(\frac{G_4}{G_3}-i\omega\right)}\right]\left(\frac{\partial\psi}{\partial x}\right)=D_\omega\left(\frac{\partial\psi}{\partial x}\right), \tag{8.56}$$

where

$$D_\omega=\left[1-\frac{i\omega\alpha T_0 R}{\left(\frac{G_4}{G_3}-i\omega\right)}\right]. \tag{8.57}$$

Again, we find real and imaginary parts of D_ω in the following way:

$$D_\omega = \left(1 + \frac{\alpha T_0 R \omega^2}{\left(\left(\frac{G_4}{G_3}\right)^2 + \omega^2\right)}\right) - i\frac{\omega \alpha T_0 R \frac{G_4}{G_3}}{\left(\left(\frac{G_4}{G_3}\right)^2 + \omega^2\right)},$$

$$\operatorname{Re}[D_\omega] = \left(1 + \frac{\alpha T_0 R \omega^2}{\left(\left(\frac{G_4}{G_3}\right)^2 + \omega^2\right)}\right), \tag{8.58}$$

$$\operatorname{Im}[D_\omega] = -\frac{\omega \alpha T_0 R \frac{G_4}{G_3}}{\left(\left(\frac{G_4}{G_3}\right)^2 + \omega^2\right)}.$$

A thermoelastic dissipation can be quantified by a ratio of the loss of mechanical energy per cycle of vibration to the general conserved energy of deformation, and it can be estimated using a required amount of energy for each cycle to keep a periodic harmonic motion [3, 82, 83]. The required amount of energy for an infinitely small beam element $d\,x$ located in point x over period $T = 2\pi/\omega$ is defined by the following formula:

$$dx \int_0^{2\pi/\omega} PN\frac{d}{dt}\left[\frac{1}{2}\left(\frac{\partial w}{\partial x}\right)^2\right] dt + dx \int_0^{2\pi/\omega} M\frac{d}{dt}\left[\left(\frac{\partial \psi}{\partial x}\right)\right] dt. \tag{8.59}$$

Substituting (8.51), (8.56) for N and M into (8.59), one obtains

$$[Energy\ loss]_{extension} = \int_0^1 \int_0^{2\pi/\omega} PN\frac{d}{dt}\left[\frac{1}{2}\left(\frac{\partial w}{\partial x}\right)^2\right] dt dx =$$

$$= \int_0^1 \int_0^{2\pi/\omega} P \operatorname{Re}[N] \operatorname{Re}\left[\frac{d}{dt}\left(\frac{1}{2}\left(\frac{\partial w}{\partial x}\right)^2\right)\right] dt\, dx = \tag{8.60}$$

$$= -\pi P \operatorname{Im}[S_\omega] \left[\int_0^1 \left(\frac{d w_0}{d x}\right)^2 dx\right]^2,$$

$$[Energy\ loss]_{bending} = \int_0^1 \int_0^{2\pi/\omega} M \frac{d}{dt}\left(\frac{\partial \psi}{\partial x}\right) dt dx =$$
$$= \int_0^1 \int_0^{2\pi/\omega} \mathrm{Re}\,[M]\, \mathrm{Re}\left[\frac{d}{dt}\left(\frac{\partial \psi}{\partial x}\right)\right] dt\, dx = -\pi\, \mathrm{Im}\,[D_\omega] \int_0^1 \left(\frac{d\,\psi_0}{d\,x}\right)^2 dx\ , \tag{8.61}$$

and it is clear that imaginary parts of the complex axial (S_ω) and bending (D_ω) stiffness define the mechanical energy loss. General generated energy of deformation due to extension bending, shear deformation and higher order moments is given by the following non-dimensional formula:

$$U = \frac{1}{2}\left\{ P\, \mathrm{Re}\,[S_\omega] \left[\int_0^1 \frac{1}{2}\left(\frac{d\,w_0}{d\,x}\right)^2 dx\right]^2 + \mathrm{Re}\,[D_\omega] \int_0^1 \left(\frac{d\,\psi_0}{d\,x}\right)^2 dx + \right.$$
$$\left. + H \int_0^1 \left(\frac{d\,w_0}{d\,x} + \psi_0\right)^2 dx + F \int_0^1 \left(\frac{d^2\,w_0}{d\,x^2} - \frac{d\psi_0}{dx}\right)^2 dx \right\}. \tag{8.62}$$

Therefore, employing equations (8.60)–(8.62), general formula describing thermoelastic dissipation or inverse quality factor Q is as follows:

$$\left(\frac{1}{Q}\right)_{Non-linear} = \frac{-P\, \mathrm{Im}\,[S_\omega] \left[\int_0^1 \left(\frac{d\,w_0}{d\,x}\right)^2 dx\right]^2 - \mathrm{Im}\,[D_\omega] \int_0^1 \left(\frac{d\,\psi_0}{d\,x}\right)^2 dx}{2U}\ . \tag{8.63}$$

8.5 Numerical and Analytical Results for a Rectangular Cross-Sectional Beam Resonator

We study the beam resonator with a rectangular cross section shown in Fig. 8.1. Functions $f_1(z)$, $f_2(z)$ that occur in (8.29) can be yielded by the adiabatic condition applied to the upper and lower microbeam surfaces, i.e. we have $d\,f/d\,z = 0$ for $z = \pm 1/2$ in the non-dimensional coordinates. Note that the adiabatic condition yields $G_2 = 0$, which is a consequence of the general theorem on divergence that holds for arbitrary forms of the beam cross section. As we have already mentioned, function $f_1(z)$ representing the temperature field generated by mechanical extension should be an even function of z. If one develops $f_1(z)$ into a series with regard to z up to the second order, it is visible that a coefficient standing by the second order should be equal to zero to fit the adiabatic state of the beam surface. Therefore, since deformation along extension does not depend on z, it implies $f_1(z) = const$.

It follows from the definition of $f_1(z)$ that the constant can be chosen arbitrarily, and hence it is allowed to take $f_1(z) = 1$.

Therefore, we obtain the values of coefficients in Eq. (8.43)

$$G_1 = 1, \quad G_2 = 0. \tag{8.64}$$

In the case of function $f_2(z)$ describing the change in the temperature field generated by the bending deformation, which is odd, for the rectangular cross section one gets

$$f_2(z) = z - \frac{4z^3}{3}, \quad G_3 = \frac{1}{15}, \quad G_4 = \frac{2}{3}. \tag{8.65}$$

Temperature profile $f_2(z)$ along beam thickness is described by the following formula [20]:

$$f_2(z) = z + \frac{1}{2\pi} \sin(2\pi z)$$

and

$$G_3 = \frac{\pi^2 + 3}{24\pi^2} \approx \frac{2}{37}, \quad G_4 = 1.$$

In further computations we take into account (8.65).

In order to study the influence of the length scalar parameter on TED, we consider beams made from gold, which are extensively employed in MEMS. Material and geometric properties of the resonator are given in Table 8.1. The values of the length material parameter for different beam thicknesses can be found in references [99, 100].

Table 8.1 Fundamental microbeam parameters

Parameter	Gold
Length (L)	200μm
Thickness (h)	2μm
Young modulus (E)	79 GPa
Poisson's coefficient (ν)	0.44
Heat transfer coefficient (λ) Density (ρ)	318 W/mK 19320 kg/m
Specific heat under constant volume (c_v)	129 J/kgK
Linear size-dependent coefficient (α)	14.2 $10^{-6} K^{-1}$
Temperature of the surrounding medium (T_0)	300 K
Scalar length material parameter (l)	0.47 μm for $0,5 < h < 1$ μm 0.73 μm for $1 < h < 2$ μm 1.05 μm for $h > 2$ μm

8.5.1 Analytical Solution for the Linear Size-Dependent Beam

We take into account undamped modes of beam vibrations as the eigenfunctions (see [101]). It is assumed that the effect of large deflection in the vibrational regime plays the secondary role, and therefore is not meaningful and the vibrational beam regime can be approximated in a linear way modelled by a beam clamped from both sides [20, 102, 103].

Substituting (8.47) into (8.45) for $\theta_1(x) = \theta_2(x) = 0$ and $N = 0$ (linear case), the following system of governing PDEs is obtained

$$\begin{aligned} &H\left(\frac{\partial^2 w}{\partial x^2} + \frac{\partial \psi}{\partial x}\right) - F\left(\frac{\partial^4 w}{\partial x^4} - \frac{\partial^3 \psi}{\partial x^3}\right) = -\omega^2 w, \\ &\frac{\partial^2 \psi}{\partial x^2} - H\left(\frac{\partial w}{\partial x} + \psi\right) - F\left(\frac{\partial^3 w}{\partial x^3} - \frac{\partial^2 \psi}{\partial x^2}\right) = \frac{-\omega^2}{P}\psi . \end{aligned} \tag{8.66}$$

The latter system of PDEs can be recast to the following counterpart form:

$$\begin{aligned} &F\frac{\partial^6 w}{\partial x^6} + \left(-4FH + F\frac{\omega^2}{P} - H\right)\frac{\partial^4 w}{\partial x^4} + \left(-H\frac{\omega^2}{P} - F\omega^2 - \omega^2\right)\frac{\partial^2 w}{\partial x^2} + \\ &+\left(-\frac{\omega^4}{P} + H\omega^2\right) w = 0, \end{aligned} \tag{8.67}$$

$$\begin{aligned} &F\frac{\partial^6 \psi}{\partial x^6} + \left(-4FH + F\frac{\omega^2}{P} - H\right)\frac{\partial^4 \psi}{\partial x^4} + \left(-H\frac{\omega^2}{P} - F\omega^2 - \omega^2\right)\frac{\partial^2 \psi}{\partial x^2} + \\ &+\left(-\frac{\omega^4}{P} + H\omega^2\right) \psi = 0. \end{aligned} \tag{8.68}$$

The following notation is introduced:

$$\begin{aligned} &a_1 = F, \quad a_2 = -4FH + F\frac{\omega^2}{P} - H, \\ &a_3 = \left(-H\frac{\omega^2}{P} - F\omega^2 - \omega^2\right), \quad a_4 = -\frac{\omega^2}{P} + H\omega^2. \end{aligned} \tag{8.69}$$

In the case of classical ($F = 0$) beam clamped on its both sides, Eq. (8.70) follow from (8.67)–(8.69)

$$a_2\frac{\partial^4 w}{\partial x^4} + a_2\frac{\partial^2 w}{\partial x^2} + a_3 w = 0, \qquad a_2\frac{\partial^4 \psi}{\partial x^4} + a_3\frac{\partial^2 \psi}{\partial x^2} + a_4\psi = 0. \tag{8.70}$$

Solutions to (8.70) in the following forms are searched:

$$W = A_1 ch\alpha x + A_2 sh\alpha x + A_3 \cos\beta x + A_4 \sin\beta x, \tag{8.71}$$

$$\psi = A_1 M_1 sh\alpha x + A_2 M_1 ch\alpha x + A_3 M_2 \sin\beta x - A_4 M_2 \cos\beta x. \tag{8.72}$$

Formula (8.66) yields

$$M_1 = \frac{-H\alpha^2 - \omega^2}{H\alpha}, \qquad M_2 = \frac{H\beta^2 - \omega^2}{H\beta}. \tag{8.73}$$

The characteristic equation

$$q^4 + \frac{a_3}{a_2}q^2 - \frac{a_4}{a_2} = 0 \tag{8.74}$$

has the following roots

$$q_{1,2} = \pm i\left[(a_3 + \frac{\sqrt{a_3{}^2 + 4a_2a_4}}{2a_2})\right]^{\frac{1}{2}}, \quad q_{3,4} = \pm\left[(-a_3 + \frac{\sqrt{a_3{}^2 + 4a_2a_4}}{2a_2})\right]^{\frac{1}{2}}, \tag{8.75}$$

and α and β can be easily found (see (8.71) and (8.72))

$$(\alpha, \beta) = \left[(\mp a_3 + \frac{\sqrt{a_3{}^2 + 4a_2a_4}}{2a_2})\right]^{\frac{1}{2}} \tag{8.76}$$

or equivalently

$$\alpha = \sqrt{\left[-\left(\frac{H}{P}+1\right)\omega^2 + \sqrt{\left(\frac{H}{P}+1\right)^2\omega^4 - 4H\left(\frac{\omega^4}{P} - H\omega^2\right)}\right]/2H}\,. \tag{8.77}$$

$$\beta = \sqrt{\left[\left(\frac{H}{P}+1\right)\omega^2 + \sqrt{\left(\frac{H}{P}+1\right)^2\omega^4 - 4H\left(\frac{\omega^4}{P} - H\omega^2\right)}\right]/2H}\,. \tag{8.78}$$

Boundary conditions (8.46) for the clamped beam (*câc*) follow

$$w|_{x=0,1} = \psi|_{x=0,1} = 0.$$

Hence, the following eigenfunctions are derived:

$$w_0 = \frac{W_0}{a}\left[ch\alpha x - \cos\beta x + \delta\left(\frac{sh\alpha x}{M_1} + \frac{\sin\beta x}{M_2}\right)\right],$$
$$\psi_0 = \frac{W_0}{a}\left[M_1 sh\alpha x - M_2 \sin\beta x + \delta\left(ch\alpha x - \cos\beta x\right)\right], \tag{8.79}$$

where

$$\delta = \frac{M_2\sin\beta - M_1 sh\alpha}{(ch\alpha - \cos\beta)}, \quad M_1 = \frac{-(H\alpha^2 + \omega^2)}{H\alpha}, \quad M_2 = \frac{(H\beta^2 - \omega^2)}{H\beta}, \tag{8.80}$$

and the frequency equation is as follows:

$$2 - 2ch\alpha \cdot \cos\beta + \left(\frac{M_2^2 - M_1^2}{M_1 M_2}\right) sh\alpha \cdot \sin\beta = 0. \tag{8.81}$$

Quantity a in (8.79) is equal to the maximum value of the eigenfunction in square brackets of the first formula of (8.79), whereas W_0 is the value of non-dimensional amplitude of beam deflection. Solving equation (8.81) and defining a, functions $w_0(x)$, $\psi_0(x)$ in (8.79) are substituted to (8.63) in order to get quality factor Q of the resonator.

As a case study, we consider a beam made from gold (see Table 7.1).

Graphs of the fundamental vibration mode $w_0(x)$, $\psi_0(x)$ for $W_0 = a = 1.589$ for the case of beam clamped on both sides ($c\hat{a}c$) for $L/h = 100$ are shown in Figs. 8.2 and 8.3.

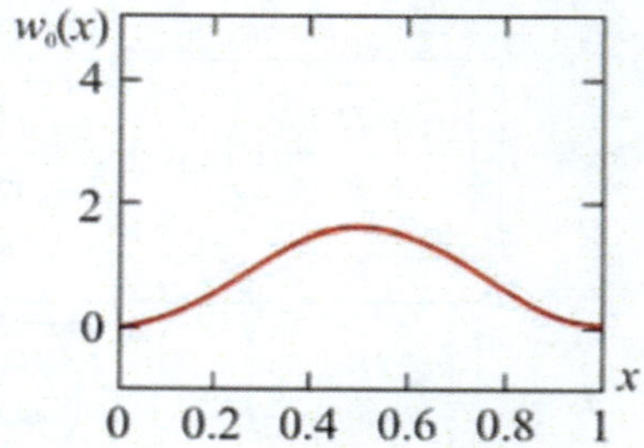

Fig. 8.2 Fundamental mode $w_0(x)$

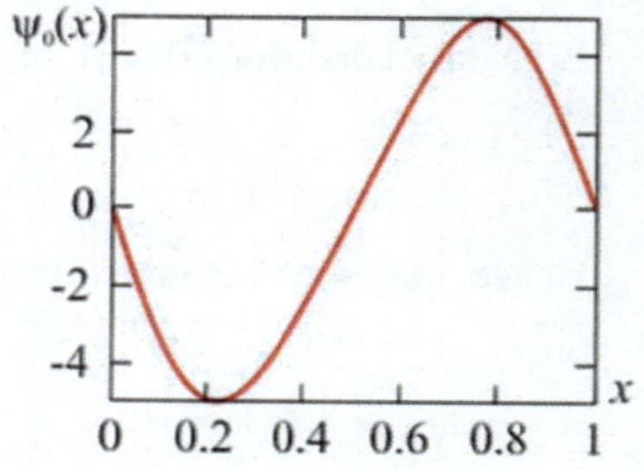

Fig. 8.3 Fundamental mode $\psi_0(x)$

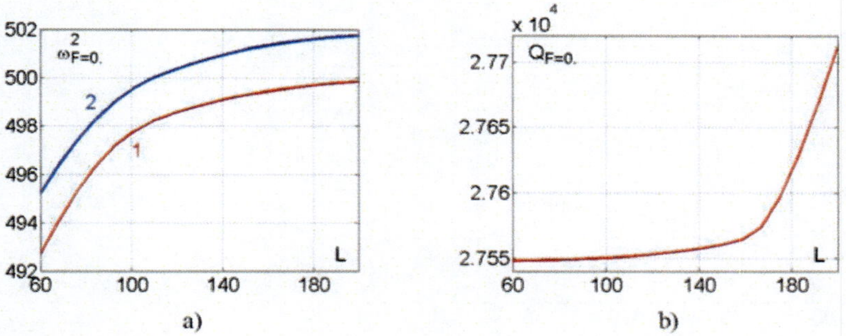

Fig. 8.4 Beam with boundary conditions $(s-s)$: **a** dependence of linear eigenfrequencies for $F=0$: 1—analytical solution; 2—numerical solution; **b** quality factor Q versus beam length L

PDE (8.81) is solved using MathCad. Figure 8.4a presents the dependence of linear non-dimensional frequencies of the fundamental vibration mode for the classical linear Timoshenko beam made from gold ($F=0$) for boundary conditions $(c-c)$, and for different values of its length L as well as for thickness $h=2\ \mu\text{m}$ versus the beam length: 1—analytical solution of equation (8.81); 2—numerical solution. Reduction of the problem of infinite dimension is reduced to the Cauchy problem based on the method of finite differences of the second-order accuracy. The Cauchy problem is solved using the sixth-order Runge-Kutta method.

Figure 8.4b shows the dependence of quality factor of the beam resonator estimated by formulas (8.63) for $W_0=10^{-6}$ (linear case) versus beam length L.

In the case of simply supported classical beam $(s-s)$, the boundary conditions take the following form:

$$w|_{x=0,1}=w''|_{x=0,1}=0.$$

The frequency equation is obtained based on (8.71), (8.72) and takes the following form $\sin\beta=0$.

In the case of the fundamental mode $\beta=\pi$. It follows from (8.80) that the eigenfunctions take the following form:

$$M_2=(H\pi^2-\omega^2)/H\pi\,,\qquad W=\sin\pi x\,,\qquad \psi=M_2\cos\pi x.$$

Formula (8.66) for $F=0$ yields

$$\pi^2M_2+H(M_2-\pi)=\frac{\omega^2M_2}{P},$$

which allows us to define the frequency of the fundamental mode of vibrations $\omega^2_{F=0}$ for $(s-s)$ Timoshenko beam.

The dependence of linear non-dimensional frequencies of the fundamental mode of the classical linear Timoshenko beam ($F=0$) for the $(s-s)$ boundary condition

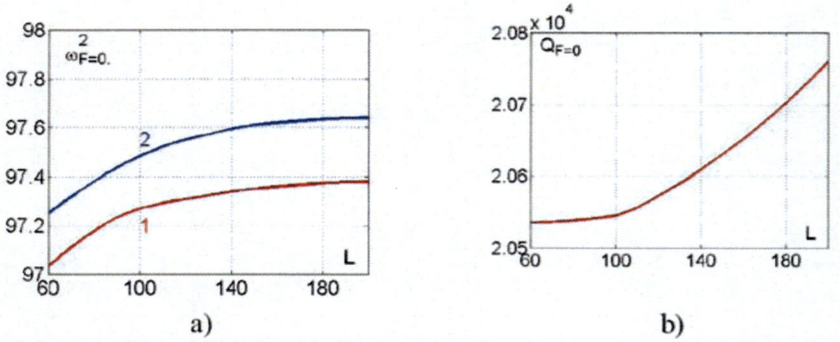

Fig. 8.5 Beam with boundary conditions $(s - s)$: **a** dependence of the linear non-dimensional eigenfrequencies for $F = 0$, where 1/2 is the analytical/numerical solution; **b** dependence of the quality factor of the resonator of length L

in dependence of its length is shown in Fig. 8.5a where analytical/numerical solution is marked by 1/2. Figure 8.5b illustrates the dependence of beam quality factor versus beam length L obtained from formulas (8.63) for the linear case.

In order to obtain linear frequencies in the modified couple stress theory ($F \neq 0$) we employ equation (8.66) and notations (8.69) to get the following sixth-order characteristic equation

$$a_1 q^6 + a_2 q^4 + a_3 q^2 + a_4 = 0. \tag{8.82}$$

The equation can be recast to the following form [104]:

$$\left(q^4 + s_1 q^2 + s_2\right)\left(s_3 q^2 + s_4\right) = 0, \tag{8.83}$$

where

$$s_1 = (a_3 - s_2 a_1)/s_4\,, \qquad s_3 = a_1\,, \qquad s_4 = a_4/s_2\,, \tag{8.84}$$

and s_2 is the solution to the following cubic equation:

$$a_1^2 s_2^3 - a_1 a_3 s_2^2 + a_4 a_2 s_2 - a_4^2 = 0. \tag{8.85}$$

The analysis of coefficients $F = \frac{1}{4}\frac{l^2 GA}{EI}$ in (8.41) leads to a conclusion that in the case of gold it has the order of 10^{-1} for $l = 10^{-6} m$, whereas the remaining coefficients P and H are of several thousand times larger in the interval $L/h = 25 - 100$. If we neglect the first term in Eq. (8.85), since $a_1^2 \approx 10^{-2}$, hence they are much smaller than coefficients $a_1 a_3\,,\ a_4 a_2\,,\ a_4^2$ standing by other terms. So, one gets

$$s_2 = \frac{1}{2}\frac{a_4}{a_1 a_3}\left(a_2 \pm \sqrt{a_2^2 - 4 a_1 a_3}\right). \tag{8.86}$$

Finally, the first six roots of equation (8.82) are as follows:

$$q_{1,2} = \pm\frac{\sqrt{2}}{2}\left(-s_1 + \sqrt{s_1^2 - 4s_2}\right)^{1/2}, \quad q_{3,4} = \pm\frac{\sqrt{2}}{2}\left(-s_1 - \sqrt{s_1^2 - 4s_2}\right)^{1/2},$$
$$q_{5,6} = \pm\sqrt{-s_4/s_2}\ . \tag{8.87}$$

The analysis shows that $q_{1,2}$ are real, $q_{3,4}$ are complex conjugated, and $q_{5,6}$ are real. Therefore

$$(\alpha,\ \beta) = \frac{\sqrt{2}}{2}\left(\mp s_1 + \sqrt{s_1^2 - 4s_2}\right)^{1/2}, \quad \gamma = \sqrt{-s_4/s_2}\ . \tag{8.88}$$

We assume the following form of the unknown functions:

$$\begin{aligned} w &= A_1 ch\alpha x + A_2 sh\alpha x + A_3\cos\beta x + A_4\sin\beta x + A_5 ch\gamma x + A_6 sh\gamma x,\\ \psi &= A_1 M_1 sh\alpha x + A_2 M_1 ch\alpha x + A_3 M_2\sin\beta x -\\ &\quad - A_4 M_2\cos\beta x + M_3 A_5 ch\gamma x + M_3 A_6 sh\gamma x. \end{aligned} \tag{8.89}$$

Substituting (8.89) into (8.66), the following coefficients are defined:

$$M_1 = \frac{F\alpha^4 - H\alpha^2 - \omega^2}{F\alpha^3 + H\alpha}, \quad M_2 = \frac{F\beta^4 + H\beta^2 - \omega^2}{-F\beta^3 + H\beta}, \quad M_3 = \frac{F\gamma^4 - H\gamma^2 - \omega^2}{F\gamma^3 + H\gamma}. \tag{8.90}$$

We consider as earlier two types of the boundary conditions: clamped-clamped beam $(c - c)$ with boundary conditions $w|_{x=0,\,1} = w'\big|_{x=0,\,1} = \psi|_{0,\,1} = 0$ and the simple-simple supported beam (s-s) with boundary conditions $w|_{x=0,\,1} = w''\big|_{x=0,\,1} = \psi'\big|_{0,\,1} = 0$.

In the case of $(c - c)$ boundary conditions, the associated frequency equation takes the following form:

$$\begin{aligned} &C_1 \mathrm{sh}\alpha\sin\beta\mathrm{sh}\gamma + C_2\mathrm{sh}\gamma\,(\mathrm{ch}\alpha\ \cos\beta - 1) + C_3\mathrm{sh}\alpha\,(\mathrm{ch}\gamma\ \cos\beta - 1) +\\ &C_4\sin\beta\,(\mathrm{ch}\alpha\mathrm{ch}\gamma - 1) = 0, \end{aligned} \tag{8.91}$$

where

$$\begin{aligned} C_1 &= -(\beta M_3 + \gamma M_2)^2 - (\alpha M_2 + \beta M_1)^2 + (\alpha M_3 - \gamma M_1)^2,\\ C_2 &= -2\,(\beta M_3 + \gamma M_2)\,(\alpha M_3 - \gamma M_1),\\ C_3 &= 2\,(\alpha M_2 + \beta M_1)\,(\alpha M_3 - \gamma M_1),\\ C_4 &= 2\,(\alpha M_2 + \beta M_1)\,(\beta M_3 + \gamma M_2)\,. \end{aligned} \tag{8.92}$$

The obtained Eq. (8.91) is strongly transcendent and to find its solution is not an easy task. Therefore, in order to estimate the frequencies for the modified couple stress theory ($F \neq 0$), we employ an approximate method. Namely, using modes (8.79) as the vibration modes for $F = 0$ one may also approximately compute fundamental frequency $F \neq 0$ via the Rayleigh method. The eigenfrequency is defined by comparison of the formulas governing kinetic and potential energies of the vibrating beam. As following from (8.62), under the lack of temperature field $\mathrm{Re}\,[S_\omega] = \mathrm{Re}\,[D_\omega] = 1$ and the values of kinetic and potential energies of the linear beams carrying out vibrations on the fundamental frequencies $\omega_{F\neq 0}$ are defined by the following formulas:

$$K = \frac{\omega_{F\neq 0}^2}{2}\int_0^1 \left(w^2(x) + \frac{1}{P}\psi^2(x)\right)dx, \tag{8.93}$$

$$U_L = \frac{1}{2}\left\{\int_0^1 \left(\frac{d\psi}{dx}\right)^2 dx + H\int_0^1 \left(\frac{dw}{dx} + \psi\right)^2 dx + F\int_0^1 \left(\frac{d^2 w}{dx^2} - \frac{d\psi}{dx}\right)^2 dx\right\}, \tag{8.94}$$

where functions $w(x)$, $\psi(x)$ are defined by (8.89) and $(c - c)$ beam boundary conditions. The exact value of frequency $\omega_{F\neq 0}$, owing to the Rayleigh relation, is as follows:

$$\omega_{F\neq 0}^2 = \left\{\int_0^1 \left(\frac{d\psi}{dx}\right)^2 + H\int_0^1 \left(\frac{dw}{dx} + \psi\right)^2 dx + \right. \\ \left. + F\int_0^1 \left(\frac{d^2 w}{dx^2} - \frac{d\psi}{dx}\right)^2 dx\right\} \Big/ \int_0^1 \left(w^2(x) + \frac{1}{P}\psi^2(x)\right)dx\,. \tag{8.95}$$

If we substitute $w_0(x)$, $\psi_0(x)$ from (8.79) for $F = 0$, instead of $w(x)$, $\psi(x)$, we obtain

$$\frac{\omega_{F\neq 0}^2}{\omega_{F=0}^2} \approx 1 + \frac{F\int_0^1 \left(\frac{d^2 w_0}{dx^2} - \frac{d\psi_0}{dx}\right)^2 dx}{\int_0^1 \left(\frac{d\psi_0}{dx}\right)^2 dx + H\int_0^1 \left(\frac{dw_0}{dx} + \psi_0\right)^2 dx}. \tag{8.96}$$

In other words, the eigenfrequencies for $F \neq 0$ are computed directly from the classical case for $F = 0$. In reference [105], a similar approach has been employed to find the approximate value $\omega_{F\neq 0}^2$ using the Bernoulli hypothesis $\partial w_0/\partial x = -\psi_0$. If we neglect Poisson's effect, (8.96) yields

$$\omega_{F\neq 0}^2 \approx (1 + 4F)\,\omega_{F=0}^2. \tag{8.97}$$

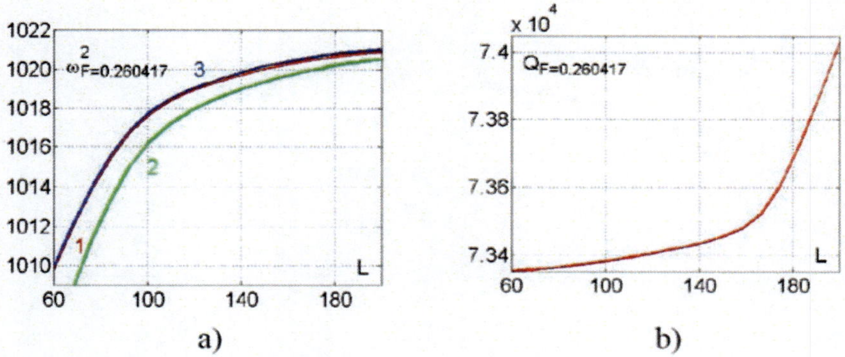

Fig. 8.6 Beam with boundary conditions $(s - s)$: **a** dependence of linear non-dimensional eigenfrequencies for $F = 0.260417$: 1—analytical solution yielded by (8.96); 2—analytical solution yielded by (8.97); 3—numerical method; **b** quality factor of the beam resonator (formula (8.63)) and frequencies (formula (8.96))

However, in this case, the shear deformations disappear, which does not hold for the Timoshenko beam. Relation (8.96) is free from this statement, and allows us to get more exact values for the Timoshenko beam using functions (8.79).

Figure 8.6a illustrates the dependence of linear non-dimensional eigenfrequencies of the fundamental vibrational mode for the non-classical linear Timoshenko beam ($F = 0.260417$) for the boundary condition $(s - s)$ versus its length L: 1—analytical solution obtained from (8.96); 2—analytical solution yielded by (8.97); 3—numerical method. Figure 8.6b shows the dependence of the quality factor of the beam resonator on the beam length estimated via formulas (8.63) (linear case).

Figure 8.6a shows that the values of $\omega^2_{F \neq 0}$ obtained from Eqs. (8.96) and (8.97) practically coincide only for both values of L, i.e. if the Timoshenko model is close to the Bernoulli-Euler model. Therefore, we employ formula (8.96) to detect the size-dependent frequencies $\omega^2_{F \neq 0}$.

In the case of the simply supported beam the analysis is simpler. If $F \neq 0$, then only the form of the coefficient $M_2 = \left(F\pi^4 + H\pi^2 - \omega^2\right)/\left(-F\pi^3 + H\pi\right)$ is changed. Frequency equation for $F \neq 0$ takes the following form:

$$\pi^2 M_2 + H(M_2 - \pi) + F\pi^2(\pi + M_2) = \frac{\omega^2 M_2}{P}. \tag{8.98}$$

After defining the frequencies and substituting the eigenfunctions $w = \sin \pi x$, $\psi = M_2 \cos \pi x$ into relations (8.63), the quality factor Q is defined in the considered linear case.

Figure 8.7a presents the dependence of the linear non-dimensional eigenfrequencies of the fundamental mode for the non-classical linear $(s - s)$ Timoshenko beam versus its length L: 1—analytical solution based on (8.98); 2—numerical solution. Figure 8.7b shows the dependence of the beam quality resonator obtained by formulas (8.63) in the linear case and for the frequencies estimated from Eq. (8.98).

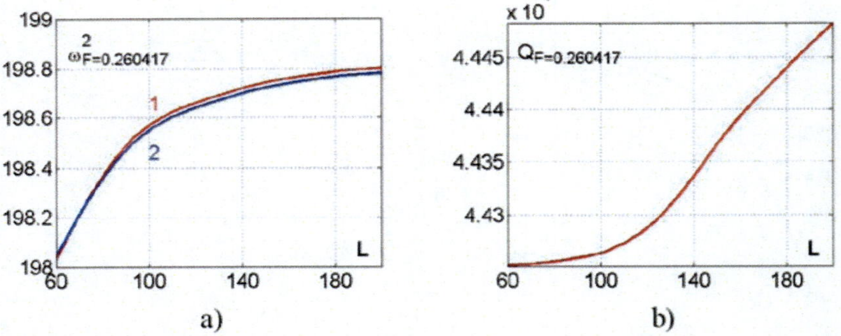

Fig. 8.7 Beam with boundary conditions $(s - s)$: **a** dependence of linear non-dimensional eigenfrequencies for $F = 0.260417$ versus L: 1—analytical solution obtained from (8.98); 2—numerical solution; **b** quality factor of the beam resonator estimated based on formula (8.63) for the linear case and frequencies obtained from (8.98) versus L

8.5.2 *Analytical Solution for the Nonlinear Size-Dependent Beam*

The influence of vibration amplitude on the resonance frequencies can be estimated from Eq. (8.45) with regard to TED. For this purpose, we employ the Rayleigh ratio which is analogous to that (see (8.93)–(8.95)) used to find approximate values of the frequencies in MMT. In the case of nonlinear beam and lack of TED, we have $\mathrm{Re}\,[S_\omega] = \mathrm{Re}\,[D_\omega] = 1$, then

$$U_{NL} = \frac{1}{2}\left\{ P\left[\int_0^1 \frac{1}{2}\left(\frac{dw}{dx}\right)^2 dx\right]^2 + \int_0^1 \left(\frac{d\psi}{dx}\right)^2 dx + \right.$$
$$\left. + H\int_0^1 \left(\frac{dw}{dx} + \psi\right)^2 dx + F\int_0^1 \left(\frac{d^2w}{dx^2} - \frac{d\psi}{dx}\right)^2 dx \right\}, \tag{8.99}$$

$$K = \frac{\omega_{NL}^2}{2}\int_0^1 \left(w^2(x) + \frac{1}{P}\psi^2(x)\right) dx. \tag{8.100}$$

If functions $w(x)$, $\psi(x)$ are the exact eigenfunctions for the first vibration mode for the given boundary conditions, the Rayleigh ratio yields an exact value of nonlinear frequency ω_{NL}, which follows

$$\omega_{NL}^2 = \frac{U_{NL}}{K}. \tag{8.101}$$

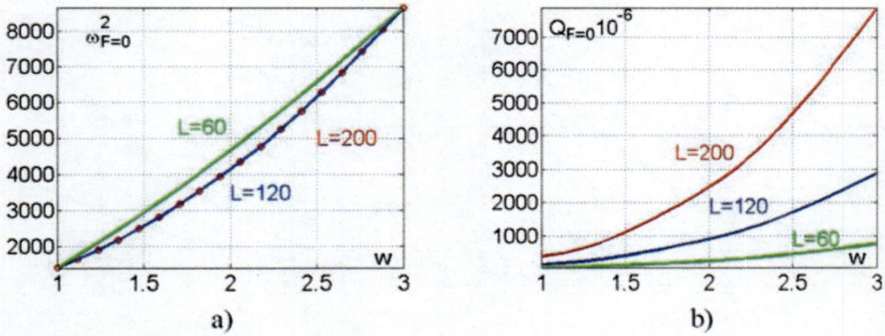

Fig. 8.8 Beam with boundary conditions $(s-s)$ for $F=0$: **a** ω^2_{NL} versus vibration amplitude w; **b** quality factor Q_{NL} versus beam vibration amplitude w

Let us start with the boundary conditions $(s-s)$. Let ω^2_L denote the square of the fundamental vibration frequency of the linear Timoshenko beam. If we consider the case $F=0$ and we substitute $w(x)$, $\psi(x)$ by w_0, $\psi_0$in Eqs. (8.99), (8.100), (8.101) the following functions are obtained:

$$w_0=\left[\mathrm{ch}\alpha x-\cos\beta x+\delta\left(\frac{\mathrm{sh}\alpha x}{M_1}+\frac{\sin\beta x}{M_2}\right)\right],$$
$$\psi_0=\left[M_1\mathrm{sh}\alpha x-M_2\sin\beta x+\delta\left(\mathrm{ch}\alpha x-\cos\beta x\right)\right],\tag{8.102}$$

where

$$\delta=\frac{M_2\sin\beta-M_1\mathrm{sh}\alpha}{(\mathrm{ch}\alpha-\cos\beta)},\quad M_1=\frac{F\alpha^4-H\alpha^2-\omega_L^2}{F\alpha^3+H\alpha},\quad M_2=\frac{F\beta^4+H\beta^2-\omega_L^2}{-F\beta^3+H\beta},\tag{8.103}$$

for $F\neq 0$, then one may estimate that

$$\frac{\omega^2_{NL}}{\omega^2_L}\approx 1+\frac{P\left(\frac{h}{L}\right)^2\left[\frac{1}{2}\int\limits_0^1\left(\frac{dw_0}{dx}\right)^2dx\right]^2\left(\frac{W_0}{a}\right)^2}{\int\limits_0^1\left(\frac{d\psi_0}{dx}\right)^2dx+H\int\limits_0^1\left(\frac{dw_0}{dx}+\psi_0\right)^2dx+F\int\limits_0^1\left(\frac{d^2w_0}{dx^2}-\frac{d\psi_0}{dx}\right)^2dx}.\tag{8.104}$$

Here, as before, W_0/a plays the role of vibration amplitude. Therefore, nonlinear eigenfrequencies can be approximately obtained from linear frequencies using linear modes of vibrations (8.102). The quality factor of the beam resonator is estimated based on formulas (8.63) for the nonlinear case and the frequencies obtained from (8.104).

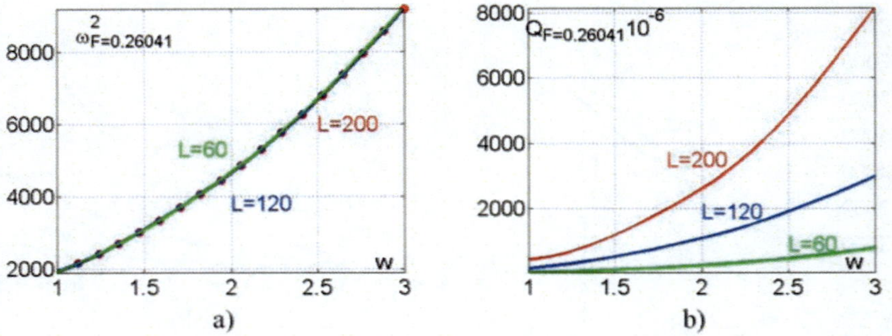

Fig. 8.9 Beam with boundary conditions $(s-s)$ for $F=0.260417$: **a** ω_{NL}^2 versus amplitude of vibrations w; **b** quality factor Q_{NL} versus beam vibration amplitude w

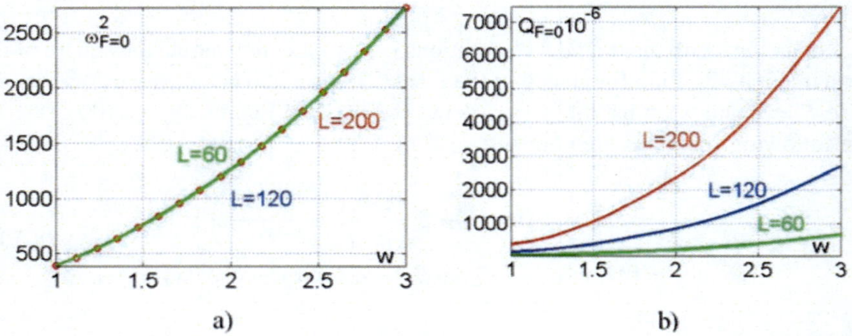

Fig. 8.10 Beam under boundary conditions $(s-s)$ and for $F=0$: **a** frequency ω_{NL}^2 versus the vibration amplitude w; **b** quality factor Q_{NL} versus the beam vibration amplitude w

Dependence w of nonlinear frequencies ω_{NL}^2 obtained approximately from formula (8.104) versus amplitude w is shown in Fig. 8.8a. Figure 8.8b presents quality factor Q_{NL} of the beam resonator under the lack of the size beam dependence $(F=0)$ versus w for three values of beam length L of the clamped beam.

Analogous results for $F=0.260417$ are given for different w in Fig. 8.9.

For the $(s-s)$ beam, the fundamental linear frequency ω_L^2 is estimated based on Eq. (8.98) and the corresponding linear vibration mode takes the following form:

$$w_0=\sin\pi x\,,\qquad \psi_0=M_2\cos\pi x, \tag{8.105}$$

where

$$M_2=\left(F\pi^4+H\pi^2-\omega^2\right)/\left(-F\pi^3+H\pi\right). \tag{8.106}$$

Substituting (8.105) into (8.104) for $a=1$ one may find the nonlinear frequency ω_{NL}^2 for different values of the vibration amplitude W_0. The quality factor of the beam resonator is computed by formula (8.63) based on the obtained frequencies ω_{NL}^2 and

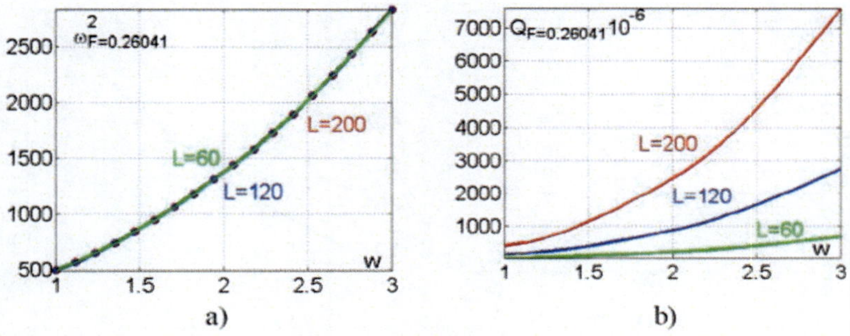

Fig. 8.11 Beam with boundary conditions $(s-s)$ for $F = 0.260417$: **a** dependence of ω_{NL}^2 on vibration amplitude w; **b** quality factor Q_{NL} versus beam vibration amplitude w

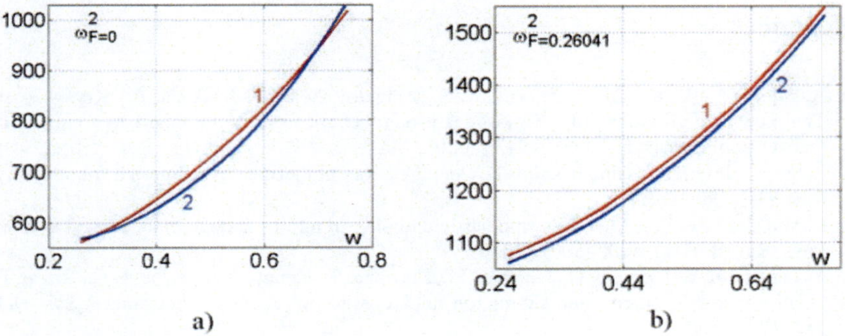

Fig. 8.12 Comparison of the eigenfrequencies for the $(s-s)$ beam: **a** $F = 0$; **b** $F = 0.260417$

the linear vibration mode (8.105). Figure 8.10a presents the dependence of nonlinear frequencies ω_{NL}^2 obtained via formula (8.104) versus the vibration amplitude w, and in Fig. 8.10b the quality factor Q_{NL} of the beam resonator with the lack of the size dependence ($F = 0$) is given for three values of the beam length.

Analogous results for $F = 0.260417$ are presented in Fig. 8.11.

In order to check the reliability of computation of eigenfrequencies, the obtained analytical results have been compared with numerical solutions.

Figure 8.12 presents the results of comparison for the beam $(s-s)$ for $F = 0$ (a) and for $F = 0.260417$ (b), i.e. taking into account the size-dependent behaviour obtained numerically (8.2) and analytically (8.1).

Figure 8.13 presents the results of comparison of the eigenfrequencies computed for the $(s-s)$ beam for $F = 0$ (a) (without the size-dependent behaviour) and for $F = 0.260417$ (b) with the size-dependent behaviour.

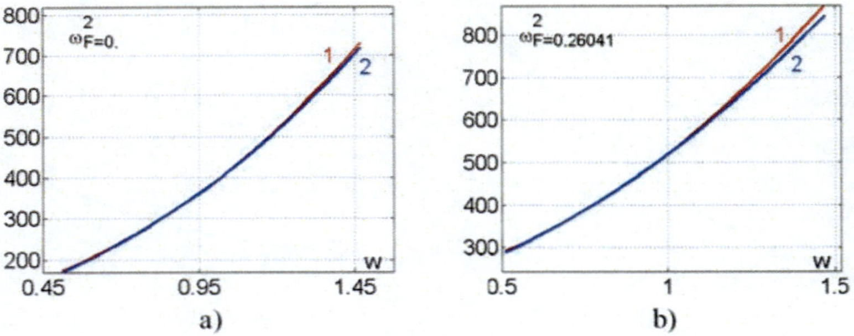

Fig. 8.13 Comparison of the eigenfrequencies for the $(s-s)$ beam: **a** $F=0$; **b** $F=0.260417$

References

1. Awrejcewicz, J., Krysko, V.A., Pavlov, S., Zhigalov, M.V., Kalutsky, L.A., Krysko, A.V.: Thermoelastic vibrations of a Timoshenko microbeam based on the modified couple stress theory. Nonlinear Dyn. **99**, 919–943 (2020)
2. Zener, C.: Internal friction in solids II. General theory of thermoelastic internal friction. Phys. Rev. **53**(1), 90–99 (1938)
3. Lifshitz, R., Roukes, M.L.: Thermoelastic damping in micro- and nanomechanical systems. Phys. Rev. B **61**(8), 5600–5609 (2000)
4. Yasumura, K.Y., Stowe, T.D., Chow, E.M., Pfafman, T., Kenny, T.W., Stipe, B.C., Rugar, D.: Quality factors in micro- and submicron-thick cantilevers. J. Microelectromech. Syst. **9**(1), 117–125 (2000)
5. Khisaeva, Z.F., Ostoja-Starzewski, M.: Thermoelastic damping in nanomechanical resonators with finite wave speeds. J. Therm. Stres. **29**, 201–216 (2006)
6. Houston, B., Photiadis, D.M., Marcus, M.H., Bucaro, J.A., Liu, X., Vignola, J.F.: Thermoelastic loss in microscale oscillators. Appl. Phys. Lett. **80**, 1300–1302 (2002)
7. Yang, J., Ono, T., Esashi, M.: Energy dissipation in submicrometer thick single-crystal silicon cantilevers. J. Microelectromech. Syst. **11**(6), 775–783 (2002)
8. Duwel, A., Gorman, J., Weinstein, M., Borenstein, J., Ward, P.: Experimental study of thermoelastic damping in MEMS gyros. Sens. Actuator A **103**, 70–75 (2003)
9. Tilmans, H.A.C., Elwespoek, M., Fluitman, J.H.J.: Micro resonant force gauges. Sens. Actuator A **30**, 35–53 (1992)
10. Lothe, J.: Aspects of the theories of dislocation mobility and internal friction. Phys. Rev. **117**, 704–708 (1960)
11. Zhang, C.L., Xu, G.S., Jiang, Q.: Analysis of the air-damping effect on a micromachined beam resonator. Math. Mech. Sol. **8**, 315–325 (2003)
12. Tortonese, M., Barrett, R.C., Quate, C.F.: Atomic resolution with an atomic force microscope using piezoresistive detection. Appl. Phys. Lett. **62**(8), 834–836 (1993)
13. Barnes, J.R., Stephenson, R.J., Welland, M.E., Gerber, Ch., Gimzewski, J.K.: Photothermal spectroscopy with femtojoule sensitivity using a micromechanical device. Nature **372**, 79–81 (1994)
14. Yurke, B., Greywall, D.S., Parqellis, A.N., Busch, P.A.: Theory of amplifier-noise evasion in an oscillator employing a nonlinear resonator. Phys. Rev. A **51**, 4211–4229 (1995)
15. Cleland, A.N., Roukes, M.L.: Fabrication of high frequency nanometer scale mechanical resonators from bulk Si crystals. Appl. Phys. Lett. **69**, 2653–2655 (1996)
16. Lun, F.Y., Zhang, P., Gao, F.B., Jia, H.G.: Design and fabrication of micro-optomechanical vibration sensor. Microfab. Technol. **120**(1), 61–64 (2006)

17. Nguyen, C.T.C.: Micromechanical resonators for oscillators and filters. IEEE Ultrason. Symp. 489–499 (1995)
18. Wang, K., Nguyen, C.T.C.: High-order medium frequency micromechanical electronic filters. J. Microelectromech. Syst. **8**(4), 534–557 (1999)
19. Lepage, S.: Thermoelastic Damping in Vibrating Beam Accelerometer: A New Thermoelastic Finite Element Approach. Incaneus, Toulouse (2006)
20. Mestrom, R.M.C., Fey, R.H.B., Phan, K.L., Nijmeijer, H.: Simulations and experiments of hardening and softening resonances in a clamped-clamped beam MEMS resonator. Sens. Actuator A: Phys. **162**(2), 225–234 (2010)
21. Rezazadeh, G., Tahmasebi, A., Zubstov, M.: Application of piezoelectric layers in electrostatic MEM actuators: controlling of pull-in voltage. Microsyst. Technol. **12**(12), 1163–1170 (2006)
22. Sadeghian, H., Rezazadeh, G., Osterberg, P.M.: Application of the generalized differential quadrature method to the study of pull-in phenomena of MEMS switches. J. Microelectromech. Syst. **16**(6), 1334–1340 (2007)
23. Sadeghian, H., Rezazadeh, G.: Comparison of generalized differential quadrature and Galerkin methods for the analysis of micro-electro-mechanical coupled systems. Commun. Nonlinear Sci. Num. Simul. **14**(6), 2807–2816 (2009)
24. Vahdat, A.S., Rezazadeh, G.: Effect of axial and residual stresses on thermoelastic damping in capacitive micro-beam resonators. J. Franklin Inst. **348**, 622–639 (2011)
25. Kim, S.B., Kim, J.H.: Quality factors for the nano-mechanical tubes with thermoelastic damping and initial stress. J. Sound Vibr. **330**, 1393–1402 (2011)
26. Evoy, S., Oikhovets, A., Sekaric, L., Parpia, J.M., Craighead, H.G., Carr, D.W.: Temperature-dependent internal friction in silicon nano-electromechanical systems. Appl. Phys. Lett. **77**, 2397–2399 (2000)
27. Zener, C.: Internal friction in solids I. Theory of internal friction in reeds. Phys. Rev. **52**(3), 230–235 (1937)
28. Zener, C., Otis, W., Nuckolls, R.: Internal friction in solids III. Experimental demonstration of thermoelastic internal friction. Phys. Rev. **53**(1), 100–101 (1938)
29. Landau, L.D., Lifshitz, E.M.: Theory of Elasticity. Pergamon Press, Oxford (1959)
30. Nayfeh, A.H., Younis, M.I.: Modeling and simulations of thermoelastic damping in microplates. J. Micromech. Microeng. **14**, 1711–1717 (2004)
31. Sun, Y., Saka, M.: Vibrations of microscale circular plates induced by ultra-fast lasers. Int. J. Mech. Sci. **50**(9), 1365–1371 (2008)
32. Sun, Y., Tohmyoh, H.: Thermoelastic damping of the axisymmetric vibration of circular plate resonators. J. Sound Vibr. **319**, 392–405 (2009)
33. Sun, Y., Saka, M.: Thermoelastic damping in micro-scale circular plate resonators. J. Sound Vibr. **329**, 328–337 (2010)
34. Sharma, J.N., Sharma, R.: Damping in micro-scale generalized thermoelastic circular plate resonators. Ultrason **51**, 352–358 (2011)
35. Srinivasa, A.R., Reddy, J.N.: A model for a constrained, finitely deforming, elastic solid with rotation gradient dependent strain energy, and its specialization to von Kármán plates and beams. J. Mech. Phys. Sol. **61**(3), 873–885 (2013)
36. Berry, B.S.: Precise investigation of the theory of damping by transverse thermal currents. J. Appl. Phys. **26**(10), 1221–1224 (1955)
37. Zhang, W., Turner, K.L.: Thermoelastic damping in the longitudinal vibration: analysis and simulation. IMECE 145–149 (2004)
38. Vengallator, S.: Analysis of thermoelastic damping in laminated composite micromechanical beam resonators. J. Micromech. Microeng. **15**, 2398–2404 (2005)
39. Prabhakar, S., Vengallatore, S.: Thermoelastic damping in bilayered micromechanical beam resonators. J. Micromech. Microeng. **17**, 532–538 (2007)
40. Yi, Y.B.: Geometric effects on thermoelastic damping in MEMS resonators. J. Sound Vib. **309**, 588–599 (2008)
41. Hajnayeb, A., Khadem, S.E., Zamanian, M.: Thermoelastic damping of a double-walled carbon nanotube under electrostatic force. Micro. Nano Lett. **6**(8), 698–703 (2011)

42. Massalas, C.V., Kalpakidis, V.K.: Coupled thermoelastic vibrations of a simply supported beam. J. Sound Vib. **88**(3), 425–429 (1983)
43. Shieh, R.C.: Thermoelastic vibration and damping for circular Timoshenko beams. J. Appl. Mech. **42**(2), 405–410 (1975)
44. Parayila, D.V., Kulkarnia, S.S., Pawaskara, D.N.: Analytical and numerical solutions for thick beams with thermoelastic damping. Int. J. Mech. Sci. **94–95**, 10–19 (2015)
45. Prevost, J.H., Tao, D.: Finite element analysis of dynamic coupled thermoelasticity problems with relaxation times. J. Appl. Mech. **50**(4), 817–822 (1983)
46. Prabhakar, S., Vengallatore, S.: Theory of thermoelastic damping in micromechanical resonators with two-dimensional heat conduction. J. Microelectromech. Sys. **17**(2), 494–502 (2008)
47. Silver, M.J., Peterson, L.D., Erwin, R.S.: Predictive elastothermodynamic damping in finite element models using a perturbation formulation. AIAA **43**(12), 2646–2653 (2005)
48. Sun, Y., Fang, D., Soh, A.K.: Thermoelastic damping in micro-beam resonators. Int. J. Sol. Struct. **43**, 3213–3229 (2006)
49. Lepage, S.: Stochastic finite element method for the modeling of thermoelastic damping in micro-resonators, Ph.D. thesis, Vibrations et Identification des Structures, Département d'Aérospatiale et Méchanique. Université de Liége, Liége (2006)
50. Guo, X., Yi, Y.B., Pourkamali, S.: A finite element analysis of thermoelastic damping in vented MEMS beam resonators. Int. J. Mech. Sci. **74**, 73–82 (2013)
51. De, S.K., Aluru, N.R.: Theory of thermoelastic damping in electrostatically actuated microstructures. Phys. Rev. B **74**, 14305 (2006)
52. Fleck, N.A., Muller, G.M., Ashby, M.F., Hutchinson, J.W.: Strain gradient plasticity: theory and experiments. Acta Metall. Mater. **42**, 475–487 (1994)
53. Lam, D.C.C., Yang, F., Chong, A.C.M., Wang, J., Tong, P.: Experiments and theory in strain gradient elasticity. J. Mech. Phys. Soli **51**, 1477–1508 (2003)
54. Liu, H.K., Pan, C.H., Liu, P.P.: Dimension effect on mechanical behavior of silicon microcantilever beams. Measurement **41**, 885–895 (2008)
55. Tsiatas, G.C.: A new Kirchhoff plate model based on a modified couple stress theory. Int. J. Sol. Struct. **46**(13), 2757–2764 (2009)
56. Wang, F.C., Yang, F.Q., Zhao, Y.P.: Size effect on the coalescence-induced self-propelled droplet. Appl. Phys. Lett. **98**(5), 053112 (2011)
57. Brezny, R., Green, D.J.: Characterization of edge effects in cellular materials. J. Mater. Sci. **25**, 4571–4578 (1990)
58. Onck, P.R., Andrews, E.W., Gibson, L.J.: Size effects in ductile cellular solids. Part I: modeling. Int. J. Mech. Sci. **43**(3), 681–699 (2001)
59. Cosserat, E., Cosserat, F.: Théorie des Corps Déformables. Hermann et Fils, Paris (1909)
60. Wang, B., Zhao, J., Zhou, S.: A micro scale Timoshenko beam model based on strain gradient elasticity theory. Euro. J. Mech. A/Sol. **29**, 591–599 (2010)
61. Mindlin, R.D., Tiersten, H.F.: Effects of couple-stresses in linear elasticity. Arch. Rat. Mech. Anal. **11**, 415–448 (1962)
62. Yang, F., Chong, M., Lam, D.C.C., Tong, P.: Couple stress based strain gradient theory for elasticity. Int. J. Solids Struct. **39**, 2731–2743 (2002)
63. Park, S.K., Gao, X.L.: Bernoulli-Euler beam model based on a modified couple stress theory. Micromech. Microeng. **16**(11), 2355–2359 (2006)
64. Kong, S., Zhou, S., Nie, Z., Wang, K.: The size-dependent natural frequency of Bernoulli–Euler micro-beams. Int. J. Eng. Sci. **46**, 427–437 (2008)
65. Kong, S., Zhou, S., Nie, Z., Wang, K.: Static and dynamic analysis of micro beams based on strain gradient elasticity theory. Int. J. Eng. Sci. **47**, 487–498 (2009)
66. Kahrobaiyan, M.H., Asghari, M., Rahaeifard, M., Ahmadian, M.T.: Investigation of the size-dependent dynamic characteristics of atomic force microscope microcantilevers based on the modified couple stress theory. Int. J. Eng. Sci. **48**(12), 1985–1994 (2010)
67. Asghari, M., Ahmadian, M.T., Kahrobaiyan, M.H., Rahaeifard, M.: On the size-dependent behavior of functionally graded micro-beams. Mater. Des. **31**, 2324–2329 (2010)

68. Ma, H.M., Gao, X.-L., Reddy, J.N.: A microstructure-dependent Timoshenko beam model based on a modified couple stress theory. J. Mech. Phys. Sol. **56**, 3379–3396 (2008)
69. Ma, H.M., Gao, X.L., Reddy, J.N.: A non-classical Mindlin plate model based on a modified couple stress theory. Acta Mech. **220**, 217–235 (2011)
70. Reddy, J.N.: Microstructure-dependent couple stress theories of functionally graded beams. Mech. Phys. Sol. **59**, 2382–2399 (2011)
71. Reddy, J.N., Kim, J.: A nonlinear modified couple stress-based third-order theory of functionally graded plates. Compos. Struct. **94**, 1128–1143 (2012)
72. Reddy, J.N., Berry, J.: Nonlinear theories of axisymmetric bending of functionally graded circular plates with modified couple stress. Compos. Struct. **94**, 3664–3668 (2012)
73. Arbind, A., Reddy, J.N.: Nonlinear analysis of functionally graded microstructure-dependent beams. Compos. Struct. **98**, 272–281 (2013)
74. Kim, J., Reddy, J.N.: Analytical solutions for bending, vibration, and buckling of FGM plates using a couple stress-based third-order theory. Compos. Struct. **103**, 86–98 (2013)
75. Guo, F.L., Rogerson, G.A.: Thermoelastic coupling effect on a micro-machined beam resonator. Mech. Res. Commun. **30**, 513–518 (2003)
76. Rezazadeh, G., Vahdat, A.S., Tayefeh-Rezaei, S., Cetinkaya, C.: Thermoelastic damping in a micro-beam resonator using modified couple stress theory. Acta Mech. **223**, 1137–1152 (2012)
77. Taati, E., Molaei Najafabadi, M.M., Basirat Tabrizi, H.: Size-dependent generalized thermoelasticity model for Timoshenko microbeams. Acta Mech. **225**(7), 1823–1842 (2014)
78. Taati, E., Molaei Najafabadi, M.M., Reddy, J.N.: Size-dependent generalized thermoelasticity model for Timoshenko micro-beams based on strain gradient and non-Fourier heat conduction theories. Comput. Struct. **116**, 595–611 (2014)
79. Eom, K., Park, H.S., Yoon, D.S., Kwon, T.: Nanomechanical resonators and their applications in biological/chemical detection: nanomechanics principles. Phys. Rep. **503**, 115–163 (2011)
80. Ekinci, K.L., Roukes, M.L.: Nanoelectromechanical systems. Rev. Sci. Instr. **76**, 061101 (2005)
81. Sharma, J.N.: Thermoelastic damping and frequency shift in micro/nanoscale anisotropic beam. J. Therm. Stress. **34**, 650–666 (2011)
82. Ru, C.Q.: Thermoelastic dissipation of nanowire resonators with surface stress. Phys. E **41**, 1243–1248 (2009)
83. Tunvir, K., Ru, C.Q., Mioduchowski, A.: Thermoelastic dissipation of hollow micromechanical resonators. Phys. E **42**, 2341–2352 (2010)
84. Singh, G., Sharma, A.K., Rao, G.V.: Large-amplitude free vibrations of beams - A discussion on various formulations and assumptions. J. Sound Vibr. **142**(8), 77–85 (1990)
85. Rao, B.N.: Large-amplitude free vibrations of simply supported uniform beams with immovable ends. J. Sound Vibr. **155**(3), 523–527 (1992)
86. Rao, G.V., Raju, K.K.: Large amplitude free vibrations of beams - an energy approach. ZAMM **83**(7), 493–498 (2003)
87. Peng, H.B., Chang, C.W., Aloni, S., Yuzvinsky, T.D., Zettl, A.: Ultrahigh frequency nanotube resonators. Phys. Rev. Lett. **97**, 087203 (2006)
88. Bunch, J.S., van der Zande, A.M., Verbridge, S.S., Frank, I.W., Tanenbaum, D.M., Parpia, J.M., Craighead, H.G., McEuen, P.L.: Electromechanical resonators from graphene sheets. Science **315**, 490–493 (2007)
89. Masmanidis, S.C., Karabalin, R.B., De Vlaminck, I., Borghs, G., Freeman, M.R., Roukes, M.L.: Multifunctional nanomechanical systems via tunably coupled piezoelectric actuation. Science **317**, 780–783 (2007)
90. Hao, Z.: Thermoelastic damping in the contour-mode vibration of micro and nano-electromechanical circular thin-plate resonators. J. Sound Vib. **313**, 77–96 (2008)
91. Xie, W.C., Lee, H.P., Lim, S.P.: Non-linear dynamic analysis of MEMS switches by nonlinear modal analysis. Nonlinear Dyn. **31**(3), 243–256 (2003)
92. Méndez, C., Paquay, S., Klapka, I., Raskin, J.P.: Effect of geometrical nonlinearity on MEMS thermoelastic damping. Nonlin. Anal. Real World Appl. **10**, 1579–1588 (2009)

93. Salajeghe, S., Khadem, S.E., Rasekh, M.: Nonlinear analysis of thermoelastic damping in axisymmetric vibration of micro circular thin-plate resonators. Appl. Math. Model. **36**, 5991–6000 (2012)
94. Tunvir, K., Ru, C.Q., Mioduchowski, A.: Large-deflection effect on thermoelastic dissipation of microbeam resonators. J. Therm. Stress **35**, 1076–1094 (2012)
95. Rochus, V., Rixen, D.J., Golinval, J.C.: Electrostatic coupling of MEMS structures: transient simulations and dynamic pull-in. Nonlinear Anal. Theor. Meth. Appl. **63**, 1619–1633 (2005)
96. Krysko, A.V., Awrejcewicz, J., Pavlov, S.P., Bodyagina, K.S., Zhigalov, M.V., Krysko, V.A.: Non-linear dynamics of size-dependent Euler-Bernoulli beams with topologically optimized microstructure and subjected to temperature field. Int. J. Non-Lin. Mech. **104**, 75–86 (2018)
97. Awrejcewicz, J., Pavlov, S.P., Bodyagina, K.S., Zhigalov, M.V., Krysko, V.A.: Design of composite structures with extremal elastic properties in the presence of technological constraints. Compos. Struct. **174**, 19–25 (2017)
98. Timoshenko, S.P., Goodier, J.N.: Theory of Elasticity, 2nd edn. McGraw-Hill, New York (1951)
99. Feng, J.T., Zhao, Y.P.: Influence of different amount of Au on the wetting behavior of PDMS membrane. Biomed. Microdev. **10**(1), 65–72 (2008)
100. Voyiadjis, G.Z., Abu Al-Rub, R.K.: Gradient plasticity theory with a variable length scale parameter. Int. J. Sol. Struct. **42**(14), 3998–4029 (2005)
101. Han, S.M., Benaroya, H., Wei, T.: Dynamics of transversely vibrating beams using four engineering theories. J. Sound Vib. **225**(5), 935–988 (1999)
102. Gorman, D.J.: Free Vibration Analysis of Beams and Shafts. Wiley, New York (1975)
103. Hao, Z., Erbil, A., Ayazi, F.: An analytical model for support loss in micromachined beam resonators with in-plane flexural vibrations. Sens. Actuators A **109**, 156–164 (2003)
104. Mahi, A., Adda Bedia, E.A., Tounsi, A., Machab, I.: An analytical method for temperature-dependent free vibration analysis of functionally graded beams with general boundary conditions. Comput. Struct. **92**, 1877–1887 (2010)
105. Araujo dos Santos, J.V., Reddy, J.N.: Free vibration and buckling analysis of beams with a modified couple-stress theory. Int. J. Appl. Mech. **4**(3), 1250026 (2012)

Chapter 9
Vibrations of Size-Dependent Beams Under Topologic Optimization and Temperature Field

9.1 Introduction

In this chapter, problems associated with multifunctional requirements with respect to effective characteristics of composites consisting of two components as well as composites with holes or technological inclusions have been studied. In the process of investigation, a strong dependence of the optimal topology of the distribution of materials in the microstructure of composites on the form of the target functions has been detected. The study of transformations of the optimal topology of the composite microstructure with a change in the weight coefficient from $\omega = 0$ (maximization of the heat transfer) up to $\omega = 1$ (maximization of the mechanical moduli) has been conducted. Moreover, a set of alternatives optimal in the Pareto sense has been constructed. The considered examples clearly indicate the inability to achieve the best/required properties simultaneously in both cases, which is caused by conflicting criteria in the target function. In addition, nonlinear dynamics of the size-dependent Euler-Bernoulli beams embedded into temperature field with topologically optimized microstructure is studied. The following new results are presented:

(i) We have developed a mathematical model of the size-dependent nonlinear beam, taking into account the topological optimization under the criterion of maximum stiffness. The mathematical model is based on the Bernoulli-Euler, von Kármán and Duhamel-Newmann hypotheses. Also, an algorithm and a computer program for numerical computations of the optimized beam microstructure for the given boundary conditions, the form of external load and the temperature have been developed (both static and dynamic problems have been considered).

(ii) Reliability of the results was confirmed by investigating the convergence along the spatial variable as well as by examining the solution to the Cauchy problems, and investigating Lyapunov exponents and time evolution of the frequency, obtained using wavelet analysis.

J. Awrejcewicz et al., *Mathematical Modelling and Numerical Analysis of Size-Dependent Structural Members in Temperature Fields*, Advanced Structured Materials 142,
https://doi.org/10.1007/978-3-030-55993-9_9

(iii) The analysis of the reliability of the results obtained for different numbers of spatial partitions was carried out based on the analysis of the power spectrum for chaotic system states. The reliability of chaos was validated by computing LLEs using four different methods.
(iv) The Cauchy problem was solved by numerous methods and the fourth-order Runge-Kutta method was chosen as the most efficient. The optimal step was chosen using the Runge principle.
(v) Based on the tests carried out for numerous wavelets (Daubechies, Gauss, Haar and Morlet), the Morlet wavelets were chosen as the most feasible for our problem.
(vi) The static analysis was carried out for three values of temperature and two values of size-dependent parameter. The investigated 'frequency-deflection' dependency exhibits different results for homogeneous and non-homogeneous (optimized) beams for all values of the length-dependent parameter.
(vii) The use of beams with the optimized microstructure allows for an increase in the range of working loads regimes compared to homogeneous beams for which the vibration regimes are either periodic or quasi-periodic.
(viii) The analysis of the scenarios of transition from periodic to chaotic vibrations was carried out. In all cases, the transition into chaotic vibrations followed a scenario similar to the classical Pomeau-Manneville scenario (a few exceptions were observed for some values of temperature and the length-dependent parameter).

9.2 Literature Review

Functionally graded materials (FGM) are fabricated based on the multi-phase composites with volume part phases and with mechanical/thermomechanical properties being changed along the chosen directions in order to achieve engineering required characteristics. Those new fabricated materials possess the efficient strength characteristics without harmful concentration of stresses of some of their parts. Owing to the mentioned properties, the beams made from FGM found wide application in numerous engineering constructions including wings of gas and wind turbines, rotor blades of helicopters, propeller blades as well as stator blades, feathering paddles, wicket gates and others (see for instance [1]).

A change of the material property along a given direction can be incorporated by either stepwise or exponential rules, which are widely used for the description of the variable FGM properties, towards the material thickness, its longitudinal direction or towards both directions simultaneously.

The classical (standard) way to change the elastic properties of a material element, i.e. its microstructure, is introduced by a design engineer (constructor). The majority of investigations devoted to FGM beams' design are focused on tailoring the beams' dynamical properties along their thickness.

In addition, the investigations are focused in determination of the free frequencies corresponding to different modes of the beam vibrations [2–8]. It should be emphasized that analysis of free vibrations of beams with non-homogenous transverse cross sections and with tapered material properties along the beam longitudinal direction is more complex and difficult to study than those with homogenous cross sections and homogenous material [9, 10]. This is due to the occurrence of variable coefficients in the governing equations, which requires specific and sometimes novel approaches for modelling the system dynamics and for solving the derived differential equations. Only in rare cases, analytical solutions can be obtained for the case of tapered beams in their axial direction [11, 12]. Therefore, majority of case studies were carried out with the help of the numerical methods. Hemmatnezhad et al. [13] studied nonlinear behaviour of functionally graded beams with an account of Kármán nonlinearity and with the help of finite element method (FEM). Kien [14] employed FEM to analyze tapered cantilever beams made of FGM. Alshorbagy et al. [15] investigated free vibrations of FGM beams using FEM. Shaba et al. [10, 16, 17] analyzed stability and vibration of FGM tapered Timoshenko and Euler-Bernoulli beams using a few numerical approaches. Ke et al. [18] investigated free vibrations of geometrical FGM beams using Galerkin procedure in its first-order approximation.

In order to study a tapered beam made from homogenous/non-homogenous material various mathematical models based on either Euler-Bernoulli [10, 19–24] or Timoshenko [17, 25–28] theories were constructed. Free vibrations as well as dynamic characteristics of a beam made from FGM and with tapered and/or non-uniform cross section were analyzed in references [29–31].

On the other hand, in order to investigate FGM mechanical structural systems with an account of the size-dependent behaviour, the following novel theories were developed: modified couple stress theory [32–38], nonlocal polar theory of elasticity [39], strain gradient theory [40] and the general theory of curved deformable interfaces [41].

Arbind and Reddy [42] carried out nonlinear study of the functionally graded microstructure-dependent beams made from non-homogenous material properties along the beams thickness and with the von Kármán non-linearity. The modified couple stress theory was used to follow the size-dependent beam behavior. The derived governing PDEs are based on the Euler-Bernoulli and Timoshenko theories. The static beam deflections versus the length size-dependent parameter, the character of change of beam thickness properties, the shear deformations and geometric nonlinearity were investigated. Arbind et al. [43] employed Hamilton principle and the modified couple stress-based third-order theory to study nonlinear behaviour of FGM beams.

More recently the beam-type structures are strengthened by composite materials tailored by the methods of topological optimization. Topological optimization of composite structures is widely used while fabricating materials with improved physical characteristics. In order to average the complex dynamical phenomena and make it computationally feasible, the complex microstructure behaviour of an elastic medium undergoes the so-called asymptotic homogenization [44–46]. In addition, in the case of design of microstructure materials other methods are employed including the method of solid isotropic material with penalization (SIMP) [47], the method of evolutionary structural optimization (ESO) [38] and the level set methods [48–51]. An important role is played in the matching of SIMP with other methods, and in particular with the method of homogenization. The main idea of the latter approach relies on the determination of the element parameters as the design variables, whereas in the optimization process the appropriate parameters are estimated by establishing links between the local parameters (for instance, density) and the global required physical material properties (for instance, Young modulus and/or heat transfer coefficients). Various algorithms are then employed to solve the optimization problem like the method of optimal criterion [52], the method of neighbourhood search based on efficient sensitivity calculations [53] or the method of moving asymptotes [54]. It should be noticed that materials in the optimization domain are graded in a way to achieve the required functional characteristics of the given constraints. Two main directions in the FGM research can be distinguished: (i) problems dealing with constructions optimization [38, 44, 50, 51] and (ii) problems of achieving effective functional material properties [55–60]. The design of composite structures requires precise modelling of elastic, thermal and thermoelastic material properties. The associated widely popular elasticity criteria take into account only mechanical load [44], whereas thermal criteria rely on the heat load [61, 62], and finally the thermoelastic criterion includes both mechanical and heat loads [63, 64].

In the case of the problems of topological optimization, there are two main approaches aimed on the inclusion of mechanical and thermal criteria. The first class considers the given material thermoelastic properties but without a full coupling of heat and mechanic fields in the topological sense, i.e. usually the temperature field is given and the thermal load is transformed to the mechanical load [63, 64]. In this statement, the given construction is controlled via mechanical characteristics. The second class takes into account both heat and mechanical fields. The problem is solved in the following way: first, the temperature field is calculated, and then it is matched with the field of deformations. In the latter approach, the microstructure material can be constructed with the help of the mechanical target function [63] or with the help of the coupled mechanical and heat functions simultaneously [65].

The qualitative radical change in the design of composite materials was introduced by Sigmund [58, 60] who proposed the method of inverse homogenization. He assumed that the composites are made from identical periodically distributed cells. The topological optimization was employed for the appropriate localization of either one-component [57] or multi-component [55, 59] materials in order to achieve the required physical properties of the constructed composite. It is well known that a direct homogenization allows consideration of the global composite properties based on the structure of one elementary cell or on a basis of the representative cell volume [66]. In contrary to this standard approach, the method of inverse homogenization allows to estimate the appropriate/required material distribution in the introduced model of an elementary cell [57]. A difference between the global homogenized material properties and the required (target) values is controlled by finding a minimum with a help of the methods of topological optimization [55, 60], However, in the latter approach, the elastic and heat material properties are usually treated separately.

More recently, the optimal microstructure of materials has been worked out in the problems of achieving maximum value and shear moduli [44, 67]. The mentioned methods allow to design novel composites with negative Poisson's coefficient [57] as well as materials with extreme or even negative heat transfer coefficient [60], which demonstrates a huge ability of the mentioned method while designing either elastic or thermoelastic composites. It is evident that in many cases heat transfer and elastic material properties are coupled [68]. Therefore, it is difficult to achieve the composite material with high heat transfer and stiffness ability simultaneously. This is why it is worthy to develop a novel procedure working effectively while designing the composite materials with a few inversely required targets. However, the multi-criteria optimization is difficult and sometimes impossible to be realized in order to obtain a global optimum for all design targets. In the optimization process, one may observe that the improvement of one criterion requires compromise with regard to another criterion. A set of the obtained solutions is defined by the Pareto space consisting of a series of points in the solutions space, where each of the included solutions cannot be improved via the used target function without deterioration of the remaining target functions [69]. This means that it is required to gather a sufficient choice of the best solutions. Torquato et al. [70] proposed an optimal design of manufacturable 3D composites with multifunctional characteristics including heat transfer and electricity transfer with the same weights [70].

Yoo and Lee [71] employed various weight coefficients while searching for optimal topology of the Pareto solution with regard to the stiffness and frequency criteria mainly using a swing arm type actuator by computing the response surface method. For many years, as it follows from the literature overview, the microstructure design of the beam-type structural members was based on the designer intuition and experience with the support of the optimization methods.

In this chapter, we solve a multistep optimization problem of reinforcement of a beam made from composite material to improve its resistance to the acting loads. The approach consists of a few steps. In the first step, using the methods of topological optimization, a microstructure of two-component composite being optimized with regards to stiffness and heat transfer is obtained. The second step relies on topological optimization of the reinforced beam structure under the criterion of minimum flexibility (maximum stiffness) for the given boundary conditions and loading conditions of the beam embedded into temperature field. In each of the considered cases, the optimal structure of the reinforced beam becomes original for the given conditions of loading and the employed boundary conditions. In result, the designed beam with the optimal microstructure exhibits non-homogeneity along the two directions, i.e. its thickness and length.

The third step is aimed at investigation of the static and dynamic beam behaviour with its mathematical model derived on a basis of the Euler-Bernoulli hypotheses, the modified couple stress theory and with an account of the geometric von Kármán non-linearity. The influence of the temperature fields is taken into account by using the Duhamel-Neumann relations.

Besides, it should be mentioned, that our research fulfils a gap in the existing research, since there are no reports of the results of chaotic vibrations of the size-dependent FGM in the available literature.

9.3 Topological Optimization Problem

The topological optimization stands for a process of determination of coupling and possible forms of localization of inclusions inside a given space occupied by an elastic material. It can be understood as the determination of the best distribution of the materials inside of the given space [72]. It guarantees a large freedom for designers in comparison to the process of optimization of the size and space forms. The latter takes into account the variables like fitness and surface of the cross sections of the being constructed elements (sizes) and the geometric peculiarities (forms) of the given design elements. Therefore, optimization of a topology plays a crucial role in both preliminary design studies where changes exhibit essential influence on the final effectiveness of the phase components. Owing to the mentioned possibility of finding the optimal material microstructure, the topological optimization belongs to the perspective tool for the thermoelastic design. There exist two types of topological optimizations: discrete and continuous [73, 74]. In what follows, we consider only continuous optimization. There are available a few different methods based on finite element for the topological optimization of a continuum. The general problem consists of determination of either existence or a lack of material in a given subspace of the space undergoing the design process. In what follows, we briefly describe the nowadays employed methods of topological optimization.

9.3.1 Methods Based on Density

Nowadays the most widely employed methods of topological optimization of structures are based on the explicit parameterization also known as methods based on density. Those methods operate on the space divided into finite elements. However, instead of characterization of elastic properties of microstructure, each of the finite elements contains only one design variable. The latter one is often understood as the material element density ρ_e. Properties of each material element, for instance, the Young moduli and the heat transfer coefficient are the functions of the variable density through the given interpolating functions. The most known are widely used in the solid isotropic material with penalization method (SIMP). The fundamental mathematical formulations of the problem of the topological optimization based on the density consist of target functions, a set of constraints (they often include constraints because of the material use) and the finite-dimensional representation of the physical system, for instance, on a basis of the method of finite element (FEM). The general statement being based on the linear static analysis of FEM can be presented in the following way

$$\begin{gathered} \min f(\mathbf{x},\, \mathbf{U}) \quad in \quad \Omega_{opt}, \\ \text{for} \quad \mathbf{K}(\mathbf{x})\,\mathbf{U} = \mathrm{F}(\mathbf{x}), \quad g_i(\mathbf{x},\, \mathbf{U}) \leq 0, \quad 0 \leq x_e \leq 1, \end{gathered} \tag{9.1}$$

where f-target function; $\mathbf{x}$-vector of design variables coupled with density ρ_e, $\mathbf{U}$-vector of displacement, $\mathbf{K}$-global matrix of stiffness, $\mathbf{F}$-force vector and g_i stands for the constraints. Observe that stiffness matrix $\mathbf{K}$ and sometimes the force vector $\mathbf{F}$ depend explicitly on the design variables. In the frame of the introduced general statement of the problem, we consider a series of case problems for various target functions including flexibility, stress state, frequency, eigenvalue problem, etc.

A crucial role of the methods based on the density plays the choice of the appropriate interpolating functions and penalty functions in order to match the physical parameters of the problem with continuously designed variables (for instance, density). As we have already mentioned, the variables are coupled with physical density of each finite element ρ_e. The intervals of the density changes are $0 \leq \rho_e \leq 1$, or $0 < \rho_{\min} \leq \rho_e \leq 1$, where 0 corresponds to an empty element, and 1 corresponds to the fundamental material, and $\rho_{\min}$ denotes the minimal density value, which is needed to overcome problems implied by singularity of the stiffness matrix. The mentioned difficulties include singularity of the matrices of the finite elements as well as lack of ability of the material to appear again in the spaces with zeroth density. The choice of the parameters implies the necessity of rigorous separation of the space into subspaces with $\rho = \rho_{\min}$ and subspaces with $\rho = 1$. It is usually realized by employing one of the implicit methods of a penalty, where the most popular is [47, 75]. In the SIMP method, often referred as the step rule or fictions/material model, the variables of the density are penalized with the use of stepwise principle. This approach is demonstrated through the following equation

$$E\,(\rho_e) = \rho_e^p E_0, \tag{9.2}$$

where SIMP is employed to the Young modulus of the finite element. Here $E\,(\rho_e)$-variable Young modulus, E_0-Young modulus of the input material, p-finite penalty parameter. It should be mentioned that for $0 \leq \rho_e \leq 1$ and positive p (usually it takes the value of 3), $E\,(\rho_e)$ is bounded for zero density and by E_0 for $\rho_e = 1$.

The alternative schemes of interpolation have been worked out mainly for the removement of disadvantages of the SIMP method. In particular, the developed Rational Approximation of Material Properties (RAMP) [76] offer, in contrary to SIMP method, non-zero sensitivity for the zero density. In result, it has been shown that the RAMP model of material improves some of the occurred difficulties in the problems dealing with small values of the density and under dependence of the loading parameters on the design variables.

The second alternative method to SIMP, known as the SINH method, is based on the hyperbolic sinusoidal functions [77]. In majority of the used methods, the

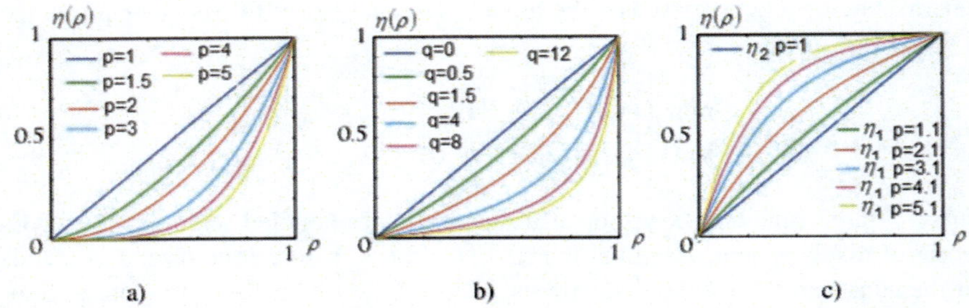

Fig. 9.1 Interpolating functions for the methods of topological optimization based on density: **a** SIMP: $\eta(\rho) = \rho^p$, **b** RAMP: $\eta(\rho) = \frac{\rho}{1+q(1-\rho)}$, **c** SINH: $\eta_1(\rho) = 1 - \frac{\sinh(p(\rho-1))}{\sinh(p)}$, $\eta_2(\rho) = \rho^p$

constraints are employed on the material parameters whereas in the SINH method the fundamental material is subjected to the constraints.

Comparison of schemes of constraints for the SIMP, RAMP and SINH methods [77] is presented in Fig. 9.1, where ρ is the physical density, p is the penalty parameter for SIMP and SINH, q is the penalty parameter for RAMP, and η_1 (ρ) and η_1 (ρ) stand for the functions of the physical and material parameters, respectively.

9.3.2 *Regularization*

Introduction of the interpolation functions, as it has been mentioned, for the purpose of parametrizations of the space of topological optimization, allows to solve the counterpart problems of numerical solutions. The latter ones refer to the so-called problems of "chess plane", where a problem of the optimal solution depends on a chosen mesh. The problems of "chess plane" refers to the occurrence of the neighbourhood elements with large and small densities located in the form of "chess plane". It has been shown that finite element of higher order in some cases allows to remove the problem of the chess plane. The dependence of the numerical results on the mesh choice can be improved by the increase of the number of the finite elements which implies increase of the mesh partition. Examples of the results of the topological optimization exhibiting effects of the chess plane depending on the mesh choice are shown in Figs. 9.2, 9.3

Fig. 9.2 Example of the chess plane effects in the process of the topological optimization

Fig. 9.3 Optimal topology versus the chosen mesh with dimensions of cells: **a** 60 × 20, **b** 90 × 30, **c** 150 × 50, **d** 300 × 100

Fig. 9.4 Results of topological optimization using filtration of density (**a**) and Heaviside projection (**b**)

In order to avoid the chess plane problem, various filtration methods are employed. The filtration process is introduced by the direct modification of two density variables or they are based on the obtained information regarding sensitivity of the neighbourhood of the surrounding elements. More recently, in addition to the commonly used projectional methods [78, 79], the counterpart procedures dealing with checkerboards, mesh-dependence and local minima [80] as well as on the density filters [81, 82] are employed. In cases where there is a need for achieving more precise boundary localization, there are methods based on that Heaviside projection and the density filter algorithms [79, 83] and methods of morphology-based operators [78], which are used to carry out projection of density 0/1 onto density on the physical density. The physical density ρ_e is equal to 1 if $x_e > 0$ and it is equal to zero if $x_e = 0$. In order to compute gradients required for sensitivity analysis while solving various problems of optimization, the so-called smoothened Heaviside function is used

$$\rho_e = 1 - e^{-\gamma x_e} + x_e e^{-\gamma}, \tag{9.3}$$

which is a continuous function in contrary to the classical Heaviside function. Parameter $\gamma \geq 0$ describes the projection curvature, which stands for a line in point $\gamma = 0$ and approaches the unit Heaviside step function for $\gamma \to \infty$. In majority of applications, a scheme of continuation is used where γ is initially set to zero and then it is gradually increased. Guest et al. [84] proposed elimination of beta-continuation from Heaviside projection. Figure 9.4 shows characteristic features of filtration and projection procedures. The occurrence of a grey transitional material along the structural boundaries is associated with the use of the density filter. This drawback is removed through the use of the Heaviside projection.

9.4 Applying Topological Cell Optimization

The topological optimization used for solving majority of application-oriented problems is restricted to design of mechanical constructions based on stiffness criteria [85, 86].

9.4.1 Flexibility Minimization

The methods of topological optimization work efficiently when a target function is defined as an integral with respect to design space. In the case of purely structural problems, a target function serves flexibility of construction with a constraint put on its volume. This requirement, in the case of linear structures, corresponds to defining an optimal stiffness tensor $E_{ijkl}(\mathbf{x})$, which is treated as variable quality in the space of optimization [72].

Let us introduce the energetic bilinear form (internal virtual work of an elastic body being in equilibrium under displacements u and under arbitrary virtual displacement v)

$$a(u, v) = \int_{\Omega} E_{ijkl}(x)\, \varepsilon_{ij}(u)\, \varepsilon_{kl}(v)\, d\Omega \tag{9.4}$$

subject to linear deformation

$$\varepsilon_{ij}(u) = \frac{1}{2}\left(\frac{\partial u_i}{\partial x_j} + \frac{\partial u_j}{\partial x_i}\right), \tag{9.5}$$

and under linear form of loading

$$c(u) = \int_{\Omega} f u\, d\Omega + \int_{\Gamma} t u d\Omega. \tag{9.6}$$

Then the problem of minimization takes the following form:

$$\min_{u \in U, E} : c(u) \tag{9.7}$$

under condition $a_E(u, v) = c(v),\ v \in U$.

The equilibrium equation (9.7) presents a weak variational form, whereas U denotes the space of kinematically allowed fields of displacement, f stand for volume forces and t describes the loads on the boundary. Solutions to the problem (9.7) found using one of the methods of topological optimization allow to achieve an optimal topology characterized by desired maximum of global stiffness. However, the problem (9.7) is solved using FEM, and hence the problem of topological optimization takes the form of (9.1). An overview of the available literature implies that the latter problem is dealt with over 85% of works devoted to topological optimization. This wide popularity of the method can be explained through two characteristic features of the target function based on flexibility: it is self-adjoint and its sensitivity to that material inclusion is negative. The mentioned two characteristics of the method allow for its simple implementation and for getting fast and reliable results, and hence it is widely used in the commercial engineering-oriented programs.

9.4.2 Thermoelastic Structures

In majority cases of realization of topological optimization under criterion of minimum flexibility and with an account of volume forces in (9.7), the problem is reduced to finding an optimal solution depending only on the boundary loads which depend only on design variables. Rodrigues and Fernandes [64] are the first researchers who solved the problem employing material based model and included the temperature influence into the flexibility form. The novel optimization is realized on minimalization through a global measurement of general displacement estimated by structure flexibility with isoparametric constraints put onto the general volume. Flexibility defined as virtual work of the given load and the generated displacement is equal to the squared energetic norm of the general displacement multiplied by two. Basing on the introduced assumptions and defining the material coefficient $\beta_{ij} = E_{ijkl}\alpha_{kl}$, where α_{kl} are coefficients of the linear heat expansion, the problem of topological optimization of flexibility of thermoelastic structures can be presented in the following form:

$$\min_{0 \le \rho \le 1} \left\{ \int_\Omega b_i u_i d\Omega + \int_\Omega \beta_{ij}(\mathbf{x})\varepsilon_{ij}(u)\Delta T d\Omega + \int_\Gamma t_i u_i d\Gamma \right\}, \tag{9.8}$$

under the condition of the isoparametric constraints put on the volume

$$\int_\Omega \rho\,(\mathbf{x})\; d\Omega \le \bar{V}, \tag{9.9}$$

where $\bar{V}$ stands for allowed volume magnitude, and under satisfaction the Duhamel-Neumann principle obeying the linear thermoelasticity expressed by the following equations:

$$\sigma_{ij}\,(\mathbf{u}) = E_{ijkl}\varepsilon_{kl}\,(\mathbf{u}) - \beta_{ij}\Delta T\,(\mathbf{x})\,. \tag{9.10}$$

Observe that **u** is allowed as a solution to the equilibrium equation in the following weak form:

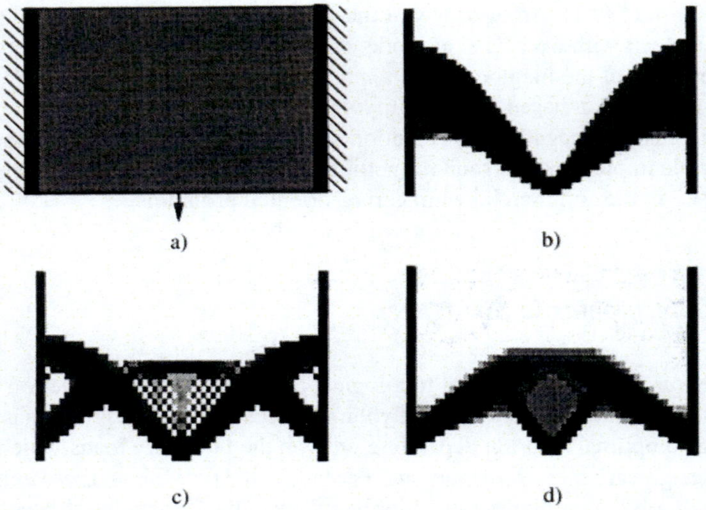

Fig. 9.5 Optimal topologies while solving problem of flexibility [64] for different magnitudes of the temperature load: **a** input construction; **b** $\Delta T = 0°\mathrm{F},\ c = 2,81$; **c** $\Delta T = 1°\mathrm{F},\ c = 4,31$; **d** $\Delta T = 4°\mathrm{F},\ c = 12,08$

$$\begin{aligned} &\int_{\Omega}\left[E_{ijkl}(\rho)\,\varepsilon_{ij}(\mathbf{u})\,\varepsilon_{kl}(\mathbf{v}) - \beta_{ij}(\rho)\varepsilon_{ij}(\mathbf{v})\Delta T - b_i v_i\right]d\Omega- \\ &-\int_{\Gamma} t_i u_i \mathrm{d}\Gamma = 0 \quad \forall \mathbf{v}. \end{aligned} \tag{9.11}$$

Figure 9.5 presents the results of the reference [64] for the case of minimalization of flexibility of a simple 2D-structure under the external discrete (point) load and subjected to uniform increase of temperature ΔT with regard to the input state. Observe that increase of ΔT yields divergence of the employed optimal topologies. Besides, increase of ΔT implies increase of the construction flexibility. This is because the amount of temperature load of the target function (9.8) depends on that design variables.

Xia et al. [87] conducted the topological optimization under the criterion of flexibility minimum and showed that dependence on the temperature load of the material may increase the structural flexibility. Many results of topological optimization carried out by the methods based on density show that there exist large spaces of separating material parts having grey colour ("grey material"). In many cases, in particular, implementation of the computed subspaces cannot be realized physically. The problem of solvability requires a precise distribution of the phases of the employed material. Therefore, with an additional computational effort, it is difficult to guarantee that the optimal topology does not include grey material when an arbitrary approach based on density consideration is employed. Gao an Zhang [88] studied the latter problem and showed that in some cases the scheme of RAMP interpretation

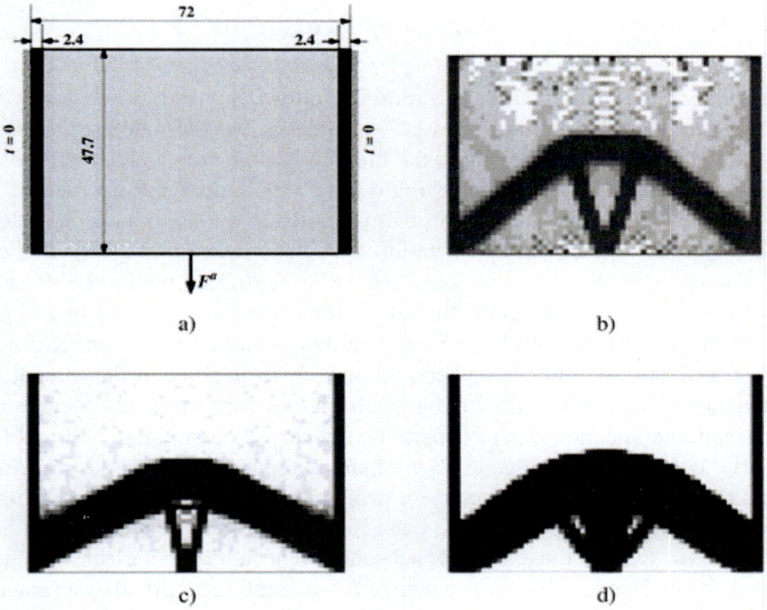

Fig. 9.6 Achieving minimum of thermoelastic flexibility (SIMP methods show grey material which is removed using RAMP): **a** input construction; **b** SIMP $\Delta T = 1°$F; **c** SIMP $\Delta T = 3°$F; **d** RAMP $\Delta T = 1°$F

allows to omit problems associated with the occurrence of the "grey material", contrary to the SIMP scheme. This is because RAMP interpolation, in contrary to SIMP, possesses different zero sensitivity form for zero density, as it has been reported in Fig. 9.1. Figure 9.6 presents an increase of accuracy of material separation while using RAMP for tests model. It should be emphasized that the problem of "grey material" is much more dependent on the magnitude of the mechanical and thermal load than on the used interpolation schemes.

There exist also other methods aimed at the increase of the accuracy of material separation in topologically optimal structures. Yang and Li [89] introduced the terminal load into topological optimization while achieving minimum dynamic flexibility of plates. In the beginning, the distribution of thermal stresses in a structure was calculated through static analysis with a given homogeneous temperature field. Then those thermal stresses were used as preliminary stresses, and then dynamic analysis was carried out with an account of harmonic load with given frequencies.

Efficient extension of the so far considered cases relies on matching problems of thermoelastic topological optimization with a given thermal load and the topological optimization of the heat transfer which yields the temperature distribution. Cho and Choi [89] proposed such an approach using coupled field at adjoint sensitivity analysis in order to take into account the temperature load dependence in the carried out structural analysis with an emphasis put on the heat exchange.

9.4.3 Topological Optimization of Stress State

While conducting topological optimization the following question arises: how essentially can we decrease thermal stresses while designing thermoelastic structures? Dusynix and Bendose [90] belong to the first few people who attack the problem of reaching minimum stress while carrying out the topological optimization of structures. After that three fundamental problems related to the use of criteria of stresses estimation in the topological optimization have been raised, i.e. a so-called occurrence of singularity, strongly nonlinear behaviour of stresses with possible changes of topology and local character of stresses [72]. The earlier carried out investigations based on stresses estimation have been abandoned and more correct and reliable criteria based on flexibility have been employed. In parallel to that recent trend, the transitional structural optimization has been based on achieving a minimum of material volume in conditions of constraints employed on stresses. The problem of singularity occurs due to the generation of singular subspaces of the n-dimensional allowed design space, i.e. some of the n projection variables are removed from the studied problem [91]. Besides, it has been found that often the global optimal construction lies in some of these singular subspaces. The latter problem has been solved by Duysinix and Bendose [90] with a help of β—relaxation modified by an analogous problem dealing with form optimization where a stress element becomes singular since a transversal cross section of the optimized space tends to be zero [92]. In a general case, the constraints employed on stresses in topological optimization can be formulated in the following way:

$$g(\rho) = \sigma(\rho) \leq \bar{\sigma}, \ \rho > 0, \tag{9.12}$$

where σ—elementary stress (for instance, von Mises stress), $\bar{\sigma}$—allowed stress, and inequality $\rho > 0$ points out existence of an element. In order to remove the condition $\rho > 0$ from constraints, one may use the following modified formula

$$g(\rho) = \rho \left(\frac{\sigma(\rho)}{\bar{\sigma}} - 1 \right) \leq 0. \tag{9.13}$$

Besides, in order to avoid singularity based on employment of ε-relaxation, the following counterpart constraints can be used

$$g(\rho) = \rho \left(\frac{\sigma(\rho)}{\bar{\sigma}} - 1 \right) \leq \varepsilon, \tag{9.14}$$

where ε stands for a given parameter. For $\varepsilon = 0$, one deals with the input problem with stress constraints. However, for arbitrary $\varepsilon > 0$, the problem is characterized by a non-singular space, i.e. the optimal solutions are distributed in subspaces of design space with non-zero ρ, and now they can be achieved by using traditional algorithms of structural optimization. In the process of optimization, the value of ε is gradually decreased and in a limit of this process the initial form (input) problem of the introduced constraints form (9.13) is obtained [90]. The so far described approach has been followed by other researchers [93–98]. Le et al. [99] proposed qualitatively different approach to remove the problem of singularity by introduction of a simple penalty based on the stresses estimation with the help of the interpolating function. Let the function interpolating stresses be $\eta_\sigma(\rho)$, then the constrained of stresses can be presented by the following weak inequality

$$g(\rho) = \eta_\sigma(\rho)\sigma(\rho) \leq \sigma_r. \tag{9.15}$$

It has been shown that using interpolation $\eta = \rho^p$ and an appropriate choice of the penalty parameter for interpolation of stresses, stiffness and volume, the singular subspaces of design space can be achieved in an effective way. Le et al. [99] proposed $\rho = 1/2,\ 3$ for stresses, stiffness and volume, respectively.

A high-order nonlinear character of stress constraints is implied by a high sensitivity of the stresses in a finite element generated by a neighbourhood changes in topology. The authors [99] recommended the use of Heaviside filters to remove the occurred problem.

The local character of stress constraints plays a crucial role in the majority of investigations within the topological optimization. In engineering problems of design, the occurred stresses should be bounded in each point of a construction. It can be easily achieved by optimizing the sizes and forms of a construction, using as engineering intuition in order to satisfy the required constraints in critical places (critical elements) on the cases of the structure geometry and employed loads. However, the process of a topological optimization does not allow to predict the elements with critical stresses, since their geometric configuration is a priori not known. Therefore, the possible solutions should satisfy the introduced stress constraints in all finite elements. On the other hand, a decrease of computational time requires a decrease of the constraints' number, and therefore it seems to be an attractive introduction of one constraint of the maximum stresses as follows

$$g(\rho) = \max_{e=1..n} (\sigma_e(\rho)) \leq \bar{\sigma}. \tag{9.16}$$

Unfortunately, the maximum operator (9.16) is not differentiable, which does not allow to describe formulas for sensitivity in an analytical way. It motivated many researchers to introduce continuous aggregating functions matching local stresses into one formula [72, 95]. The latter methods are named global ones and they usually use variation of the Kresselmeier-Steinhauser (KS) functions in the following form:

$$\sigma_{KS} = \frac{1}{p} \ln \left[\sum_{e=1}^{N_{es}} \exp\left(\frac{F(\sigma_e)}{\bar{\sigma}} \right) \right] \tag{9.17}$$

or its equivalent approximated p norm

$$\sigma_{PN} = \left[\sum_{e=1}^{N_e} \left(\frac{F(\sigma_e)}{\bar{\sigma}} \right)^p \right]^{1/p}. \tag{9.18}$$

The use of the global approach implies a drawback, since the aggregating functions offer only approximation to the maximum value in a set of the being searched values, and the quality of approximation usually decreases with the increase of the set dimension. In result, the global formulations of the problem yielded its simplification, but it cannot guarantee that the maximum stresses imply satisfaction to the stresses constraints introduced locally. The carried out review of the state of the art allows to conclude that though there is a set of the methods of topological optimization, the most suitable and mostly used are the methods based on density. There are also constraints associated with the topological optimization of thermoelastic problems, but sufficient and widely recognized target functions and proper problem statement are not defined so far, in contrary to the problems relating to getting minimum of flexibility of mechanical problems. Besides, there exist works against direct application of the target function in the form of flexibility while considering the thermoelastic problem and they call for an alternative formulation of the problem. In addition, practically there are no works aimed on the study of the thermoelastic topological optimization. Namely, only relatively simple problems regarding thermoelastic load have been studied so far dealing with cosmic heat structures. The widely used traditional problem based on achieving the flexibility minimum cannot serve as a reliable tool under thermoelastic design of structures.

It is a hope that so far the worked out criteria based on stresses in the case of mechanical loads can serve as an archetype to develop a similar approach for the thermoelastic structures and thermal stresses. It seems that the topological optimization can be an effective tool in the coming future for designing the structures resisting to heat transfer.

9.4.4 Topological Optimization in the Field of Thermoelasticity

Based on the carried out overview of the state of the art of the methods of topological optimization carried out in Sect. 9.3, our further investigations rely on the method of defining density. In spite of the lack of case studies devoted to thermoelastic structures, the fundamental methods and algorithms of optimization based on density are mostly recognized and tested. Therefore, the latter method supplemented by corresponding methods of parametrizations, and filtration can be used to extend and modify the existing algorithm of topological optimization in order to solve the problems of thermoelasticity.

A topological optimization including the thermoelastic effect is based on a typical process of a structural optimization. The process starts with an input configuration, defining a target function for minimization and the choice of constraints to be satisfied. The target function and the constraints are estimated with the help of the thermoelastic finite element analysis. Information about object sensitivity and constraints are estimated based on the algorithm of numerical optimization. The latter process is repeated iterationally unless a convergence of the target function is achieved. An important role in the mentioned process is associated with design dependence on the heat loading. In contrary to the topological optimization regarding mechanical load which does not depend on the change of constraints topology, the design heat dependence should be exhibited by sensitivity analysis. The final mathematical modelling of the topological optimization takes the following form:

$$\min f(\boldsymbol{\rho}, \mathbf{U}(\boldsymbol{\rho})), \tag{9.19}$$

under condition $\mathbf{K}(\boldsymbol{\rho})\mathbf{U} = \mathbf{F}(\boldsymbol{\rho})$, $g_i(\boldsymbol{\rho}, \mathbf{U}(\boldsymbol{\rho})) \le 0$, for the variables $0 \le x_e \le 1$, for $e = 1, 2, \ldots, N$.

The so far formulated problem is analogous of the general approach of the topological optimization (9.1) presented in the literature overview. However, the parameterization (9.19) is defined in the terms of physical variables associated with density ρ_e. The problem of optimization is still solved with regard to the projection variables x_e. Recall that the physical densities are functions of the design variables as $\rho_e(\mathbf{x})$ depending on a filter or projection, which undergo changes. The proposed algorithm includes the elementary analysis and the complex analysis of sensitivity in addition to the algorithms required for topological optimization (regularization, filtration, optimizators, etc.).

Since a chosen method regarding density for the purpose of topological optimization is composed of a system of finite elements, it should be parametrized for distribution of the design variables based on density which is taken into account in the being designed finite elements. Here we employ the parametrization based on the general thermoelastic space for design. The latter one denoted by Ω is shown for the 2D case in Fig. 9.7. It consists of the boundary conditions, external loading located on the surface parts and a given temperature distribution which can be homogenous or area dependent. The space includes holes and the optimized space Ω. Here we employ parametrization based on the general thermoelastic design space. The design space is discretized by N finite elements with N_{opt} design and N_{nopt} given elements. Each of the finite elements is associated with a design variable $0 < x_e \le 1$, for $e = 1, 2, \ldots, N_{opt}$. All of the variables define a vector $\mathbf{x}$. Design variables $\mathbf{x}$ are coupled with physical densities $\boldsymbol{\rho}$ by the choice of a standard filter as it has been described in Sect. 9.2.2.

System of equations for static space equilibrium is approximated by a choice of finite elements with an account of both mechanical and temperature loads, and it has the following form:

$$\mathbf{K}(\boldsymbol{\rho})\mathbf{U}(\boldsymbol{\rho}) = \mathbf{F}(\boldsymbol{\rho}), \tag{9.20}$$

where $\mathbf{K}(\boldsymbol{\rho})$—global matrix of stiffness, $\mathbf{U}(\boldsymbol{\rho})$—vector of node displacements and $\mathbf{F}(\boldsymbol{\rho})$ stands for a vector of nodal load. In a general case $\mathbf{F}(\boldsymbol{\rho})$ consists of a sum

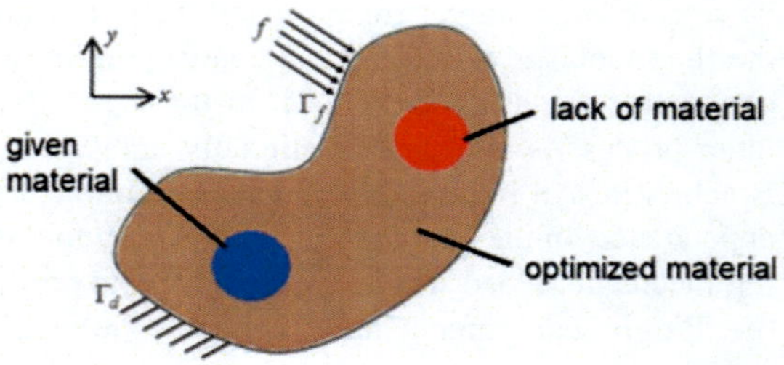

Fig. 9.7 General space of a thermoelastic design

of design-independent mechanical load $\mathbf{F}^m$ and heat-dependent load $\mathbf{F}^{th}(\boldsymbol{\rho})$, i.e. we have

$$\mathbf{F}(\boldsymbol{\rho}) = \mathbf{F}^m + \mathbf{F}^{th}(\boldsymbol{\rho}). \tag{9.21}$$

Observe that the heat load vector depends on the densities, which exhibits dependence of the heat load upon the projection configuration. Stiffness matrix $\mathbf{K}(\boldsymbol{\rho})$ consists of a sum of stiffness of finite elements

$$\mathbf{K}(\boldsymbol{\rho}) = \sum_{e=1}^{N_{opt}} \mathbf{k}_e(\rho_e), \tag{9.22}$$

where

$$\mathbf{k}_e(\rho_e) = \int_{\Omega_e} \mathbf{B}_e^T \mathbf{C}_e(\rho_e) \mathbf{B}_e d\Omega. \tag{9.23}$$

Here $\mathbf{B}_e$ is the matrix of coupling of deformation and displacement of the finite element with number e which consists of derivatives of the element form and which does not depend on the design variables, and stands for the elasticity matrix of element $\mathbf{C}_e(\rho_e)$, which in the case of isotropic materials can be presented as a linear function of the elasticity modulus in the following form:

$$\mathbf{C}_e(\rho_e) = E(\rho_e)\bar{\mathbf{C}}_e, \tag{9.24}$$

where $\bar{\mathbf{C}}_e$ does not depend in the projection variables and they are defined by material matrix; $E(\rho_e)$—elasticity modulus element e depending on the density variable. For elements belonging to non-optimized space, we take $\rho_e = 1$ in formula (9.24). The vector of mechanical load $\mathbf{F}^m$ is composed of external forces for the associated degrees of freedom.

In order to carry out the topological optimization, the vector of heat load $\mathbf{F}^{th}(\boldsymbol{\rho})$ is generalized using the coefficient of linear expansion [64, 88].

The vector of nodal loads for the optimized element e is defined in the following way:

$$\mathbf{f}_e^{th}(\rho_e) = \int_{\Omega_e} \mathbf{B}_e^T \mathbf{C}_e(\rho_e)\, \varepsilon_e^{th}(\rho_e) d\Omega, \tag{9.25}$$

where ε_e^{th} stands for the vector of thermal deformation for an element of the form

$$\varepsilon_e^{th}(\rho_e) = \alpha(\rho_e)\,\Delta T_e \boldsymbol{\varphi}^T. \tag{9.26}$$

Here the following notation is employed: $\alpha(\rho_e)$—temperature coefficient of linear expansion which depends on the element density; ΔT_e—temperature change of element e (here understood as the averaged value of mode temperatures); $\boldsymbol{\varphi}$—vector [110] or [111000] corresponding to 2D and 3D problems, respectively.

Substitution of (9.24) and (9.26) into (9.25) yields

$$\mathbf{f}_e^{th}(\rho_e) = \int_{\Omega_e} E(\rho_e)\,\alpha(\rho_e)\,\mathbf{B}_e^T \bar{\mathbf{C}}\,\Delta T_e \boldsymbol{\varphi}^T d\Omega. \tag{9.27}$$

Observe that $E(\rho_e)$ and $\alpha(\rho_e)$ in (9.27) depend on the density variables, and consequently, require interpolation of those properties. In order to simplify the problem, we unify those parameters within the following one coefficient of the heat load

$$\beta(\rho_e) = E(\rho_e)\,\alpha(\rho_e). \tag{9.28}$$

In what follows, the mentioned parameter uniquely characterizes the material property m and $\mathbf{f}_e^{th}(\rho_e)$ can be recast to the following form:

$$\mathbf{f}_e^{th}(\rho_e) = \int_{\Omega_e} \beta(\rho_e)\,\mathbf{B}_e^T \bar{\mathbf{C}} \Delta T_e \boldsymbol{\varphi}^T d\Omega. \tag{9.29}$$

Finally, the global vector of the heating load includes contributions of all elements, and hence

$$\mathbf{F}^{th}(\boldsymbol{\rho}) = \sum_{e=1}^{N} \mathbf{f}_e^{th}(\rho_e). \tag{9.30}$$

As earlier, the vector of heat load of an element for the non-optimized subspace can be obtained from the previous relations by taking $\rho_e = 1$.

9.4.5 Interpolating Schemes

As it has been mentioned in Sect. 9.3.1, in the case of the load being design depended and particularly the thermal load, there appear difficulties with proper distribution of material into phases under the SIMP interpolation scheme. Therefore, RAMP has been recommended. The stiffness $E(\rho_e)$ and heat expansion coefficient $\beta(\rho_e)$ are interpolated, respectively, by the following relations:

$$E(\rho_e) = \frac{\rho_e}{1 + R_E(1 - \rho_e)} E_0, \tag{9.31}$$

$$\beta(\rho_e) = \frac{\rho_e}{1 + R_\beta(1 - \rho_e)} E_0 \alpha_0, \tag{9.32}$$

where R_E and R_β stand for RAMP penalty parameters. Besides, E_0 and α_0 stand for the basic material properties for the elasticity modulus and heat expansion coefficient, respectively.

While working with a topological optimization, it is assumed that the target function, as well as the constraints are defined through certain functionals J depending on density vector $\boldsymbol{\rho}$, as well as the displacement vectors $\mathbf{U}(\boldsymbol{\rho})$, i.e. $J(\boldsymbol{\rho}, \mathbf{U}(\boldsymbol{\rho}))$. In order to analyze the sensitivity of the functional with regard to the densities ρ_i

$$\frac{dJ}{d\rho_i} = \frac{\partial J}{\partial \rho_i} + \frac{\partial J}{\partial \mathbf{U}} \frac{d\mathbf{U}}{d\rho_i}. \tag{9.33}$$

The derivative of the displacement vector $d\mathbf{U}/d\rho_i$ is defined by differentiating the equilibrium equations (9.20), and after a few transformations we get

$$\frac{d\mathbf{U}}{d\rho_i} = \mathbf{K}^{-1} \left(\frac{d\mathbf{F}^{th}}{d\rho_i} - \frac{d\mathbf{K}}{d\rho_i} \mathbf{U} \right). \tag{9.34}$$

Substitution of (9.34) into (9.33) yields

$$\frac{dJ}{d\rho_i} = \frac{\partial J}{\partial \rho_i} + \frac{\partial J}{\partial \mathbf{U}} \mathbf{K}^{-1} \left(\frac{d\mathbf{F}^{th}}{d\rho_i} - \frac{d\mathbf{K}}{d\rho_i} \mathbf{U} \right), \tag{9.35}$$

which defines the functional J sensitivity with regard to the density change. Now, depending on the character of the constructions optimization, formula (9.35) can be calculated using either direct or complex method. Since the problem of the topological optimization is characterized by a huge number of design variables comparing with relatively small number of constraints, formula (9.35) can be effectively computed using the complex procedure by implementing the complex variable $\mathbf{V}$ in the following way:

$$\frac{dJ}{d\rho_i} = \frac{\partial J}{\partial \rho_i} + \mathbf{V}^T \left(\frac{d\mathbf{F}^{th}}{d\rho_i} - \frac{d\mathbf{K}}{d\rho_i} \mathbf{U} \right). \tag{9.36}$$

The vector of complex variables is defined through solutions to the conjugated system of equations

$$\mathbf{KV} = \frac{\partial J}{\partial \mathbf{U}}, \tag{9.37}$$

which possesses the form similar to Eq. (9.20) for the case of direct problem. This means that its solutions can be found using the earlier defined inverse matrix $\mathbf{K}^{-1}$ while dealing with formula (9.20).

9.5 Reactions on Boundary

Reactions on boundary of the thermally loaded construction play an important role, since it can be transmitted on the neighbourhood structures. In this section, we describe the methodology aimed on getting freedom (or resultant reaction obtained through summation along a few degrees of freedom) based on the results of finite element analysis.

9.5.1 Reaction on Boundary and Their Sensitivity

After carrying out the solution procedure of FEM described in Sect. 9.2.4, the reaction force can be obtained based on the estimated displacement, and system matrix in the following way:

$$\mathbf{R}\left(\boldsymbol{\rho}\right) = \mathbf{K}\left(\boldsymbol{\rho}\right)\mathbf{U}\left(\boldsymbol{\rho}\right) - \mathbf{F}\left(\boldsymbol{\rho}\right). \tag{9.38}$$

Observe that $\mathbf{R}\left(\boldsymbol{\rho}\right)$ consists of reaction forces for all degrees of freedom. Therefore, general reactive loads spanned on a set of k degrees of freedom can be defined in the following way:

$$S\left(\boldsymbol{\rho}\right) = \sum_{i \in D} R_i\left(\boldsymbol{\rho}\right), \tag{9.39}$$

where D is a set of indices corresponding to the given degrees of freedom. Sensitivity of the reactive load S to the variable physical density ρ_j can be obtained from the following formula:

$$\frac{dS}{d\rho_j} = \sum_{i \in D} \frac{dR_i}{d\rho_j}, \tag{9.40}$$

where sensitivity of the resulting reactive load for the given degrees of freedom $dR_i/d\rho_j$ can be obtained in the way described below.

We begin with differentiation of equation (9.38)

$$\frac{d\mathbf{R}}{d\rho_j} = \frac{d}{d\rho_j}\left(\mathbf{K}\left(\boldsymbol{\rho}\right)\mathbf{U}\left(\boldsymbol{\rho}\right) - \mathbf{F}\left(\boldsymbol{\rho}\right)\right) = \frac{d\mathbf{K}}{d\rho_j}\mathbf{U} + \mathbf{K}\frac{d\mathbf{U}}{d\rho_j} - \frac{d\mathbf{F}^{th}}{d\rho_j}. \tag{9.41}$$

The derivative $d\mathbf{U}/d\rho_j$ is obtained by differentiating equilibrium in a way yielded by (9.34).

Substitution into (9.41) gives

$$\frac{d\mathbf{R}}{d\rho_j} = \frac{d\mathbf{K}}{d\rho_j}\mathbf{U} + \mathbf{K}\mathbf{K}^{-1}\left(\frac{d\mathbf{F}^{th}}{d\rho_j} - \frac{d\mathbf{K}}{d\rho_j}\mathbf{U}\right) - \frac{d\mathbf{F}^{th}}{d\rho_j}, \tag{9.42}$$

which defined sensitivity for the whole vector of reactive load. The computational requirements can be decreased if the sensitivity is computed only for a separate reaction

$$\frac{d\,R_i}{d\,\rho_j} = \frac{d\,\mathbf{K}_i}{d\,\rho_j}\mathbf{U} + \mathbf{V}_i^T\left(\frac{d\,\mathbf{F}^{th}}{d\,\rho_j} - \frac{d\,\mathbf{K}}{d\,\rho_j}\mathbf{U}\right) - \frac{d\,F_i{}^{th}}{d\,\rho_j}, \tag{9.43}$$

where the complex vector $\mathbf{V}_i$ is yielded as a solution to the conjugated system having the same form as Eq. (9.37):

$$\mathbf{K}\mathbf{V}_i = \mathbf{K}_i. \tag{9.44}$$

Here $\mathbf{K}_i$ is a column of the global stiffness matrix, and $F_i{}^{th}$ stands for the global component of the heat load corresponding to the ith degree of freedom.

9.5.2 *Formulating Problems of Topological Optimization*

We consider three methods of topological optimization. They can be employed to problems with other geometry, load conditions and target functions.

9.5.2.1 Achieving Minimum Flexibility Under Thermal Load

The first formulation is based on getting minimum flexibility (maximum of stiffness) with constraints put on the volume. As it has been already mentioned, application of the target function in the form of flexibility for topological optimization with heat loads are questionable. However, in what follows we consider this problem and we show why it cannot yield reliable results. The mathematical statement of the problem for the case of topological optimization aimed on achieving flexibility minimum is governed by the following conditions:

$$\min\ c(\boldsymbol{\rho}) = \mathbf{F}^T(\boldsymbol{\rho})\,\mathbf{U}(\boldsymbol{\rho}) = \mathbf{U}^T(\boldsymbol{\rho})\,\mathbf{K}(\boldsymbol{\rho})\,\mathbf{U}(\boldsymbol{\rho}),$$

under condition $\mathbf{K}(\boldsymbol{\rho})\,\mathbf{U}(\boldsymbol{\rho}) = \mathbf{F}^{th}(\boldsymbol{\rho})$,

$$g(\boldsymbol{\rho}) = \sum_{e=1}^{N_{opt}} \rho_e v_e \le \rho_0 V_f, \tag{9.45}$$

and the variables $0 < x_{\min} \le x_e \le 1$ for $e = 1, 2, \ldots, N_{opt}$.

Here the following notation is used: c—flexibility, ρ_e—physical density of element e (coupled with projection variable x_e via filter), v_e—volume of element e, V_f—allowed volume of fraction, N_{opt}—general number of optimized elements, and $x_{\min}$—minimum allowed value of the design variables.

Sensitivity of the flexibility function with regard to the change of physical density ρ_j takes the following form:

$$\frac{dc}{d\rho_j} = \frac{d\left(\mathbf{F}^T\mathbf{U}\right)}{d\rho_j} = \frac{d\mathbf{F}^T}{d\rho_j}\mathbf{U} + \mathbf{F}^T\frac{d\mathbf{U}}{d\rho_j}. \tag{9.46}$$

Substituting $d\mathbf{U}/d\rho_j$ obtained from (9.34) into (9.46) we get

$$\frac{dc}{d\rho_j} = \left(\frac{d\mathbf{F}^{th}}{d\rho_j}\right)^T \mathbf{U} + \left(\mathbf{F}^{th}\right)^T \mathbf{K}^{-1}\left(\frac{d\mathbf{F}^{th}}{d\rho_j} - \frac{d\mathbf{K}}{d\rho_j}\mathbf{U}\right). \tag{9.47}$$

Since the problem of flexibility functional is self-conjugated, i.e. we have $\left(\mathbf{F}^{th}\right)^T\mathbf{K}^{-1} = \mathbf{U}^T$, Eq. (9.47) is simplified and takes the following form:

$$\frac{dc}{d\rho_j} = \mathbf{U}^T\left(2\frac{d\mathbf{F}^{th}}{d\rho_j} - \frac{d\mathbf{K}}{d\rho_j}\mathbf{U}\right). \tag{9.48}$$

Observe that the sensitivity Eq. (9.48) governs changes of the physical densities ρ_j. The optimization procedure requires sensitivity estimation with regard to the projection variables x_j but it is easily computed based on the obtained physical sensitivity with the help of the filtration and projection methods.

9.5.2.2 Method of Artificial Mechanical Load

The second formulation of a topological optimization employs a different approach aimed on stiffness projection. It was firstly demonstrated by Heyni [85], who tried to model thermal effects via mechanical loads. The mathematical formulation of the problem is based on achieving the flexibility minimum under action of the mechanical load

$$\min \ c\left(\boldsymbol{\rho}\right) = \mathbf{F}^T\left(\boldsymbol{\rho}\right)\mathbf{U}\left(\boldsymbol{\rho}\right) = \mathbf{U}^T\left(\boldsymbol{\rho}\right)\mathbf{K}\left(\boldsymbol{\rho}\right)\mathbf{U}\left(\boldsymbol{\rho}\right),$$

and under condition

$$\mathbf{K}\left(\boldsymbol{\rho}\right)\mathbf{U}\left(\boldsymbol{\rho}\right) = \mathbf{F}^a, \quad g(\boldsymbol{\rho}) = \sum_{e=1}^{N_{opt}} \rho_e v_e \le \rho_0 V_f, \tag{9.49}$$

where $0 < x_{\min} \le x_e \le 1$ for $e = 1, 2, \ldots, N_{opt}$.

Here we do not have any heat load, whereas $\mathbf{F}^a$ stands for external artificial mechanical load. In order to define the points of action and direction of that load, we use the fact that the optimal structure yielded by the solution to the problem (9.6) should exhibit the required material distribution in order to most efficiently resist the applied load $\mathbf{F}^a$. In other words, the obtained construction should be stable with respect to deformation in direction of the applied load, and it should possess very small stiffness in the remaining directions. The mentioned loads should imitate reactive loads implied by thermal expansion.

It seems intuitive that the described approach may generate constructions being close to the optimal ones assuming that the artificial loads are chosen properly. In practice, the thermoelastic characteristics of the obtained constructions are sensitive to the way how the artificial loads are put to the computational model. Different configurations of the artificial load yield different thermoelastic characteristics. In spite of the mentioned drawback, the presented approach is still attractive owing to its simplicity.

9.5.2.3 Simultaneous Action of Mechanical and Thermal Load

The terminal formulation of the optimization procedure stands for extension of the previous method of artificial mechanical load, where in addition to the reactive load obtained in Sect. 9.4.1 and yielded by thermal load in the points of construction fixation, the additional constraints are supplemented. In order to achieve minimum flexibility of a target function and due to the lack of the thermal load, the artificial load is applied to the investigated structure. The thermoelastic analysis for the given temperature fields is carried out separately. Then reaction forces which appear in the supplemented constraints of the optimization problem are computed. The mathematical statement of the problem is as follows:

Find min $c(\boldsymbol{\rho}) = \mathbf{U}_1^T(\boldsymbol{\rho})\,\mathbf{K}(\boldsymbol{\rho})\,\mathbf{U}_1(\boldsymbol{\rho})$, under conditions

$$
\begin{gathered}
\mathbf{K}(\boldsymbol{\rho})\,\mathbf{U}_1(\boldsymbol{\rho}) = \mathbf{F}^m, \quad \mathbf{K}(\boldsymbol{\rho})\,\mathbf{U}_2(\boldsymbol{\rho}) = \mathbf{F}^{th}, \\
g_1(\boldsymbol{\rho}) = \sum_{e=1}^{N_{opt}} \rho_e v_e \le \rho_0 V_f, \\
g_2(\boldsymbol{\rho}) = S_L(\boldsymbol{\rho}, \mathbf{U}_2) \le R_0, \quad g_2(\boldsymbol{\rho}) = S_R(\boldsymbol{\rho}, \mathbf{U}_2) \le R_0,
\end{gathered}
\tag{9.50}
$$

with the $0 < x_{\min} \le x_e \le 1$ for $e = 1, 2, \ldots, N_{opt}$.

Relations (9.50) require solving of two problems using FEM. The first one allows to define the displacement vector $\mathbf{U}_1$ for the system of equations with artificial mechanical load $\mathbf{F}^a$, whereas the second problem yields $\mathbf{U}_2$ which stands for the displacement vector for the system of equations associated with the thermal load $\mathbf{F}^{th}(\boldsymbol{\rho})$. The target function (flexibility) is computed with regards to S_L and S_R and depends only of the artificial load. $\mathbf{U}_1$ stands for the reactive loads gathered from artificial degrees of freedom in direction x. The latter depends only for the results of computation of the thermoelastic system $\mathbf{U}_2$.

9.6 Topological Optimization of Thermoelastic Composites with Maximized Stiffness and Heat Transfer

The topological optimization of composite structures is widely used while tailoring materials to achieve the required engineering physical properties. In this section, the problem of topological optimization of the microstructure of a composite aimed at the construction of a material with most effective values of the bulk modulus of elasticity and thermal conductivity taking into account the competing mechanical and thermal properties of the materials included in the composite is defined and solved [100]. A two-phase composite consists of two base materials, one of which has a higher Young modulus but lower thermal conductivity, while the other has a lower Young modulus but higher thermal conductivity. A new class of problems for composites containing material pores or technological inclusions of different shapes is considered. Effective thermoelastic properties are obtained using the asymptotic homogenization method. A modified solid isotropic material with penalization (SIMP) model is used to regularize the problem. The problem of the isoperimetric constraints is solved by the method of moving asymptotes (MMA). The influence on the optimal topology of the composite in the presence of two competing materials, and optimality criteria using linear weight functions are investigated. Pareto spaces that provide a deep understanding of how these goals compete in achieving optimal topology are constructed.

9.6.1 Introduction

The use of new composite materials requires a theoretical background to predict microscopic/effective properties of composites based on the knowledge of their material components. It is well known and documented that cell bulk fractions, shapes and orientation are important factors in the manufacturing of composite materials with effective properties.

The topological optimization of composite structures is a widely employed method of tailoring materials, which is used to achieve their improved and required physical properties. In the recently published papers and monographs, one can find that the asymptotic homogenization approach plays a key role in achieving reliable averaging of the complex microstructural behaviour of an elastic medium [44–46, 101]. An interesting way to compute modified Young modules in both directions for a two-dimensional crystal body with a square lattice using the continuum approach of continuum mechanics is described in reference [102].

It appears that the most widely recognized and used methods for the topological optimization of structures are explicit parametrization methods, which are known as density-based methods. These methods are based on the subdivision of the analyzed mechanical object into finite elements. Instead of a set of properties, each finite element contains only one design variable, which is often understood as the finite element material density ρ_e. The main idea of this approach is to define the finite element parameters as design variables and derive a relationship between local (for instance, density) and global physical properties of the material (for instance, Young modulus or thermal conductivity) in the computational process aimed at finding parameters in the optimization process. To design microstructures composed of different materials, the solid isotropic material with penalization (SIMP) method [47], the evolutionary structural optimization (ESO) [38] and the level set method (LS) [49–51, 101] are used. In particular, the use of homogenization with SIMP is becoming increasingly popular and useful.

By virtue of numerous algorithms aimed at finding an optimum, such as the method of optimal criterion (OC) [52], successive linear programming (SLP) or the method of moving asymptotes (MMA) [54], the optimized material localization allows us to achieve the required functional material characteristics. The current research has focused on solving problems aimed at either optimizing the constructions [38, 44, 51] or getting functional properties of materials [50, 55–60]. When designing composite structures, either elastic or thermal/thermoelastic properties are usually taken into account. In the case of the elasticity criterion, only the mechanical load is considered [44, 103], whereas thermal criteria take into account the thermal load [61, 62], and thermoelastic criteria cover both mechanical and thermal loads [63, 64].

It should be emphasized that there are two approaches that include both thermal and mechanical criteria in topological optimization problems. The first one considers the material as thermoelastic without the coupling between the thermal and mechanical fields in the topological sense (the thermal load is converted into mechanical load) [63, 64]. In this case, the structure is controlled by the mechanical characteristics. On the other hand, the second approach takes into account both the heat and mechanical fields. Namely, the temperature field is considered first, and then it is matched with the field of deformations. In the mentioned cases, the design can be guided either by the mechanical target function [63] or by coupled mechanical and heat functions [65].

The qualitative seminal results in the problem of design of materials were obtained by Sigmund [58, 60] who introduced the method of inverse optimization. It is assumed that the composites are built from identical periodic base cells. The methodology of the topological optimization was employed in the case of the proper distribution of one-component [57] or multi-phase materials [55, 59] in order to achieve the required physical properties. It is known that direct homogenization makes it possible to consider global properties based on the structure of the elementary cell or the representative volume [66]. On the other hand, the inverse homogenization is aimed at searching the optimized material distributions within the elementary cell [57]. Differences between the global homogenized material properties and their target values are minimized with the help of topological optimization methods [55, 57, 60]. In this context, the elastic and heat material properties are considered separately.

Optimal microstructures of materials have been developed in the problems of maximizing bulk and shear moduli [55, 67]. Using this approach, novel composites with negative Poisson's coefficient [57] and materials with extreme or negative coefficients of thermal expansion [60] have been designed, which demonstrated wide possibilities of this method in the design of elastic and thermoelastic composites.

On the other hand, in many cases, the heat transfer and elastic properties of materials compete with each other [68]. Hence, it is often difficult to achieve both high heat transfer and high stiffness simultaneously. For this reason, it is important to develop a procedure that would work effectively in spite of the conflicting objectives. In general, the multi-target optimization is difficult or even impossible to obtain a global optimum for all expected design targets. However, in the optimization process, it can be seen that at some stage, any further improvement of one criterion requires a compromise with at least one other criterion. A set of such solutions determines the Pareto space which consists of a series of points within the solution space [69], where for each of the mentioned solutions it is impossible to further improve some target functions without worsening at least one of the remaining target functions. Thus, it is necessary to gather as much information about the Pareto space as possible to choose the best solution.

Torquato et al. [70] combined heat transfer and electrical transfer with equal weight to propose the optimal design of manufacturable 3D composites with multi-functional characteristics. Yoo and Lee [71] used various weight coefficients to find optimal topologies of the Pareto solution of a swing arm type actuator in accordance with the flexibility criteria and eigenfrequency. The mentioned works show that the use of the weight functions is the best method to obtain a set of Pareto solutions.

Seresta et al. [104] considered a wing box design optimization by employing composite laminates with blending constraints based on the inclusion of fibre orientation angle of the layers and the total thickness of the laminate as design optimization variables. They have shown that the optimum design has better continuity of the laminate lay-ups.

The multi-criteria (multi-objective) optimization of laminated composite structures under technological constraints was discussed by Bassir et al. [105]. The main method based on the NSGA-II program exhibited its efficiency in obtaining a uniform spray of the Pareto solutions.

Lee et al. [106] developed a multi-objective approach using a parallel multi-objective generic algorithm to study a stacking sequence design optimization for a multilayer composite plate. The used methodology matched a robust multi-objective evolutionary algorithm and a finite element analysis with a parallel optimization system. It has been shown analytically that Pareto optimal solutions offer a set of selections for engineers to improve the composite structure in terms of the industrial needs (cost) and mechanical properties (weight and stiffness).

Ning et al. [107] developed an experimental method of additive manufacturing using the fused deposition modelling, which affected the tensile strength, Young's modulus, yield strength, flexural stresses and toughness of reinforced thermoelastic composites.

Madeira et al. [108] proposed the optimal design of laminated composite panels with constrained layer damping aimed at minimization of weight and maximization of model damping. In particular, trade-off Pareto optimal solutions and the respective treatment configurations were obtained and analyzed.

Salem and Donaldson [109] developed and studied a methodology for a combined weight and cost optimization of sandwich plates with hybrid composite face sheets and a foam core. The used multi-objective optimization technique allowed them to minimize both the weight and the cost of the hybrid sandwich plate simultaneously. The Pareto frontier trade off the curve was generated by optimizing a sequence of combining weight and cost objective functions.

Seretis et al. [110] carried out the multi-parameter design of experiments based on the multi-objective curing cycle optimization for glass fabric/epoxy composites by using the Poisson regression and generic algorithm. The authors achieved the estimation of the curing parameters for optimum tensile and flexural performance with high accuracy, and they demonstrated that the Poisson regression theoretical model can predict the tensile and flexural response of the cured composites accurately.

Passos and Luersen [111] employed the multi-objective optimization of curvilinear fibre composites in two cases: for a square plate and a fuselage-like section. The considered problems consisted of three to twelve variables in order to detect the resulting Pareto front properly. To overcome the long computational time needed by the finite element method, the Kriging-based approaches were used.

In this section, we define and solve the problem of topological optimization of the composite microstructure in order to construct a material with the largest effective values of the Young modulus and thermal conductivity. The studied two-phase composite consists of two materials: the first one has high Young modulus and low thermal conductivity whereas the second one is characterized by low Young modulus and high thermal conductivity.

Here we study the influence of two competing material properties, i.e. mechanical and thermal, using linear weight functions. In particular, the problem of construction of the Pareto spaces yielding an insight into the role of competing material properties in the optimal topologies is addressed (see [100, 103]).

9.6.2 Determination of the Effective Tensor of Elastic Properties

The following assumptions are introduced: the composite is linearly elastic and macroscopically transversely isotropic with respect to both mechanical and heat parameters. Initial stresses are absent and the introduced inclusions are homogeneous, linearly elastic, isotropic and regularly packed. The matrix is homogeneous, linearly elastic and isotropic with respect to mechanical and thermal parameters. For composites consisting of linear elastic materials, the governing differential equations of a homogeneous and microstructural element are composed of linear elasticity equations.

In the elastic regime, the microscopic behaviour of the elementary periodic cell made from an anisotropic material can be described by the effective stress tensor $\overline{\sigma}_{ij}$ and deformation tensor $\overline{\varepsilon}_{ij}$ of a homogenized medium. The tensors are coupled by means of the effective elasticity tensor C^e_{ijkl}

$$\bar{\sigma}_{ij} = C^e_{ijkl}\bar{\varepsilon}_{kl}, \tag{9.51}$$

and C^e_{ijkl} depends on the volume fraction of the filler and the microstructure of the elementary cell.

We introduce the local coordinate system (Y) to describe the rapid changes in properties of the microstructure of the material in the global coordinate system (X) in the macro scale. The local coordinate y can be considered as a fast coordinate which is coupled with x through the relation $y = x/\varepsilon$ $(\varepsilon << 1)$. The displacement of an arbitrary material point in an elastic body can be approximated using two-scale asymptotic series [112, 113]

$$u^\varepsilon(x) = u^0(x, y) + \varepsilon u^1(x, y) + \varepsilon^2 u^2(x, y) + \cdots \tag{9.52}$$

Substituting (9.52) into the equilibrium equations, the tensor of the effective elastic properties is obtained in the following form:

$$C^e_{ijkl} = \frac{1}{|Y|}\int\limits_Y \left(C_{ijkl} - C_{ijpq}\frac{\partial \chi^{kl}_p}{\partial y_q}\right)dY, \tag{9.53}$$

where $|Y|$ stands for the area of the elementary cell, χ^{kl}_p is a periodic field of the allowed displacements in the case of kl load [112], and it satisfies the following integral equations with respect to the elementary periodic cell with periodic boundary conditions

$$\int\limits_Y C_{ijpq}\frac{\partial \chi^{kl}_p}{\partial y_q}\frac{\partial v_i}{\partial y_j}dY = \int\limits_Y C_{ijkl}\frac{\partial v_i}{\partial y_j}dY, \quad \forall \mathbf{v} \in Y. \tag{9.54}$$

Here $\mathbf{v}$ stands for the kinematically allowed arbitrary field of displacements.

The problem (9.54) is further solved on the elementary cell by the finite element method (FEM). Along the boundaries of various phases, the coupling conditions are formulated. For the problem of a plane stress-strain state, there are two independent loading cases $kl = 11, 12$. Equation (9.53) can be recast to the following form:

$$C^e_{ijkl} = \frac{1}{|Y|}\int\limits_Y \left(C_{ijkl} - C_{ijpq}\frac{\partial \chi^{kl}_p}{\partial y_q}\right)dY = \langle C_{ijkl}\rangle - \left\langle \sigma^{kl}_{ij}\right\rangle, \tag{9.55}$$

where $\langle C_{ijkl}\rangle$ denotes the averaged elastic tensor dependent on the material volume fraction, which is estimated based on the classical mixture laws; $\left\langle \sigma^{kl}_{ij}\right\rangle$ denotes the averaged stress tensor on the elementary cell in the case of kl load. It is easy to notice that $\left\langle \sigma^{kl}_{ij}\right\rangle$ stands for a correcting term representing the influence of the material microstructure of the elementary cell. Formula (9.55) can be presented in the following form:

$$C^e_{ijkl} = \frac{1}{|Y|} \int_Y C_{pqrs} \left(\varepsilon^{0(ij)}_{pq} - \varepsilon^{*(ij)}_{pq} \right) \left(\varepsilon^{0(kl)}_{rs} - \varepsilon^{*(kl)}_{rs} \right) dY, \tag{9.56}$$

where $\varepsilon^{*(ij)}_{pq} = \frac{1}{2}\left(\frac{\partial \chi^{ij}_p}{\partial y_q} + \frac{\partial \chi^{ij}_q}{\partial y_p} \right)$ and $\varepsilon^{0(ij)}_{pq}$ are linearly independent test deformations on the base cell, employed to define the characteristics of the deformations field $\varepsilon^{*(ij)}_{pq}$. Here, χ^{ij} are solutions to the following problem

$$\int_Y \left(\varepsilon^{0(ij)}_{pq} - \varepsilon^{*(ij)}_{pq} \right) C_{pqrs} \varepsilon^*_{rs} \left(v^{kl} \right) dY \quad \forall \mathbf{v} \in \mathbf{V}_Y$$

defined on the elementary periodic cell, and $\mathbf{V}_Y = \mathbf{v}(\mathbf{y})$ stands for the set of sufficiently smooth functions obtained in Y and exhibiting Y periodicity. For 2D problems, there are three test deformation fields that take the following form: $\varepsilon^{0(11)}_{pq} = [1\ 0\ 0]$, $\varepsilon^{0(22)}_{pq} = [0\ 1\ 0]$ and $\varepsilon^{0(12)}_{pq} = [0\ 0\ 0.5]$ [113] (symmetry $\varepsilon^{0(12)}_{pq} = \varepsilon^{0(21)}_{pq}$ implies reduction in the number of test fields from four to three).

Let us consider a 2D elementary periodic cell of a symmetric microstructure (Fig. 9.8a) made from an isotropic material. In the case of the plane stress-strain state, the governing equations have the following form:

$$\begin{bmatrix} \bar{\sigma}_{11} \\ \bar{\sigma}_{22} \\ \bar{\sigma}_{12} \end{bmatrix} = \begin{bmatrix} C^e_{1111} & C^e_{1122} & 0 \\ C^e_{1122} & C^e_{1111} & 0 \\ 0 & 0 & C^e_{1212} \end{bmatrix} \begin{bmatrix} \bar{\varepsilon}_{11} \\ \bar{\varepsilon}_{22} \\ 2\bar{\varepsilon}_{12} \end{bmatrix}. \tag{9.57}$$

For the isotropic tensor, the homogenized elastic tensor C^e_{ijkl} has three components C^e_{1111}, C^e_{1122}, C^e_{1212}. If initial deformation ε^0 occurs in the direction ($\varepsilon^0_{11} = 1$, $\varepsilon^0_{12} = 0$, $\varepsilon^0_{22} = 0$), then Eq. (9.56) yields

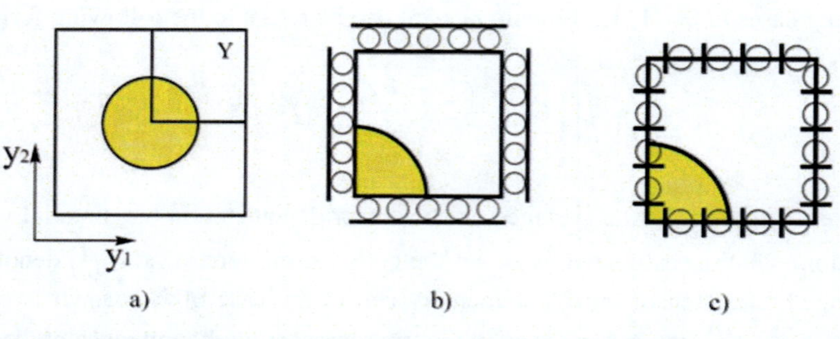

Fig. 9.8 Elementary periodic composite cell: **a** the area of topological optimization Y and boundary conditions for case **b**, **c** on the quarter of the base cell [reprinted with permission from Composites Part B publishers]

$$C^e_{1111} = \frac{1}{|Y|} \int_Y \left(C_{1111} - C_{1111}\varepsilon^{*(11)}_{11} - C_{1122}\varepsilon^{*(11)}_{22} \right) dY. \tag{9.58}$$

Taking into account initial stresses ε^0 in the direction ($\varepsilon^0_{11} = 0$, $\varepsilon^0_{12} = 0$, $\varepsilon^0_{22} = 1$), one gets

$$C^e_{1111} = \frac{1}{|Y|} \int_Y \left(C_{1122} - C_{1111}\varepsilon^{*(22)}_{11} - C_{1122}\varepsilon^{*(22)}_{22} \right) dY. \tag{9.59}$$

If we consider shear initial stresses ε^0 ($\varepsilon^0_{11} = 0$, $\varepsilon^0_{12} = 0.5$, $\varepsilon^0_{22} = 0$), then formula (9.56) yields

$$C^e_{1212} = \frac{1}{|Y|} \int_Y \left(C_{1212} - 2C_{1212}\varepsilon^{*(12)}_{12} \right) dY. \tag{9.60}$$

Now, from (9.57)–(9.60) we obtain

$$K^e = \left(C^e_{1111} + C^e_{1122} \right)/2, \qquad G^e = G^e_{1212}/2, \tag{9.61}$$

where K^e/G^e is the effective bulk/shear model.

It is necessary to impose corresponding periodic boundary conditions on the characteristic displacement fields χ. However, in the case when periodic cell exhibits symmetry, the periodicity conditions can be replaced by typical boundary conditions. Namely, if the composite material consists of isotropic components and the periodic cell has symmetry along two axes, then problem (9.54) is reduced to the problem of a quarter of the cell. In the case of the plane stress-strain state for the quarter part of the elementary cell, the mentioned boundary conditions [114] for the characteristic function can be defined as follows: (see Fig. 9.9): $i = j$ (1 or 2) on $y_1 = 0$, $y_1 = Y_1$,

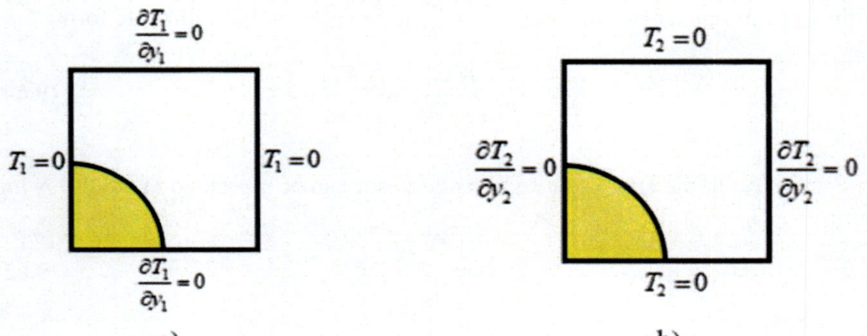

Fig. 9.9 Boundary conditions for the heat problems (**a**), (**b**) [reprinted with permission from Composites Part B publishers]

$\chi_1^{ij} = 0$ and on $y_2 = 0$, $y_2 = Y_2$, $\chi_2^{ij} = 0$ (Fig. 9.8b); in the case of load $ij = 12$ (or 21) on $y_1 = 0$, $y_1 = Y_1$, $\chi_2^{12} = 0$ and on $y_2 = 0$, $y_2 = Y_2$, $\chi_1^{12} = 0$ (Fig. 9.8c).

9.6.3 Determination of the Effective Heat Transfer Coefficient

To calculate the effective heat transfer coefficient of an elementary periodic cell, the homogenization method is applied, and the associated homogenized terms can be obtained from heat transfer equations. According to references [45, 46, 68, 112], the effective coefficient of heat transfer can be defined in the following way:

$$k_{ij}^e = \frac{1}{|Y|}\int_Y \left(k_{ij}(y) - k_{ij}(y)_j \frac{\partial \zeta_i}{\partial y_j}\right) dY = \frac{1}{|Y|}\int_Y k_{ij}(y)\left(I - \frac{\partial \zeta_i}{\partial y_j}\right) dY, \quad (9.62)$$

with a local coordinate y. In contrast to the mechanical problem (9.56), where homogenized coefficients C_{ijkl}^e are expressed in terms of initial given deformations $\varepsilon_{pq}^{0(ij}$ in the existing finite element programs, it is impossible to set initial values of temperature gradients. One of the methods proposed by the authors to calculate k_{ij}^e is as follows. Employing notation $I - \frac{\partial \zeta_i}{\partial y_j} = \frac{\partial T_i}{\partial y_j}$ [112] and using the scheme analogous to that used to interpolate the local stiffness tensor, the characteristic function T_i is yielded by the following PDE

$$\frac{\partial}{\partial y_j}\left[k_{ij}(y)\frac{\partial T_i}{\partial y_j}\right] = 0. \quad (9.63)$$

Consequently, the effective heat transfer tensor (9.62) for an isotropic material with the heat transfer coefficient $k(y_1, y_2)$ can be recast to the following form:

$$k_{ij}^e = \frac{1}{|Y|}\int_Y k(y_1, y_2)\frac{\partial T_j}{\partial y_i} dY. \quad (9.64)$$

For a 2D cell, the effective heat transfer tensor can be presented in the following form:

$$\mathbf{k}^e = \begin{bmatrix} k_{11}^e & 0 \\ 0 & k_{22}^e \end{bmatrix}. \quad (9.65)$$

Therefore, the general effective material conductance can be estimated as follows:

$$tr\left(\mathbf{k}^e\right) = k_{11}^e + k_{22}^e. \quad (9.66)$$

Now the boundary conditions [112] for the characteristic functions T_i for a representative cell which take the classical form are shown in Fig. 9.9a, b and have the following form:

for k^e_{11}

$$T_1(0, y_2) = 0, \quad T_1(1, y_2) = 1, \quad \frac{\partial T_1}{\partial y_1}(y_1, 0) = \frac{\partial T_1}{\partial y_1}(y_1, 1) = 0;$$

for k^e_{22}

$$T_2(y_1, 0) = 0, \quad T_2(y_1, 1) = 1, \quad \frac{\partial T_2}{\partial y_2}(0, y_2) = \frac{\partial T_2}{\partial y_2}(1, y_2) = 0.$$

9.6.4 Topological Optimization

The method of homogenization [44–46] is one of the most effective approaches to compute the global physical properties of a composite such as the bulk stiffness modulus, the shear modulus or the heat transfer coefficient.

Having both the elastic and thermal homogenization determined, one can define an optimization problem aimed at defining the optimal topology for thermoelastic problems. Such problems are particularly interesting when the elastic and thermal properties of materials strongly compete with each other and the optimal topologies for individual problems differ significantly from each other. To solve the problem, it is necessary to simultaneously take into account both the elastic and thermal material properties.

Let us assume that the elementary periodic cell is divided into finite elements and each finite element has its own variable density ρ_n $(n = 1, \ldots, N)$. Following the rule of SIMP [112], where the role of the control variable is played by an artificially introduced density $\rho_n(\mathbf{x})$ the corresponding Young's modulus $E(\mathbf{x})$ and heat transfer coefficient $k(\mathbf{x})$ for the two-component composite in the nth finite element can be written as

$$\begin{aligned} E_n(\mathbf{x}) &= E_1 \rho_n^p(\mathbf{x}) + E_2(1 - \rho_n^p(\mathbf{x})), \\ k_n(\mathbf{x}) &= k_1 \rho_n^p(\mathbf{x}) + k_2(1 - \rho_n^p(\mathbf{x})), \end{aligned} \tag{9.67}$$

where E_1, E_2 are Young's moduli of the materials, and k_1, k_2 are the heat transfer coefficients of the materials.

The exponent $p \geq 1$ is used as a penalty factor and the increase in p leads to a clearer decomposition of the material phases (typically $p = 4$ or $p = 5$ [113]).

Usually, for the convenience of calculations, optimization problems with respect to the maximum are reduced to the minimum.

Let M^e denote one of the mechanical quantities, i.e. the effective bulk modulus K^e or the effective shear modulus G^e. The target function is defined as follows:

$$\min\left\{(q-1)\left(\frac{\omega M^e}{M_b}+\frac{(1-\omega)tr(\mathbf{k}^e)}{k_b}\right)+q\frac{h_0h_{\max}}{A}\int\limits_Y|\nabla\rho(\mathbf{y})|^2dY\right\}, \quad (9.68)$$

where ω—weight coefficient taking into account the input of target functions of the elastic and thermal terms and M_b; k_b—given values of the effective elastic modulus (shear modulus or bulk effective modulus) and the heat transfer coefficient used for the normalization purpose. The second term is the penalty function to exclude the so-called chessboard effect in the optimization process; h_0 is the initial mesh size and h_{max} is the current mesh size. The quantity $0 \le q \le 1$ plays the role of a coefficient that makes it possible to balance the target function and penalty function. It should be noted that the search for the minimum of function (9.68) corresponds to the search of the function maximum standing next to the multiplier $(q-1)$, since $q-1 \le 0$ [68].

The isoperimetric constrains for the artificially introduced density $\rho(\mathbf{x})$ are chosen in the following way:

$$0 \le \int\limits_Y \rho(\mathbf{y})dY \le \gamma A, \quad (9.69)$$

$$0 < \delta \le \rho(\mathbf{x}) \le 1. \quad (9.70)$$

In formula (9.69) A stands for the total volume of the material of the optimized space Y in the elementary periodic cell for $\rho(\mathbf{x}) = 1$ and γ denotes the material fraction with parameters E_1, k_1.

To obtain a numerical solution, the stiffness cannot disappear entirely. Thus, in inequality (9.70), we assume that δ is sufficiently small to avoid the occurrence of singularity of the input stiffness matrix in the optimization process.

9.6.5 Numerical Results

Numerical results obey investigation of topological optimization of a composite made from two competing materials and composites with technological holes and inclusions.

9.6.5.1 Topological Optimization of a Composite Made from Two Competing Materials

We consider the elementary cell composed of two competitive materials in the ratio 1:1 ($\gamma = 0.5$), i.e. Young's modulus is greater in one material than in the second material, whereas in the second material, the heat transfer coefficient is larger than the first one. Let us take $E_1 = 1, k_1 = 5$ ($E_2 = 5, k_2 = 1$) for the first (second) material. The computations were carried out by FEM and the elementary cell was divided into 2500 elements. In order to solve the optimization problem, the method of moving asymptotes (MMA) introduced by Svanberg [54] was employed.

The MMA is based on a special type of convex approximation and it handles element sizes as design variables, shape variables and material orientation angles. In each step of the iterative process, convex approximating subproblems are generated and solved. The latter are controlled by the so-called "moving asymptotes" stabilizing and improving the process convergence.

Figure 9.10 presents the convergence of the maximization process of the bulk modulus depending on the number of iterations n. Curve 1/2 corresponds to the change in the first/second term of penalty function (9.68) for $\omega = 1$, whereas curve 3 corresponds to the integral values in inequality (9.69). At least 25 iterations are required to obtain a sufficiently accurate solution.

First, the problem dealing with the maximization of the effective heat transfer coefficients was solved, i.e. we took the weight coefficient in (9.68) $\omega = 0$. In all further reported figures, red/blue colour corresponds to the first/second material. Figure 9.11a reports the obtained optimal topology for this problem. The optimal value is $tr\left(\mathbf{k}^e\right)/2 = 2.364$. The shear modulus and bulk modulus $G^e = 1.619$, $K^e = 2.428$. For the problem dealing with the maximization of the bulk modulus ($\omega = 1$), the optimal topology is presented in Fig. 9.11b. In this case, the obtained optimal values $K^e = 3.219$, and $tr\left(\mathbf{k}^e\right)/2 = 1.725$. For the shear modulus ($\omega = 1$), the

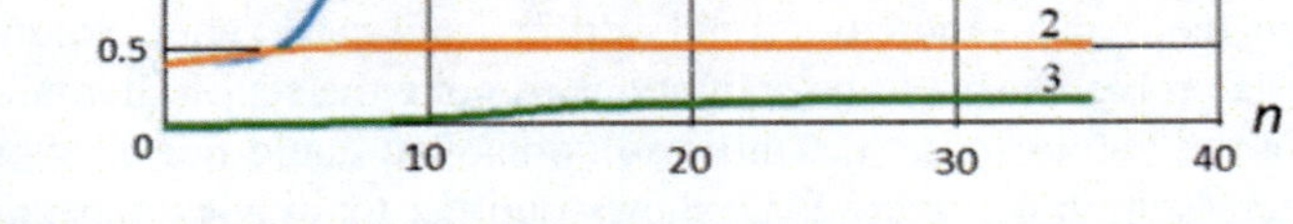

Fig. 9.10 Convergence of the maximization process of the bulk modulus (number of iterations n are marked on the horizontal axis, whereas numerical values of the bulk modulus (BM) refer to the vertical axis) [reprinted with permission from Composites Part B publishers]

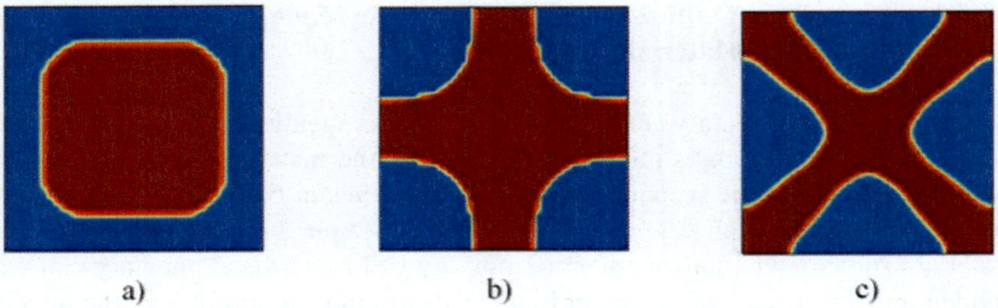

Fig. 9.11 Optimal microstructures of composites for two competing base materials: **a** maximization $tr(\mathbf{k}^e)$ $(\omega = 0)$; **b** maximization K^e $(\omega = 1)$; **c** maximization G^e $(\omega = 1)$ [reprinted with permission from Composites Part B publishers]

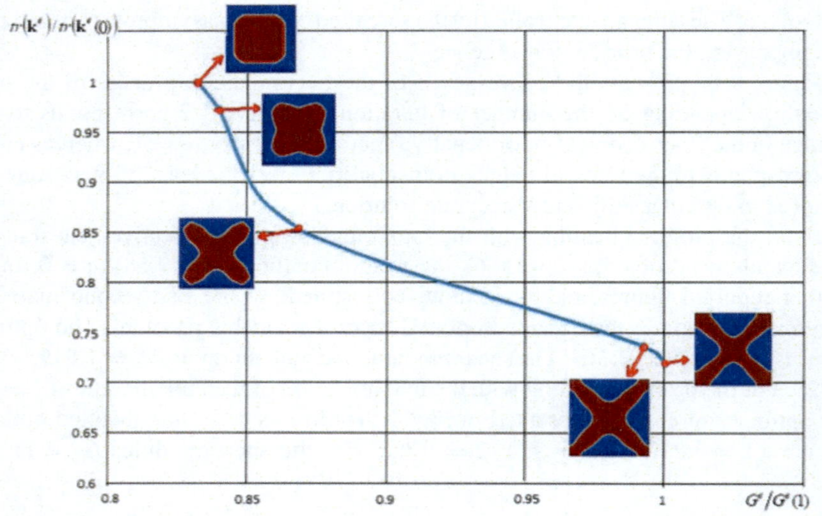

Fig. 9.12 Optimal topologies from the Pareto set (red/blue colour corresponds to the first/second material) [reprinted with permission from Composites Part B publishers]

optimal topology is shown in Fig. 9.11c (the optimal values of $G^e = 1.946$ and $tr\,(\mathbf{k}^e) = 1.633$).

As it follows from Fig. 9.11, the optimal microstructures for the maximum heat transfer and mechanical moduli strongly differ from each other. Solutions to the multi-criteria problems should be sought only among the set of alternative optimal solutions in the Pareto sense, i.e. those solutions that could not be substituted by other "better" solutions. Figure 9.12 shows optimal topologies depending on the weight coefficient for the optimization carried out with respect to two criteria, i.e. the effective shear modulus and the heat transfer coefficient belonging to the set of Pareto solutions [114]. The horizontal axis corresponds to the relative values of the shear modulus $G^e/G^e(1)$, where $G^e(1)$ denotes the optimal value of the shear

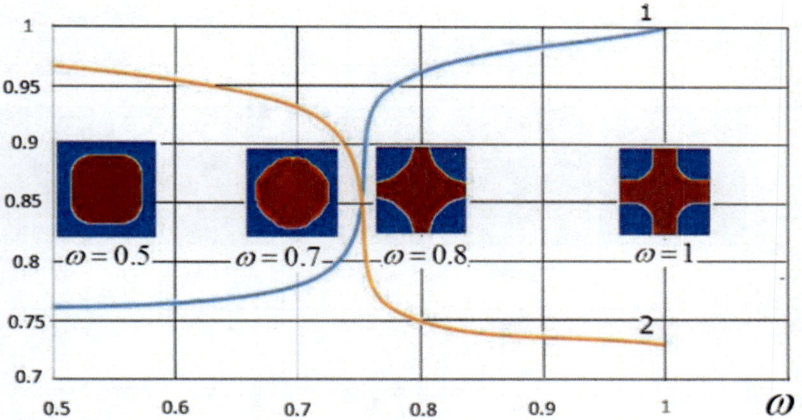

Fig. 9.13 Relative changes of the effective bulk modulus $K^e(\omega)/K^e(1)$ (1) and heat transfer coefficient $tr(k^e(\omega))/tr(k^e(0))$ (2) [reprinted with permission from Composites Part B publishers]

modulus for $\omega = 1$. The vertical axis presents the relative values of $tr(\mathbf{k}^e)/tr(\mathbf{k}^e(0))$, where $tr(\mathbf{k}^e(0))/2$ stands for the optimal value of the heat transfer coefficient for $\omega = 0$.

When the weight coefficient ω in target function (9.68) changes, there is a change in priorities of optimization from the criterion of the maximum heat transfer coefficient for $\omega = 0$ up to optimization under the criterion of the maximum bulk modulus or shear modulus.

In Fig. 9.13, curve 1 shows the relative change in the optimal effective bulk modulus $K^e(\omega)/K^e(1)$, whereas curve 2 presents the analogous value for the heat transfer coefficient.

Figure 9.13 shows that in the case study devoted to the optimization of the bulk modulus for $\omega \approx 0.75$, there is a change in priorities in the target function. For $\omega < 0.75$, the priority is given to the maximization of the heat transfer coefficient, whereas for $\omega > 0.75$, the bulk modulus is maximized. The optimal constructions obtained for ω close to the threshold point $\omega = 0.75$ strongly depend on the initial approximation, and the values of their effective moduli exhibit a sudden change when transiting through $\omega = 0.75$.

Table 9.1 presents the optimal topologies, the values of the effective mechanical moduli and heat transfer coefficients for different values of ω.

In the considered example, the balance is reached for K^e/G^e for the fixed weight coefficient $\omega \approx 0.75/\omega \approx 0.7$. One can see from Table 9.1 that a sudden change in the optimal microstructure occurs precisely in the reported points.

Table 9.1 Optimal microstructures and values of effective moduli [reprinted with permission from Composites Part B publishers]

ω	0	0.2	0.7	0.775	1
			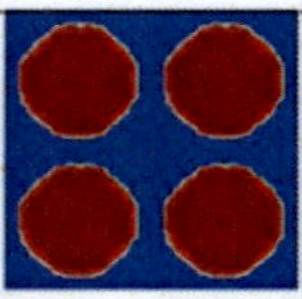		
K^e	2.4288	2.4545	2.5100	3.0513	3.2190
$tr(\mathbf{k}^e)$	2.3640	2.3401	2.2027	1.8002	1.7249
ω	0	0.2	0.5	0.715	1
		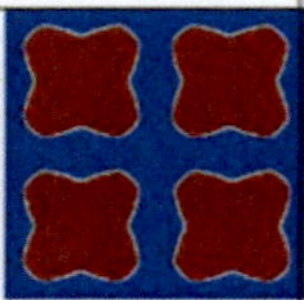	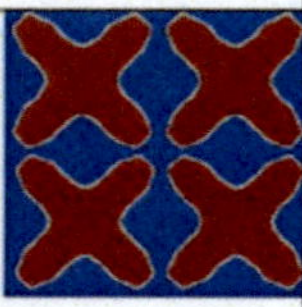		
G^e	1.6191	1.6348	1.6900	1.9314	1.9463
$tr(\mathbf{k}^e)$	2.3640	2.2099	1.9413	1.6744	1.6331

9.6.5.2 Topological Optimization of Composites with Technological Holes

In addition to continuous structures composed of two competing materials, we also studied composite structures composed of two materials with holes of given a priori forms (shapes).

Consider an elementary composite cell having a technological hole of a circular form and a radius of 0.3, or a cell with a square hole of the same area. As in the previous case, the optimized cell area is filled with two competing materials in a 1:1 ratio, with the same parameters. Table 9.2 shows the obtained microstructures of composites and the values of effective moduli for the composite with circular holes for various values of ω.

The microstructure optimal with respect to K^e has a similar form to the microstructure obtained for the cell without a hole, whereas the form of the optimal microstructure with respect to the value of $tr(\mathbf{k}^e)$ is similar to the composite without the hole but is shifted by 1/4 of the cell. Optimal topologies for the cell with a hole change more smoothly when changing ω than in the homogeneous cell optimization (the change in the target function priorities takes place without a sudden transition).

Table 9.2 Optimal microstructures and values of effective moduli for a composite with circular holes [reprinted with permission from Composites Part B publishers]

ω	0	0.2	0.5	1
			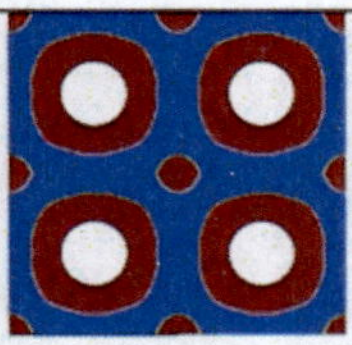	
K^e	15281	1.5809	2.1414	2.5262
$tr(\mathbf{k}^e)$	1.9783	1.9382	1.8334	1.3070
		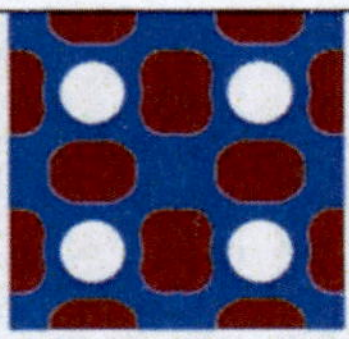		
G^e	1.2160	1.3434	1.5361	1.6338
$tr(\mathbf{k}^e)$	1.9783	1.8326	1.3626	1.2438

Table 9.3 shows the optimal microstructures as well as values of the effective moduli for the cell with a square hole.

As can be seen in Tables 9.2 and 9.3, the presence of holes in the cell significantly changes the microstructures that are optimal with respect to G^e. In contrast to the circular holes, square holes more strongly change optimal topologies for all effective moduli. The transition from the microstructures optimal with respect to heat modulus to the microstructures optimal with respect to mechanical moduli occurs more smoothly than in the case of the optimization of a homogeneous cell.

9.6.5.3 Topological Optimization of Composites with Technological Inclusions

Let us consider structures containing inclusions of a given form, made from a material different than the previous two. The shape and size of inclusions coincide with the characteristics of holes considered in the previous case. In the figures, the area of inclusions is marked in yellow. Both the shape and location of inclusions is a priori

Table 9.3 Optimal microstructures and values of effective moduli for a composite with square holes [reprinted with permission from Composites Part B publishers]

ω	0	0.2	0.5	1

K^e	1.5073	1.5634	2.0590	2.3727
$tr(\mathbf{k}^e)$	1.9035	1.7934	1.8063	1.2943
			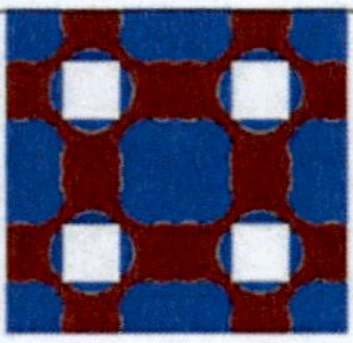	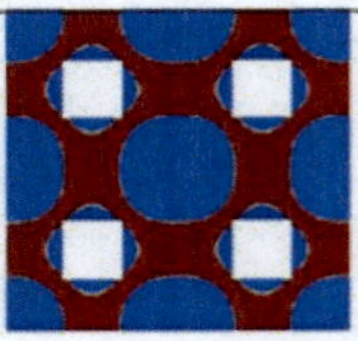
G^e	1.1788	1.2905	1.5709	1.6093
$tr(\mathbf{k}^e)$	1.9035	1.8345	1.2583	1.2422

specified, Young's modulus of inclusion $E_{inc} = 3$, and the heat transfer coefficient $k_{inc} = 3$. The ratio of the first and the second material to the remaining space part is equal to 1:1.

The obtained computational results for the composites with circular inclusions are presented in Table 9.4.

The presence of inclusions of a circular shape, in contrast to the holes, does not fundamentally change the form of topologies for the full optimization of either the mechanical or heat moduli. However, it weakens the transition with the change in the priorities in the target function and increases the values of the optimal effective moduli for all considered criteria.

Table 9.5 reports the computational results for the composites with the inclusions of a square shape.

The presence of inclusions of a square form strongly affects the topology in solving the heat problem or the problem of the maximum bulk stiffness modulus K^e.

Table 9.4 Optimal microstructures and values of effective moduli for a composite with square inclusions [reprinted with permission from Composites Part B publishers]

ω	0	0.2	0.5	1
			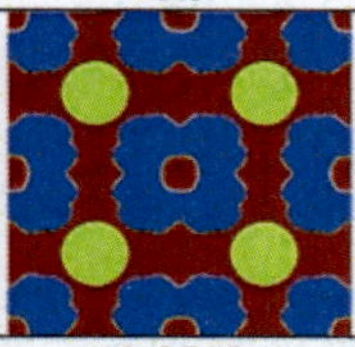	
K^e	2.6069	2.6254	3.0512	3.367
$tr(\mathbf{k}^e)$	2.6844	2.6683	1.7779	1.6626
	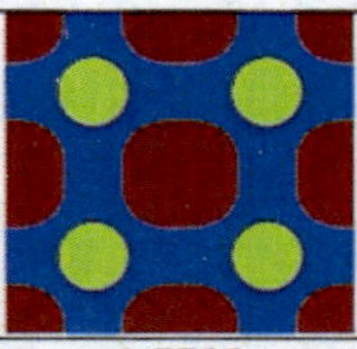			
G^e	1.7792	1.8013	1.8782	1.9416
$tr(\mathbf{k}^e)$	2.6844	2.5697	2.3977	1.7052

9.7 Non-Linear Dynamics of Size-Dependent Euler-Bernoulli Beams with Topologically Optimized Microstructure and Subjected to Temperature Field

This section is devoted to the investigation of nonlinear dynamics of non-homogeneous beams with a material optimally distributed along the height and length of the beam. The study was initiated by topological optimization for the given boundary and loading conditions, which yielded maximum stiffness of a beam microstructure. As a result, the beam with an optimized microstructure exhibiting non-homogeneity in two directions, i.e. along beam thickness and length, was obtained.

In the second step, a beam model was derived based on the kinematic Euler-Bernoulli hypotheses and the modified couple stress theory including the von Kármán geometric nonlinearity and heat flow action obeying the Duhamel-Neumann law.

Both static and dynamic behaviour of the optimized (non-homogeneous) and homogeneous beams were studied for different values of the material length-dependent parameter and temperature. Differences and peculiarities in static and dynamic problems were illustrated and discussed. In particular, the influence of the scale-size parameter on chaotic beam dynamics was investigated. Also, scenarios of transition into deterministic chaos were detected and analyzed for both homogeneous and optimized beams [115].

Table 9.5 Optimal microstructures and the effective moduli values [reprinted with permission from Composites Part B publishers]

ω	0	0.2	0.5	1
			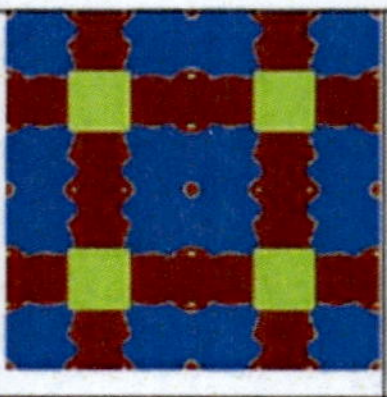	
K^e	2.6143	2.7014	3.0023	3.1615
$tr(\mathbf{k}^e)$	2.5402	2.5364	1.8873	1.8085

G^e	1.7488	1.7497	1.9005	1.9273
$tr(\mathbf{k}^e)$	2.5402	2.4476	2.4397	1.7220

9.7.1 Introduction

Non-homogeneous materials (NM) are fabricated as multi-phase composites in which material properties change along the selected direction. This feature allows achieving better/required dynamic or static characteristics without a simultaneous occurrence of the concentration of stresses. Owing to this, NM beams are widely applied in gas and wind turbines, rotor blades of helicopters, bow feathers, and in numerous cosmic and marine structures [57].

The mentioned change in the material properties can be described either by pitch-like or exponential laws. For beams, the properties change can be realized either in the length or the thickness direction or in these two directions simultaneously.

Modification of elastic properties of the material, i.e. the microstructure properties of the structural members, has been widely used by practicing engineers/designers.

The majority of scientific studies are focused on investigations of free NM beam vibrations, where the material change is introduced along with the beam thickness. The main target of the studies is to determine the beam frequencies with respect to different forms of beam vibration. A few different methods and methodologies have been employed for this purpose [2–8]. It has been observed that the problems of free vibrations of beams with non-homogeneous transverse cross sections or with material properties changing in the longitudinal direction exhibit more complexity comparing to the classical investigations of beams with homogeneous transverse cross sections and made of homogeneous materials due to the occurrence of time-dependent coefficients in the governing differential equations [9, 10]. This is why the majority of the carried out investigations rely on numerical methods [11, 12]. For instance, nonlinear behaviour of functionally graded beams with the von Kármán geometric nonlinearity has been studied in reference [13] by using finite element method (FEM). Kien [14] has considered large displacement response of the tapered cantilever beams made of axially functionally graded material. Alshorbagy et al. [15] have studied free vibration characteristics of a functionally graded beam also by using FEM. Shahba et al. [10, 16, 17] have carried out investigations of free vibration and stability of axially functionally graded tapered Euler-Bernoulli and Timoshenko beams with classical/non-classical boundary conditions by using a few numerical techniques. An analytical study of nonlinear vibrations of functionally graded beams has been carried out by Ke et al. [18] with the use of the first-order Galerkin method.

In order to analyze vibrations of non-uniform and exponentially functionally graded and tapered beams, numerous mathematical models based on the Euler-Bernoulli [10, 19–24, 116] and Timoshenko [17, 25–28] hypotheses have been employed. Authors of the references [29–31] have studied free vibrations of clamped tapered beams on linear elastic foundations, axially functionally graded tapered Euler-Bernoulli beams, and axially functionally graded Timoshenko beams having non-uniform cross sections.

It should be noted that the classical mechanics of rigid bodies do not allow one to interpret or forecast the size-dependent behaviour of the MEMS/NEMS structural members due to a lack of a size-dependent parameter responsible for quantifying micron and submicron scale processes playing a crucial role in the final global dynamic behaviour of the mentioned structural elements. So far, the following theories have been used for modelling of the scale effects occurred in a continuum: the couple stress theory of elasticity [37, 117], the nonlocal theory of elasticity [39], the strain gradient theory of elasticity [40] and the general theory of curved deformable interfaces in solids [41].

A theoretical foundation for the couple stress-based strain gradient theory for elasticity has been proposed by Yang et al. [118]. In that paper, the governing equations contain, in spite of two classical Lame constants, an additional higher order material constant. This theory has been employed and validated by numerous researchers to get a reliable interpretation of the size-dependent dynamic behaviour of microstructures [32–34, 119].

Rajabi and Ramezani [36] have derived PDEs for a geometrically nonlinear homogeneous Euler-Bernoulli beam by using the von Kármán relations. The Galerkin method in the first approximation has been employed and the influence of a size-dependent coefficient of the magnitude equal to the fundamental frequency of nonlinear vibrations has been studied.

The nonlinear forced vibrations of a microbeam have been investigated in [120], employing the strain gradient elasticity theory. The geometrically nonlinear equation of motion, the microbeam, taking into account the size effect, has been obtained employing a variational approach. Results of the frequency response of the system have been studied.

Arbind and Reddy [42] have considered functionally graded microstructure-dependent beams made from a material non-homogeneously distributed along the beam thickness, taking into account the von Kármán nonlinearity. The modified couple stress theory has been used and the counterpart governing equations for the Euler-Bernoulli and the Timoshenko beams have been derived. In particular, the influence of the scale length parameter, the law of the beam properties change along with the thickness, shear deformations, and geometric nonlinearity on the static beam deflections have been studied. Arbind et al. [43] have derived nonlinear PDEs of the size-dependent Sheremetev-Pelekh-Reddy beam made from NM by using Hamilton's principle.

In the present work, we are aimed at carrying out a comparative analysis of the static and dynamic behaviour of size-dependent beams [121, 122]. An investigation of nonlinear dynamics of micro/nanobeams, plates, and shells has been carried out in references [103, 123, 124]. The studies, which can be found in the available literature devoted to the microstructure of beams, have been conducted without using the optimization methods. Furthermore, the overviewed references did not offer a rigorous analysis of the reliability of results. Also, there is a lack of works devoted to studying the chaotic dynamics of size-dependent beams made from NM in the existing literature, which stands for the motivation of the present study. In this work, a mathematical model of a functionally graded size-dependent nonlinear Euler-Bernoulli beam is derived, the algorithms and programs for construction of a microstructure beam (working in the given conditions), and then investigation of its static/dynamic features, are developed.

At the first stage of our analysis, for the given conditions of the beam loading, boundary conditions and the temperature field, the topological optimization was introduced following the criterion of keeping the maximum stiffness. In all case studies, the achieved optimal microstructure is original and not obtained by others. As a result of the carried out optimization, the optimal microstructure beam exhibits functional grading in two directions, i.e. along its length and thickness.

The second stage of our study was aimed at analyzing static and dynamic beam behaviour. The mathematical model is constructed based on the Euler-Bernoulli hypotheses, the modified couple stress theory and the von Kármán geometric nonlinearity. The influence of the temperature field on the model follows the classical Duhamel-Neumann relations.

The carried out comparative analysis of the statics and nonlinear dynamics of homogeneous and non-homogeneous (functionally and optimally graded) beams yielded the main observation that the stress-strain states and magnitudes of the fundamental frequencies of vibrations of these structures that essentially depend on the used temperature.

9.7.2 Mathematical Background

9.7.2.1 Topological Optimization

Consider a 2D elastic beam area Ω, bounded by the closed surface $\Gamma = \Gamma_1 \cup \Gamma_2 \cup \Gamma_3$ and subjected to a plane stress state. We assume that the material of beams is linearly elastic and isotropic and that it is subjected to the temperature field $T(\mathbf{x})$, $\mathbf{x} = \{x_1, x_3\}$.

The computational model with the scheme of boundary conditions is reported in Fig. 9.14, and $\theta = T(\mathbf{x}) - T_0$ stands for the temperature change with respect to the initial temperature T_0. The boundary Γ_1 corresponds to the clamping-clamping boundary conditions, whereas Γ_2 stands for the free surface part, i.e. without loading. Boundary surface Γ_3 is under action of vertical load $t_2 = q$ acting in the direction of the axis OX_3.

In the case of the displacements field $(u_1,\ u_3)$, the governing equilibrium equation takes the following form

$$\sigma_{ij,\,j} = 0 \quad \text{in} \quad \Omega, \tag{9.71}$$

where σ_{ij} stands for the stress tensor. The relationship between linear deformations and displacements obeys the following relation

$$\varepsilon_{ij} = \frac{1}{2}\left(u_{i,\,j} + u_{j,i}\right),\ i,\ j = 1, 2. \tag{9.72}$$

The corresponding stress-strain relationship follows the Duhamel-Neumann law [125]

$$\sigma_{ij} = E(\mathbf{x})(\varepsilon_{ij} - \alpha\,\theta\,\delta_{ij}), \tag{9.73}$$

where $E(\mathbf{x})$, $\alpha(\mathbf{x})$, $\theta(\mathbf{x})$ and δ_{ij} denote Young's modulus, the temperature coefficient of linear extension of the non-homogeneous material of the beam Ω, the difference between the current and initial beam temperatures and the Kronecker symbol, respectively. Recall that the displacement and temperature fields are coupled by Eq. (9.73).

With the help of the topological optimization, the optimal beam material distribution can be found, which allows one to obtain either maximum stiffness or minimum flexibility of the beam thermoelastic body.

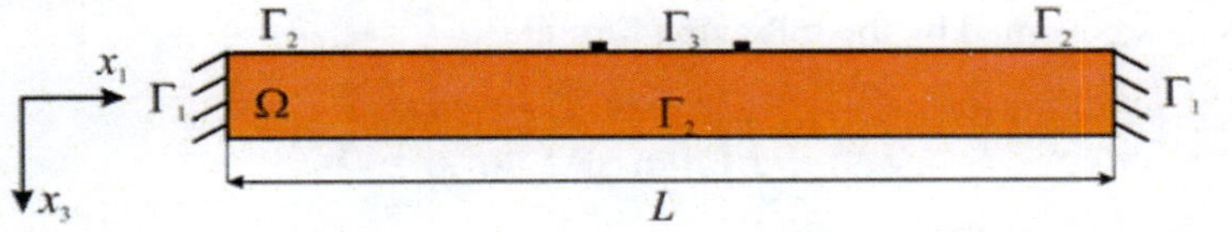

Fig. 9.14 Beam computational model [reprinted with permission from International Journal of Non-Linear Mechanics publishers]

The problem of topological optimization in its counterpart thermoelastic formulation obeys the following formal description: [126–128]

$$\min_{0\le r(x)\le 1} C = \iint_{\Omega} \beta\,\theta\, u_{i,i} d\Omega + \int_{\Gamma} t_i u_i d\Gamma \tag{9.74}$$

and is subjected to the following constraint

$$\int_{\Omega} \rho(x) d\Omega \le A(\Omega)\,\gamma, \tag{9.75}$$

where $\beta(\mathbf{x}) = \alpha(\mathbf{x})\,E(\mathbf{x})$; $\rho(\mathbf{x})$—beam material density, γ—coefficient associated with the basic material of the beam.

To carry out the topological structural optimization, the space Ω is divided into finite elements, and the beam material density $\rho(\mathbf{x})$ is assumed to be constant for each employed finite element. The projection variable $r(\mathbf{x})$ is coupled with Young moduli $E(\mathbf{x})$, with $\beta(\mathbf{x})$, and with the volume material density $\rho(\mathbf{x})$ of each element, following the scheme for minimum compliance topology optimization (the so-called ramp scheme) [76]:

$$\begin{aligned} E(\mathbf{x}) &= \frac{E_0\, r(\mathbf{x})}{(1 + p \cdot (1 - r(\mathbf{x}))}, & \beta(\mathbf{x}) &= \frac{\alpha_0\, E_0\, r(\mathbf{x})}{(1 + q \cdot (1 - r(\mathbf{x}))}, \\ \rho(x) &= \frac{\rho_0\, r(\mathbf{x})}{(1 + p \cdot (1 - r(\mathbf{x}))}, & \mathbf{x} &\in \Omega. \end{aligned} \tag{9.76}$$

where p, q—penalty parameters used to guarantee the compact material distribution; $r(\mathbf{x})$—field of the projection variables $0 < r_0 \le r(\mathbf{x}) \le 1$; r_0—small number guaranteeing non-zero stiffness of the finite elements. Observe that for $r(\mathbf{x}) = 1$ the whole space is filled with the reinforced basic material.

9.7.2.2 Size-Dependent Bernoulli-Euler Beam Model

Owing to the modified couple stress theory [118], the accumulated energy of deformation U of an elastic beam, taking into account the size-dependent behaviour of the beam, is governed by the following formula

$$U = \frac{1}{2} \iint_{\Omega} \left(\sigma_{ij}\varepsilon_{ij} + m_{ij}\chi_{ij}\right) d\Omega, \tag{9.77}$$

where σ_{ij}, ε_{ij}, m_{ij}, and χ_{ij} denote components of the symmetric part of the stress tensor σ, components of the deformation tensor ε, components of the deviatory part

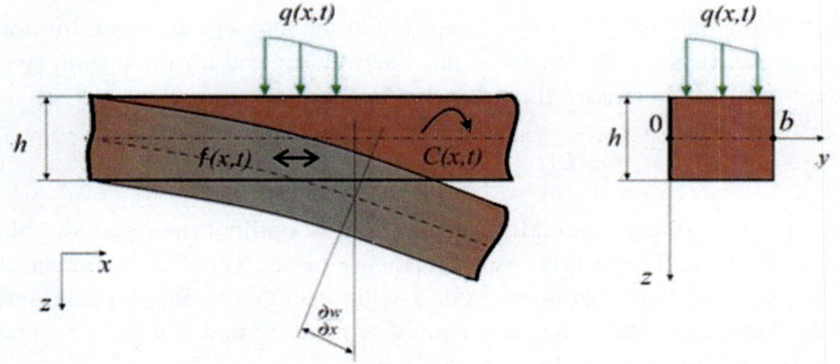

Fig. 9.15 Kinematic parameters and loads acting on the Bernoulli-Euler beam element [reprinted with permission from International Journal of Non-Linear Mechanics publishers]

of the symmetric tensor of the higher order **m** and components of the symmetric part of the curvature tensor χ, respectively. The employed system of coordinates $x_1 = x,\ x_2 = y,\ x_3 = z$, kinematic parameters and loads for the extended model of the Bernoulli-Euler beam based on the modified couple stress theory are shown in Fig. 9.15. The load $f(x,\ t)$ is coupled with the axial mass force per beam unit length, $C(x,\ t)$ stands for the moments distribution per beam unit length, and $q\ (x,\ t)$ is the intensity of the distributed transverse force per beam unit length. Observe that beam properties change along the coordinate OX (axial displacements) and along the thickness (the axis OZ).

The field of displacements in an arbitrary point of the Bernoulli-Euler microbeam is defined as follows:

$$u_1 = u\ (x,t) - zw_{,x}\ (x,\ t)\,, \qquad u_2 = 0, \qquad u_3 = w\ (x,\ t)\,, \tag{9.78}$$

where $u\ (x,\ t)\,,\ w\ (x,\ t)$ stand, respectively, for the axial and transverse deflection of the middle surface points, i.e. for $z = 0$. Employing (9.78), the nonlinear Kármán relations can be described by the displacements field in the following way:

$$\varepsilon_{xx} = u_{,x} + \frac{1}{2}\left(w_{,x}\right)^2 - zw_{,xx}. \tag{9.79}$$

On the other hand, the formula $\phi_i = (rot\ (\mathbf{u}))_i/2$ yields

$$\varphi_x = 0, \qquad \varphi_y = -w_{,x}\,, \qquad \varphi_z = 0. \tag{9.80}$$

Finally, the following relation for the non-zeroth component of the symmetric curvature tensor part is obtained

$$\chi_{xy} = \chi_{yx} = -\frac{1}{2}w_{,xx}. \tag{9.81}$$

In the case when the structure is subjected to the temperature field, for non-zero components of the symmetric part of the stress tensor and for the deviatory part of the tensor of the higher order, the following relations hold

$$\sigma_{xx} = E(\varepsilon_{xx} - \alpha\theta), \qquad m_{xy} = 2l^2 G\chi_{xy}, \tag{9.82}$$

where $E(x, z)$—Young's modulus of the beam optimal material distribution; $G(x, z) = E(x, z)/2(1 + \nu)$ shear modulus; ν—Poisson's coefficient treated as constant; l—material length parameter. For a wide beam ($b > 5h$), a plane deformation state dominates, and hence one should replace in relations (9.76) $E(x, z)$ by $E(x, z)/(1- \nu^2)$ and $\alpha(x, z)$ by $\alpha(x, z)/(1 - \nu)$.

Equations (9.77) and (9.79)–(9.82) allow for derivation of the deformation energy U of a non-homogeneous microbeam, i.e. we have

$$\begin{aligned} U &= \frac{1}{2}\int_0^L\int_A \left(\sigma_{xx}\varepsilon_{xx} + 2m_{xy}\chi_{xy}\right) dA\, dx = \\ &= \frac{1}{2}\int_0^L\int_A \left\{\sigma_{xx}\left(u_{,x} + \frac{1}{2}(w_{,x})^2\right) - \sigma_{xx} z w_{,xx} + 2m_{xy}(-\frac{1}{2}w_{,xx})\right\} dA\, dx = \\ &= \frac{1}{2}\int_0^L \left\{N\left(u_{,x} + \frac{1}{2}(w_{,x})^2\right) - M w_{,xx} + 2Y(-\frac{1}{2}w_{,xx})\right\} dx, \end{aligned} \tag{9.83}$$

where A stands for the area of the transverse beam cross section, the normal resulting force is denoted by N and the bending moment M and the higher order moment Y satisfy the following relations:

$$\begin{aligned} N &= \int_A \sigma_{xx} dA = k_1\left(u_{,x} + \frac{1}{2}(w_{,x})^2\right) - k_2 w_{,xx} - N_T\,, \\ M &= \int_A \sigma_{xx} z dA = k_2\left(u_{,x} + \frac{1}{2}(w_{,x})^2\right) - k_3 w_{,xx} - M_T\,, \\ Y &= \int_A m_{xx} dA = -k_4 w_{,xx}\,, \quad N_T = \int_A E\alpha\theta\, dA, \quad M_T = \int_A E\alpha\theta\, z\, dA. \end{aligned} \tag{9.84}$$

Kinetic beam energy is as follows:

$$
\begin{aligned}
K &= \frac{1}{2}\int_0^L\int_A \rho\left\{\left(u_{,t} - zw_{,xt}\right)^2 + w_{,t}^2\right\} dA\,dx = \\
&= \frac{1}{2}\int_0^L\int_A \rho\left\{u_{,t}^2 - 2zu_{,t}w_{,xt} + z^2 w_{,xt} + w_{,t}^2\right\} dA\,dx = \\
&= \frac{1}{2}\int_0^L \left\{b_0 u_{,t}^2 - 2b_1 u_{,t}w_{,xt} + b_2 w_{,xt}^2 + b_0 w_{,t}^2\right\} dx\ .
\end{aligned}
\tag{9.85}
$$

Stiffness components and initial terms of equations (9.84) and (9.85) are defined in the following manner:

$$
\begin{aligned}
&(k_1,\, k_2,\, k_3) = \int_A E\,(1,\, z,\, z^2) dA, \\
&k_4 = \int_A G l^2\, dA,\ (b_0,\, b_1,\, b_2) = \int_A \rho\,(1,\, z,\, z^2) dA.
\end{aligned}
\tag{9.86}
$$

The work variation δW, caused by external forces $f,\ q$, moments C, and by the stresses and moments employed on the beam borders $\bar{N},\ \bar{V},\ \bar{M}$, is defined by the following formula: [129]

$$
\begin{aligned}
\delta W &= \int_0^L \left(f\delta u + q\,\delta w + C\delta\,\varphi_y\right) dx + \\
&+ \left(\bar{N}\delta u + \bar{Q}\delta w - \bar{M}\delta\left(w_{,x}\right)\right)\Big|_{x=0}^{x=L} = \\
&= \int_0^L \left[f\delta u + \left(q + C_{,x}\right)\delta w_y\right] dx + \\
&+ \left(\bar{N}\delta u + \left(\bar{Q} - C\right)\delta w - \bar{M}\delta\left(w_{,x}\right)\right)\Big|_{x=0}^{x=L}\ .
\end{aligned}
\tag{9.87}
$$

Equations of motion of a non-homogeneous size-dependent Bernoulli-Euler beam are yielded by Hamilton's principle and take the following form:

$$
\int_0^t \left(\delta K - \delta U + \delta W\right) dt = 0. \tag{9.88}
$$

Substituting (9.83), (9.85), and (9.87) into (9.88) and carrying out the integration by parts, the following system of governing motion equations is obtained

$$N_{,x} + f = b_0 u_{,tt} - b_1 w_{,xtt}, \tag{9.89}$$

$$\left(N w_{,x}\right)_{,x} + (M_{,xx} + Y_{,xx}) + q + C_{,x} = b_0 w_{,tt} + b_1 u_{,xtt} - b_2 w_{,xxtt}, \tag{9.90}$$

and the corresponding beam boundary conditions are as follows:

$$N = \bar{N}\Big|_{x=0,\,L} = 0 \quad \text{or} \quad \delta\, u|_{x=0,\,L} = 0, \tag{9.91}$$

$$\left\{N w_{,x} + M_{,x} + Y_{,x} + C - b_1 u_{,tt} + b_2 w_{,xtt} = \bar{Q}\right\}\Big|_{x=0,\,L} \tag{9.92}$$

$$\text{or} \quad \delta\, w|_{x=0,\,L} = 0,$$

$$M + Y = \bar{M}\Big|_{x=0,\,L} \quad \text{or} \quad \delta\, w_{,x}\Big|_{x=0,\,L} = 0. \tag{9.93}$$

Equations (9.91)–(9.93) imply that there are six boundary conditions for the size-dependent (non-classical) non-homogeneous Bernoulli-Euler beam, on the contrary to the classical beam associated with four boundary conditions.

Equations (9.84), (9.89) and (9.90) allow one to derive the equations of motion in terms of displacements:

$$\left(k_1\left(u_{,x} + \frac{1}{2}(w_{,x})^2\right)\right)_{,x} - \left(k_2 w_{,xx}\right)_{,x} - N_{T,x} + f = b_0 u_{,tt} - b_1 w_{,xtt}, \tag{9.94}$$

$$\begin{gathered}\left[\left(k_1\left(u_{,x} + \frac{1}{2}(w_{,x})^2\right)\right) w_{,x}\right]_{,x} - \left[\left(k_2 w_{,xx} + N_T\right) w_{,x}\right]_{,x} + \\ + \left[k_2\left(u_{,x} + \frac{1}{2}(w_{,x})^2\right)\right]_{,xx} - \left[\left(k_3 + k_4\right) w_{,xx}\right]_{,xx} \\ - M_{T,xx} + q + C_{,x} = b_0 w_{,tt} + b_1 u_{,xtt} - b_2 w_{,xxtt}\,.\end{gathered} \tag{9.95}$$

The corresponding boundary conditions can be defined in the following way:

$$k_1\left(u_{,x} + \frac{1}{2}(w_{,x})^2\right) - k_2 w_{,xx} - N_T = \bar{N}\Bigg|_{x=0,\,L} = 0 \quad \text{or} \quad \delta\, u|_{x=0,\,L} = 0, \tag{9.96}$$

$$\begin{gathered}\left\{\left(k_1\left(u_{,x} + \frac{1}{2}(w_{,x})^2\right) - k_2 w_{,xx} - N_T\right) w_{,x} + \right. \\ + \left(k_2\left(u_{,x} + \frac{1}{2}(w_{,x})^2\right) - k_3 w_{,xx} - M_T\right)_{,x} \\ \left. -\left(k_4 w_{,xx}\right)_{,x} + C - b_1 u_{,tt} + b_2 w_{,xtt} = \bar{Q}\right\}\Bigg|_{x=0,\,L} = \bar{Q}\,, \quad \text{or } \delta\, w|_{x=0,\,L} = 0.\end{gathered} \tag{9.97}$$

$$k_2\left(u_{,x}+\frac{1}{2}\left(w_{,x}\right)^2\right)-\left(k_3+k_4\right)w_{,xx}-M_T=\bar{M}\Big|_{x=0,\,L} \tag{9.98}$$

$$\text{or}\quad \delta\, w_{,x}\big|_{x=0,\,L}=0.$$

We introduce the following non-dimensional variables and parameters

$$\begin{aligned}
&\bar{x}=\frac{x}{L},\quad \bar{u}=\frac{u}{h},\quad \bar{w}=\frac{w}{h},\quad \lambda=\frac{L}{h},\quad l_0=\frac{l}{h},\\
&(\bar{k}_1,\,\bar{k}_2,\,\bar{k}_3)=\left(\frac{k_1}{K_0},\,\frac{k_2}{K_0h},\,\frac{k_3}{K_0h^2}\right),\quad \bar{k}_4=\frac{k_4}{K_0l^2},\\
&(\bar{b}_0,\,\bar{b}_1,\,\bar{b}_2)=\left(\frac{b_0}{I_0},\,\frac{b_1}{I_0h},\,\frac{b_2}{I_0h^2}\right),\quad \bar{t}=\frac{t}{L}\sqrt{\frac{K_0}{I_0}},\\
&\bar{\omega}=\frac{L^2\,I_0}{K_0}\omega^2,\quad \bar{f}=\frac{fL\lambda^2}{K_0},\quad \bar{q}=\frac{qL\lambda}{K_0},\\
&\bar{C}=\frac{C\lambda}{K_0},\quad \bar{\theta}=\alpha_0\theta,\quad \bar{N}_T=\frac{N_T\lambda^2}{K_0},\quad \bar{M}_T=\frac{M_T\lambda}{K_0L},
\end{aligned} \tag{9.99}$$

where $K_0=AE_0$ and $I_0=A\rho_0$; the quantities k_1 and b_0 correspond to homogeneous microbeams with Young's modulus E_0; material density is denoted by ρ_0 and α_0 stands for the linear temperature extension coefficient.

The following counterpart non-dimensional equations of beam motion can be derived (bars over non-dimensional quantities are omitted)

$$\left(k_1\left(u_{,x}+\frac{1}{2}\left(w_{,x}\right)^2\right)\right)_{,x}-\left(k_2w_{,xx}\right)_{,x}-N_{T,x}+f=b_0u_{,tt}-b_1w_{,xtt}, \tag{9.100}$$

$$\begin{aligned}
&\frac{1}{\lambda^2}\left[\left(k_1\left(u_{,x}+\frac{1}{2}\left(w_{,x}\right)^2\right)\right)w_{,x}\right]_{,x}-\frac{1}{\lambda^2}\left[k_2w_{,xx}\,w_{,x}\right]_{,x}-\frac{1}{\lambda^2}\left(N_Tw_{,x}\right)_{,x}+\\
&+\frac{1}{\lambda^2}\left[k_2\left(u_{,x}+\frac{1}{2}\left(w_{,x}\right)^2\right)\right]_{,xx}-\frac{1}{\lambda^2}\left[\left(k_3+l_0^2k_4\right)w_{,xx}\right]_{,xx}-\\
&-M_{T,xx}+q+C_{,x}=b_0w_{,tt}+\frac{1}{\lambda^2}b_1u_{,xtt}-\frac{1}{\lambda^2}b_2w_{,xxtt}\,.
\end{aligned} \tag{9.101}$$

The corresponding non-dimensional boundary conditions have the following form:

(i) clamping-clamping

$$u=w=w_{,x}\big|_{x=0,\,1}=0, \tag{9.102}$$

(ii) simple support

$$u = w|_{x=0,\,1} = 0,$$
$$\frac{1}{\lambda^2}k_2\left(u_{,x} + \frac{1}{2}(w_{,x})^2\right) - \frac{1}{\lambda^2}\left(k_3 + k_4 l_0^2\right) w_{,xx} - M_T = 0\Big|_{x=0,\,1}. \tag{9.103}$$

9.7.3 *Results and Discussions*

The numerical study concerns a thermoelastic beam with $\lambda = 40$ and made from steel (E_0= 200 MPa, $\alpha_0 = 12.3 \cdot 10^{-6}$ 1/K, $\rho_0 = 7800\,\mathrm{kg}/m^3$, $\nu = 0.3$). In order to simplify the carried out numerical investigation, we take $f(x, t) = 0$, $C(x, t) = 0$, $b_1 = 0$, $b_2 = 0$. The optimal topology of the beam microstructure is obtained based on the solution to the plane problems of thermoelasticity by using FEM and the method of moving asymptotes [54]. Examples of the optimal topology of the material distribution of the beam under different temperatures θ are reported in Table 9.6 where the left half of the beam is shown (red colour corresponds to the reinforced basic material).

With the use of the relations (9.86), the obtained optimal beam microstructure allowed us to compute the values $k_1(x)$, $k_2(x)$, $k_3(x)$, which are changed along the beam length. The numerically obtained functions are shown in Table 9.7.

Next, the estimated values of the physical quantities were employed to analyze static and dynamic problems of the Bernoulli-Euler nonlinear beam.

9.7.3.1 Method of Solution

In order to reduce the problem governed by PDEs (9.100), (9.101) to a system of ODEs (the Cauchy problem) with respect to u, w, the finite difference method (FDM) regarding spatial coordinates and with approximation $O(\Delta x^2)$ is employed.

Table 9.6 Microstructures of the optimal beams [reprinted with permission from International Journal of Non-Linear Mechanics publishers]

θ=+100	
θ=0	
θ=-100	

Table 9.7 Estimated functions $k_1(x)$, $k_2(x)$ and $k_3(x)$ for different θ [reprinted with permission from International Journal of Non-Linear Mechanics publishers]

θ=-100	θ=0	θ=100
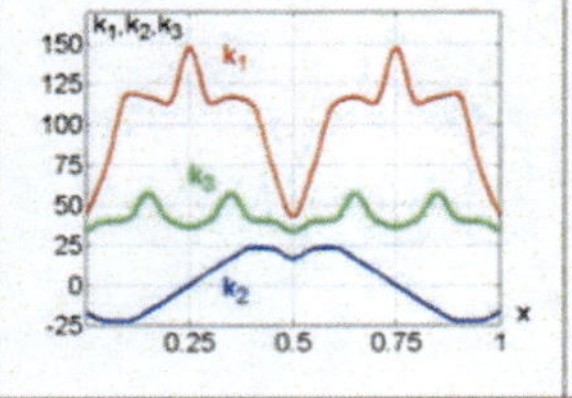	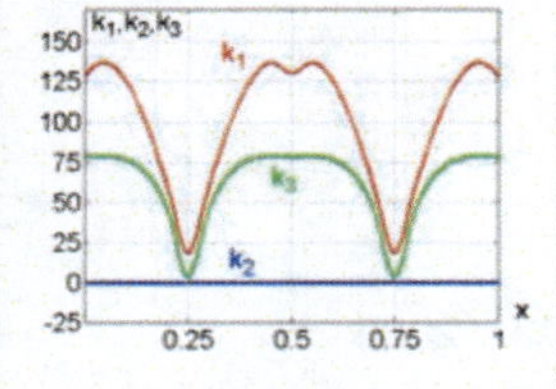	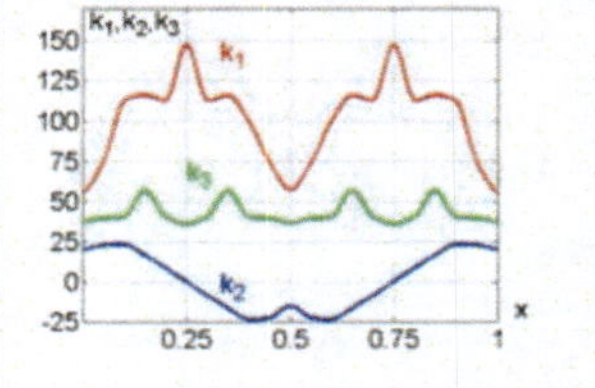

The obtained solutions to the problems (9.100)–(9.103) were compared for different numbers of beam partitions $n = 20, 40, 80, 100, 120, 160$. In the case of chaotic beam vibrations, the study was carried out with the help of time series (signal) and frequency power spectra analyses. In order to quantify the strength of chaotic dynamics, the sign of the largest Lyapunov exponent (LLE) was estimated by four qualitatively different methods, i.e. the Wolf [130], Rosenstein [131], Kantz [132] and neural network [133] methods, to validate the results. Based on the mentioned numerical study we have further employed $n = 100$.

To solve the Cauchy problems, we considered the fourth- (rk4) and second- (rk2) order Runge-Kutta methods, the fourth-order Runge-Kutta-Fehlberg method (rkf45), the fourth-order Cash-Karp method (rkck), and the eighth-order Runge-Kutta-Dormand-Prince method (rk8pd). Eventually, the fourth-order Runge-Kutta method was chosen and the optimal time step was found by the Runge principle.

Solutions to the static problems could be relatively easily obtained using the dynamic equations (9.100), (9.101) with the help of the dissipative term $\varepsilon w_{,t}$ occurring in Eq. (9.101). Such an approach is known as the setup relaxation method [122]. In other words, when the load q does not depend on time, one obtains a static solution based on the dynamic approach. In the considered static problems, the initial conditions play a role of excitation, while the term with the first derivative multiplied by the damping coefficient stands for the dissipative force. A solution to the dynamic problem can be obtained using an arbitrary method for solving the Cauchy problems. Finally, we have obtained stationary solutions, i.e. solutions to the static problems based on the dynamical approach.

9.7.3.2 Static Versus Dynamic Problems

Based on the already described methods, we considered static and free oscillations problems of the size-dependent homogeneous and non-homogeneous Euler-Bernoulli beams subjected to the temperature field for four sets of parameters: 1—without the size-dependent effect ($k_4 = 0$) for a homogeneous beam; 1*—without

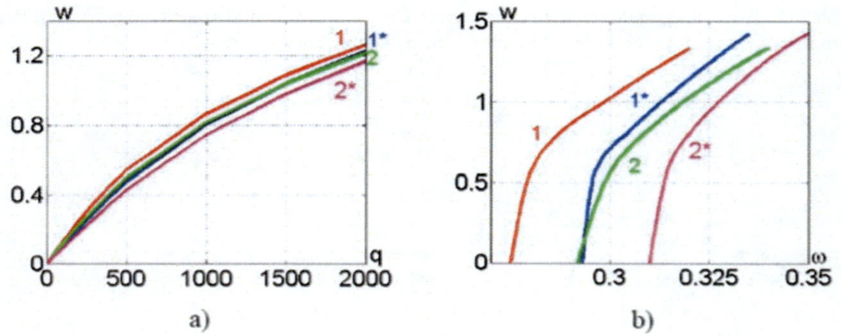

Fig. 9.16 Functions "*load-deflection*" $q(w)$ (**a**) and "*fundamental frequency-deflection*" $\omega(w)$ (**b**) for $\theta = 0$ [reprinted with permission from International Journal of Non-Linear Mechanics publishers]

the size-dependent effect ($k_4 = 0$) for the topologically optimal microstructure; 2—with the account of the size-dependent effect ($k_4 = 0.3$) of the homogeneous beam; 2*—taking into account the size-dependent effect for the topologically optimal microstructure.

Figure 9.16 shows the dependencies "load-deflection" $q(w)$ and "fundamental frequency-deflection" $\omega(w)$ for $\theta = 0$ for the nonlinear problem. The analysis of the obtained results shows that for the optimal beam, the deflection w is smaller than the counterpart homogeneous beam for the same load magnitudes, taking or not taking into account the size-dependent behaviour. In the case of presence of the temperature field, for large deflections, there is a tendency to bring closer the results for optimal and non-optimal beams. This benefit, i.e. a decrease in the maximum deflection of the topologically optimal beam, comparing with the homogeneous beam, equals 12.7% (for $\theta = -100$ and $q = 500$) and for the case without the size-dependent behaviour, it reaches 10.4%. For the case $\theta = +100$ and $q = 500$, the decrease is 13.25/11.5% for the problem without/with the account of the size-dependent parameter. For the lack of the temperature field, the results are 12.4% and 13.8%, respectively.

The analysis of changes in the fundamental frequencies for different temperature magnitudes and the scale-size length parameters yields the following results. For $\theta = 100$, the difference between the frequencies of the homogeneous and non-homogeneous (optimal) beam for the deflection 0.00001 achieves 5.7 and 3.76%, without and with the account of the size-dependent parameter, respectively. In the case of a unit (9.71) deflection of the beam, the corresponding values are 3.9 and 3.15%.

For the temperature $\theta = -100$ and for deflection 0.00001, we have 7.4% and 5.1%, respectively, and for the deflection of the magnitude 1, we have 4.4% and 4.47%, respectively. Finally, for the deflection 0.00001, we get 6.5 and 6.1%, and for the deflection 1, we obtain 4% and 3.12%, respectively.

Therefore, the *frequency-deflection* $\omega(w)$ dependency exhibits the difference of the results obtained for the optimal and homogeneous beam in all studied cases ($\theta = -100, \theta = 0, \theta = 100$).

9.7.3.3 Dynamic Problems

To investigate the influence of the topologically optimal beam microstructure on the nonlinear dynamics of the beam, Fourier and wavelet spectra, Poincaré maps, phase portraits and the largest Lyapunov exponents (LLEs) can be employed. In this study, all results were obtained for the harmonic load $q = q_0 \sin(\omega_p t)$. The dissipation coefficient was taken as $\varepsilon = 1$. In order to investigate the reliability of the computed LLEs based on the Wolf method [130], the LLEs (λ_1) were additionally estimated using the methods of Rosenstein [131]], Kantz [132] and neural networks NW [133]. We compared the results for different mother wavelets (Morlet, Daubechies, Gauss, Haar, etc.), and eventually used the Morlet wavelet, which is the most suitable for our purpose.

Tables 9.8, 9.9 and 9.10 contain the following results: (a) time series (signal) $w(0.5; t)$; (b) Fourier spectra $S(\omega)$ computed based on the FFT (Fast Fourier Transforms); (c) 2D wavelet spectra based on the Morlet wavelets; (d) phase portraits $\dot{w}\,[w(t)]$; (e) Poincaré maps $w_{t+T}\,[w_t]$; (f) values of LLEs estimated based on employment of four qualitatively different approaches. Column designations (1, 1*, 2, 2*) of the Tables 9.8, 9.9 and 9.10 correspond to the types of tasks described in Sect. 9.6.3.2.

We began with studying the beam dynamics for θ=0, focusing on estimation and tracing the changes in the frequency band. For the lack of the size-dependent behaviour ($k_4 = 0$), the following excitation parameters were fixed: $q_0 = 18500$, $\omega_p = 8.5$. For taking into account the size-dependent behaviour ($k_4 = 0.3$), the excitation amplitude $q_0 = 19000$, and its frequency $\omega_p = 8.7$.

Case study 1 ($k_4 = 0$, *homogeneous beam*). The signal (1a) exhibits complexity, the frequency spectrum band (9.71b) contains 20 fundamental frequencies separated by the interval $\omega_n - \omega_{n-1} \approx 0.42$, and a set of frequencies of small powers. One can observe two subharmonic frequencies $\omega_1 = \omega_p/2$ and $\omega_2 = \omega_p/4$ resulting from the period doubling bifurcation. The wavelet spectrum (9.71c) demonstrates non-uniform character and some frequencies are switched on and/or switched off, i.e. their births/deaths depend on the time evolution. The Poincaré section (9.71e) shows a set of points in the form of a loop. The phase portrait indicates the occurrence of a strange chaotic attractor (9.71d).

Case study 1* ($k_4 = 0$, *optimal microstructure*). Signal (9.71*a) is periodic. The Fourier spectrum (9.71b) exhibits two frequencies, i.e. ω_p and $\omega_1 = 0.6538 = \omega_p/13$. The Morlet wavelet (9.71*c) reports, besides of ω_1, also a few frequencies of small powers in the low-frequency band. Phase portrait (9.71*d) is structured and more compressed in comparison to the previous case. Poincaré map (9.71*e) presents a point proving the existence of periodic vibrations. Therefore, for the given

Table 9.8 Dynamical characteristics of the beam at temperature $\theta = 0$: **a** time series $w(0.5; t)$; (b) Fourier spectra $S(\omega)$; (c) 2D wavelet spectra; (d) phase portraits $\dot{w}\,[w(t)]$; (e) Poincaré maps $w_{t+T}\,[w_t]$; (f) values of LLEs [reprinted with permission from International Journal of Non-Linear Mechanics publishers]

	1	1*	2	2*
a)	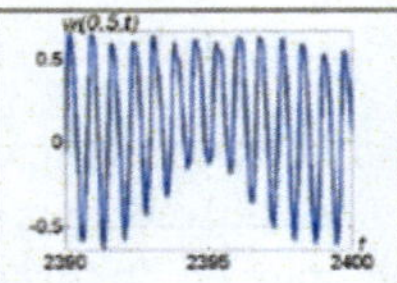	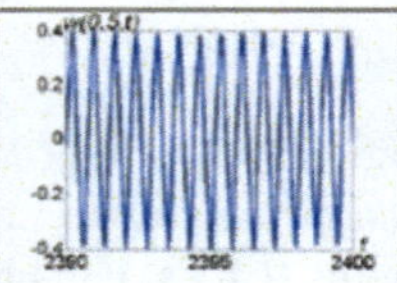	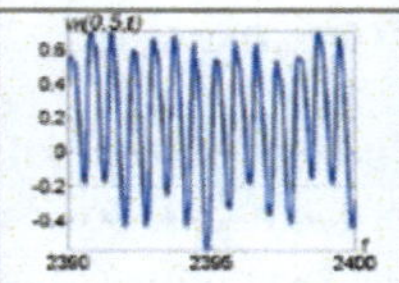	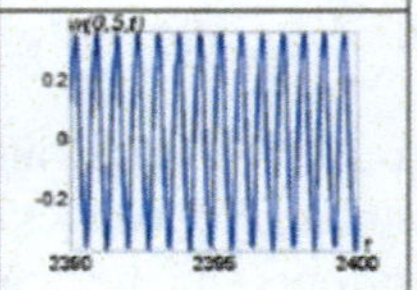
b)	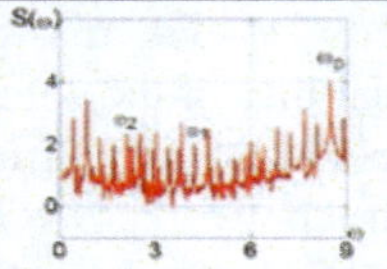	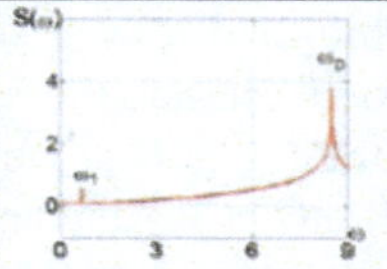	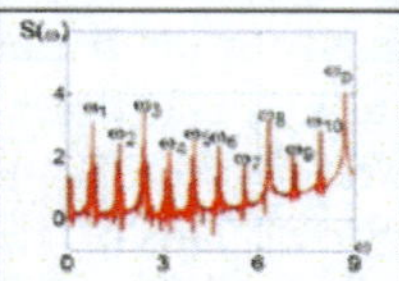	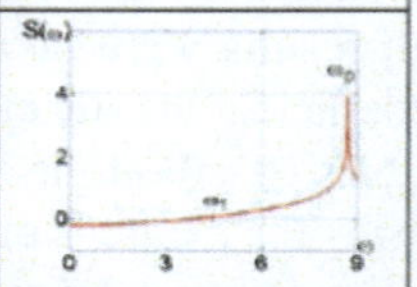
c)	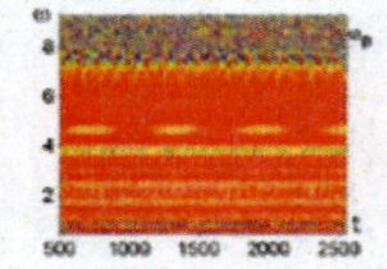	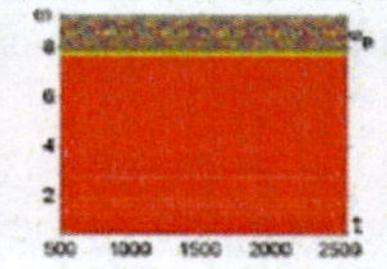	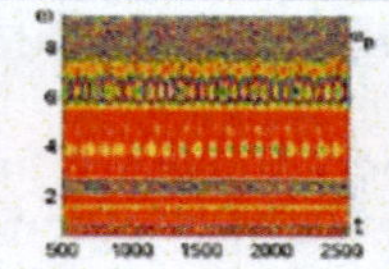	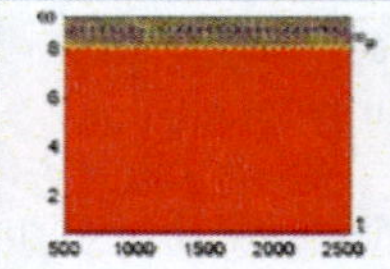
d)		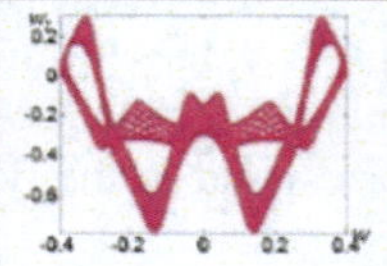	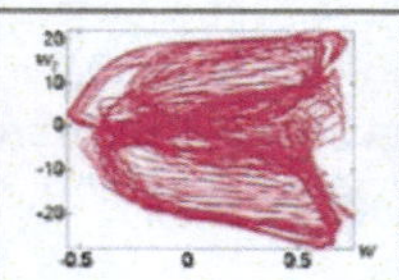	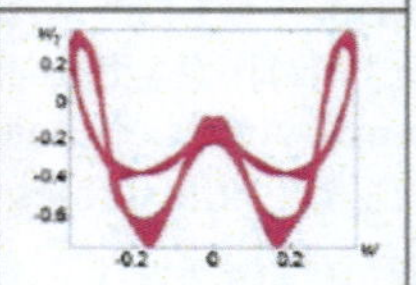
e)	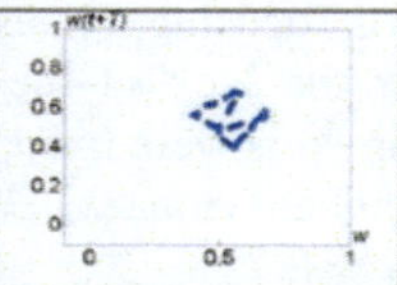	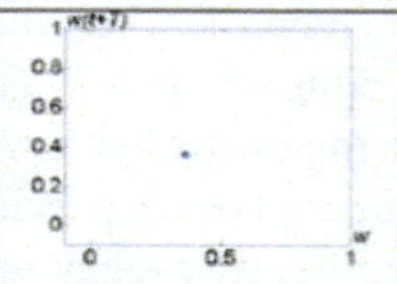	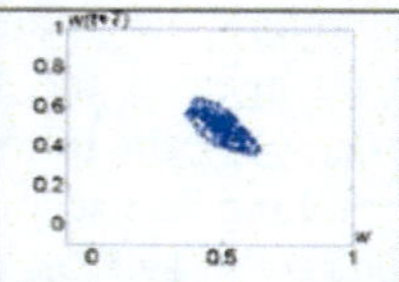	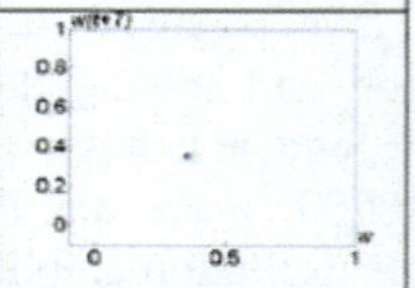
f)	Wolf [130] $\lambda_1 = 0.021$ Rosenstein [131] $\lambda_1 = 0.0087$ Kantz [132] $\lambda_1 = 0.01478$ NN [133] $\lambda_1 = 0.0065$	Wolf [130] $\lambda_1 = -0.0032$ Rosenstein [131] $\lambda_1 = -0.0009$ Kantz [132] $\lambda_1 = -0.0017$ NN [133] $\lambda_1 = -0.0005$	Wolf [130] $\lambda_1 = 0.043$ Rosenstein [131] $\lambda_1 = 0.0099$ Kantz [132] $\lambda_1 = 0.0314$ NN [133] $\lambda_1 = 0.011$	Wolf [130] $\lambda_1 = -0.0123$ Rosenstein [131] $\lambda_1 = -0.0012$ Kantz [132] $\lambda_1 = -0.00629$ NN [133] $\lambda_1 = -0.0073$

Table 9.9 Dynamical characteristics of the beam at temperature $\theta = 100$: **a** time series $w(0.5; t)$; (b) Fourier spectra $S(\omega)$; (c) 2D wavelet spectra; (d) phase portraits $\dot{w}\,[w(t)]$; (e) Poincaré maps $w_{t+T}\,[w_t]$; (f) values of LLEs [reprinted with permission from International Journal of Non-Linear Mechanics publishers]

	1	1*	2	2*
a)	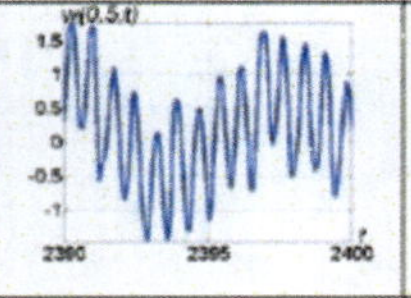	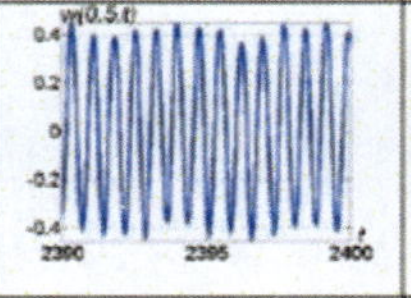	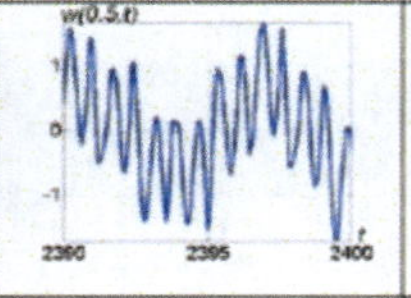	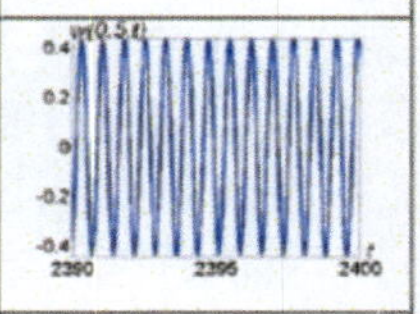
b)	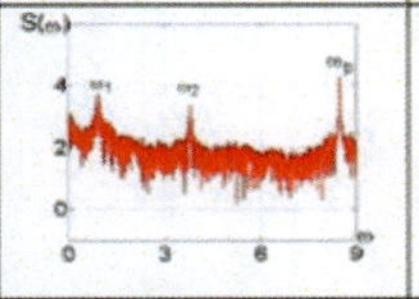	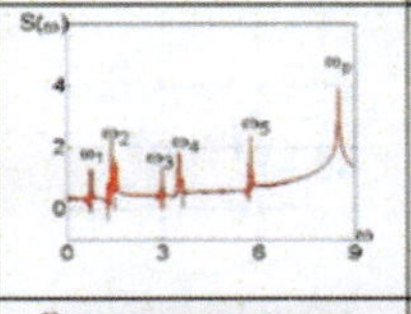	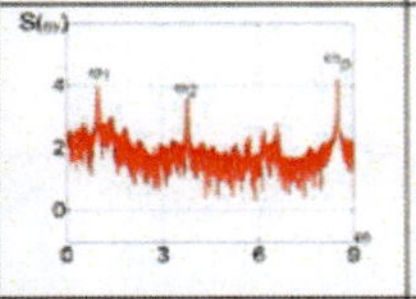	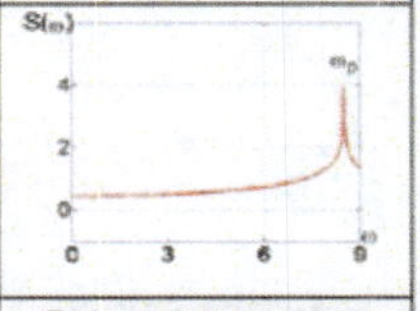
c)	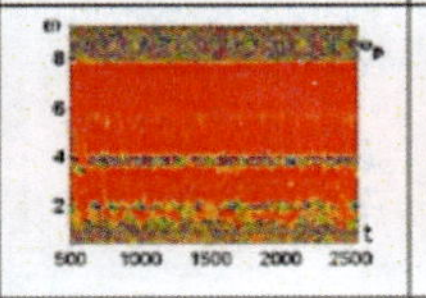	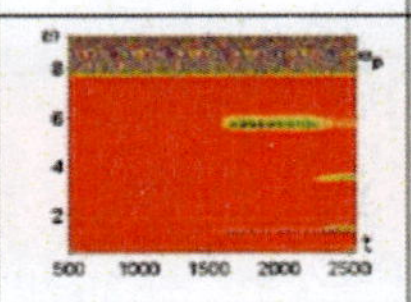	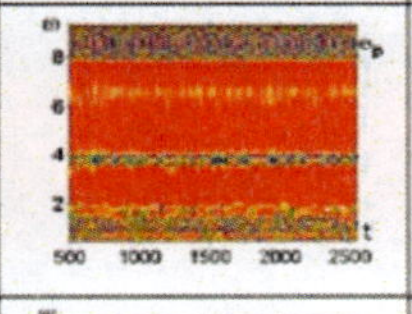	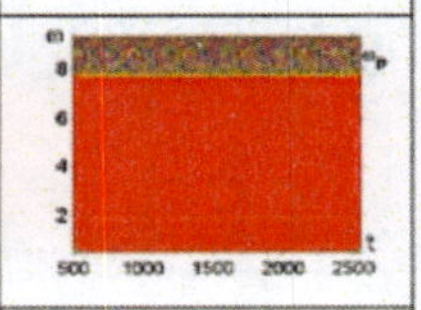
d)	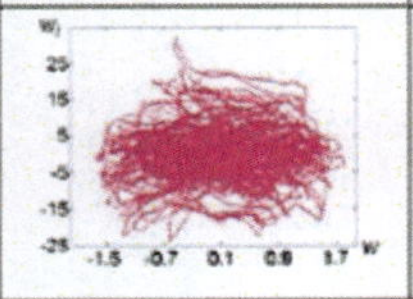	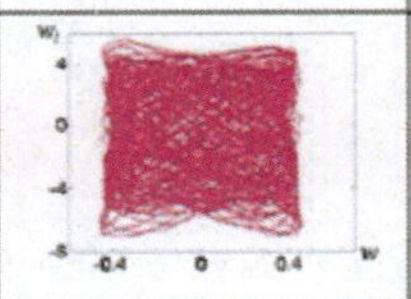	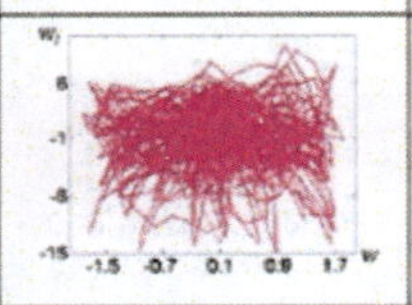	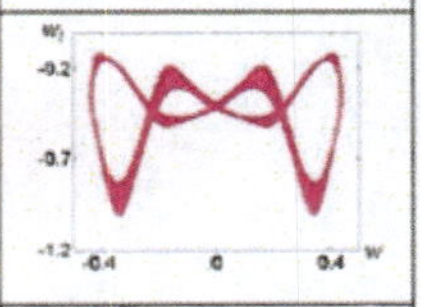
e)	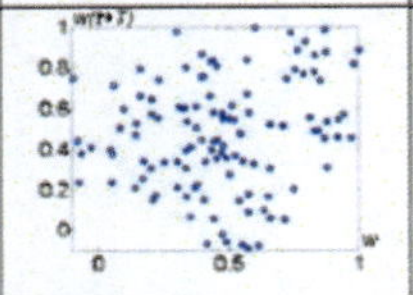	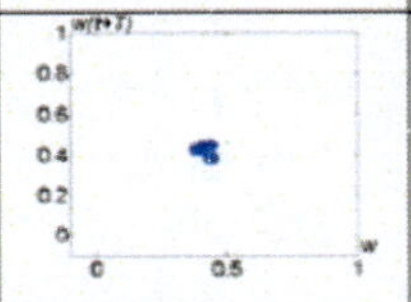	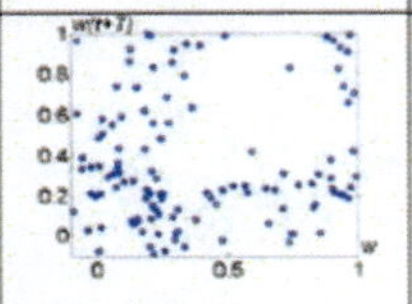	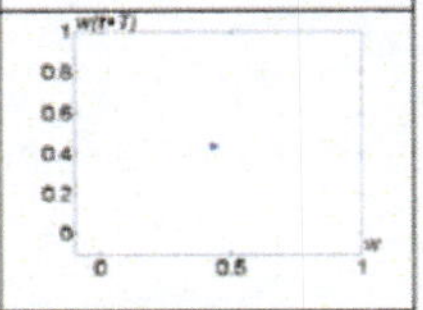
f)	Wolf [130] $\lambda_1 = 0.0341$ Rosenstein [131] $\lambda_1 = 0.017$ Kantz [132] $\lambda_1 = 0.02741$ NN [133] $\lambda_1 = 0.012$	Wolf [130] $\lambda_1 = 0.0081$ Rosenstein [131] $\lambda_1 = 0.006$ Kantz [132] $\lambda_1 = 0.0035$ NN [133] $\lambda_1 = 0.001$	Wolf [130] $\lambda_1 = 0.031$ Rosenstein [131] $\lambda_1 = 0.0168$ Kantz [132] $\lambda_1 = 0.0252$ NN [133] $\lambda_1 = 0.021$	Wolf [130] $\lambda_1 = -0.014$ Rosenstein [131] $\lambda_1 = -0.009$ Kantz [132] $\lambda_1 = -0.01$ NN [133] $\lambda_1 = -0.0076$

Table 9.10 Dynamical characteristics of the beam at temperature $\theta = -100$: (a) time series $w(0.5; t)$; (b) Fourier spectra $S(\omega)$; (c) 2D wavelet spectra; (d) phase portraits $\dot{w}\,[w(t)]$; (e) Poincaré maps $w_{t+T}\,[w_t]$; (f) values of LLEs [reprinted with permission from International Journal of Non-Linear Mechanics publishers]

	1	1*	2	2*
a)	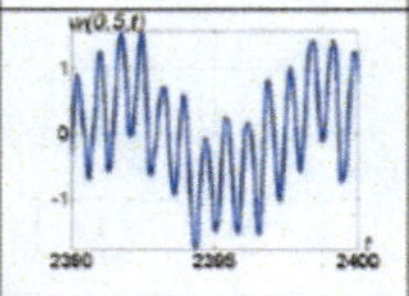	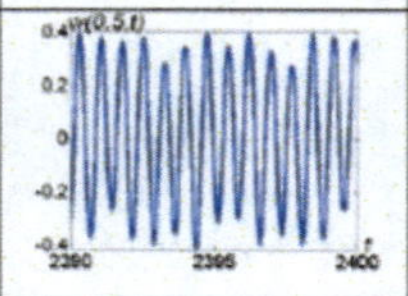	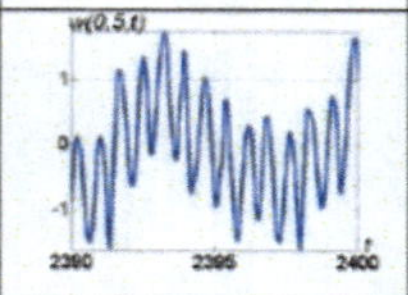	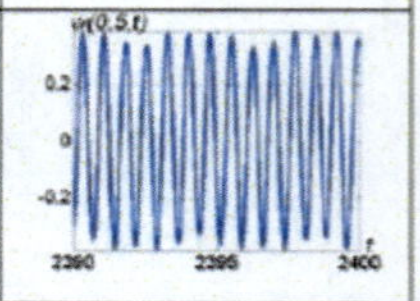
b)	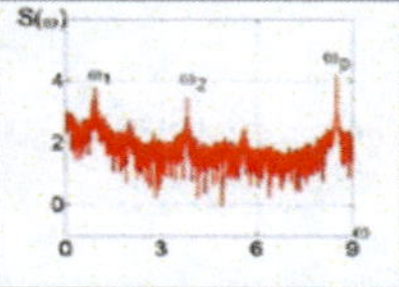	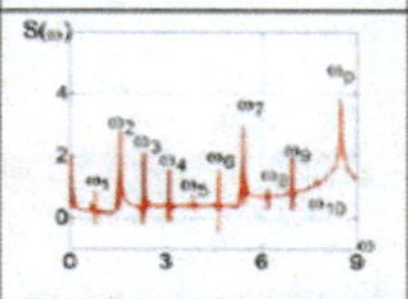	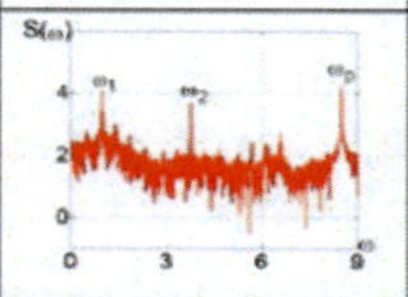	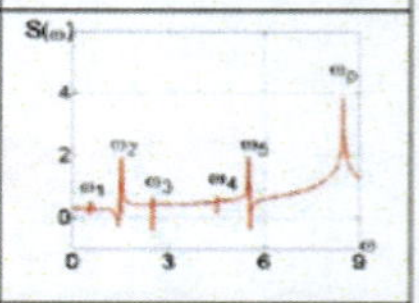
c)	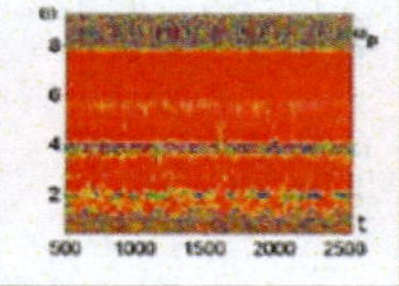	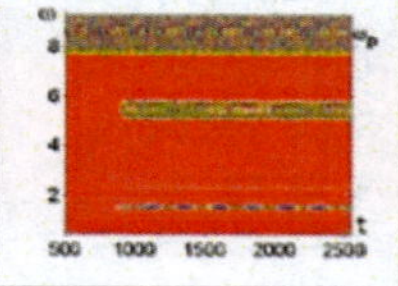	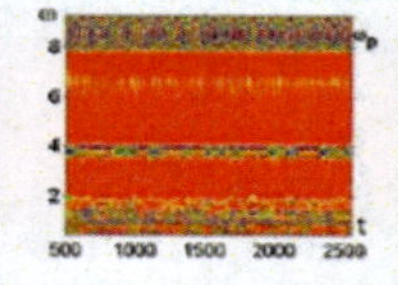	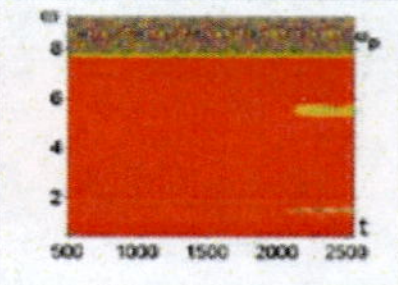
d)	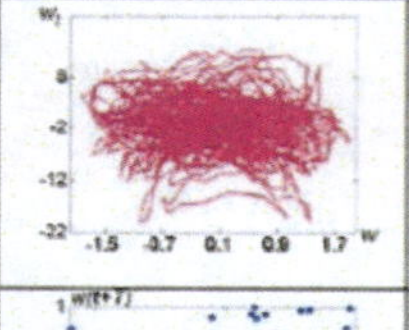	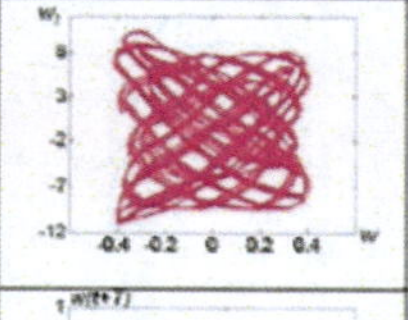	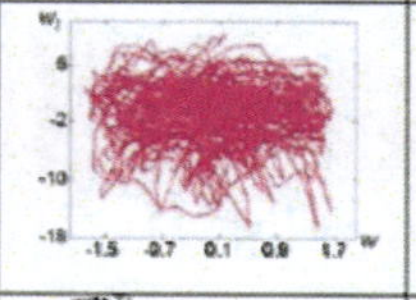	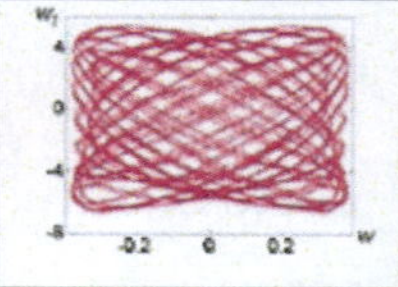
e)	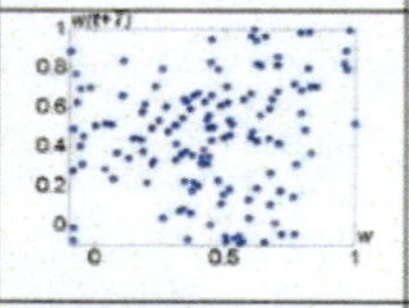	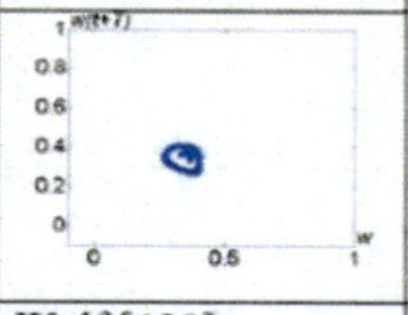	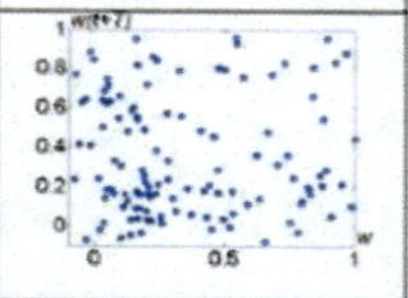	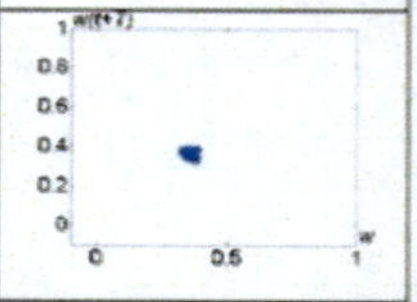
f)	Wolf [130] $\lambda_1 = 0.0756$ Rosenstein [131] $\lambda_1 = 0.06568$ Kantz [132] $\lambda_1 = 0.02982$ NN [133] $\lambda_1 = 0.0254$	Wolf [130] $\lambda_1 = 0.0012$ Rosenstein [131] $\lambda_1 = 0.0017$ Kantz [132] $\lambda_1 = 0.00264$ NN [133] $\lambda_1 = 0.00234$	Wolf [130] $\lambda_1 = 0.0812$ Rosenstein [131] $\lambda_1 = 0.06537$ Kantz [132] $\lambda_1 = 0.02753$ NN [133] $\lambda_1 = 0.0348$	Wolf [130] $\lambda_1 = -0.00023$ Rosenstein [131] $\lambda_1 = -0.0094$ Kantz [132] $\lambda_1 = -0.00091$ NN [133] $\lambda_1 = -0.00001$

frequency and amplitude of the harmonic load, the topologically optimal beam structure exhibits a transition from chaotic vibrations to two-frequency periodic vibrations.

Case study 2 ($k_4 = 0.3$, *homogeneous beam*). The presented signal (9.72a), Fourier spectrum (9.72b), and in particular Morlet wavelet spectrum (9.72c) indicate chaotic vibrations of the beam. The power spectrum comprises numerous frequencies. The order in the frequencies distribution is remarkable. Pairs of the frequencies satisfy the following resonance relationship: $\omega_3 - \omega_2 = \omega_4 - \omega_3 = \omega_6 - \omega_5 = \omega_8 - \omega_7 = \omega_9 - \omega_8 = \omega_p - \omega_{10} = 0.766$, whereas the remaining frequency pairs are governed by formulas $\omega_2 - \omega_1 = \omega_{10} - \omega_9 = 0.862$ and $\omega_5 - \omega_4 = \omega_7 - \omega_6 = 0.804$. Therefore, the frequencies of the spectrum depend on the excitation frequency ω_p. The wavelet spectrum (9.72c) exhibits non-uniform character, since certain frequencies are switched on and switched off in time. Phase portrait (9.72d) exhibits similarity to the phase portrait of the first case, but its form is more compressed with respect to velocity w_t. The Poincaré map consists of the gathered points in the form of a spiral.

Case study 2* ($k_4 = 0.3$, *optimal microstructure*). Signal (9.72*a) is periodic, but the amplitude is twice lower than in the previous case. Fourier spectrum (9.72*b), besides ω_p, exhibits the period doubling frequency $\omega_1 = 4.367 \approx \omega_p/2$. The wavelet spectrum (besides ω_1) contains a series of frequencies located in the low-frequency zone. The phase portrait (9.72*d) qualitatively repeats the phase portrait (9.71*d). The Poincaré map (9.72*e) consists of one point. Therefore, in the case of neglecting of the size-dependent behaviour, the optimized beam material essentially influences the vibration characteristics of the beams.

In what follows, we consider the dynamic characteristics of the homogeneous and optimized beam in the temperature field. All further presented results were obtained for $q_0 = 50000$, $\omega_p = 8.5$. Table 9.9 reports the vibration characteristics for the temperature ($\theta = 100C^{\circ}$).

Case study 1 ($k_4 = 0$, *homogeneous beam*). Signal (9.71a) exhibits a chaotic structure. The Fourier spectrum is typical for chaos spanned on 3 frequencies, i.e. $\omega_p = 8.5$ and two dependent frequencies $\omega_1 = 0.998$, $\omega_2 = 3.832$. The following resonance relationship between them holds $\omega_p = 3 \cdot (\omega_2 - \omega_1)$. The wavelet spectrum (9.71c), besides the frequencies ω_2, ω_1, presents noisy-type frequencies localized in the low-frequency band. The phase portrait (9.71d) and the Poincaré map also confirm the chaotic character of vibrations.

Case study 1* ($k_4 = 0$, *optimal microstructure*). Signal (9.71*a) has the form of a sinusoid embedded in a sinusoid-like envelope. The Fourier spectrum (9.71b), besides ω_p, contains the following dependent frequencies: ω_1 , ω_2, ω_3, ω_4, ω_5. The following resonance relations hold: $\omega_p - \omega_5 = \omega_5 - \omega_4$, $\omega_1 + \omega_2 + \omega_3 = \omega_p/4$. The wavelet spectrum (9.71*c) gives additional information on the results obtained by the FFT. Namely, the frequencies ω_1, ω_2, ω_4, ω_5 are born at time instants $t \approx 2380$, 1550, 2350, 1570, respectively. Therefore, up to time $t \approx 1500$, the beam vibrations exhibit only two frequencies ω_p and ω_3. The phase portrait (9.71*d) and the Poincaré map (9.71*e) exhibit the multi-frequency vibrations, and the points $w(t)$ and $w(t+T)$ are densely distributed, which indicates a chaotic structure.

Case study 2 ($k_4 = 0.3$, *homogeneous beam*). The forms of the signal (9.72a), Fourier spectrum (9.72b), and particularly the wavelet spectrum (9.72c) imply chaotic vibrations. The frequency Fourier spectrum exhibits ω_p and $\omega_1 = 0.958$, $\omega_2 = 3.791$. It can be observed on the wavelet spectrum that ω_2 has time-dependent power. The phase portrait (9.72d) and Poincaré map (9.72c) also indicate chaotic vibrations of the beam.

Case study 2* ($k_4 = 0.3$, *optimal microstructure*). The signal (9.72*a) is periodic. The Fourier spectrum (9.72*b) contains only the excitation frequency. The wavelet spectrum (9.72*c) exhibits noisy frequencies localized in the low-frequency band. The phase portrait (9.72*d) is qualitatively similar to the phase portrait (9.72*d) of the previous case (θ=0). The Poincaré map (9.72*e) shows a point indicating the existence of a periodic orbit.

Therefore, it is illustrated that in the case of absence of the temperature field, the beam optimized material has an essential influence on the dynamic characteristics of the beam.

In Table 9.10, the results for the temperature $\theta = -100C^\circ$ are shown, and they are briefly described in what follows.

Case study 1 ($k_4 = 0$, *homogeneous beam*). The signal (9.71a) is chaotic. The Fourier spectrum exhibits two of the most powerful frequencies $\omega_1 = 0.8947$, $\omega_2 = 3.813$ which satisfy the relations $\omega_1 = \omega_p/9.5$, $\omega_2 = \omega_p/2.223$. The phase portrait (9.71d) and the Poincaré map (9.71c) also manifest chaotic beam vibrations.

Case study 1* ($k_4 = 0$, *optimal microstructure*). The signal (9.71*a) presents a modified sinusoid-like vibration with the amplitude twice lower than in the case of the homogeneous beam (9.71). The Fourier spectrum (9.71b) contains the frequencies ω_1, ω_2, ω_3, ω_4, ω_5, ω_6, ω_7, ω_8, ω_9, ω_{10}, and the excitation frequency ω_p. All frequencies are separated from each other by distances approximately equal to 0.775. The wavelet spectrum (9.71*c) supplements the Fourier spectrum with respect to their time evolution. Namely, the frequencies ω_1, ω_2, ω_3 , ω_7, ω_{10} appear in the spectrum beginning with the time instant $t \approx 850$. Points of the Poincaré maps are located on a ring, implying regular vibrations.

Case study 2 ($k_4 = 0.3$, *homogeneous beam*). The forms of the signal (9.72a), Fourier spectrum (9.72b), and particularly the Morlet spectrum (9.72c) indicate chaotic vibrations. The spectrum, besides ω_p, exhibits two dominating frequencies $\omega_1 = 0.9604$, $\omega_2 = 3.777$, which satisfy the estimation $\omega_1 = \omega_p/8.85$, $\omega_2 = \omega_p/2.25$. The wavelet spectrum illustrates how ω_2 changes its power over time in a way analogous to the case 1. The phase portrait (9.72d) and Poincaré map (9.72c) also indicate beam chaotic vibrations.

Case study 2* ($k_4 = 0.3$, *optimal microstructure*). The signal (9.72*a) is composed of a sinusoid modulated regularly and having the amplitude three times lower than in the case of the homogeneous beam (9.72). The Fourier spectrum (9.72*b) contains, besides of ω_p, the dependent frequencies ω_1, ω_2, ω_3, ω_4, ω_5. The following resonance phenomena hold: $\omega_p - \omega_5 = \omega_5 - \omega_4$, $\omega_1 + \omega_2 + \omega_3 = \omega_p/4$. The wavelet spectrum (9.72*c), similar to the classical Fourier spectrum, exhibits temporal behaviour of the frequencies ω_2, ω_5, beginning from the time instants $t \approx 2050$, 2150, respectively. In other words, before reaching $t \approx 2050$, the vibrations are practically harmonic with the fundamental frequency ω_p, since the power of the frequencies ω_1, ω_3, ω_4 is extremely small. The Poincaré map consists of densely distributed points, i.e. the trajectories are not divergent and the beam vibrations are regular.

9.7.3.4 Scenarios of Transition into Chaos

Detection, construction and analysis of a scenario of transition from regular to chaotic vibrations, and vice versa, are important and fundamental tasks when investigating nonlinear dynamics of continuous mechanical systems. These transitions can be based on the classical Feigenbaum, Pomeau-Manneville, and Ruelle-Takens-Newhouse scenarios. There are papers reporting scenarios of the transition for structural members, including beams, plates, and shells, by using the Bernoulli-Euler, Timoshenko and Sheremetev-Pelekh-Reddy models without [134–136] and taking [121, 137] into account of the size-dependent behaviour. Furthermore, scenarios of transition from regular to chaotic dynamics of nonlinear non-homogeneous beams have been detected and studied with respect to their thickness for different values of the size-dependent parameter [138].

We begin the investigations with a construction and analysis of scenarios for two temperature values (θ=100, $\theta = -100$), for two cases, i.e. taking into account the size-dependent behaviour $\gamma_2 = 0.3$ or not $\gamma_2 = 0$, for a non-homogeneous (optimized) beam. The following parameters were fixed: $\omega_p = 8.5$, $\varepsilon = 1$, $\gamma_1 = 40$, $\nu = 0.3$. All detected scenarios have a similar qualitative picture, i.e. a transition into chaotic vibrations which follows the Pomeau-Manneville scenario [139], where load plays a role of the control parameter.

Chaotic vibrations are achieved after the third cycle of the process "periodic vibrations-chaotic vibrations". Each cycle is realized through different scenarios.

Let us consider the scenarios for the non-homogeneous (optimized) beam for the temperature $\theta = -100$, $\gamma_2 = 0$. The following dynamic characteristics are reported in Tables 9.11, 9.12 and 9.13: (a) signal $w(0.5; t)$; (b) Fourier spectrum $S(\omega)$ based on the FFT; (c) 2D wavelet spectrum based on the Morlet mother wavelet; (d) phase portrait $\dot{w}\,[w(t)]$; (e) Poincaré section $w_{t+T}\,[w_t]$; (f) LLEs computed by four methods.

The first cycle consist of three frequencies exhibiting the following resonance behaviour $\omega_p - \omega_3 = \omega_3 - \omega_2 = \omega_2 - \omega_1$. An increase in the amplitude of load

Table 9.11 Dynamical characteristics of the beam at temperature θ=-100: (a) time series $w(0.5;\, t)$; (b) Fourier spectra $S(\omega)$; (c) 2D wavelet spectra; (d) phase portraits $\dot{w}\,[w(t)]$; (e) Poincaré maps $w_{t+T}\,[w_t]$; (f) values of LLEs ($q = 250$) [reprinted with permission from International Journal of Non-Linear Mechanics publishers]

a) 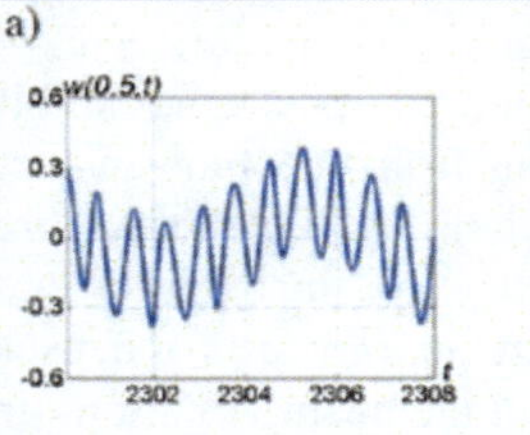	b) 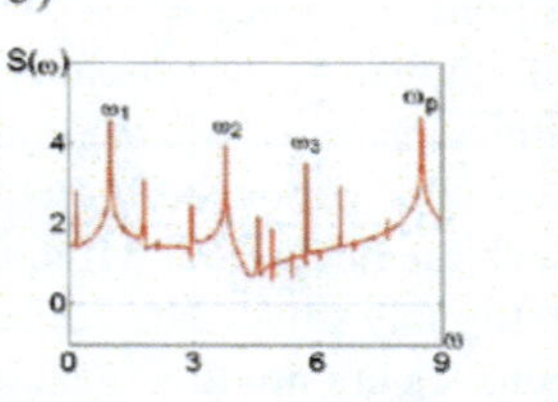	c) 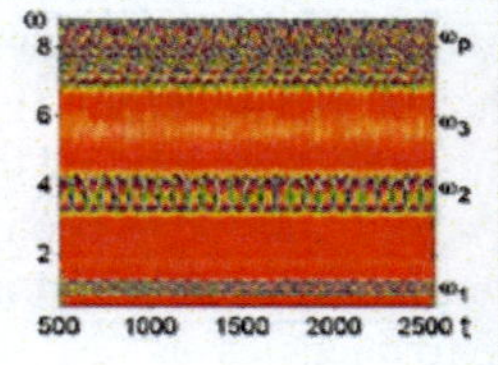
d) 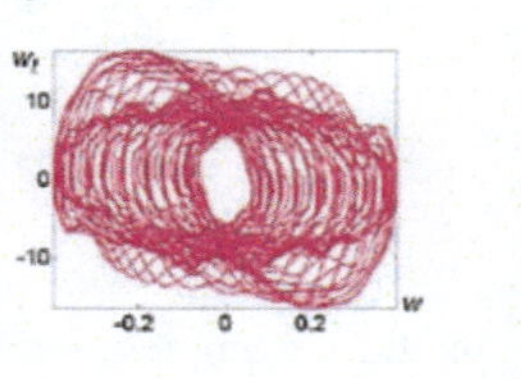	e) 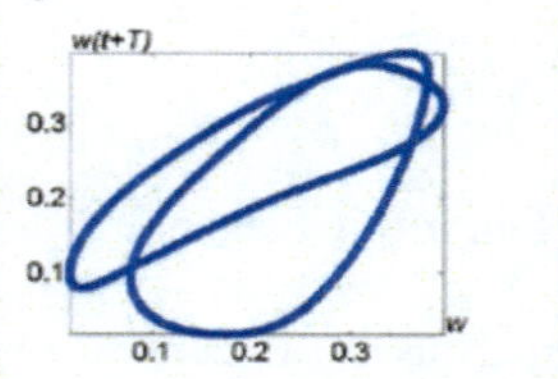	f) LE Spectrum: Wolf [130] λ_1 =0.1023 Kantz [132] λ_1 =0.02998 Rosenstein [131] λ_1 =0.10030 NN [133] λ_1 =0.00357

yields the occurrence of the so-called "pair frequencies", i.e. two frequencies, which are localized on the same distance from the fundamental frequency. It should be noted that for all obtained scenarios, the distance is the same and equals to $\approx$0.33. A further increase in the excitation amplitude yields an increase in the number of pair frequencies and chaotic orbits appear (Table 9.11). The occurrence of the chaotic state is validated by the magnitude of the LLE (f) estimated by four methods. The phase portrait (d) and, in particular, the Poincaré map (e) composed of two ellipses indicate that chaos is spanned on four frequencies $\omega_1,\ \omega_2,\ \omega_3,\ \omega_p$. This observation is also validated by both Fourier and wavelet spectra.

A further increase in the load amplitude results in the purification of the spectrum to one-frequency vibrations. Then, two frequencies, which are coupled by the resonance relation $\omega_p = 2 \cdot \omega_2 + \omega_1$, appear. The increase in the load yields also the occurrence of pair frequencies localized at a distance of $\approx$0.88 from the fundamental frequency. Further increase causes an increase in the number of pair frequencies, and a transition into chaos (Table 9.12). It should be emphasized that, as it was in the case of Tables 9.8, 9.9 and 9.10, the Fourier spectrum does not allow one to detect time instants of the birth and the death of frequencies, and hence it is necessary to employ the wavelet analysis. The wavelet spectrum is non-uniform, and the frequencies 6.44; 5.31; 3.65; 2.126; 1.595; 1.197 appear at different time instants after t=1500. Furthermore, the power of those frequencies is time dependent. The frequency $\omega = 5.17$ appears and vanishes in time. Its first occurrence exhibits the maximum power comparable with the excitation frequency. Next, this frequency appears with half of the excitation frequency power, which is also validated by the Fourier spectrum. The

Table 9.12 Dynamical characteristics of the beam at temperature θ=-100: (a) time series $w(0.5; t)$; (b) Fourier spectra $S(\omega)$; (c) 2D wavelet spectra; (d) phase portraits $\dot{w}\,[w(t)]$; (e) Poincaré maps $w_{t+T}\,[w_t]$; (f) values of LLEs ($q = 580$) [reprinted with permission from International Journal of Non-Linear Mechanics publishers]

a) 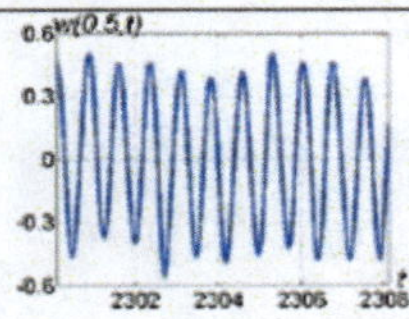	b) 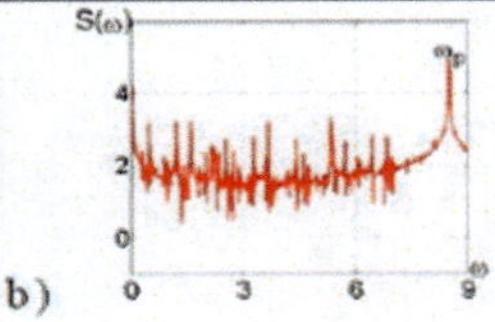	c) 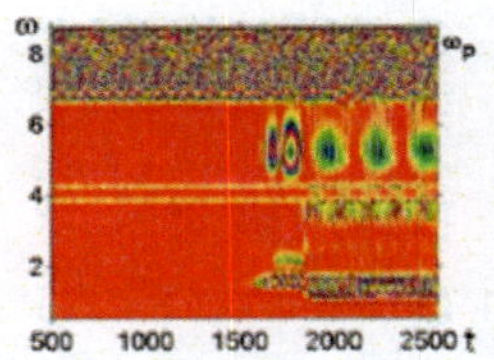
d) 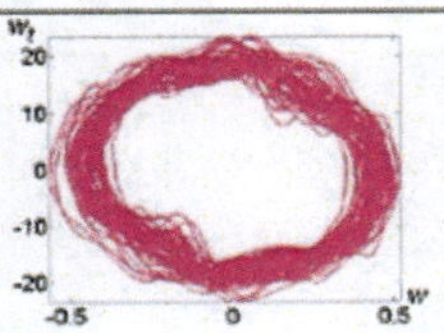	e) 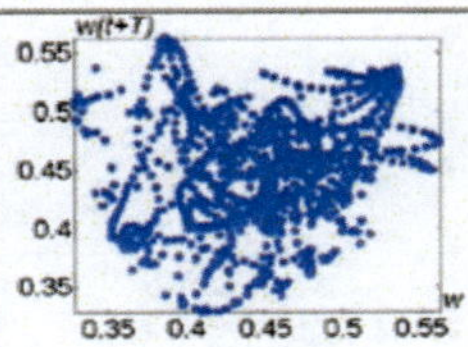	f) LE Spectrum: Wolf [130] λ_1 =0.10156 Kantz [132] λ_1 =0.03182 Rosenstein [131] λ_1 = 0.09747 NN [133] λ_1 =0.10618

lifespan of the frequency $\omega = 2.224$ is limited to the interval $t \approx [1701;\ 1829]$. The non-uniform form of the frequency spectrum influences the phase trajectories. In spite of that, the trajectories are close to each other in the phase portrait (d), the Poincaré map (e) shows a chaotic distribution of trajectories on the plane $\dot{w}\,[w(t)]$. LLEs computed by four methods are positive, indicating beam chaotic state.

The increase in the amplitude of the excitation load yields a transition into one-frequency periodic vibrations. Then, two more frequencies occur: $\omega_1 = \omega_p/4$ and $\omega_2 = 3\omega_p/4$. Further increase in the load yields the frequency $\omega_3 = \omega_p/2$, and a pair of frequencies around the existing $\omega_1, \omega_2, \omega_3$. Then, the number of frequencies increases, which results in the chaotic state (Table 9.13). All vibration characteristics reported in Table 9.13 validate the beam chaotic state.

As it has been already mentioned, scenarios of transition into chaos for all optimized (non-homogeneous) beams are qualitatively the same. There are, however, a few differences:

(i) for the temperature $\theta = -100$, the first cycle exhibits a low-band frequency $\omega_1 = 1.15$, and the increase in the load yields the frequency $\omega_1 = 3.6662$. For the temperature $\theta = 100$, the mentioned frequencies appear in a reversed manner;

(ii) for the size-dependent parameter $\gamma_2 = 0.3$, chaotic vibrations appear at a higher excitation amplitude than for $\gamma_2 = 0$.

Table 9.13 Dynamical characteristics of the beam at temperature θ=-100: (a) time series $w(0.5; t)$; (b) Fourier spectra $S(\omega)$; (c) 2D wavelet spectra; (d) phase portraits $\dot{w}\,[w(t)]$; (e) Poincaré maps $w_{t+T}\,[w_t]$; (f) values of LLEs ($q = 642$) [reprinted with permission from International Journal of Non-Linear Mechanics publishers]

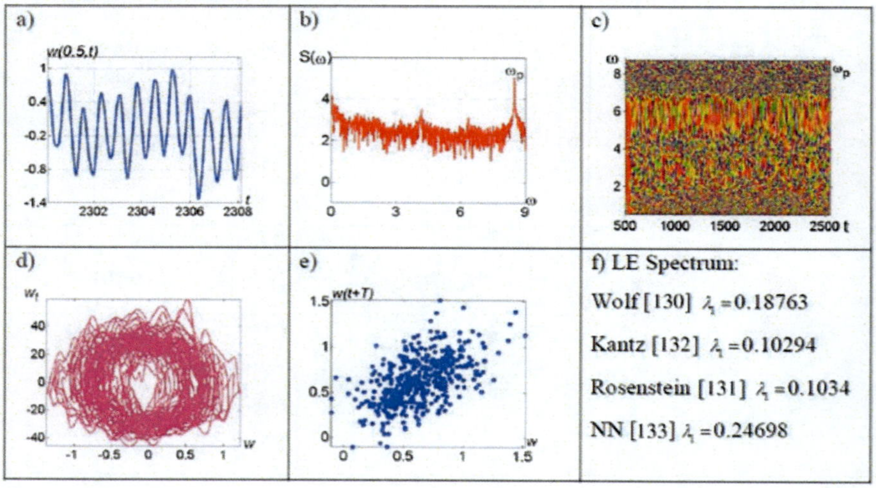

References

1. Gorman, D.J.: Free Vibration Analysis of Beams and Shafts. Wiley, New York (1975)
2. Aydogdu, M., Taskin, V.: Free vibration analysis of functionally graded beams with simply supported edges. Mater. Des. **28**, 1651–1656 (2007)
3. Kapuria, S., Bhattacharyya, M., Kumar, A.N.: Bending and free vibration response of layered functionally graded beams: a theoretical model and its experimental validation. Compos. Struct. **82**(3), 390–402 (2008)
4. Pradhan, K.-K., Chakraverty, S.: Free vibration of Euler and Timoshenko functionally graded beams by Rayleigh-Ritz method. Compos. Part B: Eng. **51**(4), 175–184 (2013)
5. Simsek, M., Kocaturk, T., Akbas, S.D.: Static bending of a functionally graded microscale Timoshenko beam based on the modified couple stress theory. Compos. Struct. **95**, 740–747 (2013)
6. Simsek, M.: Fundamental frequency analysis of functionally graded beams by using different higher-order beam theories. Nucl. Eng. Design. **240**(4), 697–705 (2010)
7. Sina, S.A., Navazi, H.M., Haddadpour, H.: An analytical method for free vibration analysis of functionally graded beams. Mater. Design. **30**(3), 741–747 (2009)
8. Thai, H.T., Vo, T.P.: Bending and free vibration of functionally graded beams using various higher order shear deformation beam theories. Int. J. Mech. Sci. **62**, 57–66 (2012)
9. Rajasekaran, S.: Differential transformation and differential quadrature methods for centrifugally stiffened axially functionally graded tapered beams. Int. J. Mech. Sci. **74**(3), 15–31 (2013)
10. Shahba, A., Rajasekaran, S.: Free vibration and stability of tapered Euler-Bernoulli beams made of axially functionally graded materials. Appl. Math. Model. **36**(7), 3094–3111 (2012)
11. Calio, I., Elishakoff, I.: Closed-form solutions for axially graded beam-columns. J. Sound Vib. **280**(3–5), 1083–1094 (2009)

12. Elishakoff, I., Johnson, V.: Apparently the first closed-form solution of vibrating inhomogeneous beam with a tip mass. J. Sound Vib. **286**(4–5), 1057–1066 (2009)
13. Hemmatnezhad, M., Ansari, R., Rahimi, G.H.: Large-amplitude free vibrations of functionally graded beams by means of a finite element formulation. Appl. Math. Model. **37**, 8495–8504 (2013)
14. Kien, N.D.: Large displacement response of tapered cantilever beams made of axially functionally graded material. Compos. Part B **55**, 298–309 (2013)
15. Alshorbagy, A.E., Eltaher, M.A., Mahmoud, F.F.: Free vibration characteristics of a functionally graded beam by finite element method. Appl. Math. Model. **35**(1), 412–425 (2011)
16. Shahba, A., Attarnejad, R., Hajilar, S.: Free vibration and stability of axially functionally graded tapered Euler-Bernoulli beams. Shock Vib. **18**(5), 683–696 (2011)
17. Shahba, A., Attarnejad, R., Tavanaie, M.M., Hajilar, S.: Free vibration and stability analysis of axially functionally graded tapered Timoshenko beams with classical and non-classical boundary conditions. Compos. Part B: Eng. **42**(4), 0801–0808 (2011)
18. Ke, L.L., Yang, J., Kitipornchai, S.: An analytical study on the nonlinear vibration of functionally graded beams. Meccanica **45**, 743–752 (2010)
19. Akgoz, B., Civalek, O.: Free vibration analysis of axially functionally graded tapered Bernoulli-Euler microbeams based on the modified couple stress theory. Compos. Struct. **98**, 314–322 (2013)
20. Banerjee, J.R., Su, H., Jackson, D.R.: Free vibration of rotating tapered beams using the dynamic stiffness method. J. Sound Vib. **298**(4–5), 1034–1054 (2006)
21. Hsu, J.-C., Lai, H.-Y., Chen, C.-K.: Free vibration of nonuniform Euler-Bernoulli beams with general elastically end constraints using Adomian modified decomposition method. J. Sound Vib. **318**(4–5), 965–981 (2008)
22. Huang, Y., Li, X.-F.: A new approach for free vibration of axially functionally graded beams with non-uniform crosssection. J. Sound Vib. **329**(11), 2291–2303 (2010)
23. Li, X.-F., Kang, Y.-A., Wu, J.-X.: Exact frequency equations of free vibration of exponentially functionally graded beams. Appl. Acoust. **74**(3), 413–420 (2013)
24. Mao, Q., Pietrzko, S.: Free vibration analysis of a type of tapered beams by using Adomian decomposition method. Appl. Math. Comput. **219**(6), 3264–271 (2012)
25. Ozgumus, O.-O., Kaya, M.O.: Flapwise bending vibration analysis of a rotating double-tapered Timoshenko beam. Arch. Appl. Mech. **78**(5), 379–392 (2008)
26. Rajasekaran, S.: Free vibration of centrifugally stiffened axially functionally graded tapered Timoshenko beams using differential transformation and quadrature methods. Appl. Math. Model. **37**(6), 4440–4463 (2013)
27. Rajasekaran, S., Norouzzadeh, T.E.: Free vibration analysis of axially functionally graded tapered Timoshenko beams using differential transformation element method and differential quadrature element method of lowest-order. Meccan. **49**(4), 995–1009 (2014)
28. Yardimoglu, B.: Vibration analysis of rotating tapered Timoshenko beams by a new finite element model. Shock Vib. **13**(2), 117–126 (2006)
29. Baghani, M., Mazaheri, H., Salarieh, H.: Analysis of large amplitude free vibrations of clamped tapered beams on a nonlinear elastic foundation. Appl. Math. Model. **38**(3), 1176–1186 (2013)
30. Huang, Y., Yang, L.E., Luo, Q.Z.: Free vibration of axially functionally graded Timoshenko beams with non-uniform cross-section. Compos. Part B: Eng. **45**, 1493–1498 (2013)
31. Shahba, A., Attarnejad, R., Hajilar, S.: A Mechanical-Based Solution for Axially Functionally Graded Tapered Euler-Bernoulli Beams. Mech. Adv. Mater. Struct. **20**, 696–707 (2012)
32. Asghari, M., Ahmadian, M.T., Kahrobaiyan, M.H., Rahaeifard, M.: On the size-dependent behavior of functionally graded micro-beams. Mater. Des. **31**, 2324–2329 (2010)
33. Asghari, M., Kahrobaiyan, M.H., Ahmadian, M.T.: A nonlinear Timoshenko beam formulation based on the modified couple stress theory. Int. J. Eng. Sci. **48**, 1749–1766 (2010)
34. Lazopoulos, K.A., Lazopoulos, A.K.: Bending and buckling of thin strain gradient elastic beams. Eur. J. Mech. A-Sol. **29**, 837–843 (2010)

35. Mindlin, R.D., Tiersten, H.F.: Effects of couple-stresses in linear elasticity. Arch. Rat. Mech. Anal. **11**, 415–448 (1962)
36. Rajabi, F., Ramezani, S.: A nonlinear microbeam model based on strain gradient elasticity theory with surface energy. Arch. Appl. Mech. **82**, 363–376 (2012)
37. Toupin, R.A.: Elastic materials with couple-stresses. Arch. Ration. Mech. Anal. **11**, 385–414 (1962)
38. Xie, Y., Steven, G.: A simple evolutionary procedure for structural optimization. Comput. Struct. **49**, 885–896 (1993)
39. Eringen, A.C.: Nonlocal polar elastic continua. Int. J. Eng. Sci. **10**, 1–16 (1972)
40. Aifantis, E.C.: Strain gradient interpretation of size effects. Int. J. Fract. **95**, 299–314 (1999)
41. Gurtin, M.E., Weissmuller, J., Larche, F.: The general theory of curved deformable inter-faces in solids at equilibrium. Philos. Mag. A **78**, 1093–1109 (1998)
42. Arbind, A., Reddy, J.N.: Nonlinear analysis of functionally graded microstructure-dependent beams. Compos. Struct. **98**, 272–281 (2013)
43. Arbind, A., Reddy, J.N., Srinivasa, A.R.: Modified couple stress-based third-order theory for nonlinear analysis of functionally graded beams Lat. Am. J. Sol. Struct. **11**, 459–487 (2014)
44. Bendsoe, M., Kikuchi, N.: Generating optimal topologies in optimal design using a homogenization method. Comput. Meth. Appl. Mech. Eng. **71**, 197–224 (1988)
45. Bensoussan, A., Lions, J., Papanicolaou, G.: Asymptotic Analysis for Periodic Structures. North-Holland, Amsterdam (1978)
46. Sanchez-Palencia, E.: Non-Homogeneous Media and Vibration Theory. Springer, Berlin (1980)
47. Bendsoe, M.: Optimal shape design as a material distribution problem. Struct. Optimiz. **1**, 193–202 (1989)
48. Allaire, G., Jouve, F., Toader, A.: Structural optimization using sensitivity analysis and a level-set method. J. Comput. Phys. **194**, 363–393 (2004)
49. Belytschko, T., Xiao, S., Parimi, C.: Topology optimization with implicit functions and regularization. Int. J. Num. Meth. Eng. **57**(8), 1177–1196 (2003)
50. Sethian, J., Wiegmann, A.: Structural boundary design via level set and immersed interface methods. J. Comput. Phys. **163**(2), 489–528 (2000)
51. Wang, M., Wang, X., Guo, D.: A level set method for structural topology optimization. Comput. Meth. Appl. Eng. **192**, 227–246 (2003)
52. Zhou, M., Rozvany, G.: Dcoc: an optimality criteria method for large systems, part i: theory. Struct. Optimiz. **5**, 12–29 (1993)
53. Svanberg, K., Werme, M.: Topology optimization by neighbour-hood search method based on efficient sensitivity calculations. Int. J. Numer. Meth. Eng. **67**, 1670–1699 (2006)
54. Svanberg, K.: The method of moving asymptotes - a new method for structural optimization. Int. J. Num. Meth. Eng. **24**, 359–373 (1987)
55. Gibiansky, L., Sigmund, O.: Multiphase elastic composites with extremal bulk modulus. J. Mech. Phys. Sol. **48**, 461–498 (2000)
56. Larsen, U., Sigmund, O., Bouwstra, S.: Design and fabrication of compliant micro mechanisms and structures with negative Poisson's ratio. J. Microelectromech. Sys. **6**(2), 99–106 (1997)
57. Sigmund, O.: Materials with prescribed constitutive parameters: an inverse homogenization problem. Int. J. Sol. Struct. **31**, 2313–2329 (1994)
58. Sigmund, O.: Tailoring materials with prescribed elastic properties. Mech. Mater. **20**, 351–368 (1999)
59. Sigmund, O.: A new class of extremal composites. J. Mech. Phys. Sol. **48**, 397–428 (2000)
60. Sigmund, O., Torquato, S.: Design of materials with extreme thermal expansion using a three-phase topology optimization method. J. Mech. Phys. Sol. **45**, 1037–1067 (1997)
61. Jonsmann, J., Sigmund, O., Bouwstra, S.: Compliant thermal microactuators. Sens. Actuat. **76**, 463–469 (1999)
62. Li, Q., Steven, G., Xie, Y., Qurein, O.: Shape and topology design for heat conduction by evolutionary structural optimization. Int. J. Heat Mass Trans. **42**, 3361–3371 (1999)

63. Li, Q., Steven, G., Xie, Y.: Thermoelastic topology optimization for problems with varying temperature fields. J. Therm. Stres. **24**, 347–366 (2001)
64. Rodrigues, H., Fernandes, P.: A material based model for topology optimization of thermoelastic structures. Int. J. Num. Meth. Eng. **38**(12), 1951–1969 (1999)
65. Cho, S., Choi, J.: Efficient topology optimization of thermo-elasticity problems using coupled field adjoint sensitivity analysis method. Fin. Elem. Anal. Des. **41**(15), 1481–1499 (2009)
66. Chen, X., Mai, Y.: Micromechanics of rubber-toughened polymers. J. Mater. Sci. **33**(14), 3529–3539 (1998)
67. Hashin, Z., Shtrikman, S.: A variational approach to the theory of the elastic behaviour of multiphase materials. J. Mech. Phys. Sol. **11**, 127–140 (1963)
68. Guest, J., Prevost, J.: Optimizing multifunctional materials: design of microstructures for maximized stiffness and fluid permeability. Int. J. Sol. Struct. **43**, 7028–7047 (2006)
69. Prager, W., Taylor, J.: Problems of optimal structural design. J. Appl. Mech. **35**, 102–106 (1968)
70. Torquato, S., Hyun, S., Donev, A.: Optimal design of manufacturable three-dimensional composites with multifunctional characteristics. J. Appl. Phys. **94**(9), 5748–5759 (2003)
71. Yoo, J., Lee, C.: Topology optimization of a swing arm type actuator using the response surface method. Microsys. Technol. Micro Nanosys. Inf. Stor. Proc. Syst. **13**(1), 21–31 (2007)
72. Bendsoe, M.P., Sigmund, O.: Topology Optimization. Theory, Methods and Applications. Springer, Berlin (2003)
73. Eschenauer, H.A., Olhoff, N.: Topology optimization of continuum structures: a review. Appl. Mech. Rev. **54**(4), 331–390 (2001)
74. Rozvany, G.I.N.: Aims, scope, methods, history and unified terminology of computer-aided topology optimization in structural mechanics. Struct. Multidiscip. Optim. **21**(2), 90–108 (2001)
75. Rozvany, G.I.N., Zhou, M., Birker, T.: Generalized shape optimization without homogenenization. Struct. Optim. **4**(3–4), 250–254 (1992)
76. Stolpe, M., Svanberg, K.: An alternative interpolation scheme for minimum compliance topology optimization. Struct. Multidiscip. Optim. **22**(2), 116–124 (2001)
77. Bruns, T.E.: A reevaluation of the SIMP method with filtering and an alternative formulation for solid-void topology optimization. Struct. Multidiscip. Optim. **30**(6), 428–436 (2009)
78. Guest, J.K., Prevost, J.H., Belytschko, T.: Achieving minimum length scale in topology optimization using nodal design variables and projection functions. Int. J. Num. Meth. Eng. **61**(2), 238–254 (2004)
79. Sigmund, O.: Morphology-based black and white filters for topology optimization. Struct. Multidiscip. Optim. **33**(4–5), 401–424 (2007)
80. Sigmund, O., Petersson, J.: Numerical instabilities in topology optimization: a survey on procedures dealing with checkerboards, mesh-dependencies, and local minima. Struct. Optim. **16**(1), 68–75 (1998)
81. Bourdin, B.: Filters in topology optimization. Int. J. Num. Meth. Eng. **50**(9), 2143–2158 (2001)
82. Bruns, T.E., Tortorelli, D.A.: Topology optimization of non-linear elastic structures and compliant mechanisms. Comput. Meth. Appl. Mech. Eng. **190**(26–27), 3443–3459 (2001)
83. Wang, F., Lazarov, B.S., Sigmund, O.: On projection methods, convergence and robust formulations in topology optimization. Struct. Multidiscip. Optim. **43**(6), 767–784 (2011)
84. Guest, J.K., Asadpoure, A., Ha, S.-H.: Eliminating beta-continuation from Heaviside projection and density filter algorithms. Struct. Multidiscip. Optim. **44**(4), 443–453 (2011)
85. Deaton, J.D., Grandhi, R.V.: Stiffening of restrained thermal structures via topology optimization. Struct. Multidiscip. Optim. **48**(4), 731–745 (2013)
86. Sigmund, O., Maute, K.: Topology optimization approaches. Struct. Multidiscip. Optim. **48**(6), 1031–1055 (2013)
87. Xia, Q., Wang, M.Y.: Topology optimization of thermoelastic structures using level set method. Comput. Mech. **42**(6), 837–857 (2008)

88. Gao, T., Zhang, W.: Topology optimization involving thermo-elastic stress loads. Struct. Multidiscip. Optim. **42**(5), 725–738 (2010)
89. Yang, X., Li, Y.: Topology optimization to minimize the dynamic compliance of a bi-material plate in a thermal environment. Struct. Multidiscip. Optim. **47**(3), 399–408 (2013)
90. Duysinx, P., Bendsoe, M.P.: Topology optimization of continuum structures with local stress constraints. Int. J. Num. Meth. Eng. **43**(8), 1453–1478 (1998)
91. Cheng, G., Jiang, Z.: Study on topology optimization with stress constraints. Eng. Optim. **20**(2), 129–148 (1992)
92. Cheng, G., Guo, X.: Epsilon-relaxed approach in structural topology optimization. Struct. Optim. **13**, 258–266 (1997)
93. Bruggi, M.: On an alternative approach to stress constraints relaxation in topology optimization. Struct. Multidiscip. Optim. **36**(2), 125–141 (2008)
94. Bruggi, M., Venini, P.: A mixed FEM approach to stress-constrained topology optimization. Int. J. Num. Meth. Eng. **73**(12), 1693–1714 (2008)
95. Guilherme, C.E.M., Fonseca, C.E.M.: Topology optimization of continuum structures with epsilon-relaxed stress constraints. In: Alves, M., Da Costa Mattos, H. (eds.) International Symposium on Solid Mechanics, vol. 1, pp. 239–250 (2007)
96. Reddy, J.N., Kim, J.: A nonlinear modified couple stress-based third-order theory of functionally graded plates. Compos. Struct. **94**, 1128–1143 (2012)
97. Paris, P., Navarrina, F., Colominas, I., Casteleiro, M.: Topology optimization of continuum structures with local and global stress constraints. Struct. Multidiscip. Optim. **39**(4), 419–437 (2009)
98. Pereira, J., Fancello, E., Barcellos, C.: Topology optimization of continuum structures with material failure constraints. Struct. Multidiscip. Optim. **26**(1–2), 50–66 (2004)
99. Le, C., Norato, J., Bruns, T.E., Ha, C., Tortorelli, D.A.: Stress-based topology optimization for continua. Struct. Multidiscip. Optim. **41**(4), 605–620 (2010)
100. Krysko, A.V., Awrejcewicz, J., Pavlov, S.P., Bodyagina, K.S., Krysko, V.A.: Topological optimization of thermoelastic composites with maximized stiffness and heat transfer. Compos. Part B **158**, 319–327 (2019)
101. Andrianov, I.V., Awrejcewicz, J., Danishevskyy, V.V.: Asymptotical Mechanics of Composites. Springer, Berlin (2018)
102. Alizadeh, A.N., Sofiyev, A.H.: Modified Young's moduli of nano-materials taking into account the scale effects and vacancies. Meccanica **46**, 915–920 (2011)
103. Awrejcewicz, J., Pavlov, S.P., Bodyagina, K.S., Zhigalov, M.V., Krysko, V.A.: Design of composite structures with extremal elastic properties in the presence of technological constraints. Compos. Struct. **174**, 19–25 (2017)
104. Seresta, O., Gürdal, Z., Adams, D.B., Watson, L.T.: Optimal design of composite wing structures with blended laminates. Comput. Part B **38**, 469–480 (2007)
105. Bassir, D.H., Irisarri, F.X., Maire, J.F., Carrere, N.: Incorporating industrial constraints for multiobjective optimization of composite laminates using a GA. Int. J. Simul. Multidisci. Des. Optim. **2**, 101–106 (2008)
106. Lee, D.S., Morillo, C., Bugeda, G., Oller, S., Onate, E.: Multilayered composite structure design optimisation using distributed/parallel multi-objective evolutionary algorithms. Compos. Struct. **94**(3), 1087–1096 (2012)
107. Ning, F., Cong, W., Qiu, J., Wei, J., Wang, S.: Additive manufacturing of carbon fiber reinforced thermoplastic composites using fused deposition modelling. Compos. Part B **80**, 369–378 (2015)
108. Madeira, J.F.A., Araujo, A.L., Mota Soares, C.M.: Multiobjective optimization of constrained layer damping treatments in composite plate structures. Mech. Adv. Mater. Struc. **24**(5) 427–436 (2017)
109. Salem, A.I., Donaldson, S.L.: Weight and cost multi-objective optimization of hybrid composite sandwich structures. Int. J. Comput. Meth. Exp. Meas. **5**(2), 200–210 (2017)
110. Seretis, G., Kouzilos, G., Manolakos, D., Provatidis, Ch.: Multi-objective curing cycle optimization for glass fabric/epoxy composites using Poisson regression and genetic algorithm. Mater. Res. **21**(2), e20170815 (2018)

111. Passos, A.G., Luersen, M.A.: Multiobjective optimization of laminated composite parts with curvilinear fibers using Kriging-based approaches. J. Struct. Multidisci. Optim. **57**(3), 1115–1127 (2018)
112. Bakhvalov, N.S., Panasenko, G.: Homogenisation: Averaging Processes in Periodic Media. Springer, Netherlands (1989)
113. Andrianov, I.V., Awrejcewicz, J., Danishevs'kyy, V.V., Ivankov, A.O.: Asymptotic Methods in the Theory of Plates with Mixed Boundary Conditions. Wiley, New York (2014)
114. Pareto, V.: Manual of Political Economy. Macmillan, New York (1971)
115. Krysko, A.V., Awrejcewicz, J., Pavlov, S.P., Bodyagina, K.S., Zhigalov, M.V., Krysko, V.A.: Non-linear dynamics of size-dependent Euler-Bernoulli beams with topologically optimized microstructure and subjected to temperature field. Int. J. Non-Lin. Mech. **104**, 75–86 (2018)
116. Ozdemir, O., Kaya, M.O.: Flapwise bending vibration analysis of a rotating tapered cantilever Bernoulli-Euler beam by differential transform method. J. Sound Vib. **289**(1–2), 413–420 (2006)
117. Mindlin, R.D., Tiersten, H.F.: Effects of couple-stresses in linear elasticity. Arch. Ration. Mech. Anal. **11**, 415–448 (1962)
118. Yang, F., Chong, M., Lam, D.C.C., Tong, P.: Couple stress based strain gradient theory for elasticity. Int. J. Solids Struct. **39**, 2731–2743 (2002)
119. Farokhi, H., Ghayesh, M.H., Amabili, M.: Nonlinear dynamics of a geometrically imperfect microbeam based on the modified couple stress theory. Int. J. Eng. Sci. **68**, 11–23 (2013)
120. Ghayesh, M.H., Amabili, M., Farokhi, H.: Nonlinear forced vibrations of a microbeam based on the strain gradient elasticity theory. Int. J. Eng. Sci. **63**, 52–60 (2013)
121. Krysko, A.V., Awrejcewicz, J., Zhigalov, M.V., Pavlov, S.P., Krysko, V.A.: Nonlinear behaviour of different flexible size-dependent beams models based on the modified couple stress theory. Part 2. Chaotic dynamics of flexible beams. Int. J. Non-Lin. Mech. **93**, 106–121 (2017)
122. Krysko, A.V., Awrejcewicz, J., Zhigalov, M.V., Pavlov, S.P., Krysko, V.A.: Nonlinear behaviour of different flexible size-dependent beams models based on the modified couple stress theory. Part 1. Governing equations and static analysis of flexible beams. Int. J. Non-Lin. Mech. **93**, 96–105 (2017)
123. Krysko, A.V., Awrejcewicz, J., Pavlov, S.P., Zhigalov, M.V., Krysko, V.A.: Mathematical model of a three-layer micro- and nano-beams based on the hypotheses of the Grigolyuk-Chulkov and the modified couple stress theory. Int. J. Sol. Struct. **117**, 39–50 (2017)
124. Awrejcewicz, J., Krysko, V.A., Sopenko, A.A., Zhigalov, M.V., Kirichenko, A.V., Krysko, A.V.: Mathematical modelling of physically/geometrically non-linear micro-shells with account of coupling of temperature and deformation fields. Chaos Sol. Fract. **104**, 635–654 (2017)
125. Duhamel, J.M.C.: Second memoire sur les phenomenes thermoomecaniques. de l'Ecole Polytechnique, 15 (1837)
126. Deaton, J.D., Grandhi, R.V.: Stress-based topology optimization of thermal structures. In: 10th World Congress on Structural and Multidisciplinary Optimization, May 19–24, Orlando (2013)
127. Choa, S., Choib, J.-Y.: Efficient topology optimization of thermo-elasticity problems using coupled field adjoint sensitivity analysis method. Fin. Elem. Anal. Des. **41**, 1481–1499 (2005)
128. Rodrigues, H., Fernandes, P.: Topology optimal design of thermoelastic structures using a homogenization method. Control Cyber. **23**(3), 553–563 (1994)
129. Asghari, M., Rahaeifard, M., Kahrobaiyan, M.H., Ahmadian, M.T.: The modified couple stress functionally graded Timoshenko beam formulation. Mater. Des. **32**, 1435–1443 (2011)
130. Wolf, A., Swift, J.B., Swinney, H.L., Vastano, J.A.: Determining Lyapunov exponents from a time series. Phys. D **16**, 285–317 (1985)
131. Rosenstein, M.T., Collins, J.J., De Luca, C.J.: A practical method for calculating largest Lyapunov exponents from small data sets. Phys. D **65**, 117–134 (1993)
132. Kantz, H.: A robust method to estimate the maximal Lyapunov exponent of a time series. Phys. Lett. A **185**, 77–87 (1994)

133. Awrejcewicz, J., Krysko, V.A., Dobriyan, V., Papkova, I.V. and Krysko, A.V.: On the Lyapunov exponents computation of coupled non-linear Euler-Bernoulli beams. In: Proceedings of the Fourteenth International Conference on Civil, Structural and Environmental Engineering Computing. Civil-Comp Press, Stirlingshire, UK, 2013, Paper 53
134. Awrejcewicz, J., Krysko, V.A.: Feigenbaum scenario exhibited by thin plate dynamics. Nonlinear Dyn. **24**, 373–398 (2006)
135. Awrejcewicz, J.A., Krysko, V.A., Zhigalov, M.V., Saltykova, O.A., Krysko, A.V.: Chaotic vibrations in flexible multilayered Bernoulli-Euler and Timoshenko type beams. Lat. Am. J. Sol. Struct. **5**(4), 319–363 (2008)
136. Krysko, V.A., Awrejcewicz, J., Kutepov, I.E., Zagniboroda, N.A., Serebryakov, A.V., Krysko, A.V., Papkova, I.V.: Chaotic dynamics of flexible beams with piezoelectric and temperature phenomena. Phys. Lett. A **377**, 2058–2061 (2013)
137. Krysko, A.V., Awrejcewicz, J., Pavlov, S.P., Zhigalov, M.V., Krysko, V.A.: Chaotic dynamics of the size-dependent non-linear micro-beam model. Commun. Nonlin. Sci. Num. Simul. **50**, 16–28 (2017)
138. Awrejcewicz, J., Krysko, A.V., Pavlov, S.P., Zhigalov, M.V., Krysko, V.A.: Chaotic dynamics of size dependent Timoshenko beams with functionally graded properties along their thickness. Mech. Sys. Sig. Proces. **93**, 415–430 (2017)
139. Pomeau, Y., Manneville, P.: Intermittent transition to turbulence in dissipative dynamical systems. Commun. Math. Phys. **174**(2), 189–198 (1980)